# Lectures on Tight Closure and Its Applications

CONTEMPORARY MATHEMATICS

831

# Lectures on Tight Closure and Its Applications

International Graduate Course on Tight Closure and Its Applications
May 16, 2022–July 25, 2022
Virtual

School on Commutative Algebra and Algebraic Geometry in Prime Characteristics
May 2–5, 2023
Trieste, Italy

Irena Swanson
Kei-ichi Watanabe
Editors

AMS AMERICAN MATHEMATICAL SOCIETY

2020 *Mathematics Subject Classification*. Primary 13A35, 13C14, 13H10, 13H15.

---

**Library of Congress Cataloging-in-Publication Data**

Contemporary Mathematics ISSN: 0271-4132 (print); ISSN: 1098-3627 (online)
DOI: https://doi.org/10.1090/conm/831

---



∞ The paper used in this book is acid-free and falls within the guidelines
established to ensure permanence and durability.
Visit the AMS home page at https://www.ams.org/

10 9 8 7 6 5 4 3 2 1    30 29 28 27 26 25

# Contents

# Preface

The volume consists of edited lecture notes from two events at the International Centre for Theoretical Physics (ICTP) in Trieste, Italy, organized in honor of Melvin Hochster and Craig Huneke. The two events were the online International Graduate course on Tight Closure and Its Applications, spread over summer 2022, and the School on Commutative Algebra and Algebraic Geometry in Prime Characteristics in May 2023. The unifying basis of these events was the theory of tight closure, the brainchild of Melvin Hochster and Craig Huneke in the late 1980s, which has had a dramatic effect on the field of commutative algebra, giving unified proofs and strong generalizations of many major theorems in commutative algebra, and stimulating much research, including recent proofs of long-standing conjectures. The aim of the two events as well as of this volume is to provide training in the foundations of commutative algebra and algebraic geometry in prime characteristic and to present some of the exciting recent developments in and beyond tight closure. The lecture notes for the online school also come with exercises and solutions.

The topics in the volume include characteristic p methods, test ideals, direct summands, singularities, Hilbert-Kunz multiplicity, Briançon-Skoda theorems, big Cohen-Macaulay algebras, the localization problem, uniform Artin-Rees results, vector bundles and tight closure, singularity invariants. The intended audience is graduate students learning the material as well as researchers wanting the latest advances in one reference.

We thank the organizers of the two events: Kyle Maddox, Kriti Goel, Maria Evelina Rossi, Viet Trung Ngo, Maria Evelina Rossi, Jugal Verma, and Lotthar Goettsche.

Just like the two events, this volume is dedicated to Melvin Hochster and Craig Huneke.

# Online International Graduate course on Tight Closure and Its Applications, summer 2022

Contemporary Mathematics
Volume **831**, 2025
https://doi.org/10.1090/conm/831/16628

# An introduction to the tight closure operation

Neil Epstein, with exercise solutions by Kriti Goel

## 1. Introduction

These notes arose from a series of three lectures given online, as part of the summer school "Graduate Course on Tight Closure of Ideals and its Applications", in association with the Abdus Salam International Centre for Theoretical Physics, May 16 and 18, 2022. There were ten such series (and several one-offs), but mine was the first, so my goal was to introduce the tight closure operation concretely and in context of other operations in mathematics, so as to set the stage for the rest of the instructors. There have of course been many introductions to tight closure theory; mine is bound to have a lot of overlap with them, though my own spin should also be apparent. However, almost none of the definitions and results herein are due to me. I have drawn liberally from the sources [**HH93**, **Hun96**, **Hun98**, **Hoc04a**, **Hoc04b**, **Eps12**], among others.

## 2. What is a closure operation?

It turns out that the definition for closure operations that we use in commutative algebra arises from analysis, over a century ago.

Definition 2.1 ([**Moo10**]). Let $M$ be a set and let $\mathcal{S}$ be a collection of subsets of $M$ that is closed under arbitrary intersection. Assume also $M \in \mathcal{S}$. A *closure operation* on $\mathcal{S}$ is a function $(-)^{\mathrm{c}} : \mathcal{S} \to \mathcal{S}$ that satisfies the following properties:

(1) [Extensiveness] For any $L \in \mathcal{S}$, $L \subseteq L^{\mathrm{c}}$.
(2) [Order-preservation] For any $K, L \in \mathcal{S}$ with $K \subseteq L$, we have $K^{\mathrm{c}} \subseteq L^{\mathrm{c}}$.
(3) [Idempotence] For any $L \in \mathcal{S}$, $L^{\mathrm{cc}} = L^{\mathrm{c}}$.

If $L \in \mathcal{S}$ with $L = L^{\mathrm{c}}$, we say $L$ is *closed* with respect to c, or c-*closed.*

Remark 2.2. The condition that $M \in \mathcal{S}$ is superfluous if we apply the convention that among subsets of $M$, we have $\bigcap \varnothing = \bigcap\{L \subseteq M \mid L \in \varnothing\} = M$.[1]

Remark 2.3. Note that for any closure operation c, the set $\mathcal{C} := \{L \in \mathcal{S} \mid L = L^{\mathrm{c}}\}$ is itself closed under arbitrary intersections. That is, for any $\mathcal{A} \subseteq \mathcal{C}$, we have $\bigcap \mathcal{A} \in \mathcal{C}$. To see this, let $L = \bigcap \mathcal{A}$. Then for any $K \in \mathcal{A}$, we have $L \subseteq K$, so by

[1] Here I use the convention that if $\mathcal{S}$ is a collection of subsets of a ground set $X$, then $\bigcap \mathcal{S}$ limits the elements of $X$ to those that are in all members of $\mathcal{S}$. Hence $\bigcap \varnothing = X$, as there is then no limitation on the elements selected from $X$.

order-preservation we have $L^c \subseteq K^c = K$ (since $K \in \mathcal{C}$). Hence, $L^c \subseteq \bigcap \mathcal{A} = L$. But $L \subseteq L^c$ by extensiveness, so $L = L^c$.

Conversely, let $\mathcal{C}$ be any subset of $\mathcal{S}$. Define $(-)^c : \mathcal{S} \to \mathcal{S}$ by the rule $L^c = \bigcap\{K \in \mathcal{C} \mid L \subseteq K\}$, where we use the convention on empty sets from Remark 2.2. Then c is a closure operation. Extensiveness holds since $L^c$ is an intersection of sets that contain $L$. Order-preservation holds for a similar reason. To see idempotence, let $K \in \mathcal{C}$ with $L \subseteq K$. Then $K \supseteq \bigcap\{U \in \mathcal{C} \mid L \subseteq U\} = L^c$. Conversely, if $L^c \subseteq K$, then since $L \subseteq L^c$, we have $L \subseteq K$. That is, if $K \in \mathcal{C}$, then $L \subseteq K \iff L^c \subseteq K$. Thus, $(L^c)^c = \bigcap\{K \in \mathcal{C} \mid L^c \subseteq K\} = \bigcap\{K \in \mathcal{C} \mid L \subseteq K\} = L^c$.

Hence, we have the following:

PROPOSITION 2.4. *Let $M$ be a set and $\mathcal{S}$ a collection of subsets of $M$ closed under arbitrary intersection. Then there is a one-to-one correspondence between closure operations on $\mathcal{S}$ and subcollections of $\mathcal{S}$ that are closed under arbitrary intersection. Namely, one assigns to any closure operation* c *the collection $\{L \in \mathcal{S} \mid L = L^c\}$, and one assigns to any intersection-closed subcollection $\mathcal{C}$ of $\mathcal{S}$ the operation $L \mapsto \bigcap\{K \in \mathcal{C} \mid L \subseteq K\}$.*

REMARK 2.5. Suppose $M$ is an $R$-module and $\mathcal{S} = 2^M$. Let $\mathcal{C}$ = all $R$-submodules of $M$. Then the corresponding closure operation sends $X \mapsto$ the $R$-submodule of $M$ generated by $X$.

Similarly, "the ideal/subring/subgroup/subalgebra generated by..." can all be seen as closure operations on the power set of a ring/ring/group/algebra.

REMARK 2.6. However, in what follows, $M$ will be a ring (or a module over a ring), and then $\mathcal{S}$ will be all ideals (resp. all submodules). In this case we will say c is a closure operation *on* $M$.

DEFINITION 2.7 ([**Pet64**]). A closure operation c on a ring $R$ is *semiprime* if for all ideals $I, J \subseteq R$, we have $J \cdot I^c \subseteq (JI)^c$.

LEMMA 2.8. *Let* c *be a semiprime closure operation on a ring $R$. Then for any ideals $I, J$ of $R$, we have $(I : J)^c \subseteq (I^c : J)$. Hence, if $I$ is* c*-closed, so is $(I : J)$. It follows that the isolated primary components of a decomposable closed ideal (e.g. any closed ideal in a Noetherian ring) are closed.*

PROOF. By semiprime-ness, $J \cdot (I : J)^c \subseteq (J(I : J))^c \subseteq I^c$. Hence, $(I : J)^c \subseteq (I^c : J)$. If $I$ is c-closed (i.e. $I = I^c$), we then have $(I : J) \subseteq (I : J)^c \subseteq (I^c : J) = (I : J)$, whence $(I : J) = (I : J)^c$.

For the final statement, let $I$ be a decomposable ideal that is c-closed, and let $\mathfrak{q}$ be an isolated primary component of $I$. Then by [**AM69**, Chapter 4, Exercise 15] and prime avoidance, there is some $f \in R$ with $\mathfrak{q} = I : f$, which is c-closed by the first part of the proof. □

DEFINITION 2.9. A *closure operation* cl on a *class* $\mathcal{M}$ of $R$-modules assigns to each $M \in \mathcal{M}$ a closure operation $(-)^{\mathrm{cl}}_M$ on $M$. A closure operation on a *category* of modules is just a closure operation on the underlying class of objects.

DEFINITION 2.10. A closure operation cl on a category $\mathcal{M}$ of $R$-modules is *functorial* if for any morphism $g : M \to N$ in $\mathcal{M}$ and any submodule $L \subseteq M$, we have $g(L^{\mathrm{cl}}_M) \subseteq g(L)^{\mathrm{cl}}_N$.

LEMMA 2.11. *Any closure operation* cl *that is functorial on $\mathcal{M}$, where $\mathcal{M}$ is either the category of modules or the category of finitely generated $R$-modules, has* *the property* $I^{\mathrm{cl}}_R L^{\mathrm{cl}}_M \subseteq (IL)^{\mathrm{cl}}_M$ *for any ideal* $I$ *and any* $M \in \mathcal{M}$. *In particular,* cl *is semiprime on* $R$.

PROOF. Let $x \in I$ and let $\mu_x : M \to M$ be the homothety map. Then $x \cdot L^{\mathrm{cl}}_M = \mu_x(L^{\mathrm{cl}}_M) \subseteq (\mu_x(L))^{\mathrm{cl}}_M = (xL)^{\mathrm{cl}}_M$. Hence, $I \cdot L^{\mathrm{cl}}_M = \sum_{x\in I} x \cdot L^{\mathrm{cl}}_M \subseteq \sum_{x\in I}(xL)^{\mathrm{cl}}_M \subseteq (IL)^{\mathrm{cl}}_M$.

Similarly, for any $z \in L$, define the $R$-linear map $\nu_z : R \to M$ by $r \mapsto rz$. Then $I^{\mathrm{cl}}_R \cdot z = \nu_z(I^{\mathrm{cl}}_R) \subseteq (\nu_z(I))^{\mathrm{cl}}_M = (Iz)^{\mathrm{cl}}_M$. Hence, $I^{\mathrm{cl}}_R L = \sum_{z\in L} I^{\mathrm{cl}}_R z \subseteq \sum_{z\in L}(Iz)^{\mathrm{cl}}_M \subseteq (IL)^{\mathrm{cl}}_M$.

Combining the two statements, we have $I^{\mathrm{cl}}_R L^{\mathrm{cl}}_M \subseteq (I^{\mathrm{cl}}_R L)^{\mathrm{cl}}_M \subseteq \left((IL)^{\mathrm{cl}}_M\right)^{\mathrm{cl}}_M = (IL)^{\mathrm{cl}}_M$. □

**First examples.** We will be building examples of closure operations throughout these lectures, with tight closure taking central place. However, we mention the following here:

- *The identity closure*: Define d on ideals of $R$ (resp. submodules of a module $M$) via $I^{\mathrm{d}} = I$ (resp. $L^{\mathrm{d}}_M := L$) This is manifestly a closure operation, a functorial closure operation, a semiprime operation, and the smallest possible example of any of these.
- *Radical*: The operation $\sqrt{-}$ on ideals is a semiprime closure operation. It will be larger than any of the other closures we care about here.

## 3. The Frobenius endomorphism and Frobenius closure

*Convention*: From here on, all rings are commutative, Noetherian, and of a fixed positive prime characteristic $p > 0$. The letter $q$ will always denote $p^e$, where $e$ is a positive integer. It is true that "reduction mod $p$" is an important reason to study tight closure theory, and allows definition and use of tight closure in equal characteristic zero. However, I will omit this topic; see [**Hun96**, Appendix by Mel Hochster] for a good introduction to the subject.

**The Frobenius endomorphism.** Consider the function $F : R \to R$ given by $r \mapsto r^p$. This is a ring homomorphism (called the *Frobenius*), because of the following:

LEMMA 3.1 (Freshman's dream). *Let* $R$ *be a ring of characteristic* $p$ *and let* $q = p^e$ *for some* $e \in \mathbb{N}$. *Let* $a, b \in R$. *Then* $(a + b)^q = a^q + b^q$.

PROOF. It is enough to show this for $e = 1$, so $q = p$; then a quick induction finishes the proof.

By the binomial theorem, we have

$$(a + b)^p = a^p + b^p + \sum_{i=1}^{p-1} \binom{p}{i} a^{p-i} b^i.$$

Now, for each $1 \le i \le p-1$, the numerator of $\binom{p}{i} = \frac{p!}{(p-i)!i!}$ has $p$ as a prime factor, but all the factors of the denominator are strictly less than $p$, hence they do not have $p$ as a prime factor. Thus, $p \mid \binom{p}{i}$. Write $\binom{p}{i} = n_i p$, $n_i \in \mathbb{N}$. Then the $i$th summand in the sum above becomes $n_i \cdot p \cdot a^{p-i} b^i = n_i \cdot 0 = 0$, since $R$ has characteristic $p$. □

**The functors $F^{e*}$ and $F^e_*$.** Let $g : R \to S$ be any ring homomorphism. Then one has the *extension of scalars* functor $g^* : {}_R\text{Mod} \to {}_S\text{Mod}$, given by $S \otimes_R -$, and the *restriction of scalars* functor $g_* : {}_S\text{Mod} \to {}_R\text{Mod}$, given by considering all $S$-modules and $S$-linear maps as $R$-modules and $R$-linear maps via the elementwise action $r \cdot z := g(r)z$.[2] These are adjoint functors, with $g^*$ left-adjoint to $g_*$. Hence, for any $M \in {}_R\text{Mod}$ there is a natural $R$-linear map $M \to g_*g^*M$ (the *unit* of the adjunction) given by $z \mapsto 1 \otimes z$. Note that $g_*$ is exact, but $g^*$ is in general only right exact.

When one applies this to the iterated Frobenius functors, one obtains $F^{e*}$ and $F^e_*$.[3] When a ring $A$ needs to be specified, we write $F_A$ for the Frobenius morphism on $A$. Concretely, if $M$ is a module presented by a (possibly transfinite) matrix $(a_{ij})$ of ring elements, then $F^{e*}(M)$ is presented by the matrix $(a_{ij}^{p^e})$. Thus for any free module $G$, we may identify $F^{e*}(G)$ with $G$.

On the other hand, when one applies this to quotient maps, i.e. when $R = S/I$ and $\pi : S \twoheadrightarrow R$ is the canonical map, we have $\pi^*(M) = M/IM$ for an $S$-module $M$.

LEMMA 3.2. *Let $S$ be a ring, $I$ an ideal, and $R = S/I$. Then for any $S$-module $M$, we have $F_S^{e*}(M)/IF_S^{e*}(M) = F_R^{e*}(M/IM)$ as $R$-modules.*

PROOF. Let $\pi : S \twoheadrightarrow R$ be the canonical map. Then

$$\begin{aligned} F_S^{e*}(M)/IF_S^{e*}(M) &= \pi^*(F_S^{e*}(M)) = (\pi \circ F_S^e)^*(M) \\ &= (F_R^e \circ \pi)^*(M) = F_R^{e*}(\pi^*(M)) = F_R^{e*}(M/IM). \end{aligned}$$

The equality from the first line to the second holds because $\pi \circ F_S^e = F_R^e \circ \pi$. That is, for any $x \in S$, we have

$$\pi(F_S^e(x)) = \pi(x^{p^e}) = \overline{x^{p^e}} = \bar{x}^{p^e} = F_R^e(\pi(x)).$$ □

**The $q$th root extension.** Another way to see the Frobenius endomorphism, in the case of a *reduced* ring, is as follows: Suppose $R$ is a reduced Noetherian ring of characteristic $p$. Let $K$ be the product of the algebraic closures of the fraction fields of $R/\mathfrak{p}$, where $\mathfrak{p}$ ranges over the (finitely many) minimal primes of $R$. Then for any power $q = p^e$ of $p$, let $R^{1/q} := \{\alpha \in K \mid \alpha^q \in R\}$. Then $R^{1/q}$ *is isomorphic (as a ring) to* $R$, via the map $\alpha \mapsto \alpha^q$. In particular, another description of $R^{1/q}$ is $\{r^{1/q} \mid r \in R\}$, as any element of $R$ has a *unique* $q$th root in $K$. To see this, first note that for any $r \in R$, the algebraic equation $x^q - r = 0$ must have a solution in $K$. On the other hand, suppose $x, y \in K$ with $x^q = y^q$. Then $(x-y)^q = x^q - y^q = 0$ by Freshman's dream, whence $x - y = 0$ since $K$ is reduced. Now, consider the inclusion map $i : R \hookrightarrow R^{1/q}$, and let $\varphi : R^{1/q} \to R$ be the isomorphism given above. Then $\varphi \circ i = F_R^e$. Hence, *up to isomorphism*, the Frobenius endomorphism can be seen as the inclusion map $R \hookrightarrow R^{1/p}$. In this way, we can think of $F^e_*(R)$ as $R^{1/p^e}$, as the two are isomorphic as $R$-modules.

LEMMA 3.3. *Let $R$ be a reduced Noetherian ring of characteristic $p$, and let $W \subseteq R$ be a multiplicative set. Then we have a ring isomorphism*

$$W^{-1}(R^{1/p}) \cong (W^{-1}R)^{1/p}.$$

[2]To the categorically minded, this notation might seem backward, as upper star (resp. lower star) classically represents contravariance (resp. covariance). But the convention exists this way because it arose in algebraic geometry, where maps of schemes are in the opposite direction from maps of rings.

[3]An older convention was to write $F^{e*}(M)$ as $F^e(M)$, and $F^e_*(M)$ as ${}^eM$.

PROOF IDEA. Let $\varphi: W^{-1}(R^{1/p}) \to (W^{-1}R)^{1/p}$ be the map $r^{1/p}/w \mapsto (r/w^p)^{1/p}$ and $\psi: (W^{-1}R)^{1/p} \to W^{-1}(R^{1/p})$ the map $(r/w)^{1/p} \mapsto (rw^{p-1})^{1/p}/w$. It is routine to verify that $\varphi$ and $\psi$ are well-defined, that they are ring homomorphisms, and that they are inverses of each other. □

**Bracket powers of ideals and submodules.** Let $I$ be an ideal and $q$ a power of $p$. Then the *qth bracket power* $I^{[q]}$ of $I$ is defined by any of the following equivalent methods:

(1) $I^{[q]}$ is the ideal generated by the set $\{a^q \mid a \in I\}$. (It may not be *equal* to that set, as the set is generally not itself an ideal.)
(2) Suppose $I = (a_1, \dots, a_n)$. Then $I^{[q]} = (a_1^q, \dots, a_n^q)$. (The equivalence of (1) and (2) is due to Freshman's Dream.)
(3) $I^{[q]}$ is the annihilator of the cyclic $R$-module $F^{e*}(R/I)$. That is, $F^{e*}(R/I) \cong R/I^{[q]}$.
(4) If $R$ is reduced, then $I \cdot R^{1/q} = (I^{[q]})^{1/q}$.

More generally, let $M$ be a module and $L \subseteq M$ a submodule. Then for any $e \in \mathbb{N}$ and $q = p^e$, the inclusion map $i : L \hookrightarrow M$ induces an $R$-module map $F^{e*}(i) : F^{e*}(L) \to F^{e*}(M)$. We set $L_M^{[q]} :=$ the image of $F^{e*}(i)$. Note that if $M = R$, the two notions above are consistent, in that $I_R^{[q]} = I^{[q]}$ when we identify $F^{e*}(R)$ with $R$.

Now if $z \in M$, and $u : M \to F^{e*}(M)$ is the unit of the adjunction above, we set $z_M^{p^e} := u(z)$. The notation is again consistent, in that if $z \in R$, then $z_R^q = z^q \in R = F^{e*}(R)$. Also note that if $z \in M$ and $i : Rz \to M$ is the inclusion, then $(Rz)_M^{[q]} = R \cdot z_M^q$ as $R$-submodules of $F_R^{e*}(M)$.

The results of the following exercise will be used freely in the sequel.

EXERCISE 1 (The algebra of bracket powers). Let $R$ be a ring, let $q, q'$ be powers of $p$, let $n$ be a positive integer, let $I, J$ be ideals, let $M$ be a finitely generated $R$-module and let $K, L$ be submodules. Prove the following identities:

(1) $(I^{[q]})^{[q']} = I^{[qq']}$.
(2) $L_M^{[qq']} = \left(L_M^{[q]}\right)_{F^{e*}(M)}^{[q']}$.
(3) $(K + L)_M^{[q]} = K_M^{[q]} + L_M^{[q]}$ (Hence in particular, $(I + J)^{[q]} = I^{[q]} + J^{[q]}$).
(4) If $K \subseteq L$ then $K_M^{[q]} \subseteq L_M^{[q]}$.
(5) For any $R$-linear map $g : M \to N$, $F^{e*}(g)(L_M^{[q]}) = g(L)_N^{[q]}$.
(6) $(I^n)^{[q]} = (I^{[q]})^n$.

We are now in position to introduce a new closure operation and a new type of ring.

DEFINITION 3.4. Let $x \in R$ and $I$ an ideal of $R$. We say $x$ is in the *Frobenius closure of* $R$, or $x \in I^{\mathrm{F}}$, if there is some power $q$ of $p$ with $x^q \in I^{[q]}$.

More generally, if $M$ is a module, $L \subseteq M$ a submodule, and $z \in M$, we say $z$ is in the *Frobenius closure of* $L$ *in* $M$, or $z \in L_M^{\mathrm{F}}$, if there is some $q$ with $z_M^q \in L_M^{[q]}$.

LEMMA 3.5. *Frobenius closure is a functorial closure operation on the category of* $R$*-modules.*

Proof. Let $L$ be a submodule of $M$. First we need to see that $L_M^{\mathrm{F}}$ is a submodule of $M$. Let $x, y \in L_M^{\mathrm{F}}$ and $r \in R$. Then there exist $q$ and $q'$ with $x_M^q \in L_M^{[q]}$ and $y_M^{q'} \in L_M^{[q']}$. Then $(x+ry)_M^{[qq']} = (x_M^q)^{q'}_{F^{e*}(M)} + r^{qq'}(y_M^{q'})^q_{F^{e'*}(M)} \in (L_M^{[q]})_M^{[q']} + (L_M^{[q']})_M^{[q]} = L_M^{[qq']}$. Thus, $x + ry \in L_M^{\mathrm{F}}$, which is therefore a submodule of $M$.

Since 1 is a power of $p$, for any $z \in M$, if $z \in L$ then $z_M^1 = z \in L = L_M^{[1]}$, so $z \in L_M^{\mathrm{F}}$.

If $K \subseteq L \subseteq M$ are submodules and $z \in K_M^{\mathrm{F}}$, then for some $q$ we have $z_M^q \in K_M^{[q]}$. But $K_M^{[q]} \subseteq L_M^{[q]}$, so $z_M^q \in L_M^{[q]}$ and $z \in L_M^{\mathrm{F}}$.

For idempotence, suppose $z \in (L_M^{\mathrm{F}})_M^{\mathrm{F}}$. Then for some $q$, $z_M^q \in (L_M^{\mathrm{F}})_M^{[q]}$. This means that there are elements $y_1, \dots, y_t \in L_M^{\mathrm{F}}$ and $r_1, \dots, r_t \in R$ with $z_M^q = \sum_{i=1}^t r_i (y_i)_M^q$. But then for each $i$, there is some $q_i$ such that $y_i^{q_i} \in L_M^{[q_i]}$. Let $q' = \max\{q_i \mid 1 \le i \le t\}$. Then $y_i^{q'} \in L_M^{[q']}$ for all $i$. Then

$$z_M^{qq'} = (z_M^q)^{q'}_{F^{e*}(M)} = \sum_{i=1}^t r_i^{q'} (y_i)_M^{qq'} = \sum_{i=1}^t r_i^{q'} ((y_i)_M^{q'})^q_{F^{e'*}(M)} \in (L_M^{[q']})^{[q]}_{F^{e'*}(M)} = L_M^{[qq']}.$$

Hence, $z \in L_M^{\mathrm{F}}$.

Finally, we show functoriality. Let $M$ be an $R$-module, $L \subseteq M$ and $g : M \to N$ an $R$-linear map. Let $z \in L_M^{\mathrm{F}}$. Then there is some power $q = p^e$ of $p$ with $z^q \in L_M^{[q]}$. Hence $g(z)_N^q = F^{e*}(g)(z_M^q) \in F^{e*}(g)(L_M^{[q]}) \subseteq g(L)_N^{[q]}$ by Exercise 1. Thus, $g(z) \in g(L)_N^{\mathrm{F}}$. □

For the next example, recall that a ring $R$ is F-*pure* if for any $R$-module $M$, the natural map $M \to F_R^*(M)$ is injective. It is equivalent to say that Frobenius closure on $R$-modules is trivial.

Example 3.6. Let $R$ be any Stanley-Reisner ring – i.e. a polynomial ring over a field, modulo a squarefree monomial ideal. For instance, $R = k[x, y]/(xy)$. Then every ideal is Frobenius closed. This is because such rings are F-pure by Fedder's criterion. See Theorem 9.2 and Example 9.3.

## 4. Regularity and flatness of Frobenius

Recall that a Noetherian local ring $(R, \mathfrak{m})$ of dimension $d$ is *regular* if $\mathfrak{m}$ can be generated by $d$ elements. A Noetherian ring is *regular* if $R_\mathfrak{m}$ is regular for all maximal ideals $\mathfrak{m}$ of $R$. A polynomial ring in finitely many variables over a field is always regular, as is a power series ring in finitely many variables over a field.

Theorem 4.1 (Kunz, 1969). *Let $R$ be a reduced Noetherian ring of characteristic $p$. Then $R$ is regular if and only if $R^{1/p}$ is flat as an $R$-module.*

We only prove the easy direction, as it is all we need here.

Proof of the "$\Longrightarrow$" direction. We prove it first for polynomial algebras over a field. Let $R = k[x_1, \dots, x_d]$. Then the inclusion $i : R \to R^{1/p} = k^{1/p}[x_1^{1/p}, \dots, x_d^{1/p}]$ is the composition of the inclusions $j_1 : k[x_1, \dots, x_d] \to k[x_1^{1/p}, \dots, x_d^{1/p}]$ and $j_2 : k[x_1^{1/p}, \dots, x_d^{1/p}] \to k^{1/p}[x_1^{1/p}, \dots, x_d^{1/p}]$. In each case, we have a free module. Namely, the target of $j_1$ is free over the source with basis given by $\{\prod_{j=1}^d x_j^{i_j} \mid 0 \le i_j < p\}$, and the target of $j_2$ is free over the source with basis given by a vector space basis of the field extension $k^{1/p}/k$. Thus, $R^{1/p}$ is a free, hence flat, $R$-module. Setting

$\mathfrak{m} := (x_1, \dots, x_d)R$ and $\mathfrak{m}^{1/p} := (x_1^{1/p}, \dots, x_d^{1/p})R^{1/p}$, we have $\mathfrak{m}^{1/p} \cap R = \mathfrak{m}$, so that $R_\mathfrak{m} \to (R^{1/p})_{\mathfrak{m}^{1/p}}$ is a flat local homomorphism. Hence, the same holds for the completion $k[\![x_1, \dots, x_d]\!] = \widehat{R_\mathfrak{m}} \to \widehat{(R^{1/p})}_{\mathfrak{m}^{1/p}} = k^{1/p}[\![x_1^{1/p}, \dots, x_d^{1/p}]\!] = \widehat{R_\mathfrak{m}}^{1/p}$.

Now we handle the complete local case. By the Cohen Structure Theorem, if $R$ is a complete regular local ring of characteristic $p$, then $R \cong k[\![x_1, \dots, x_d]\!]$. But then the above paragraph shows that $R^{1/p}$ is flat as an $R$-module.

Next we handle the local case. If $L \subseteq M$ is an inclusion of $R$-modules, then since $\hat{R}$ is flat over $R$ and $\hat{R}^{1/p}$ is flat over $\hat{R}$, we have the following commutative diagram of injections and isomorphisms:

$$\begin{array}{ccc}
(L \otimes_R \hat{R}) \otimes_{\hat{R}} \hat{R}^{1/p} & \hookrightarrow & (M \otimes_R \hat{R}) \otimes_{\hat{R}} \hat{R}^{1/p} \\
\| & & \| \\
(L \otimes_R \hat{R}) \otimes_{\hat{R}} \widehat{R^{1/p}} & \hookrightarrow & (M \otimes_R \hat{R}) \otimes_{\hat{R}} \widehat{R^{1/p}} \\
\| & & \| \\
(L \otimes_R R^{1/p}) \otimes_{R^{1/p}} \widehat{R^{1/p}} & \hookrightarrow & (M \otimes_R R^{1/p}) \otimes_{R^{1/p}} \widehat{R^{1/p}},
\end{array}$$

where "=" means isomorphism. Then since $\widehat{R^{1/p}}$ is faithfully flat over $R^{1/p}$, we get the inclusion $L \otimes_R R^{1/p} \hookrightarrow M \otimes_R R^{1/p}$, proving flatness.

For the general case, let $\mathfrak{m}$ be a maximal ideal of $R$. Then $(R^{1/p})_\mathfrak{m} \cong (R_\mathfrak{m})^{1/p}$ is flat over $R_\mathfrak{m}$ by the local case. Thus, $R^{1/p}$ is flat over $R$. □

COROLLARY 4.2. *Let $R$ be a regular ring and $I$, $J$ ideals of $R$. Then for any power $q$ of $p$, we have $I^{[q]} : J^{[q]} = (I : J)^{[q]}$.*

PROOF. It is elementary that for any flat $R$-module $M$, one has $IM :_M J = (I :_R J)M$. (Indeed, by [**Bou72**, Exercise I.2.22], this is *equivalent* to flatness.)

Moreover, note that by induction on the exponent $e$ in $q = p^e$, it is enough to prove the result for $q = p^1 = p$.

To see that $(I : J)^{[p]} \subseteq (I^{[p]} : J^{[p]})$, let $a \in (I : J)$. Then $aJ \subseteq I$, whence $a^p J^{[p]} = (aJ)^{[p]} \subseteq I^{[p]}$, so that $a^p \in (I^{[p]} : J^{[p]})$. Since $(I : J)^{[p]}$ is generated by all such $a^p$, the inclusion follows.

Finally, let $a \in (I^{[p]} : J^{[p]})$. Then $aJ^{[p]} \subseteq I^{[p]}$. Taking $p$th roots, we have $a^{1/p} J \subseteq a^{1/p} J R^{1/p} \subseteq I R^{1/p}$, so that $a^{1/p} \in (IR^{1/p} :_{R^{1/p}} J) = (I :_R J)R^{1/p}$ by the aforementioned Bourbaki criterion for flatness. Then taking $p$th powers, we have $a = (a^{1/p})^p \in (I :_R J)^{[p]}$. □

## 5. The tight closure operation

At long last, we are ready to meet the tight closure operation.

DEFINITION 5.1. Set $R^\circ :=$ the complement of the union of the minimal primes of $R$.

Note that $R^\circ$ is always a multiplicatively closed set. Note also that if $R$ is a domain, $R^\circ = R \smallsetminus \{0\}$.

DEFINITION 5.2. Let $I$ be an ideal and $x \in R$. We say $x$ is in the *tight closure of $I$*, or $x \in I^*$, if there is some $c \in R^\circ$ and some power $q_0$ of $p$ such that for all $q \geq q_0$, $cx^q \in I^{[q]}$.

More generally, if $L \subseteq M$ are $R$-modules and $z \in M$, we say $z$ is in the *tight closure of* $L$ *in* $M$, or $z \in L^*_M$, if there is some $c \in R^\circ$ and some power $q_0$ of $p$ such that for all $q \geq q_0$, $cz^q_M \in L^{[q]}_M$.

PROPOSITION 5.3. *Tight closure is a unctorial closure operation on the category of finitely generated R-modules. Moreover, we always have* $L^{\mathrm{F}}_M \subseteq L^*_M$.

PROOF. First we must see that $L^*_M$ is a submodule of $M$ whenever $L \subseteq M$. Let $x, y \in L^*_M$ and $r \in R$. There exist $c, d \in R^\circ$ and powers $q_0$, $q_1$ of $p$ such that for all $q \geq q_0$ (resp. $q \geq q_1$), we have $cx^q \in L^{[q]}_M$ (resp. $dy^q \in L^{[q]}_M$). Since $R^\circ$ is a multiplicative set, $cd \in R^\circ$, and for all $q \geq \max\{q_0, q_1\}$, we have $cd(x + ry)^q = d(cx^q) + r^q c(dy^q) \in L^{[q]}_M$. Hence, $x + ry \in L^*_M$, which is thus a submodule.

By letting $c = 1$, we see that $L \subseteq L^*_M$. Also, since $K \subseteq L$ implies $K^{[q]}_M \subseteq L^{[q]}_M$, it follows that $K^*_M \subseteq L^*_M$.

Now let $L \subseteq M$ and $z \in (L^*_M)^*_M$. Then there is some $c \in R^\circ$ and power $q_0$ of $p$ with $cz^q \in (L^*_M)^{[q]}_M$ for all $q \geq q_0$. Let $z_1, \ldots, z_n$ be a generating set for the finitely generated module $L^*_M$. Then there exist $c_1, \ldots, c_n \in R^\circ$ and powers $q_1, \ldots, q_n$ of $p$ such that for each $i$, we have $c_i z_i^{q_i} \in L^{[q]}_M$ whenever $q \geq q_i$. Letting $d = \prod_{i=1}^n c_i$ and $q' := \max\{q_i \mid 1 \leq i \leq n\}$, it follows that $d(L^*_M)^{[q]}_M \subseteq L^{[q]}_M$ for all $q \geq q'$. Hence, for all $q \geq \max\{q_0, q'\}$, we have $cdz^q_M \in d(L^*_M)^{[q]}_M \subseteq L^{[q]}_M$, and hence $z \in L^*_M$.

For the Frobenius closure claim, if $z \in L^{\mathrm{F}}_M$, there is some $q_0 = p^{e_0}$ with $z^{q_0}_M \in L^{[q_0]}_M$, whence for any $q \geq q_0$, we have $z^q = (z^{q_0})^{q/q_0} \in (L^{[q_0]}_M)^{[q/q_0]}_{F^{e_0 *}(M)} = L^{[q]}_M$. Then taking $c = 1$ in the definition of tight closure, we have $z \in L^*_M$.

For functoriality, let $L \subseteq M$ and $N$ be finitely generated $R$-modules, let $g : M \to N$ be $R$-linear, and let $z \in L^*_M$. Then one can choose $c \in R^\circ$ and a power $q_0$ of $p$ such that $cz^q_M \in L^{[q]}_M$ for all $q \geq q_0$. Hence, $c \cdot g(z)^q_N = c \cdot F^{e*}(g)(z^q_M) = F^{e*}(g)(cz^q_M) \in F^{e*}(g)(L^{[q]}_M) \subseteq g(L)^{[q]}_N$ for all such $q$. Thus, $g(z) \in g(L)^*_N$. □

REMARK 5.4. Note that the only place we needed $M$ to be finitely generated in the above was in showing tight closure was idempotent. Later on, (big) test elements will allow us to generalize to non-finitely generated modules (see Proposition 8.4).

LEMMA 5.5. *The tight closure and the Frobenius closure of the zero ideal both coincide with the nilradical* $J$ *of the ring.*

PROOF. Let $x \in (0)^*$. Then there is some $c \in R^\circ$ with $cx^q = 0$ for all large $q$. But for any minimal prime $\mathfrak{p}$, $c \notin \mathfrak{p}$; thus, $x^q \in \mathfrak{p}$, so $x \in \mathfrak{p}$. Hence, $x \in J$, the intersection of all such $\mathfrak{p}$.

On the other hand, suppose $x \in J$. Then for $q \gg 0$, $x^q = 0 = (0)^{[q]}$. Hence $x \in (0)^{\mathrm{F}}$.

Thus, $(0)^* \subseteq J \subseteq (0)^{\mathrm{F}}$. But since $(0)^{\mathrm{F}} \subseteq (0)^*$ by Proposition 5.3, all three ideals coincide. □

For the next properties of tight closure, we need the following "Prime Avoidance Containment (PAC)" result.

LEMMA 5.6 (PAC lemma). *Let* $R$ *be a Noetherian ring, let* $\mathfrak{p}_1, \ldots, \mathfrak{p}_n$ *be prime ideals such that for all* $i \neq j$, $\mathfrak{p}_i \nsubseteq \mathfrak{p}_j$. *Let* $W = R \smallsetminus \bigcup_{i=1}^n \mathfrak{p}_i$. *For each* $1 \leq i \leq n$, *let* $0 \neq d_i \in R/\mathfrak{p}_i$. *Then for each* $i$, *there is some* $c_i \in W$ *with* $c_i + \mathfrak{p}_i = d_i$. *Moreover, there exist elements* $t_i \in \left(\bigcap_{j \neq i} \mathfrak{p}_j\right) \smallsetminus \mathfrak{p}_i$, *and for any such choice we have* $c := \sum_{i=1}^n t_i c_i \in W$.

PROOF. Fix $i$. Let $\widetilde{d_i}$ be a lift of $d_i$ to $R$. Let $S := \{j \mid 1 \le j \le n,\ \widetilde{d_i} \in \mathfrak{p}_j\}$, and $T := \{1, \dots, n\} \setminus S$. We have $\bigcap_{j\in T} \mathfrak{p}_j \not\subseteq \bigcup_{j\in S} \mathfrak{p}_j$ by prime avoidance, the non-containment condition on the $\mathfrak{p}_j$, and the fact that all the $\mathfrak{p}_j$ are prime. Choose $e_i \in \bigcap_{j\in T} \mathfrak{p}_j \setminus \bigcup_{j\in S} \mathfrak{p}_j$. Set $c_i := \widetilde{d_i} + e_i$. For $j \in S$, we have $\widetilde{d_i} \in \mathfrak{p}_j$ and $e_i \notin \mathfrak{p}_j$, whence $c_i \notin \mathfrak{p}_j$. For $j \in T$, we have $\widetilde{d_i} \notin \mathfrak{p}_j$ and $e_i \in \mathfrak{p}_j$, so $c_i \notin \mathfrak{p}_j$. Hence, $c_i \in W$. Moreover, $c_i - \widetilde{d_i} = e_i \in \mathfrak{p}_i$ (since $i \in T$), so $c_i$ is a lift of $d_i$ to $R$.

The existence of the elements $t_i$ is immediate. Moreover, if $c \in \mathfrak{p}_i$ for some $i$, then since $t_j \in \mathfrak{p}_i$ for all $j \neq i$, it follows that $t_i c_i \in \mathfrak{p}_i$. But $t_i \notin \mathfrak{p}_i$ and $c_i \notin \mathfrak{p}_i$, contradicting the fact that $\mathfrak{p}_i$ is prime. Thus $c \in W$. □

PROPOSITION 5.7 (Reducing to the domain case). *Let $L \subseteq M$ be $R$-modules and $z \in M$. Then $z \in L^*_M$ if and only if for each minimal prime $\mathfrak{p}$ of $R$, $\bar{z} \in (L + \mathfrak{p}M/\mathfrak{p}M)^*_{M/\mathfrak{p}M}$ as modules over $R/\mathfrak{p}$, where $\bar{z}$ is the image of $z$ in $M/\mathfrak{p}M$.*

PROOF. By Lemma 3.2, for any ideal $J$, we have

$$(L + JM/JM)^{[q]}_{M/JM} = \frac{(L + JM)^{[q]}_M + JF^{e*}(M)}{JF^{e*}_R(M)} = \frac{L^{[q]}_M + JF^{e*}(M)}{JF^{e*}(M)}.$$

Let $z \in L^*_M$. Then there is some $c \in R^\circ$ and power $q_0$ of $p$ with $cz^q_M \in L^{[q]}_M$ for all $q \ge q_0$. Going mod $\mathfrak{p}$, we then have $\bar{c}\bar{z}^q_{M/\mathfrak{p}M} \in (L^{[q]}_M + \mathfrak{p}F^{e*}(M))/\mathfrak{p}F^{e*}(M) \subseteq (L + \mathfrak{p}M/\mathfrak{p}M)^{[q]}_{M/\mathfrak{p}M}$. Since $\bar{c} \neq 0$ in the domain $R/\mathfrak{p}$, it follows that $\bar{z} \in (L + \mathfrak{p}M/\mathfrak{p}M)^*_{M/\mathfrak{p}M}$.

Conversely, let $\mathfrak{p}_1, \dots, \mathfrak{p}_n$ be the minimal primes of $R$, and suppose $\bar{z} \in (L + \mathfrak{p}_iM/\mathfrak{p}_iM)^*_{M/\mathfrak{p}_iM}$ for each $i$. Then there exist $0 \neq d_i \in R/\mathfrak{p}_i$ and a power $q_0$ of $p$ such that for all $q \ge q_0$ and all $i$, $d_i\bar{z}^q_{M/\mathfrak{p}_iM} \in (L + \mathfrak{p}_iM/\mathfrak{p}_iM)^{[q]}_{M/\mathfrak{p}_iM}$. Choose $t_i, c_i$, and $c$ as in the PAC lemma. Then for each $i$, we have $c_iz^q_M \in L^{[q]}_M + \mathfrak{p}_iF^{e*}(M)$ for all $q \ge q_0$. Thus, $t_ic_iz^q_M \in \left(\bigcap_{j\neq i} \mathfrak{p}_j\right)(L^{[q]}_M + \mathfrak{p}_iF^{e*}(M)) \subseteq L^{[q]}_M + JF^{e*}(M)$, where $J$ is the nilradical of $R$. Choose a power $q_1 = p^{e_1}$ of $p$ such that $J^{[q_1]} = 0$. Then for any $q \ge q_0$,

$$c^{q_1}z^{qq_1}_M \in L^{[qq_1]}_M + J^{[q_1]}F^{(e+e_1)*}_R(M) = L^{[qq_1]}_M.$$

Hence, $z \in L^*_M$. □

EXAMPLE 5.8. It follows that the tight closure of the ideal $(x+y)$ in $k[x,y]/(xy)$ contains $x$, and hence $(x,y)$. By Example 3.6, it then follows that Frobenius closure *cannot* be checked mod minimal primes, and in particular in this ring $(x+y)^{\mathrm{F}} \neq (x+y)^*$.

Our next result shows that what holds for all *large* $q$ often also holds for *all* $q$.

PROPOSITION 5.9. *Let $R$ be a ring, $I$ an ideal, and $x \in I^*$. If $\operatorname{ht} I > 0$ or $R$ is reduced, then there is some $c \in R^\circ$ such that $cx^q \in I^{[q]}$ for all powers $q$ of $p$.*

PROOF. First suppose $\operatorname{ht} I > 0$. Then there is some $b \in R^\circ$ and power $q_0$ of $p$ with $bx^q \in I^{[q]}$ for all $q \ge q_0$. By Prime Avoidance, there is some $a \in I^{[q_0]} \cap R^\circ$. Let $c = ab$. Then $cx^q \in I^{[q]}$ for all $q$.

Now suppose $R$ is reduced. Let $\mathfrak{p}_1, \dots, \mathfrak{p}_n$ be the minimal primes of $R$. Then for each $1 \le i \le n$, either $IR/\mathfrak{p}_i = 0$ (in which case $\bar{x} \in (0)^* = (0)$ in $R/\mathfrak{p}_i$) or $\operatorname{ht}(IR/\mathfrak{p}_i) > 0$. Then by the first paragraph of the proof, for every $i$ there exists

$0 \neq d_i \in R/\mathfrak{p}_i$ such that $d_i \bar{x}^q \in (IR/\mathfrak{p}_i)^{[q]}$ for all powers $q$ of $p$. Choose $c_i, t_i$, and $c$ as in the PAC lemma. Then for each $i$, since $\bigcap_{j=1}^n \mathfrak{p}_j = 0$, we have for all $q$:

$$t_i c_i x^q \in \left(\bigcap_{j \neq i} \mathfrak{p}_j\right)(I^{[q]} + \mathfrak{p}_i) \subseteq I^{[q]} + \bigcap_{j=1}^{n} \mathfrak{p}_j = I^{[q]}.$$

Thus $cx^q \in I^{[q]}$ for all $q$, and $c \in R^\circ$. □

Next, we show the relationship between the tight closure and the radical of an ideal.

PROPOSITION 5.10. *Let $I$ be any ideal of $R$. Then $I^* \subseteq \sqrt{I}$.*

PROOF. For our first, step, we show that prime ideals are tightly closed. Suppose this holds for integral domains. Let $\mathfrak{p} \in \operatorname{Spec} R$ and let $\mathfrak{q}$ be a minimal prime ideal contained in $\mathfrak{p}$. If $x \in \mathfrak{p}^*$, then by Proposition 5.7, $x + \mathfrak{q} = \bar{x} \in (\mathfrak{p}R/\mathfrak{q})^* = \mathfrak{p}R/\mathfrak{q} = (\mathfrak{p} + \mathfrak{q})/\mathfrak{q} = \mathfrak{p}/\mathfrak{q}$. Hence $x \in \mathfrak{p}$.

Thus, we may assume $R$ is an integral domain. Let $\mathfrak{p} \in \operatorname{Spec} R$, $x \in \mathfrak{p}^*$, and choose a DVR $(V, \mathfrak{m})$ between $R$ and its fraction field such that $\mathfrak{m} \cap R = \mathfrak{p}$.[4] Then there is some $0 \neq c \in R$ such that for all $q \gg 0$, $cx^q \in \mathfrak{p}^{[q]} \subseteq \mathfrak{m}^{[q]} \subseteq \mathfrak{m}^q$. Applying the $\mathbb{Z}$-valued valuation $v$ associated to $V$, we have $v(c) + qv(x) = v(cx^q) \geq v(\mathfrak{m}^q) = qv(\mathfrak{m}) = q$. Dividing by $q$, follows that $v(x) + \frac{v(c)}{q} \geq 1$. As this holds in particular for $q > v(c)$, then since $v(x)$ is an integer, we have $v(x) \geq 1$. Hence, $x \in \mathfrak{m} \cap R = \mathfrak{p}$.

Since prime ideals are tightly closed, radical ideals are intersections of prime ideals, and intersections of tightly closed ideals are tightly closed, it follows that all radical ideals are tightly closed. Hence for any ideal $I$, since $I \subseteq \sqrt{I}$, it follows that $I^* \subseteq (\sqrt{I})^* = \sqrt{I}$. □

REMARK 5.11. The above actually shows more. It shows that the tight closure of an ideal is contained in the *integral* closure of it. See Ian Aberbach's lectures later in this volume for more on this.

DEFINITION 5.12. We say $R$ is *weakly F-regular* if every ideal is tightly closed, or *F-regular* if for every prime ideal $\mathfrak{p}$, $R_\mathfrak{p}$ is weakly F-regular.[5] We say a local ring $(R, \mathfrak{m})$ is *F-rational* if any ideal generated by a system of parameters is tightly closed.

Note that any weakly F-regular local ring is F-rational. For more on F-rationality, see Florian Enescu's lectures later in this volume.

We next give a result that explains "regular" part of the term "F-regular".

THEOREM 5.13. *Any regular ring is F-regular.*

PROOF. Let $R$ be regular. By the Auslander-Buchsbaum-Serre theorem, $R_\mathfrak{p}$ is a regular local ring for all prime ideals $\mathfrak{p}$. Hence, it suffices to show that a regular *local* ring is *weakly* F-regular.

---

[4]Construction: Let $A = R_\mathfrak{p}$, and $\mathfrak{p}' := \mathfrak{p}A = \mathfrak{p}A_\mathfrak{p}$. Let $x_1 \dots, x_n \in A$ be a generating set for $\mathfrak{p}'$. Set $S := A[\frac{x_2}{x_1}, \dots, \frac{x_n}{x_1}]$. Then $\mathfrak{p}'S = x_1 S$. Let $\mathfrak{q}$ be a minimal prime of $x_1 S$. By the Principal Ideal Theorem, $\operatorname{ht} \mathfrak{q} = 1$, so $S_\mathfrak{q}$ is a 1-dimensional Noetherian domain. Let $T$ be the integral closure of $S_\mathfrak{q}$. Then $T$ is integral over $S_\mathfrak{q}$, and Noetherian by the Krull-Akizuki theorem. Let $Q \in \operatorname{Spec} T$ lie over $\mathfrak{q}S_\mathfrak{q}$. Then $\operatorname{ht} Q = 1$ by integrality. Then $V := T_Q$ is a DVR (by Serre's normality criterion), with maximal ideal $\mathfrak{m} = QT_Q$. Moreover, $\mathfrak{m} \cap R = \mathfrak{p}$.

[5]These are known to be equivalent in many cases. Moreover, no counterexamples are known.

Accordingly, let $(R, \mathfrak{m})$ be a regular local ring (hence, a domain). Let $I$ be an ideal and let $x \in I^*$. Then there is some $c \neq 0$ and $q_0$ such that for all $q \geq q_0$, $cx^q \in I^{[q]}$. Then by Corollary 4.2, $c \in \bigcap_{q \geq q_0}(I^{[q]} : x^q) = \bigcap_{q \geq q_0}(I :_R x)^{[q]}$. If $x \notin I$, then $(I : x) \subseteq \mathfrak{m}$, so $0 \neq c \in \bigcap_q \mathfrak{m}^{[q]} = 0$ (by the Krull intersection theorem), a contradiction. Hence $x \in I$. Thus, $I^* = I$. □

PROPOSITION 5.14.

(1) *If $(R, \mathfrak{m})$ is F-rational, then any ideal generated by* part *of a system of parameters is tightly closed. In particular, the zero ideal is tightly closed, so $R$ is reduced.*
(2) *If for every maximal ideal $\mathfrak{m}$ of $R$, every $\mathfrak{m}$-primary ideal is tightly closed, then $R$ is weakly F-regular.*

PROOF.

(1) Let $x_1, \ldots, x_d$ be a system of parameters and $I = (x_1, \ldots, x_t)$, where $0 \leq t < d$. For each positive integer $u$, $x_1, \ldots, x_t, x_{t+1}^u, \ldots, x_d^u$ is a system of parameters. Hence, we have

$$I^* \subseteq \bigcap_u (x_1, \ldots, x_t, x_{t+1}^u, \ldots, x_d^u)^* = \bigcap_u (x_1, \ldots, x_t, x_{t+1}^u, \ldots, x_d^u)$$
$$\subseteq \bigcap_u (I + \mathfrak{m}^u) = I,$$

with the last equality by the Krull intersection theorem.

The zero ideal is the ideal generated by the empty subset of any system of parameters. Its tight closure is the nilradical of the ring, hence $R$ is reduced.

(2) Let $I$ be an ideal, $x \in I^*$, and $\mathfrak{m}$ a maximal ideal. If $I \nsubseteq \mathfrak{m}$, then $IR_\mathfrak{m} = R_\mathfrak{m}$, whence $x \in IR_\mathfrak{m}$. Otherwise, for any positive integer $u$ we have $\mathfrak{m}^u \subseteq I + \mathfrak{m}^u \subseteq \mathfrak{m}$, so that $I + \mathfrak{m}^u$ is $\mathfrak{m}$-primary. It follows that $x \in I^* \subseteq (I + \mathfrak{m}^u)^* = I + \mathfrak{m}^u$, so that

$$x/1 \in \bigcap_u (I + \mathfrak{m}^u)R_\mathfrak{m} = \bigcap_u \left(IR_\mathfrak{m} + (\mathfrak{m}R_\mathfrak{m})^u\right) = IR_\mathfrak{m}$$

by the Krull intersection theorem. So for every $\mathfrak{m}$, $x/1 \in IR_\mathfrak{m}$. Thus $x \in I$. □

EXERCISE 2. Let $\mathfrak{m}$ be a maximal ideal of $R$ and $I$ an $\mathfrak{m}$-primary ideal. Prove that $(I_\mathfrak{m})^* = (I^*)_\mathfrak{m}$. (*Hint:* For the forward direction, first prove it in the domain case, then keep track of which minimal primes lie inside $\mathfrak{m}$ or not.) Conclude that a ring $R$ is weakly F-regular if and only if $R_\mathfrak{m}$ is weakly F-regular for every maximal ideal $\mathfrak{m}$.

## 6. Module-finite extensions and plus closure

DEFINITION 6.1. Let $R$ be a domain, $x \in R$, and $I$ an ideal. We say $x$ is in the *plus closure* of $I$, written $x \in I^+$, if there is some injective ring homomorphism $g : R \to S$, where $S$ is also a domain, such that $g$ makes $S$ a finitely generated $R$-module and $g(x) \in IS$. (We write "$x \in IS \cap R$.")

PROPOSITION 6.2. *Plus closure is a closure operation on any integral domain $R$.*

PROOF. Let $I$ be an ideal. Let $x, y \in I^+$ and $r \in R$. Then there are module-finite extension domains $R \hookrightarrow S$ and $R \hookrightarrow T$ with $x \in IS \cap R$ and $y \in IT \cap R$. Then there is some module-finite extension domain $U$ of $R$ that dominates both $S$ and $T$.[6] Hence, $x, y \in IU \cap R$, so $x + ry \in IU \cap R$ as well. Hence, $x + ry \in I^+$. Thus, $I^+$ is an ideal.

The extension and order-preservation properties of a closure operation are obvious. Now let $x \in (I^+)^+$. Then there is a module-finite extension domain $S$ of $R$ with $x \in (I^+)S$. Thus, there exist $y_1, \ldots, y_k \in I^+$ such that $x \in (y_1, \ldots, y_k)S$. Then there exist module-finite extension domains $T_j$ of $R$ with $y_j \in IT_j$. Find a module finite extension domain $U$ of $R$ that dominates $S$ and all the $T_j$. Then $x \in IU \cap R \subseteq I^+$. □

PROPOSITION 6.3. *Let $R \to S$ be a module-finite extension of Noetherian domains. Let $0 \neq d \in S$. Then there is some $R$-linear map $g : S \to R$ such that $g(d) \neq 0$.*

PROOF. Let $W = R \smallsetminus \{0\}$, so that $Q := \operatorname{Frac} R = R_W$. Then for any $w \in W$, we have $wd \neq 0$, so $d/1 \neq 0$ in $S_W$. Hence, $S_W$ is a $Q$-vector space with $d/1 \neq 0$, so there is some $Q$-basis of $S_W$ of which $d/1$ is a member. Then by projection, there is a nonzero $Q$-linear map $\varphi : W^{-1}S \to Q$ sending $d/1 \mapsto 1$.

Since $S$ is finitely presented as an $R$-module, we have

$$\operatorname{Hom}_Q(S_W, Q) = \operatorname{Hom}_{R_W}(S_W, R_W) \cong \operatorname{Hom}_R(S, R)_W.$$

It follows that there is some $R$-linear map $g : S \to R$ and some $c \in W$ with $\varphi = g/c$. Thus,

$$\frac{g(d)}{c} = \varphi(d/1) = 1/1 = \frac{c}{c}.$$

Thus, $g(d) = c$. □

This allows us to prove the following:

THEOREM 6.4. *Let $g : R \hookrightarrow S$ be a module-finite extension of Noetherian domains, and $I$ an ideal of $R$. Then $I^* = (IS)^* \cap R$. Hence, $I^+ \subseteq I^*$.*

PROOF. Let $x \in I^*$. Then there is some $0 \neq c \in R$ such that for all large $q$, $cx^q \in I^{[q]}$. Then $g(c)g(x)^q \in I^{[q]}S = (IS)^{[q]}$, so since $g(c) \neq 0$, $g(x) \in (IS)^*$. That is, $x \in (IS)^* \cap R$.

Conversely, let $x \in (IS)^* \cap R$. Then there is some $0 \neq d \in S$ with $dx^q \in (IS)^{[q]} = I^{[q]}S$ for $q \gg 0$. By Proposition 6.3, there is an $R$-linear map $\varphi : S \to R$ with $\varphi(d) \neq 0$. Then for all large $q$, the $R$-linearity of $\varphi$ gives

$$\varphi(d)x^q = \varphi(dx^q) \in \varphi(I^{[q]}S) = I^{[q]}\varphi(S) \subseteq I^{[q]}.$$

Hence, $x \in I^*$.

For the final statement, let $x \in I^+$. Then there is a module-finite domain extension $R \hookrightarrow S$ with $x \in IS \cap R \subseteq (IS)^* \cap R = I^*$. □

REMARK 6.5. At this point, it would be natural to give a simple example showing the above inclusion can be strict. Such an example can be found in [**BM10**]. But it took decades to find any example at all! More was said of this in Wenliang Zhang's lectures, detailed later in this volume.

---

[6]Let $Q$ be a prime ideal of $S \otimes_R T$ that lies over the zero ideal of $R$. Then set $U = S \otimes_R T/Q$.

Another consequence of Proposition 6.3 is that F-rational rings are normal.

PROPOSITION 6.6. *Let $R$ be an F-rational local ring. Then $R$ is an integrally closed domain.*

PROOF. First we prove that $R$ is a domain. Let $\mathfrak{p}_1, \dots, \mathfrak{p}_n$ be the minimal primes of $R$, and suppose $n \geq 2$. Then by Prime Avoidance we have $\mathfrak{p}_1 + \bigcap_{j=2}^n \mathfrak{p}_j \not\subseteq \bigcup_{i=1}^n \mathfrak{p}_i$, which means there exists $x \in \mathfrak{p}_1$ and $y \in \bigcap_{j=2}^n \mathfrak{p}_j$ with $x + y \in R^\circ$. It follows that $x \notin \bigcup_{j=2}^n \mathfrak{p}_j$ and that $y \notin \mathfrak{p}_1$. Set $I := (x+y)R$. We have $\bar{x} = \bar{0} \in IR/\mathfrak{p}_1$, and for any $j \geq 2$, we have $\bar{x} = \overline{x+y} \in IR/\mathfrak{p}_j$. So by Proposition 5.7, $x \in I^*$, even though one sees easily that $x \notin I$, so that $I$ is not tightly closed. But as $x + y$ is not in any minimal prime, it is part of a system of parameters, so Proposition 5.14 provides the desired contradiction. Therefore $R$ can have only one minimal prime. But by Proposition 5.14 again, $R$ is reduced. Hence it is an integral domain.

Let $Q$ be the fraction field of $R$. Let $\alpha \in Q$ be integral over $R$. Then $S = R[\alpha]$ is module-finite over $R$. Say $\alpha = a/b$, $a, b \in R$, $b \neq 0$. Then $a = b\alpha$, so $a \in (b)S \cap R \subseteq (b)^* = (b)$ by Proposition 5.14, since any nonzero element of $R$ is part of a system of parameters. Thus, $a = bc$ for some $c \in R$, whence $c = \alpha \in R$. □

## 7. Colon capturing

Much of the power of tight closure theory comes from "colon capturing" results, which allow one to treat systems of parameters as if they were regular sequences. That is, tight closures "captures the failure of the ring to be Cohen-Macaulay" [**AHS96**].

THEOREM 7.1. *Let $R$ be an equidimensional ring that is a quotient of a Cohen-Macaulay local ring. Let $\mathbf{y} = y_1, \dots, y_d$ be a system of parameters for $R$. Then*

(1) *For all $j < d$, $(y_1, \dots, y_j) : y_{j+1} \subseteq (y_1, \dots, y_j)^*$, and*
(2) *For all $t \in \mathbb{N}$, we have $(y_1^t, \dots, y_d^t) : (y_1 \cdots y_d)^{t-1} \subseteq (y_1, \dots, y_d)^*$.*

PROOF. We first reduce to the domain case. Let $\mathfrak{p}_1, \dots, \mathfrak{p}_k$ be the minimal primes of $R$. Then by equidimensionality, $\dim R/\mathfrak{p}_i = d$ for all $i$, and the image of $\mathbf{y}$ mod $\mathfrak{p}_i$ is a system of parameters for $R/\mathfrak{p}_i$ for each $i$. To prove (1), let $a \in (y_1, \dots, y_j) :_R y_{j+1}$. Then $ay_{j+1} \in (y_1, \dots, y_j)$, so for each $i$ we have $ay_{j+1} + \mathfrak{p}_i \in (y_1, \dots, y_j) + \mathfrak{p}_i$. It will then follow from the domain case that $a + \mathfrak{p}_i \in ((y_1, \dots, y_j) + \mathfrak{p}_i/\mathfrak{p}_i)^*$ in $R/\mathfrak{p}_i$. By Proposition 5.7, $a \in (y_1, \dots, y_j)^*$. The reduction of (2) to the domain case is entirely analogous.

Hence, we may assume $R$ is a domain. Write $R = S/J$, where $S$ is a Cohen-Macaulay local ring. Then $J$ is prime. Let $t = \dim S$. Then $h := \operatorname{ht} J = t - d$. Let $x_1, \dots, x_h \in S$ localize to a system of parameters for the Cohen-Macaulay local ring $S_J$. Then $(x_1, \dots, x_h)S_J$ is $JS_J$-primary, so there is some $n$ with $(JS_J)^n \subseteq (x_1, \dots, x_h)S_J$. It follows that there is some $c \in S \setminus J$ such that $cJ^n \subseteq (x_1, \dots, x_h)S$. Choose a power $q_0 \geq n$ of $p$. Let $z_1, \dots, z_d$ be liftings of $y_1, \dots, y_d$ respectively to $S$.

Then $x_1, \dots, x_h, z_1, \dots, z_d$ is a system of parameters for $S$, hence a regular sequence. Indeed for any power $q$ of $p$, $x_1, \dots, x_h, z_1^q, \dots, z_d^q$ is a permutable regular sequence.

To prove (1), let $a \in S$ with $\bar{a} \in (y_1, \dots, y_j) :_R y_{j+1}$. Then $z_{j+1}a \in (z_1, \dots, z_j) + J$. Then for any $q \geq q_0$, we have $cz_{j+1}^q a^q \in (z_1, \dots, z_j)^{[q]} + cJ^n \subseteq (z_1^q, \dots, z_j^q, x_1, \dots, x_h)$. Thus, $ca^q \in (z_1^q, \dots, z_j^q, x_1, \dots, x_h) \subseteq (z_1, \dots, z_j)^{[q]} + J$, so that $\bar{c}\bar{a}^q \in (y_1, \dots, y_j)^{[q]}$. Since $c \notin J$, we have $\bar{c} \neq 0$ in the domain $R$, whence $\bar{a} \in (y_1, \dots, y_j)^*$.

The proof of (2) is similar. Let $b \in S$ with $\bar{b} \in (y_1^t, \dots, y_d^t) : (y_1 \cdots y_d)^{t-1}$. Then $b \cdot (z_1 \cdots z_d)^{t-1} \in (z_1^t, \dots, z_d^t) + J$. Hence for any $q \geq q_0$, we have $cb^q (z_1^q \cdots z_d^q)^{t-1} \in (z_1^t, \dots, z_d^t)^{[q]} + cJ^n \subseteq (z_1^{tq}, \dots, z_d^{tq}, x_1, \dots, x_h)$. Then by Exercise 3,

$$cb^q \in (z_1^q, \dots, z_d^q, x_1, \dots, x_h) \subseteq (z_1, \dots, z_d)^{[q]} + J.$$

Hence, $\bar{c}\bar{b}^q \in (y_1, \dots, y_d)^{[q]}$, and since $\bar{c} \neq 0$ and the above holds for all $q \geq q_0$, we have $\bar{b} \in (y_1, \dots, y_d)^*$. □

EXERCISE 3. Let $x_1, \dots, x_n$ be a regular sequence in a local ring, let $s \leq n$, and let $t$ be a positive integer. Then $(x_1^t, \dots, x_s^t, x_{s+1}, \dots, x_n) : (x_1 \cdots x_s)^{t-1} = (x_1, \dots, x_n)$. (Hint: Use induction on $s$, and the fact that regular sequences in this context are permutable.)

REMARK 7.2. Kawasaki [**Kaw02**, Corollary 1.2] proved that any excellent local ring is the homomorphic image of a Cohen-Macaulay local ring. Hence, the conclusions of Theorem 7.1 hold for all equidimensional excellent local rings.

COROLLARY 7.3 (The Monomial Theorem in characteristic p). *Let $(R, \mathfrak{m})$ be a Noetherian local ring of characteristic p that is a homomorphic image of a Cohen-Macaulay local ring, and $x_1, \dots, x_d$ a system of parameters. Then for any $t$, we have $(x_1 \cdots x_d)^{t-1} \notin (x_1^t, \dots, x_d^t)$.*

PROOF. We have $1 \notin \mathfrak{m} = \mathfrak{m}^* \supseteq (x_1, \dots, x_d)^*$ by Proposition 5.10. Then use Theorem 7.1(2). □

REMARK 7.4. The above theorem was first proved for rings containing a field in 1973 [**Hoc73**], albeit with a much more difficult proof than the later, tight closure proof from the 1980s. However, a proof in *mixed* characteristic did not arrive until 2016 [**And18**]!

COROLLARY 7.5. *Let $R$ be F-rational and the homomorphic image of a Cohen-Macaulay local ring. Then $R$ is Cohen-Macaulay.*

PROOF. By Proposition 6.6, $R$ is a normal domain, hence equidimensional. Let $\mathbf{y} = y_1, \dots, y_d$ be a system of parameters. Then for any $j < d$, we have $(y_1, \dots, y_j) : y_{j+1} \subseteq (y_1, \dots, y_j)^* = (y_1, \dots, y_j)$ by F-rationality. Hence $\mathbf{y}$ is a regular sequence. □

## 8. Test elements and the persistence of tight closure

DEFINITION 8.1. Let $c \in R^\circ$. We say $c$ is a *test element* if for all ideals $I$ and all $x \in I^*$, we have $cx^q \in I^{[q]}$ for all powers $q$ of $p$. If $q_0$ is a fixed power of $p$, we say that $c$ is a *$q_0$-weak test element* if the above holds for all $q \geq q_0$. We say $c$ is a ***big*** *($q_0$-weak) test element* if for all $R$-modules $M$, all submodules $L \subseteq M$, and all $z \in L_M^*$, we have $cz_M^q \in L_M^{[q]}$ for all powers $q$ of $p$ (resp. for all $q \geq q_0$).

The above might seem a lot to ask for. It amounts to switching the order of quantifiers in the definition of tight closure, and yields uniform annihilation results. However, big test elements exist in a large variety of cases.

THEOREM 8.2 ([**HH89, HH94, Sha10**]). *Let $R$ be a reduced ring that is either essentially of finite type over an excellent local ring, or excellent and F-pure, or such that $R^{1/p}$ is a finitely generated $R$-module. Let $c \in R^\circ$ such that $R_c$ is regular. Then some power of $c$ is a big test element.*

In Kevin Tucker's lectures later in this volume (Theorem 7, Definition 8, and Exercise 11), you can find a modern proof of the existence of big test elements over F-finite complete local domains. But proving existence of (big) test elements for domains implies existence of (big) weak test elements in general, as laid out below:

PROPOSITION 8.3. *Suppose $R$ is a ring such that $R/\mathfrak{p}$ has a (big) test element for each minimal prime $\mathfrak{p}$. If $q_0 = p^{e_0}$ is a power of $p$ such that $J^{[q_0]} = 0$, where $J$ is the nilradical of $R$, then $R$ has a $q_0$-weak (big) test element.*

*In particular, if $R$ satisfies the above conditions and is reduced, then it has a (big) test element.*

PROOF. Let $\mathfrak{p}_1, \dots, \mathfrak{p}_n$ be the minimal primes of $R$. For each $1 \le i \le n$, let $d_i \in R/\mathfrak{p}_i$ be a (big) test element. Choose $c_i, t_i$, and $c$ as in the PAC lemma (Lemma 5.6). Then $c^{q_0}$ is the desired element. (When $R$ is reduced, choose $q_0 = 1$.)

To see this in the case of test elements, let $I$ be an ideal and $x \in I^*$. Then for each $i$, $\bar{x} \in (IR/\mathfrak{p}_i)^*$ by Proposition 5.7, so for all $q$ we have $d_i\bar{x}^q \in (IR/\mathfrak{p}_i)^{[q]} = (I^{[q]} + \mathfrak{p}_i)/\mathfrak{p}_i$. Hence, $c_i x^q \in I^{[q]} + \mathfrak{p}_i$. Then we have $t_i c_i x^q \in (\bigcap_{j \ne i} \mathfrak{p}_j)(I^{[q]} + \mathfrak{p}_i) \subseteq I^{[q]} + J$, so that $cx^q \in I^{[q]} + J$ for all $q$, whence $c^{q_0} x^{qq_0} \in I^{[qq_0]} + J^{[q_0]} = I^{[qq_0]}$. Since $c \in R^\circ$ we are done.

In the case of big test elements, let $L \subseteq M$ be $R$-modules and $z \in L^*_M$. Then for each $i$, $\bar{z} \in ((L + \mathfrak{p}_i M)/\mathfrak{p}_i M)^*_{M/\mathfrak{p}_i M}$ by Proposition 5.7, so for all $q$ we have $d_i \bar{z}^q_{M/\mathfrak{p}_i M} \in ((L + \mathfrak{p}_i M)/\mathfrak{p}_i M)^{[q]}_{M/\mathfrak{p}_i M} = \frac{L^{[q]}_M + \mathfrak{p}_i F^{e*}(M)}{\mathfrak{p}_i F^{e*}(M)}$. Hence, $c_i z^q_M \in L^{[q]}_M + \mathfrak{p}_i F^{e*}(M)$. Then we have $t_i c_i z^q_M \in (\bigcap_{j \ne i} \mathfrak{p}_j)(L^{[q]}_M + \mathfrak{p}_i F^{e*}(M)) \subseteq L^{[q]}_M + JF^{e*}(M)$. Thus, $cz^q_M \in L^{[q]}_M + JF^{e*}(M)$, so that $c^{q_0} z^{qq_0}_M \in L^{[qq_0]}_M + J^{[q_0]} F^{(e+e_0)*}(M) = L^{[qq_0]}_M$. □

As a corollary to Theorem 8.2, we show that in cases where the theorem applies, tight closure is a closure operation, even for non-finitely generated modules.

PROPOSITION 8.4. *Let $R$ be a ring that admits a big weak test element (e.g. any ring that satisfies the conditions of Theorem 8.2). Then for any $R$-module $M$, tight closure is a closure operation on $M$.*

PROOF. All we need to establish after the proof of Proposition 5.3 is idempotence. Let $L \subseteq M$ be a submodule and $z \in (L^*_M)^*_M$. Let $c \in R$ be a big $q_0$-weak test element. Then for any $q \ge q_0$, $cz^{[q]} \in (L^*_M)^{[q]}_M$. But for any $y \in L^*_M$, we have $cy^q \subset L^{[q]}_M$ for any $q \ge q_0$. Hence $c(L^*_M)^{[q]}_M \subseteq L^{[q]}_M$ for any such $q$, so $c^2 z^q_M \in c(L^*_M)^{[q]}_M \subseteq L^{[q]}_M$ for all $q \ge q_0$. Thus, $z \in L^*_M$. □

The next two exercises are abstracted from [**Eps05**], though the conclusion to Exercise 4 was present in early work of Hochster and Huneke (see [**Hun96**, Theorem 1.5(4)]).

EXERCISE 4 (Tight closure Krull intersection theorem). Suppose $(R, \mathfrak{m})$ is local and admits a test element. Then for any ideals $I$ and $J$, we have $\bigcap_{t \in \mathbb{N}} (J + \mathfrak{m}^t I)^* = J^*$. Conclude that every tightly closed ideal is an intersection of $\mathfrak{m}$-primary tightly closed ideals.

EXERCISE 5 (Tight closure Nakayama lemma). Let $(R, \mathfrak{m})$ be local and $I, J$ ideals such that $J \subseteq I \subseteq (J + \mathfrak{m} I)^*$. Show that $I \subseteq \bigcap_{t \in \mathbb{N}} (J + \mathfrak{m}^t I)^*$. Thus, if $R$ has a test element, conclude from Exercise 4 that $I^* = J^*$.

**Persistence of tight closure.**

DEFINITION 8.5. Let $g : R \to S$ be a ring homomorphism. We say that tight closure *persists* across $g$ (for ideals) if for any ideal $I$ of $R$, we have $g(I^*) \subseteq (IS)^*$.

For modules, if $M$ is an $R$-module and $L \subseteq M$, write $LS = \operatorname{im}(L \otimes_R S \to M \otimes_R S)$. We say tight closure persists across $g$ for *modules* if for any submodule inclusion $L \subseteq M$, we have $L^*_M S \subseteq (LS)^*_{M \otimes_R S}$.

EXERCISE 6. Let $g : R \to S$ be a ring map. Suppose either that $g(R^\circ) \subseteq S^\circ$, or $R$ has a weak (resp. big weak) test element $c$ such that $g(c) \in S^\circ$. Then tight closure persists across $g$ for ideals (resp. modules). In particular, persistence always holds for inclusions of domains.

The existence of test elements as in Theorem 8.2 allows us to conclude persistence of tight closure in a wide variety of cases. We prove it for ideals below. The proof for modules is similar, though the notation is fussier so we omit it.

THEOREM 8.6. *Let $\varphi : R \to S$ be a ring map such that $R$ satisfies the assumptions of Theorem* 8.2. *Then tight closure persists along $\varphi$, both for ideals and for modules.*

PROOF FOR IDEALS. We begin with the case where $R, S$ are domains, $\varphi$ is surjective, and $P := \ker \varphi$ is a height one prime. Let $R'$ be the integral closure of $R$ in its fraction field. By excellence of $R$, $R'$ is finitely generated as an $R$-module. By integrality, we can choose $P' \in \operatorname{Spec} R'$ that lies over $P$. By integrality, $P'$ also has height 1, so $R'_{P'}$ is a DVR by Serre's criterion for normality. By excellence of $R'$ (as a finitely generated $R$-algebra), the regular locus of $R'$ is open. Hence there is some $c \in R' \smallsetminus P'$ such that $R'_c$ is regular. Moreover, $R'$ satisfies the conditions of Theorem 8.2 (as a finitely generated $R$-module), so $c$ has a power that is a big test element for $R'$. By replacing $c$ by that power, we may assume $c$ itself is a big test element for $R'$.

Now let $I$ be an ideal of $R$ and $x \in I^*$. Then $x \in (IR')^*$ by Exercise 6. Thus for all $q$, we have $cx^q \in (IR')^{[q]}$. Going mod $P'$, we have $\bar{c}\bar{x}^q \in (IR'/P')^{[q]}$ for all $q$. Since $\bar{c} \neq 0$ in $R'/P'$, it follows that $\bar{x} \in (IR'/P')^*$. But $R'/P'$ is a module-finite extension domain of $R/P$. Hence, $\varphi(x) = \bar{x} \in (IR'/P')^* \cap (R/P) = (IR/P)^* \subseteq (IS)^*$ by Theorem 6.4 and Exercise 6.

Now suppose only that $S$ is a domain and $\varphi$ is surjective. Let $P = \ker \varphi$. Then we can make a saturated chain $P_0 \subset P_1 \subset \dots \subset P_h = P$, where $h = \operatorname{ht} P$. If $h = 0$, then $P = P_0$ is a minimal prime and tight closure persists along $\varphi$ by Proposition 5.7. If $h > 0$, we may assume by induction on $h$ that tight closure persists along all canonical surjections $R \twoheadrightarrow R/Q$ such that $Q$ is a prime with $\operatorname{ht} Q < h$. In particular, tight closure persists along $R \twoheadrightarrow R/P_{h-1}$, and also tight closure persists along $R/P_{h-1} \twoheadrightarrow R/P$ by the first two paragraphs of the proof, since $\operatorname{ht} P/P_{h-1} = 1$. Thus tight closure persists along $\varphi$.

If we only assume $S$ is a domain, then letting $P = \ker \varphi$, $\varphi$ factors as $R \twoheadrightarrow R/P \hookrightarrow S$. Tight closure persists along both maps in the composition by the above and by Exercise 6.

Finally, we handle the general case. Let $\mathfrak{q}$ be a minimal prime of $S$. Let $I$ be an ideal and $x \in I^*$. Then by the above, $\overline{\varphi(x)} \in (IS/\mathfrak{q})^*$ in $S/\mathfrak{q}$. Since $\mathfrak{q}$ was arbitrary, $\varphi(x) \in (IS)^*$ by Proposition 5.7. $\square$

As a corollary, we obtain the following:

COROLLARY 8.7. *Let $S$ be a complete weakly F-regular local ring, and let $R$ be a complete local subring such that for any ideal $I$ of $R$, $IS \cap R = I$ (e.g. if $R$ is a direct summand of $S$). Then $R$ is weakly F-regular.*

PROOF. Let $I$ be an ideal of $R$. Then by persistence of the map and weak F-regularity of $S$, $I \subseteq I^* \subseteq (IS)^* \cap R = IS \cap R = I$. Hence $I = I^*$. □

As a corollary to the above, we obtain the following:

PROPOSITION 8.8. *Let $S$ be a regular ring and $R$ a Noetherian subring of $S$, such that the inclusion map $R \to S$ is split as a map of $R$-modules. Then $R$ is Cohen-Macaulay. If $R$ is locally excellent, then $R$ is also normal.*

PROOF. First we prove the result when $R$, $S$ are complete local rings, in which case the inclusion map is a local homomorphism. Then since $S$ is regular, it is weakly F-regular. Thus by Corollary 8.7, $R$ is weakly F-regular, and hence F-rational. Thus by Corollary 7.5 (and the Cohen structure theorem), $R$ is Cohen-Macaulay, and by Proposition 6.6, $R$ is normal.

The conditions and the conclusions are local; hence we may assume $R \to S$ is a local homomorphism. Moreover, $R$ is Cohen-Macaulay iff $\hat{R}$ is. Finally, let $\mathbf{x}$ be a system of parameters for $R$, and $I = (\mathbf{x})$. Then $\mathbf{x}$ is also a system of parameters for $\hat{R}$, and by faithful flatness we have $I\hat{R} \cap R = I$. But $\hat{R}$ is F-rational. Since $R$ is an excellent local ring, persistence holds for the completion map by Theorem 8.6. Thus, $I \subseteq I^* \subseteq (I\hat{R})^* \cap R = I\hat{R} \cap R = I$, whence $I = I^*$. Therefore $R$ is F-rational, and hence normal by Proposition 6.6. □

We conclude with an application of tight closure to prove a theorem of invariant theory.

THEOREM 8.9 ([**HR74**]). *Let $S = k[x_1, \ldots, x_n]$, with $k$ a field. Let $G$ be a linearly reductive group acting $k$-rationally on $S$. Then the ring of invariants $S^G$ is Cohen-Macaulay.*

PROOF. The assumptions on $G$ imply that $S^G$ is a direct summand of $S$ as an $S^G$-module, via use of a so-called "Reynolds operator". Also, $S^G$ is a finitely generated $k$-algebra, hence Noetherian. The result then follows from Proposition 8.8. □

## 9. Examples

We start with the classical example to give in introductions to tight closure: the cubical Fermat surface.

EXAMPLE 9.1. Let $R = k[x,y,z]/(x^3 + y^3 + z^3)$, where $k$ is a field. Then:

$$(y,z)^F = \begin{cases} (y,z), & \text{if } p \equiv 1 \mod 3 \\ (x^2,y,z), & \text{if } p \equiv 2 \mod 3 \\ (x,y,z), & \text{if } p = 3. \end{cases}$$

$$(y,z)^* = \begin{cases} (x^2,y,z), & \text{if } p \neq 3, \\ (x,y,z), & \text{if } p = 3, \end{cases}$$

and

$$\sqrt{(y,z)} = (x,y,z).$$

Here's why. First, set $A := k[y,z]$, and note that $R$ is a free $A$-module on the basis $\{1, x, x^2\}$. Hence for any $q$, we have $(y,z)^{[q]} = (y,z)^{[q]}R = (y,z)^{[q]}A \oplus (y,z)^{[q]}Ax \oplus (y,z)^{[q]}Ax^2$. It further follows that $R$ has a basis as a vector space over $k$ given by the monomials $\{x^i y^j z^k \mid j, k \in \mathbb{N}_{\geq 0},\ i \in \{0,1,2\}\}$.

If $p \equiv 1 \mod 3$, then for any power $q = p^n$ of $p$, we also have $q \equiv 1 \mod 3$. Say $q = 3a+1$ and $p = 3b+1$, $a, b \in \mathbb{N}$. Then we have $(x^2)^q = x^{2q} = x^{6a+2} = (x^3)^{2a}x^2 = (y^3+z^3)^{2a}x^2 = \sum_{i=0}^{2a}\binom{2a}{i}x^2(y^3)^i(z^3)^{2a-i}$. Now, for $i < a$, $2a - i \geq a+1$, so the $i$th term is a multiple of $(z^3)^{a+1} = z^{q+2} \in (y,z)^{[q]}$. Similarly, for $i > a$, the $i$th term of the above sum is a multiple of $(y^3)^{a+1} = y^{q+2} \in (y,z)^{[q]}$. Thus, $(x^2)^q \equiv \binom{2a}{a}x^2y^{3a}z^{3a} \mod (y,z)^{[q]}$. But by Lucas's theorem, $\binom{2a}{a}$ is a unit in $\mathbb{F}_p$ and hence in $R$, and $x^2y^{3a}z^{3a} \notin (y,z)^{[q]}Ax^2$, since $3a < q$. Thus, $\binom{2a}{a}x^2y^{3a}z^{3a} \notin (y,z)^{[q]}$, whence, $(x^2)^q \notin (y,z)^{[q]}$. Since $q$ was arbitrary, $x^2 \notin (y,z)^F$.

However, in the above if we multiply the left hand side by $yz$, then every term in the corresponding sum is in $(y,z)^{[q]}$. That is, $yz(x^2)^q \in (y,z)^{[q]}$ for all powers $q$ of $p$. Since $R$ is a domain and $yz \neq 0$, it follows that $x^2 \in (y,z)^*$.

Now suppose $p \equiv 2 \mod 3$. Let $q$ be an *odd* power of $p$. Then $q \equiv 2 \mod 3$ as well. Say $q = 3a+2$. Then $(x^2)^q = x^{6a+4} = (x^3)^{2a+1}x = (y^3+z^3)^{2a+1}x = \sum_{i=0}^{2a+1}\binom{2a+1}{i}x(y^3)^i(z^3)^{2a+1-i}$. When $i \leq a$ we have $2a+1-i \geq a+1$, so that the $i$th term is a multiple of $(z^3)^{a+1} = z^{3a+3} = zz^q \in (y,z)^{[q]}$. On the other hand, when $i > a$, then the $i$th term is a multiple of $(y^3)^{a+1} = y^{3a+3} = yy^q \in (y,z)^{[q]}$. Hence $x^2 \in (y,z)^F$. Since Frobenius closure is contained in tight closure, we also have $x^2 \in (y,z)^*$.

However, I claim that $x \notin (y,z)^*$ as long as $p \neq 3$. To see this, note that since $R$ is a normal graded domain of dimension 2, it has an isolated singularity at the homogeneous maximal ideal. Hence, $R_y$ is regular. It then follows from Theorem 8.2 that some power $y^t$ of $y$ is a test element for $R$. Now, suppose $x \in (y,z)^*$. Let $q$ be a power of $p$ with $q > t$. Write $q = 3a+j$, with $j = 1$ or $2$. Then by definition of test element, we have $y^tx^q \in (y,z)^{[q]}$. Thus, $y^tx^q = y^t(x^3)^ax^j = \pm y^t(y^3+z^3)^ax^j = \pm\sum_{i=0}^{a}\binom{a}{i}x^jy^{3i+t}z^{3(a-i)} \in Ay^qx^j + Az^qx^j$. But this is a contradiction to the above representation of $R$ as a $k$-vector space, since the $i = 0$ summand of the above is $\pm y^tz^{3a}$, and both $t$ and $3a$ are $< q$. Since $(y,z)^*$ must be a graded ideal and $\mathfrak{m}/(x^2,y,z)$ has length 1, it follows that $(y,z)^* = (x^2,y,z)$.

On the other hand, $x^3 = -y^3 - z^3 \in (y,z)$, so that $x \in \sqrt{(y,z)}$. Hence, $\sqrt{(y,z)} = (x,y,z)$.

Finally, if $p = 3$, then $x^p = x^3 = -y^3 - z^3 \in (y,z)^{[3]} = (y,z)^{[p]}$. Hence, $x \in (y,z)^F$, whence $(y,z)^F = (x,y,z)$. Since Frobenius closure lives in tight closure, we also have $(y,z)^* = (x,y,z)$ in that case.

Our next example will show that in a Stanley-Reisner ring, though all ideals are Frobenius closed, even parameter ideals are typically very far from being tightly closed. First, we need to recall some results.

THEOREM 9.2 (Fedder's criterion [**Fed83**]). *Let $S$ be a regular ring of characteristic $p$, $I$ an ideal, and $R = S/I$. Assume either that $(S, \mathfrak{n})$ is local, or that $S$ is standard graded over a field, $\mathfrak{n}$ is its unique graded maximal ideal, and $I$ is graded. Suppose $(I^{[p]} : I) \nsubseteq \mathfrak{n}^{[p]}$. Then every (homogeneous) ideal of $R$ is Frobenius closed.*

Note: The converse holds too, under mild conditions on $R$ (see [**Hoc77**][7]).

PROOF. Let $\mathfrak{a}$ be a (homogeneous) ideal of $R$ and $g \in R$ (homogeneous) with $g^p \in \mathfrak{a}^{[p]}$. Let $\pi : S \twoheadrightarrow R$ be the canonical surjection, $h \in \pi^{-1}(\{g\})$ (homogeneous), and $J = \pi^{-1}(\mathfrak{a})$. Then $h^p \in J^{[p]} + I$. Let $s \in (I^{[p]} : I) \setminus \mathfrak{n}^{[p]}$ (and in the homogeneous case, make sure to pick $s$ to be a homogeneous element). Then $sh^p \in J^{[p]} + I^{[p]} = J^{[p]}$. Thus, $s \in (J^{[p]} :_S h^p)$. But by Corollary 4.2, $(J^{[p]} :_S h^p) = (J :_S h)^{[p]}$. If $h \notin J$, we then have $(J : h) \subseteq \mathfrak{n}$, so $s \in (J :_S h)^{[p]} \subseteq \mathfrak{n}^{[p]}$, a contradiction. Thus, $h \in J$, whence $g = \bar{h} \in \bar{J} = \mathfrak{a}$.

Now let $\mathfrak{a}$ be a (homogeneous) ideal and $g \in R$ (homogeneous) with $g \in \mathfrak{a}^{\mathrm{F}}$. Let $e$ be minimal with $g^{p^e} \in \mathfrak{a}^{[p^e]}$. If $e > 0$, we have $(g^{p^{e-1}})^p = g^{p^e} \in \mathfrak{a}^{[p^e]} = (\mathfrak{a}^{[p^{e-1}]})^{[p]}$. Then by the first paragraph of the proof, $g^{p^{e-1}} \in \mathfrak{a}^{[p^{e-1}]}$, contrary to our choice of $e$. Hence $e = 0$, which means that $g \in \mathfrak{a}$. □

EXAMPLE 9.3. Let $R = S/I$, where $S = k[x_1, \ldots, x_n]$, $k$ an infinite field of characteristic $p$ (e.g. $\mathbb{F}_p(t)$, or the algebraic closure of $\mathbb{F}_p$), and $I$ a nonzero ideal generated by squarefree nonlinear monomials in the $x_i$. Then every homogeneous ideal of $R$ is Frobenius closed, but many parameter ideals generated by linear forms are not tightly closed.

For the Frobenius closure statement, let $s = \prod_{i=1}^n x_i^{p-1}$. Let $m = m(x_1, \ldots, x_n)$ be a monomial generator of $I$. Then it is easily seen that $sm$ is a multiple of $m(x_1^p, \ldots, x_n^p) = m^p$. Hence, $sI \subseteq I^{[p]}$. However, $s \notin (x_1^p, x_2^p, \ldots, x_n^p) = \mathfrak{n}^{[p]}$. Therefore $(I^{[p]} : I) \nsubseteq \mathfrak{n}^{[p]}$, so by Theorem 9.2, every homogeneous ideal of $R$ is Frobenius closed.

For the tight closure statement, we quote the following result.

THEOREM 9.4 ([**Ale18, GVM20**]). *Let $R$ be as in Example 9.3, and $d = \dim R$. Then there is a nonempty Zariski-open subset $U$ of $k^{d \times n}$ such that whenever $A = (a_{ij}) \in U$, setting $f_i = \sum_{j=1}^n a_{ij} x_j$ for $1 \le i \le d$ creates a graded system of parameters $f_1, \ldots, f_d$ of $R$ such that $(f_1, \ldots, f_d)^* = (x_1, \ldots, x_n)$. In doing so, one obtains precisely all systems of parameters for $R$ that consist of linear forms.*

PROOF IDEA. Each minimal prime $\mathfrak{p}$ of $R$ is generated by the homomorphic images of a subset of the variables of $S$. Hence, $R/\mathfrak{p}$ is isomorphic to a polynomial ring in the remaining variables over $k$, hence regular. Therefore, for any ideal $I$ of $R$, $I^* = \bigcap_{\mathfrak{p} \text{ a minimal prime of } R} (I + \mathfrak{p})$. If $I$ is generated by $d$ linear polynomials, we then have to solve certain linear systems of equations to arrange that $IR/\mathfrak{p}$ is the maximal ideal of $R/\mathfrak{p}$ for each $\mathfrak{p}$. Hence, $U$ can be taken to be the complement of a finite union of finite intersections of vanishing loci of certain minors of the matrix $A$.

The statement on systems of parameters follows from the Krull height theorem, along with the fact that in a polynomial ring, any linear system of parameters generates the graded maximal ideal. □

---

[7]The condition that every ideal is Frobenius closed is equivalent to the cyclic purity of the Frobenius endomorphism. But under the mild condition of being approximately Gorenstein, cyclic purity is equivalent to the *a priori* stronger condition of purity [**Hoc77**]. Fedder's article concerns the purity of the Frobenius endomorphism.

## SOLUTIONS TO EXERCISES

by KRITI GOEL

EXERCISE 1 (The algebra of bracket powers). Let $R$ be a ring, let $q, q'$ be powers of $p$, let $n$ be a positive integer, let $I, J$ be ideals, let $M$ be a finitely generated $R$-module and let $K, L$ be submodules. Prove the following identities:

(1) $(I^{[q]})^{[q']} = I^{[qq']}$.
(2) $L_M^{[qq']} = (L_M^{[q]})^{[q']}_{F^{e*}(M)}$.
(3) $(K + L)_M^{[q]} = K_M^{[q]} + L_M^{[q]}$. (Hence, in particular, $(I + J)^{[q]} = I^{[q]} + J^{[q]}$).
(4) If $K \subseteq L$, then $K_M^{[q]} \subseteq L_M^{[q]}$.
(5) For any $R$-linear map $g : M \to N$, $F^{e*}(g)(L_M^{[q]}) \subseteq g(L)_N^{[q]}$.
(6) $(I^n)^{[q]} = (I^{[q]})^n$.

PROOF. Recall that for any submodule $L \subseteq M$ of an $R$-module $M$, $L_M^{[p^e]} = \operatorname{im}(F^{e*}(i))$ where $i : L \hookrightarrow M$ is an inclusion and $F^{e*}(i) : F^{e*}(L) \to F^{e*}(M)$. In particular, for any ideal $I$ in a ring $R$, $F^{e*}(i)(I) = I^{[p^e]}$.

(1) The ideal $I^{[q]}$ is generated by the set $\{a^q \mid a \in I\}$. Then the ideal $(I^{[q]})^{[q']}$ may be generated by the $q'$th powers of these generators. That is, $(I^{[q]})^{[q']}$ is generated by the set $\{(a^q)^{q'} \mid a \in I\} = \{(a^{qq'} \mid a \in I\}$. But this later description also gives a generating set for $I^{[qq']}$. Hence, the two ideals are equal.

Alternately, the statement is a special case of (2) by taking $M = R$ and $L = I$.

(2) Let $i : L \hookrightarrow M$ be the inclusion map. Let $j : L_M^{[q]} \to F^{e*}(M)$ be the inclusion map. Then we have the following commutative diagram.

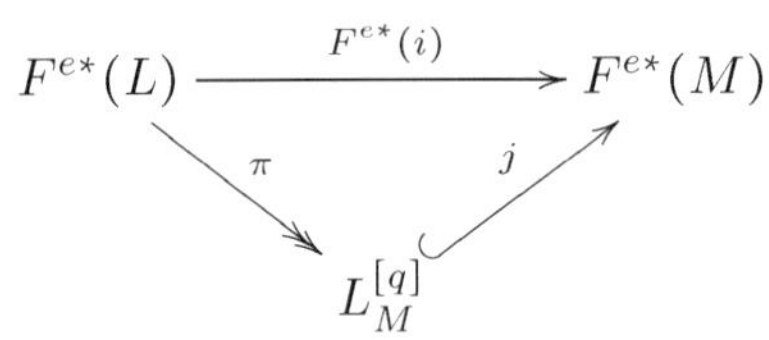


Since $F^{(e+e')*} = F^{e'*} \circ F^{e*}$ as functors, we then get the following commutative diagram:

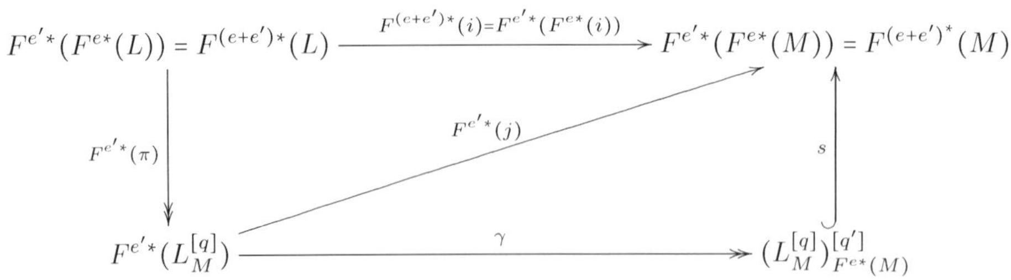


Since $s$ is an inclusion map and $\gamma \circ F^{e'*}(\pi)$ is surjective, it follows that $(L_M^{[q]})^{[q']}_{F^{e*}(M)}$ is the image of $F^{(e+e')*}(i)$, i.e., $L_M^{[qq']}$.

(3) Using the following diagram,

$$\begin{array}{ccccc} K & \overset{i_1}{\hookrightarrow} & K+L & \xrightarrow{i} & M \\ & & \nearrow_{i_2} & & \\ L & & & & \end{array}$$

we get $(K+L)_M^{[q]} = F^{e*}(i)(K+L) = F^{e*}(i \circ i_1)(K) + F^{e*}(i \circ i_2)(L) = K_M^{[q]} + L_M^{[q]}$.

(4) From the following commutative diagram

$$\begin{array}{ccc} K & \overset{i_K}{\hookrightarrow} & M \\ & {}_i\searrow \quad \nearrow_{i_L} & \\ & L & \end{array}$$

we get $K_M^{[q]} = \operatorname{im}(F^{e*}(i_K)) = \operatorname{im}(F^{e*}(i_L \circ i)) \subseteq \operatorname{im}(F^{e*}(i_L)) = L_M^{[q]}$.

(5) Consider the following commutative diagrams.

$$\begin{array}{ccc} L & \overset{i}{\hookrightarrow} & M \\ \gamma\downarrow & & \downarrow g \\ g(L) & \overset{j}{\hookrightarrow} & N \end{array} \qquad \begin{array}{ccc} F^{e*}(L) & \overset{F^{e*}(i)}{\hookrightarrow} & F^{e*}(M) \\ F^{e*}(\gamma)\downarrow & & \downarrow F^{e*}(g) \\ F^{e*}(g(L)) & \overset{F^{e*}(j)}{\hookrightarrow} & F^{e*}(N) \end{array}$$

Note that $\gamma = g|_L$ is a surjective map. Then

$$\begin{aligned} F^{e*}(g)(L_M^{[q]}) = F^{e*}(g)(\operatorname{im} F^{e*}(i)) &= \operatorname{im}(F^{e*}(g) \circ F^{e*}(i)) \\ &= \operatorname{im}(F^{e*}(g \circ i)) \quad [\text{since } F^{e*} \text{ is a functor}] \\ &= \operatorname{im}(F^{e*}(j \circ \gamma)) = F^{e*}(j)(\operatorname{im} F^{e*}(\gamma)). \end{aligned}$$

Since $F^{e*}$ is right exact and $\gamma$ is surjective, $F^{e*}(\gamma)$ must be surjective as well. Hence $F^{e*}(g)(L_M^{[q]}) = \operatorname{im} F^{e*}(j) = g(L)_N^{[q]}$ by definition of bracket power.

(6) Since $I^{[q]} = (\{a^q \mid a \in I\})$, we have

$$\begin{aligned} (I^{[q]})^n &= (\{a_1^{q\gamma_1} \cdots a_g^{q\gamma_g} \mid a_1, \ldots, a_g \in I,\ \gamma_1 + \cdots + \gamma_g = n\}) \\ &= (\{a_1^{\gamma_1} \cdots a_g^{\gamma_g} \mid a_1, \ldots, a_g \in I,\ \gamma_1 + \cdots + \gamma_g = n\})^{[q]} = (I^n)^{[q]}. \end{aligned}$$

Or, show that $(IJ)^{[q]} = I^{[q]} J^{[q]}$ and use induction on $n$. □

EXERCISE 2. Let $\mathfrak{m}$ be a maximal ideal of $R$ and $I$ an $\mathfrak{m}$-primary ideal. Prove that $(I_\mathfrak{m})^* = (I^*)_\mathfrak{m}$. Conclude that a ring $R$ is weakly $F$-regular if and only if $R_\mathfrak{m}$ is weakly $F$-regular for every maximal ideal $\mathfrak{m}$.

PROOF. We first show that $I^* R_\mathfrak{m} \subseteq (IR_\mathfrak{m})^*$. Let $x/1 \in I^* R_\mathfrak{m}$ such that $x \in I^*$. Then there exists $c \in R^\circ$ such that $cx^q \in I^{[q]}$ for all large $q$. This implies that $(c/1)(x/1)^q \in I^{[q]} R_\mathfrak{m} = (IR_\mathfrak{m})^{[q]}$ for all large $q$. Since $(c/1) \in R_\mathfrak{m}^\circ$, we get $x/1 \in (IR_\mathfrak{m})^*$.

Now, let $x/s \in (IR_\mathfrak{m})^*$. Then there exists $c/c' \in R_\mathfrak{m}^\circ$ such that $cx^q/c's^q \in (IR_\mathfrak{m})^{[q]} = I^{[q]} R_\mathfrak{m}$ for all large $q$. This implies that there exists $t_q \in R \smallsetminus \mathfrak{m}$ such that

$t_q c x^q \in I^{[q]}$ for all large $q$. Since $I$ is an $\mathfrak{m}$-primary ideal, we get $cx^q \in (I^{[q]} : t_q) = I^{[q]}$ for all large $q$. Thus, $cx^q/1 \in I^{[q]}R_{\mathfrak{m}}$ for all large $q$. Note that

$$c \in (R_{\mathfrak{m}})^\circ \bigcap R = R \setminus \left(\bigcup\{P \mid P \in \operatorname{Min}(R), P \subseteq \mathfrak{m}\}\right).$$

Let

$$d \in \left(\bigcap_{\substack{P \in \operatorname{Min}(R) \\ c \notin P}} P\right) \setminus \left(\bigcup_{\substack{P \in \operatorname{Min}(R) \\ c \in P}} P\right)$$

be a non-zero element (using prime avoidance lemma). Then $d/1$ is a nilpotent element in $R_{\mathfrak{m}}$ (as for all $P \in \operatorname{Min}(R)$ such that $P \subseteq \mathfrak{m}$, $c \notin P$ and thus $d \in P$) and by replacing it by its own power we may assume that $d/1$ is actually zero in $R_{\mathfrak{m}}$. Then $c' = c + d$ is an element of $R^\circ$ and $c'x^q \in I^{[q]}R_{\mathfrak{m}} \cap R$. Since $I$ is $\mathfrak{m}$-primary, $I^{[q]}R_{\mathfrak{m}} \cap R = I^{[q]}$. But this says that $x \in I^*$. Thus, $x/s \in I^*R_{\mathfrak{m}}$ and so, $(IR_{\mathfrak{m}})^* \subseteq I^*R_{\mathfrak{m}}$.

Let $R$ be weakly $F$-regular and let $\mathfrak{m}$ be a maximal ideal in $R$. We need to show that $R_{\mathfrak{m}}$ is weakly $F$-regular. Let $J$ be an $\mathfrak{m}R_{\mathfrak{m}}$-primary ideal in $R_{\mathfrak{m}}$. Using Proposition 5.14(2), it is sufficient to show that $J$ is tightly closed. Observe that there exists an ideal $I$ of $R$ such that $J = IR_{\mathfrak{m}}$. Since $J$ is an $\mathfrak{m}R_{\mathfrak{m}}$-primary ideal, we get that $I$ is $\mathfrak{m}$-primary. Thus, $J^* = (IR_{\mathfrak{m}})^* = I^*R_{\mathfrak{m}} = IR_{\mathfrak{m}} = J$. Hence, $J$ is tightly closed.

Now let $R_{\mathfrak{m}}$ be weakly $F$-regular, for all maximal ideals $\mathfrak{m}$ in $R$. Using Proposition 5.14, $R$ is weakly $F$-regular if every ideal primary to a maximal ideal is tightly closed in $R$. Let $\mathfrak{m}$ be a maximal ideal and $I$ be an $\mathfrak{m}$-primary ideal. Then $I^*R_{\mathfrak{m}} = (IR_{\mathfrak{m}})^* = IR_{\mathfrak{m}}$. Since this is true for all the maximal ideals, we get $I^* = I$.

**Alternate proof of $(I_{\mathfrak{m}})^* \subseteq (I^*)_{\mathfrak{m}}$:** Suppose $R$ is a domain. Let $\mathfrak{m}$ be a maximal ideal and $I \subseteq \mathfrak{m}$ an $\mathfrak{m}$-primary ideal. Let $x/s \in (I_{\mathfrak{m}})^*$. Then there is some $c \in (R_{\mathfrak{m}})^\circ$ with $cx^q/s^q \in I^{[q]}R_{\mathfrak{m}}$ for all $q \gg 0$. Then for each such $q$, there is some $t_q \in R \setminus \mathfrak{m}$ with $t_q c x^q \in I^{[q]}$. But since $I^{[q]}$ is $\mathfrak{m}$-primary, it follows that $cx^q \in I^{[q]}$. The fact that $c/1 \in (R_{\mathfrak{m}})^\circ$ means $c/1 \neq 0$ in $R_{\mathfrak{m}}$, which implies $c \neq 0$ in $R$ since $R$ is a domain. Thus $x \in I^*$, so $x/s \in (I^*)_{\mathfrak{m}}$.

Now drop the assumption that $R$ is a domain. Let $x/s \in (IR_{\mathfrak{m}})^*$. The minimal primes of $R_{\mathfrak{m}}$ are the ideals of the form $\mathfrak{p}R_{\mathfrak{m}}$, where $\mathfrak{p} \subseteq \mathfrak{m}$ and $\mathfrak{p}$ is a minimal prime of $R$. By Proposition 5.7, for each such $\mathfrak{p}$ we have $\overline{x/s} \in (I(R_{\mathfrak{m}}/\mathfrak{p}R_{\mathfrak{m}}))^* = ((IR/\mathfrak{p})_{\mathfrak{m}})^* = ((IR/\mathfrak{p})^*)_{\mathfrak{m}}$ by the domain case. Thus, there is some $d \in R \setminus \mathfrak{m}$ such that $\overline{dx} \in (IR/\mathfrak{p})^*$ for each such $\mathfrak{p}$. On the other hand, one can pick $e \in R$ such that $e \notin \mathfrak{m}$ but $e$ is in every minimal prime of $R$ that is not contained in $\mathfrak{m}$. Then $\overline{ex} = 0 \in (IR/\mathfrak{p})$ for each such prime $\mathfrak{p}$. Thus, $\overline{dex} \in (IR/\mathfrak{p})^*$ for all minimal primes $\mathfrak{p}$ of $R$, whence by Proposition 5.7 we have $dex \in I^*$. Thus, $x/s \in (I^*)_{\mathfrak{m}}$. □

EXERCISE 3. Let $x_1, \dots, x_n$ be a regular sequence in a local ring, let $s \leq n$, and let $t$ be a positive integer. Then $(x_1^t, \dots, x_s^t, x_{s+1}, \dots, x_n) : (x_1 \cdots x_s)^{t-1} = (x_1, \dots, x_n)$.

PROOF. Clearly, $(x_1, \dots, x_n) \subseteq (x_1^t, \dots, x_s^t, x_{s+1}, \dots, x_n) : (x_1 \cdots x_s)^{t-1}$. We use induction on $s$ to prove the other containment. Let $s = 1$ and let $a \in (x_1^t, x_2, \dots, x_n) : x_1^{t-1}$. Write $ax_1^{t-1} = c_1x_1^t + c_2x_2 + \cdots + c_nx_n$, for some $c_1, \dots, c_n \in R$. Then

$$(a - c_1x_1)x_1^{t-1} - c_2x_2 - \cdots - c_nx_n = 0.$$

Since $x_1^{t-1}, x_2, \dots, x_n$ is a regular sequence, we get $a - c_1x_1 \in (x_1^{t-1}, x_2, \dots, x_n)$ and hence $a \in (x_1, \dots, x_n)$. We now prove that $(x_1^t, \dots, x_{s+1}^t, x_{s+2}, \dots, x_n) : (x_1 \cdots x_{s+1})^{t-1} \subseteq (x_1, \dots, x_n)$. Let $a \in R$ such that $a(x_1 \cdots x_{s+1})^{t-1} \in (x_1^t, \dots, x_{s+1}^t, x_{s+2}, \dots, x_n)$. In other words,

$$ax_{s+1}^{t-1}(x_1 \cdots x_s)^{t-1} \in (x_{s+1}^t, x_1^t, \dots, x_s^t, x_{s+2}, \dots, x_n).$$

Since $R$ is a local ring, $x_{s+1}, x_1^t, \dots, x_s^t, x_{s+2}, \dots, x_n$ is a regular sequence and hence using $s = 1$ case, we get $a(x_1 \cdots x_s)^{t-1} \in (x_1^t, \dots, x_s^t, x_{s+1}, \dots, x_n)$ implying that $a \in (x_1, \dots, x_n)$ using induction hypothesis. □

EXERCISE 4 (Tight closure Krull intersection theorem). Suppose $(R, \mathfrak{m})$ is local and admits a test element. Then for any ideals $I$ and $J$, we have $\bigcap_{t \in \mathbb{N}} (J + \mathfrak{m}^t I)^* = J^*$. Conclude that every tightly closed ideal is an intersection of $\mathfrak{m}$-primary tightly closed ideals.

PROOF. Clearly, $J^* \subseteq \bigcap_{t \in \mathbb{N}} (J + \mathfrak{m}^t I)^*$. Let $c$ be a test element of $R$ and let $a \in \bigcap_{t \in \mathbb{N}} (J + \mathfrak{m}^t I)^*$. Then $ca^q \in \bigcap_{t \in \mathbb{N}} (J + \mathfrak{m}^t I)^{[q]} = \bigcap_{t \in \mathbb{N}} (J^{[q]} + \mathfrak{m}^{t[q]} I^{[q]})$ for all $q$. By Krull Intersection theorem, $\bigcap_{t \in \mathbb{N}} (J^{[q]} + \mathfrak{m}^{t[q]} I^{[q]}) = J^{[q]}$ and hence $a \in J^*$.

Take $I = R$. Then for any tightly closed ideal $J$, we have $J = J^* = \bigcap_{t \in \mathbb{N}} (J + \mathfrak{m}^t)^*$, where $(J + \mathfrak{m}^t)^*$ are tightly closed $\mathfrak{m}$-primary ideals. □

EXERCISE 5 (Tight closure Nakayama lemma). Let $(R, \mathfrak{m})$ be a Noetherian local ring and $I, J$ ideals such that $J \subseteq I \subseteq (J + \mathfrak{m}I)^*$. Show that $I \subseteq \bigcap_{t \in \mathbb{N}} (J + \mathfrak{m}^t I)^*$. Thus, if $R$ has a test element, conclude from Exercise 4 that $I^* = J^*$.

PROOF. Since $R$ is Noetherian, say $I = (a_1, \dots, a_g)$. Then for every $a_i$, $i = 1, \dots, g$, there exists $c_i \in R^\circ$ and $q_i$ such that $c_i a_i^q \in (J + \mathfrak{m}I)^{[q]}$ for all $q \geq q_i$. Take $c = c_1 \cdots c_g$ and $q_0 = \max_i q_i$. Then

$$cI^{[q]} \subseteq (J + \mathfrak{m}I)^{[q]} = J^{[q]} + \mathfrak{m}^{[q]} I^{[q]} \quad \text{for all } q \geq q_0. \tag{1}$$

We show that $I \subseteq (J + \mathfrak{m}^t I)^*$ for all $t \in \mathbb{N}$ and we prove this by induction on $t$. The result holds for $t = 1$. Assume that $I \subseteq (J + \mathfrak{m}^t I)^*$. Let $x \in I$. Then there exists $c' \in R^\circ$ and $q'$ such that $c'x^q \in (J + \mathfrak{m}^t I)^{[q]}$ for all $q \geq q'$. Hence, for all $q \geq \max\{q_0, q'\}$, we get

$$\begin{aligned} cc'x^q \in c\big(J^{[q]} + (\mathfrak{m}^t)^{[q]} I^{[q]}\big) &= cJ^{[q]} + (\mathfrak{m}^t)^{[q]} (cI^{[q]}) \\ &\subseteq cJ^{[q]} + (\mathfrak{m}^t)^{[q]} (J^{[q]} + \mathfrak{m}^{[q]} I^{[q]}) \\ &\subseteq J^{[q]} + (\mathfrak{m}^{t+1})^{[q]} I^{[q]}, \end{aligned}$$

implying that $x \in (J + \mathfrak{m}^{t+1} I)^*$. Thus, for all $t$, we have $I \subseteq (J + \mathfrak{m}^t I)^*$ and hence $I^* \subseteq ((J + \mathfrak{m}^t I)^*)^* = (J + \mathfrak{m}^t I)^*$. Hence $I^* \subseteq \bigcap_{t \in \mathbb{N}} (J + \mathfrak{m}^t I)^*$. If $R$ has a test element, then using Exercise 4 we get $J^* \subseteq I^* \subseteq \bigcap_{t \in \mathbb{N}} (J + \mathfrak{m}^t I)^* = J^*$ and hence $I^* = J^*$. □

EXERCISE 6. Let $g : R \to S$ be a ring map. Suppose either that $g(R^\circ) \subseteq S^\circ$, or $R$ has a weak (resp. big weak) test element $c$ such that $g(c) \in S^\circ$. Then tight closure persists across $g$ for ideals (resp. modules). In particular, persistence always holds for inclusions of domains.

PROOF. We prove the ideal case first. Let $I$ be an $R$-ideal. We need to show that $g(I^*) \subseteq (IS)^*$. First, suppose that $g(R^\circ) \subseteq S^\circ$. Let $x \in I^*$. Then there exists $c \in R^\circ$ such that $cx^q \in I^{[q]}$ for all large $q$. This implies that $g(c)g(x)^q = g(cx^q) \in g(I^{[q]}) \subseteq (IS)^{[q]}$ for all large $q$. Since $g(c) \in S^\circ$, we get $g(x) \in (IS)^*$.

Now, let $c$ be a $q_0$-weak test element of $R$ such that $g(c) \in S^\circ$. Then for any $x \in I^*$, $cx^q \in I^{[q]}$ for all $q \geq q_0$. Hence, $g(c)g(x)^q = g(cx^q) \in g(I^{[q]}) \subseteq (IS)^{[q]}$ for all $q \geq q_0$. Since $g(c) \in S^\circ$, we get $g(x) \in (IS)^*$.

For the module case, let $M$ be an $R$-module and $L \subseteq M$ be an $R$-submodule of $M$. Consider the following commutative diagrams.

$$\begin{array}{ccc} L = L \otimes_R R & \xrightarrow{i} & M = M \otimes_R R \\ \downarrow & & \downarrow {\scriptstyle 1\otimes g} \\ L \otimes_R S & \xrightarrow{i\otimes S} & M \otimes_R S \end{array} \qquad \begin{array}{ccc} F^{e*}(L) & \xrightarrow{F^{e*}(i)} & F^{e*}(M) \\ \downarrow & & \downarrow {\scriptstyle g'} \\ F^{e*}(L \otimes_R S) & \xrightarrow{F^{e*}(i\otimes S)} & F^{e*}(M \otimes_R S) \end{array}$$

Note that $g'(L_M^{[q]}) = g'(\operatorname{im} F^{e*}(i)) \subseteq \operatorname{im} F^{e*}(i \otimes S) \subseteq (LS)^{[q]}_{M\otimes_R S}$. We need to show that $(1 \otimes g)(L^*_M) \subseteq (LS)^*_{M\otimes_R S}$.

First, suppose that $g(R^\circ) \subseteq S^\circ$. Let $x \in L^*_M$. Then there exists $c \in R^\circ$ such that $cx^q \in L_M^{[q]}$ for all large $q$. This implies that $g(c)((1 \otimes g)(x))^q = g'(cx^q) \in g'(L_M^{[q]}) \subseteq (LS)^{[q]}_{M\otimes_R S}$ for all large $q$. Since $g(c) \in S^\circ$, we get $(1 \otimes g)(L^*_M) \subseteq (LS)^*_{M\otimes_R S}$.

Now, let $c$ be a big $q_0$-weak test element of $R$ such that $g(c) \in S^\circ$. Let $x \in L^*_M$. Then $cx^q \in L_M^{[q]}$ for all $q \geq q_0$. This implies that $g(c)((1 \otimes g)(x))^q = g'(cx^q) \in g'(L_M^{[q]}) \subseteq (LS)^{[q]}_{M\otimes_R S}$ for all $q \geq q_0$. Since $g(c) \in S^\circ$, we get $(1 \otimes g)(L^*_M) \subseteq (LS)^*_{M\otimes_R S}$. □

## References


[AHS96] Ian M. Aberbach, Craig Huneke, and Karen E. Smith, *A tight closure approach to arithmetic Macaulayfication*, Illinois J. Math. **40** (1996), no. 2, 310–329. MR1398097

[Ale18] Thomas Ales, *The *-core of the graded maximal ideal in a Stanley-Reisner ring*, arXiv:1810.05896 [math.AC], 2018.

[And18] Yves André, *La conjecture du facteur direct* (French, with French summary), Publ. Math. Inst. Hautes Études Sci. **127** (2018), 71–93, DOI 10.1007/s10240-017-0097-9. MR3814651

[AM69] M. F. Atiyah and I. G. Macdonald, *Introduction to commutative algebra*, Addison-Wesley Publishing Co., Reading, Mass.-London-Don Mills, Ont., 1969. MR242802

[BM10] Holger Brenner and Paul Monsky, *Tight closure does not commute with localization*, Ann. of Math. (2) **171** (2010), no. 1, 571–588, DOI 10.4007/annals.2010.171.571. MR2630050

[Bou72] Nicolas Bourbaki, *Elements of mathematics. Commutative algebra*, Hermann, Paris; Addison-Wesley Publishing Co., Reading, MA, 1972. Translated from the French. MR360549

[Eps05] Neil M. Epstein, *A tight closure analogue of analytic spread*, Math. Proc. Cambridge Philos. Soc. **139** (2005), no. 2, 371–383, DOI 10.1017/S0305004105008546. MR2168094

[Eps12] Neil Epstein, *A guide to closure operations in commutative algebra*, Progress in commutative algebra 2, Walter de Gruyter, Berlin, 2012, pp. 1–37. MR2932590

[Fed83] Richard Fedder, *F-purity and rational singularity*, Trans. Amer. Math. Soc. **278** (1983), no. 2, 461–480, DOI 10.2307/1999165. MR701505

[GVM20] Kriti Goel, J. K. Verma, and Vivek Mukundan, *Tight closure of powers of ideals and tight Hilbert polynomials*, Math. Proc. Cambridge Philos. Soc. **169** (2020), no. 2, 335–355, DOI 10.1017/s0305004119000215. MR4138926

[HH89] Melvin Hochster and Craig Huneke, *Tight closure and strong F-regularity*, Mém. Soc. Math. France (N.S.) **38** (1989), 119–133. Colloque en l'honneur de Pierre Samuel (Orsay, 1987). MR1044348

[HH93] Melvin Hochster and Craig Huneke, *Phantom homology*, Mem. Amer. Math. Soc. **103** (1993), no. 490, vi+91.

[HH94] Melvin Hochster and Craig Huneke, *F-regularity, test elements, and smooth base change*, Trans. Amer. Math. Soc. **346** (1994), no. 1, 1–62, DOI 10.2307/2154942. MR1273534

[Hoc73] Melvin Hochster, *Contracted ideals from integral extensions of regular rings*, Nagoya Math. J. **51** (1973), 25–43. MR349656

[Hoc77] Melvin Hochster, *Cyclic purity versus purity in excellent Noetherian rings*, Trans. Amer. Math. Soc. **231** (1977), no. 2, 463–488, DOI 10.2307/1997914. MR463152

[Hoc04a] Melvin Hochster, *Current state of the homological conjectures*, downloadable at `http://www.math.utah.edu/vigre/minicourses/algebra/hochster.pdf`, 2004.

[Hoc04b] Melvin Hochster, *Tight closure theory and characteristic p methods*, Trends in commutative algebra, Math. Sci. Res. Inst. Publ., vol. 51, Cambridge Univ. Press, Cambridge, 2004, pp. 181–210, DOI 10.1017/CBO9780511756382.007. With an appendix by Graham J. Leuschke. MR2132652

[HR74] Melvin Hochster and Joel L. Roberts, *Rings of invariants of reductive groups acting on regular rings are Cohen-Macaulay*, Advances in Math. **13** (1974), 115–175, DOI 10.1016/0001-8708(74)90067-X. MR347810

[Hun96] Craig Huneke, *Tight closure and its applications*, CBMS Regional Conference Series in Mathematics, vol. 88, Published for the Conference Board of the Mathematical Sciences, Washington, DC; by the American Mathematical Society, Providence, RI, 1996. With an appendix by Melvin Hochster, DOI 10.1016/0167-4889(95)00136-0. MR1377268

[Hun98] Craig Huneke, *Tight closure, parameter ideals, and geometry [MR1648666]*, Six lectures on commutative algebra, Mod. Birkhäuser Class., Birkhäuser Verlag, Basel, 2010, pp. 187–239, DOI 10.1007/978-3-0346-0329-4_3. MR2641238

[Kaw02] Takesi Kawasaki, *On arithmetic Macaulayfication of Noetherian rings*, Trans. Amer. Math. Soc. **354** (2002), no. 1, 123–149, DOI 10.1090/S0002-9947-01-02817-3. MR1859029

[Moo10] Eliakim H. Moore, *Introduction to a form of general analysis*, The New Haven Mathematical Colloquium, Yale University Press, New Haven, 1910, pp. 1–150.

[Pet64] J. W. Petro, *Some results on the asymptotic completion of an ideal*, Proc. Amer. Math. Soc. **15** (1964), 519–524, DOI 10.2307/2034739. MR162814

[Sha10] Rodney Y. Sharp, *An excellent F-pure ring of prime characteristic has a big tight closure test element*, Trans. Amer. Math. Soc. **362** (2010), no. 10, 5455–5481, DOI 10.1090/S0002-9947-10-05166-4. MR2657687

Department of Mathematical Sciences, George Mason University, Fairfax, Virginia 22030

Indian Institute of Technology, Jodhpur, India

Contemporary Mathematics
Volume **831**, 2025
https://doi.org/10.1090/conm/831/16629

# F-rational rings and rational singularities

Florian Enescu, with exercise solutions by Kyle Maddox

ABSTRACT. These lectures provide an introduction to the notion of F-rationality by following the fundamental work on the subject by Hochster, Huneke, and Smith, and including some of the contributions and ideas of Vélez, Watanabe, Hara, Fedder and Aberbach, among others. The notes start with a careful development of the main concept and grow into a presentation of the main results of the theory, some with proofs sketched, while some included without proof. The connection to rational singularities, an essential feature of the theory, is discussed in the last part of the notes.

CONTENTS



## 1. Terminology, basic facts

Let $R$ be a Noetherian ring of prime characteristic $p$.

We will denote by $\mathrm{Min}(R)$, respectively $\mathrm{Ass}(R)$, the collection of minimal primes, respectively associated primes of $R$. In addition, let $\mathrm{Minh}(R) = \{P \in \mathrm{Spec}(R) : \dim(P) = \dim(R/P)\}$. We say that $R$ is *equidimensional* if $\dim(R/P) = \dim(R) < \infty$ for all minimal primes of $R$. Moreover, we say that $R$ is *unmixed* if $\mathrm{Minh}(R) = \mathrm{Ass}(R)$. We will let $R^o$ denote the complement of the union of the minimal primes of $R$. A ring is *catenary* if, for every pair of prime ideals $P \subseteq Q$ in $R$, the height of $Q/P$, $\mathrm{ht}(Q/P)$, is finite and all maximal chains of prime ideals of $R$ starting with $P$ and ending with $Q$ have the same length. A Noetherian ring $R$ is *universally catenary* if every finite type algebra over $R$ is catenary.

Let $F^e : R \to R$ denote the $e$th iteration of the Frobenius map, that is, $F^e(r) = r^q$, where $q = p^e, e \in \mathbb{N}$. This defines a new $R$-module structure on $R$ denoted here by $R^{(e)}$ (this is denoted in lectures by other authors by $F^e_* R$). For an $R$-module $M$,

2020 *Mathematics Subject Classification.* Primary 13A35, 14B05, 14B15.

let $F^e_R(M) = R^{(e)} \otimes_R M$. For a submodule $N \subseteq M$ we denote $N^{[q]}_M = Im(F^e_R(N) \to F^e_R(M))$ and for $x \in M$, we let $x^q$ denote the image of $1 \otimes x$ in $F^e_R(M)$.

DEFINITION 1.1. Let $I$ be an ideal of $R$. Then the *tight closure* of $I$ is the ideal

$$I^* = \{x \in R: \text{ there exists } c \in R^o \text{ such that } cx^q \in I^{[q]}, \text{ for all } q = p^e \gg 0\}.$$

The *Frobenius closure* of an ideal $I$ is $I^F = \{x \in R : x^q \in I^{[q]}, \text{ for some } q = p^e\}$. Clearly, $I \subseteq I^F \subseteq I^*$.

For an $R$-module $M$ and a submodule $N$ in $M$

$$N^*_M = \{x \in M: \text{ there exists } c \in R^o \text{ such that } cx^q \in N^{[q]}_M, \text{ for all } q = p^e \gg 0\}.$$

We simply will write $N^*$ when $M$ is understood from the context. In general, it is easy to check that, for a multiplicative set $S$ in $R$, we have $S^{-1}(N^*_M) \subseteq (S^{-1}N)^*_{S^{-1}M}$.

Let $(R, \mathfrak{m}, K)$ be a local Noetherian ring and let $\dim(R) = d$. A *system of parameters* is a sequence of elements $x_1, \ldots, x_d$ such that $Rad(x_1, \ldots, x_d) = \mathfrak{m}$. A *parameter ideal* $I$ is an ideal generated by a sequence of elements $x_1, \ldots, x_k$ such that $\text{ht}(I) \geq k$. This condition is equivalent to the images of $x_1, \ldots, x_k$ form a part of system of parameters in any local ring $R_P$, where $P$ is a prime ideal such that $(x_1, \ldots, x_k) \subseteq P$. In this case, $x_1, \ldots, x_k$ are called *parameters*.

PROPOSITION 1.2. *Let $(R, \mathfrak{m})$ be a local, equidimensional and catenary ring, then $x_1, \ldots, x_k$ are parameters in $R$ if and only if they are part of a system of parameters in $R$.*

PROOF. If $x_1, \ldots, x_k$ are parameters, then their images form a part of a system of parameters in $R_{\mathfrak{m}} = R$. Conversely let $x_1, \ldots, x_d$ be a system of parameters for $R$ where $d = \dim(R)$. We will show that $x_1, \ldots, x_k$ generate an ideal of height at least $k$. Assume that $\text{ht}(x_1, \ldots, x_k) = l < k$. So, there exists a minimal prime ideal $Q$ of $(x_1, \ldots, x_k)$ of height $l$. There exists then a minimal prime ideal $P$ of $R$ such that there is a saturated chain of prime ideals of maximal length equal to $l$.

But $\dim(R/(x_1, \ldots, x_k)) = d - k$ per Theorem 14.1 in [**28**], so the length of a maximal saturated between $Q$ and $\mathfrak{m}$ is at most $n - k$. Therefore, there is a saturated chain of prime ideals between $P$ and $\mathfrak{m}$ on length $l + (d - k) < d$. But because $R$ is equidimensional and catenary, the length of this chain should be $d$, a contradiction. □

In the ring $R = K[X, Y, Z]/(XY, XZ) = K[x, y, z]$, with $K$ field, the elements $y, x + z$ form a system of parameters for $R$, yet $\text{ht}((y)) = 0$. Of course, the issue is that the ring $R$ is not equidimensional.

THEOREM 1.3 (Hochster-Huneke; Colon-Capturing Theorem, see for example Theorem 10.1.9 in [**4**]). *Let $R$ be an equidimensional ring, homomorphic image of a Cohen-Macaulay ring and $x_1, \ldots, x_d$ a system of parameters for $R$.*

*Then*

$$(x_1, \ldots, x_i) : x_{i+1} \subseteq (x_1, \ldots, x_i)^*,$$

*for all $i = 0, \ldots, d - 1$.*

The theory of excellent rings plays an important role in this theory. This has a lot to do with Theorem 2.4 below. Recall that for a local ring $R$, the *formal fiber* of $R$ at a prime ideal $P$ of $R$ is the ring $k(P) \otimes_R \widehat{R}$, where $k(P)$ is the residue field of $R_P$ and $\widehat{R}$ is the completion of $R$ at its maximal ideal. Let $K$ be a field. A $K$-algebra $S$ is *geometrically regular* if $L \otimes_K S$ is regular for every finite extension $L$ of $K$. For a ring $R$ the *regular locus* is the set $\{P \in \operatorname{Spec}(R) : R_P \text{ is regular }\}$.

DEFINITION 1.4. A Noetherian ring $R$ is excellent if $R$ is universally catenary, the formal fibers of every localization $R_P$ at a prime ideal $P$ of $R$ are geometrically regular, and, for any finitely generated $R$-algebra $S$, the regular locus of $S$ is Zariski open.

It is known that if $R$ is excellent then every localization of $R$, every homomorphic image of $R$, and every finitely generated $R$-algebra is excellent. Also, complete local rings are excellent. For these claims regarding excellent rings, please see [**27**], Sections 34.A and 34.B.

Moreover, every $F$-finite ring is excellent and every $F$-finite ring is a homomorphic image of a $F$-finite regular ring (these are two non-trivial results due to Kunz [**25**] and, respectively, Gabber [**10**]).

For a local excellent ring, the properties of equidimensionality (see 7.8.3 (x) in [**11**]), reducedness (see page 177 item (7) in [**37**], for example), and normality (see Lemma 15.52. in [**36**]) are preserved under completion.

## 2. F-rationality: definition, main properties

The theory of F-rational rings was developed by Hochster and Huneke in the influential paper [**17**] and, naturally, our approach follows their treatment.

DEFINITION 2.1. Let $R$ be a local ring, homomorphic image of a Cohen-Macaulay ring. We say that $R$ is *F-rational* if and only if every ideal generated by a system of parameters in $R$ is tightly closed. Let $R$ be a homomorphic image of a Cohen-Macaulay ring. Then we say that $R$ is F-rational if and only if $R_{\mathfrak{m}}$ is F-rational, for any maximal ideal $\mathfrak{m}$ of $R$.

THEOREM 2.2. *Let $R$ be an F-rational ring. Then every parameter ideal is tightly closed.*

PROOF. Let $I$ be a parameter ideal and $\mathfrak{m}$ a maximal ideal. Then $IR_{\mathfrak{m}} \subseteq I^*R_{\mathfrak{m}} \subseteq (IR_{\mathfrak{m}})^*$. It is enough to show $IR_{\mathfrak{m}}$ is tightly closed for any arbitrary maximal ideal $\mathfrak{m}$ containing $I$. This is so, because then $IR_{\mathfrak{m}} = I^*R_{\mathfrak{m}}$ for all maximal ideals $\mathfrak{m}$ containing $I$, which implies that $I = I^*$. So, we can localize and assume that $R$ is local. We need to show that every ideal generated by a part of a system of parameters is tightly closed.

Let $x_1, \ldots, x_k$ be such parameters and complete it to a full s.o.p. $x_1, \ldots, x_k, \ldots, x_d$ where $d = \dim(R)$. Then for any $t \geq 0$ we have

$$(x_1, \ldots, x_k) \subseteq (x_1, \ldots, x_k, x_{k+1}^t \cdots, x_d^t)$$

and, hence by taking the tight closure in both sides

$$(x_1, \ldots, x_k)^* \subseteq (x_1, \ldots, x_k, x_{k+1}^t \cdots, x_d^t)^* \subseteq (x_1, \ldots, x_k, x_{k+1}^t \cdots, x_d^t).$$

Therefore $(x_1, \ldots, x_k)^* \subseteq \cap_t (x_1, \ldots, x_k, x_{k+1}^t \cdots, x_d^t) = (x_1, \ldots, x_k)$, by the Krull's intersection theorem. This proves that the ideal $(x_1, \ldots, x_k)$ is tightly closed. □

The following result is due to Hochster and Huneke, see Theorem 4.2 in [**17**].

THEOREM 2.3. *An F-rational ring is normal and Cohen-Macaulay. In particular an F-rational local ring is a normal Cohen-Macaulay domain.*

PROOF. Normality, Cohen-Macaulayness, and F-rationality are local properties, so we can assume that $R$ is a local ring. If $R$ is zero-dimensional, then the zero ideal is tightly closed and hence the nilpotent radical of $R$ is zero. So $R$ is a field. Now assume that $\dim(R) > 0$. A principal ideal of height one is tightly closed in an F-rational ring and hence integrally closed, due to Corollary 5.8 in [**14**]. This condition implies that $R$ is normal by Lemma 5.9 in [**14**], hence $R$ is a local normal domain, and in consequence also equidimensional.

Let $x_1, \ldots, x_d$ be a system of parameters. We will show that they form a regular sequence on $R$. By colon-capturing, as in Theorem 1.3, we have $(x_1, \ldots, x_i) : x_{i+1} \subseteq (x_1, \ldots, x_i)^*$, for all $i = 0, \ldots, d-1$. So $(x_1, \ldots, x_i) : x_{i+1} \subseteq (x_1, \ldots, x_i)$, for all $i = 0, \ldots, d-1$ and this proves that we have a regular sequence. □

EXERCISE 1. Let $R$ be a Noetherian ring, $M$ be a finitely generated $R$-module and $x_1, \ldots, x_n$ and $y_1, \ldots, y_n$ two regular sequences on $M$ that generate the ideals $I$ and, respectively, $J$, with $J \subseteq I$. Write $y_i = \sum_{j=1}^{n} a_{ij} x_j$, for $i = 1, \ldots, n$, with $a_{ij} \in R$ for all $j = 1, \ldots, n$. Denote $A = (a_{ij})_{1 \leq i,j \leq n}$ and $\Delta = \det(A)$. Then

(1) $\Delta I \subseteq J$.
(2) The map

$$M/IM \overset{\cdot\Delta}{\to} M/JM$$

is injective.

The concept of test elements is of major importance in tight closure theory. Let us define them now. We say that $c \in R^o$ is a *test element* for $R$ if for every finitely generated $R$-module $M$ and finitely generated submodule $N$ of $M$ an element $x \in M$ belongs to $N_M^*$ if and only if $cx^q \in N_M^{[q]}$, for all $q = p^e$. A completely stable test element for $R$ is an element $c \in R^o$ such that its image is a test element in every completion of a local ring of $R$.

A very important and difficult result due to Hochster and Huneke is the following theorem on the existence of test elements in tight closure theory, see [**17**] Theorem 6.1.

THEOREM 2.4 (Hochster-Huneke, 1994). *Let $R$ be a reduced ring that is either F-finite or essentially of finite type over a local excellent ring. Assume that $c \in R^o$ such that $R_c$ is weakly F-regular and Gorensteing (e.g. regular). Then a power of $c$ is a test element for $R$.*

Test elements appear naturally in the tight closure computations. Often times in practice, it is very useful to identify concrete elements that are test elements. Another hard result, due to Hochster and Huneke based on work of Lipman and Sathaye, provides such elements. We refer the reader to Theorem 3.3 in [**18**].

THEOREM 2.5. *Let $R$ be a geometrically reduced, equidimensional algebra finitely generated over a field $K$ of prime characteristic $p > 0$. Then the elements of the Jacobian ideal $J(R/K)$ that are in $R^o$ are (completely stable big) test elements for $R$.*

EXERCISE 2 (see Lemma 4.6 in [**17**]). Let $R$ be a Noetherian ring, $I$ an ideal of $R$ and $S$ a multiplicative closed set in $R$.

(1) Then there exists $s \in S$ such that

$$\cup_{w \in S} (I^m : w) = I^m : s^m$$

for all integers $m \geq 0$.

(2) If characteristic of $R$ is $p > 0$, with $p$ prime, and $I$ is generated by a regular sequence of length $n$, then for $s$ chosen as above in (1) we have

$$\cup_{w \in S}(I^{[q]} : w) = I^{[q]} : s^{(n+1)q}$$

for all integers $q = p^e$.

EXERCISE 3. Let $R$ be a Noetherian ring and $W$ a multiplicatively closed set in $R$. Then every element in $(W^{-1}R)^o$ can be written in the form $c/t$ with $c \in R^o$ and $t \in W$.

The following result is Theorem 4.5 in [**17**].

PROPOSITION 2.6. *Let $R$ be a Noetherian ring of characteristic $p$, $p$ prime. Let $I$ be an ideal generated by a regular sequence and $W \subset R$ a multiplicatively closed set. Then $W^{-1}(I^*) = (W^{-1}I)^*$.*

PROOF. Let $u = x/s$, where $s \in W$, $x \in I^*$. Then there exists $c \in R^o$ such that $cx^q \in I^{[q]}$, for large $q$. But then $c(x/s)^q \in (W^{-1}I)^{[q]}$ and $c \in (W^{-1}R)^o$. So, $x/s \in (W^{-1}I)^*$.

Let $I$ be an ideal generated by a regular sequence $x_1, \ldots, x_n$. Assume $x/s \in (W^{-1}I)^*$ and so there exists $c/t \in (W^{-1}R)^o$ such that $(c/t)(x/s)^q \in (W^{-1}I)^{[q]}$ for large enough $q$ and $s, t \in W$. But then we can assume $c \in R^o$ (see Exercise 3 above). There exists $w_q \in W$ such that

$$w_q c x^q \in I^{[q]}.$$

Therefore,

$$cx^q \in \cup_{w \in W}(I^{[q]} : w) = I^{[q]} : s_0^{(n+1)q}$$

for all integers $q$, for some $s_0 \in W$ that is independently chosen of $q$, as in Exercise 2.

This gives

$$c(s_0^{n+1}x)^q \in I^{[q]},$$

and hence $s_0^{n+1}x \in I^*$, for all large $q$. Therefore $u = \frac{x}{s} = \frac{s_0^{n+1}x}{s_0^{n+1}s} \in W^{-1}I^*$. □

EXERCISE 4. Let $(R, \mathfrak{m}, K)$ be an excellent Noetherian local ring of characteristic $p$, $p$ prime. Let $I$ be an $\mathfrak{m}$-primary ideal. Then $I^*\hat{R} = (I\hat{R})^*$.

The following result discusses the behavior of F-rationality under completion, see Lemma 1.4 in [**34**], which is based on ideas from [**17**].

THEOREM 2.7. *If $R$ is local and excellent, then $R$ is F-rational if and only if $\widehat{R}$ is F-rational.*

PROOF. Let $I$ be an ideal generated by system of parameters for $\widehat{R}$. Note that $\mathfrak{m}\widehat{R}$ is the maximal ideal of $\widehat{R}$ and for any element in $y \in \widehat{R}$ and any positive integer $N$ one can find $x \in R$ such that $y - x \in \mathfrak{m}^N\widehat{R}$. These two observations can be used to show that there exists an ideal $J$ generated by a system of parameters in $R$ such that $I = J\widehat{R}$. We leave these details to the reader.

Let $x \in I^* = (J\widehat{R})^* = J^*\widehat{R}$. So, if $J$ is tightly closed in $R$, then $I$ is tightly closed in $\widehat{R}$, which shows that the F-rationality of $R$ implies the F-rationality of $\widehat{R}$.

Conversely, assume $\widehat{R}$ is F-rational. Let $I$ be an ideal generated by a system of parameters in $R$, so $I\widehat{R}$ is an ideal generated by a system of parameters in $\widehat{R}$. But

$I^* = I^*\widehat{R} \cap R = (I\widehat{R})^* \cap R = I\widehat{R} \cap R = I$, a string of equalities where we use that $R \to \widehat{R}$ is faithfully flat. □

THEOREM 2.8 (per Theorem 4.2 (e) in [**17**]). *Assume that $R$ is a homomorphic image of a Cohen-Macaulay ring. The ring $R$ is F-rational if and only if every localization at a maximal ideal of $R$ is F-rational if and only if every localization is F-rational.*

PROOF. We need to show that if $R$ is local F-rational, then $R_P$ is F-rational for any prime ideal $P$ of $R$. Choose $x_1, \ldots, x_k$ such that their images form a system of parameters in $R_P$. Recall that $R$ is Cohen-Macaulay by Theorem 2.3, and so $R_P$ is also Cohen-Macaulay. Let $J$ be the parameter ideal they generate in $R$, so we know that $J = J^*$ by Theorem 2.2. But then $(JR_P)^* = J^*R_P = JR_P$, where the first equality is due to Proposition 2.6. This finishes our proof. □

Note that if $R$ has a test element $c$, then $c(I^{[q]})^* \subseteq I^{[q]}$, for all ideals $I$ and $q$. Contrast this statement to the one below, formulated as an exercise.

EXERCISE 5. Let $I$ be an ideal in a Noetherian ring $R$. Then there exists $c \in R^o$ such that $c(I^*)^{[q]} \subseteq I^{[q]}$, for all $q$ large enough. ( Hint: to solve the exercise, the reader should start by using the definition of tight closure for every element in a generating set of $I^*$.)

The next result was shown by Hochster and Huneke in Theorem 4.3 in [**17**] and our proof follows their approach.

PROPOSITION 2.9. *Let $(R, \mathfrak{m})$ be an equidimensional local ring that is a homomorphic image of a Cohen-Macaulay ring. Let $x_1, \ldots, x_d$ be a system of parameters for $R$, where $d = \dim(R)$. If $R$ admits a test element (for example when $R$ is reduced and excellent), then*

(1) *for any $i = 1, \ldots, d-1$*

$$(x_1, \ldots, x_k)^* : x_{k+1} = (x_1, \ldots, x_k)^*.$$

(2) *If $x_1, \ldots, x_d$ generate a tightly closed ideal, then $(x_1, \ldots, x_k)$ is tightly closed for all $1 \leq k \leq d$.*

(3) *If $x_1, \ldots, x_d$ generate a tightly closed ideal, then they form a regular sequence on $R$ and hence $R$ is Cohen-Macaulay.*

(4) *If $x_1, \ldots, x_d$ generate a tightly closed ideal, then $R$ is F-rational.*

PROOF. (1) One inclusion is clear. So, let us consider $z \in R$ such that $x_{k+1}z \in (x_1, \ldots, x_k)^*$. Hence, there exists $c \in R^o$ such that $c(x_{k+1}z)^q \in (x_1^q, \ldots, x_k^q)$, for $q$ large enough. Therefore

$$cz^q \in (x_1^q, \ldots, x_k^q) : x_{k+1}^q \subseteq (x_1^q, \ldots, x_k^q)^*,$$

where the last inclusion following from Theorem 1.3.

Let $c'$ be a test element for $R$, then

$$c'cz^q \in c'(x_1^q, \ldots, x_k^q)^* \subseteq (x_1^q, \ldots, x_k^q),$$

for large enough $q$ and so $z \in (x_1, \ldots, x_k)^*$.

(2) We have the statement for $k = d$. Assume it is true that $(x_1, \ldots, x_{k+1})$ is tightly closed and let us prove that $(x_1, \ldots, x_k)$ is tightly closed.

Let $z \in (x_1, \ldots, x_k)^* \subseteq (x_1, \ldots, x_{k+1})^* = (x_1, \ldots, x_k) + Rx_{k+1}$. So, we can write $z = x + rx_{k+1}$ such that $x \in (x_1, \ldots, x_k)$. Further we conclude that $rx_{k+1} = z - x \in (x_1, \ldots, x_k)^*$. So,

$$r \in (x_1, \ldots, x_k)^* : x_{k+1} = (x_1, \ldots, x_k)^*,$$

by the first part, which was just established above.

This shows that

$$(x_1, \ldots, x_k)^* = (x_1, \ldots, x_k) + \mathfrak{m} \cdot (x_1, \ldots, x_k)^*,$$

and by NAK we conclude that $(x_1, \ldots, x_k)^* = (x_1, \ldots, x_k)$. Reverse induction concludes the claim for every $1 \leq k \leq d$.

(3) We have just proved that $(x_1, \ldots, x_k)$ are tightly closed for $k = 1, \ldots, d$, if $(x_1, \ldots, x_d$ is tightly closed. Now Theorem 1.3 immediately implies that

$$(x_1, \ldots, x_k)^* : x_{k+1} \subseteq (x_1, \ldots, x_k)^* = (x_1, \ldots, x_k),$$

for all $0 \leq k \leq d-1$, and hence $x_1, \ldots, x_d$ form a regular sequence on $R$.

(4) Let $I_t = (x_1^t, \ldots, x_d^t)$, for any integer $t \geq 1$. We will first show that $I_t$ is tightly closed for every $t \geq 1$. Fix $t \geq 1$. Assume not and hence $I_t^* \not\subseteq I_t$ which means that there exists an element $v$ in $I_t^*$ with nonzero image in the socle of $R/I_t$. The reader can check that the map $R/I \to R/I_t$, given by multiplication by $\mathbf{x}^{t-1}$, where $\mathbf{x} = x_1 \cdots x_d$ provides an isomorphism on socles. So, we can write $v = \mathbf{x}^{t-1} \cdot u$ such that the image of $u$ belongs to the socle of $R/I$ and it is nonzero.

There exists $c \in R^o$ such that $c(\mathbf{x}^{t-1} \cdot u)^q \in I_t^{[q]}$, for all large $q$. This implies that $cu^q \in I_{tq} : \mathbf{x}^{q(t-1)} \subseteq (x_1^q, \ldots x_d^q)$, for all large $q$, where the last inclusion from the regular sequence property for $x_1, \ldots, x_d$ proved in (3). But then we can conclude that $u \in I^*$ and hence $u \in I$, as $I$ is tightly closed, but this contradicts that the image of $u$ is nonzero in $R/I$.

Now, for an arbitrary ideal $J$ generated by a system of parameters, we take a large enough $t$ such that $I_t \subseteq J$ and by Exercise 1, we know that there exists an injection

$$0 \to R/J \to R/I_t,$$

in other words, there exists $c_0 \in R$ such that $(I_t : c_0) = J$. If we take $x \in J^*$ then there exists $c \in R^o$ such that $cx^q \in J^{[q]}$, for large enough $q$. So, $c(c_0x)^q \in I_t^{[q]}$ and hence $c_0x \in I_t^* = I_t$, so $x \in J$.

(Alternatively, one could use Proposition 3.2 instead of the above exercise to produce an inclusion

$$0 \to R/J \to R/I_t$$

and finish then just like above.) □

It would be useful to provide an example of an F-rational ring capturing some of the intricacies of the theory and highlighting the technical difficulties that can appear in practice. As we develop more theory, we will see that there are results in the literature that allow one to give examples more easily than the one that follows below.

EXAMPLE 2.10. Let $K$ be a field of characteristic $p > 0$, $p$ prime greater than 5. Let us show that $R = K[X, Y, Z]/(X^2 + Y^3 + Z^5)$ is F-rational. In what follows, we denote freely by $x, y, z$ the images of $X, Y, Z$, respectively, in $R$. For $p$ greater than 5, the Jacobian ideal contains $x$ and hence $x$ can be used as a completely stable test element. Localizing and completing $R$ at its homogenous maximal ideal we can assume that $R = K[[X,Y,Z]]/(X^2 + Y^3 + Z^5)$, a local complete hypersurface. The elements $y, z$ form a system of parameters in $R$. We plan to show that $(y, z) = (y, z)^*$. If not, the tight closure of $I = (y, z)$ modulo $I$ must intersect the socle of $R/I$ nontrivially. But the socle is 1-dimensional and generated by $x$, which, for simplicity, will also denote the image of $X$ in $R/I$. Hence, if $R$ is not F-rational, then $x \in (y, z)^*$. Since we know that $x$ is a completely stable test element, then we must have

$$xx^p \in (y^p, z^p).$$

But since $p$ is odd, we can write $p+1 = 2k$, for some integer $k$. Then in $R$, $x^{p+1} = x^{2k} = (-y^3 - z^5)^k = (-1)^k \sum_{i=0}^{k} \binom{k}{i} y^{3i} z^{5(k-i)}$, where the binomial coefficient is never zero since $k < p$. Moreover, $(-y^3 - z^5)^k \in (y^p, z^p)$ can be regarded in $K[[y, z]]$, which shows that all terms of the expansion should be in $(y^p, z^p)$.

But, one can show that there exists an integer $0 \leq j \leq k$ such that both $3j < p, 5(k-j) < p$, so the monomial $x^{3j} z^{5(k-j)}$ is not in $(y^p, z^p)$, which is a contradiction.

We invite the reader to study the following ring along the same reasoning lines.

EXERCISE 6. Let $K$ be a field of characteristic $p > 0$, $p$ prime greater than 3. Show that $R = K[X,Y,Z]/(X^2 + Y^3 + Z^7)$ is not F-rational, by showing that $x \cdot x^q \in (y^q, z^q)$, for all $q = p^e$.

DEFINITION 2.11 (Strong F-rationality).

(1) Let $R$ be a reduced Noetherian ring of characteristic $p > 0$, $d \in R^o$, $q$ denote $p^e, e \in \mathbb{N}$ and let $\phi_{d,q} : R \to R^{1/q}$ defined by $\phi_{d,q}(r) = d^{1/q} r$.

We say that $\phi_{d,q}$ is *quasi-split* if it is injective and stays injective upon tensoring with any quotient $R/I$, where $I$ is a parameter ideal in $R$. Note that $R/I \otimes_R R^{1/q} \simeq R^{1/q}/IR^{1/q}$.

(2) We say that $R$ is *strongly F-rational* if and only if for each $d \in R^o$ there exists $q_0$ such that $\phi_{d,q}$ is quasi-split for $q \geq q_0$.

THEOREM 2.12. *Let $R$ be a strongly F-rational ring which is a homomorphic image of a Cohen-Macaulay ring. Then $R$ is F-rational.*

PROOF. If $R$ is strongly F-rational then $R_P$ is strongly F-rational for any prime ideal $P$ in $R$. The argument is as follows: let $P$ be a prime ideal and say that $IR_P$ is generated by parameters in $R_P$, so we can assume that $I = (x_1, \ldots, x_k)$ is a parameter ideal of $R$ contained in $P$. Let $d = a/s$ not in any minimal prime of $R_P$. This means that $a$ is in $R^o$. Let $z = x/t \in R_P$ such that $dz \in I(R/P)^{1/q}$. Lift this relation back to $R$ and obtain that $ax \in IR^{1/q}$. Since $R$ is strongly F-rational, then it follows that there exists $q_0$ such that $\phi_{a,q}$ is quasi-split for $q \geq q_0$ and so $x \in I$ and hence $z \in IR_P$. This shows that $\phi_{d,q}$ is quasi-split for all $q \geq q_0$. Hence $R_P$ is strongly F-rational.

So, we can assume that $R$ is local and let $I$ be an ideal generated by a system of parameters. Assume that $x \in I^*$, so there exists $d \in R^o$ such that $dx^q \in I^{[q]}$ for all large enough $q$. But then $d^{1/q} x \in IR^{1/q}$ and then by the definition of strong F-rationality we get $x \in I$. $\square$

In our next section, we will establish the following theorem due to Vélez, see [**38**] Proposition 1.4. This will be achieved by combining Corollary 3.6 together with Theorem 2.12.

THEOREM 2.13. *Let $R$ be a reduced local ring that is a homomorphic image of a Cohen-Macaulay ring. Then $R$ is F-rational if and only if $R$ is strongly F-rational.*

An application of the characterization given in Theorem 2.13 is in the proof of the next theorem, also due to Vélez. We will only state the result, for completeness, and refer the reader to the details of the proof in [**38**].

THEOREM 2.14 (Vélez). *Let $R$ be a reduced F-finite ring. The set $\{P \in \operatorname{Spec}(R) \mid R_P \text{ is } F - \text{rational}\}$ is Zariski open.*

We close this section with an easy result on the deformation property of F-rational rings. The reader should be aware that the simplicity of this proof is in contrast with studying the deformation property of other classes of rings related to tight closure theory.

PROPOSITION 2.15 (see Theorem 4.7 (c) in [**17**]). *Let $R$ be a local ring, homomorphic image of a Cohen-Macaulay ring. Assume that $z$ is a nonzerodivisor on $R$ and $R/zR$ is F-rational. Then $R$ is F-rational.*

PROOF. Since $R/zR$ is F-rational then it is Cohen-Macaulay and an integral domain by Theorem 2.3. So, $zR$ is a prime ideal in $R$. Additionally, we can complete $z$ to a full system of parameters on $R$ say $z, x_2, \dots, x_d$ which also form a regular sequence on $R$.

We will show that $(z, x_2, \dots, x_d)$ is tightly closed in $R$. Let $cx^q \in (z^q, x_2^q, \dots, x_d^q)$, for some $c \in R^o$ and all large $q$. We can write $c = c'z^t$, for a large $t$ such that $c' \notin zR$.

For $q > t$, we get $c'z^t x^q \in (z^q, x_2^q, \dots, x_d^q)$ which implies, by the regular sequence properties, that $c'x^q \in (z^{q-t}, x_2^q, \dots, x_d^q)$. Now, go modulo $zR$ and use that $R/zR$ is F-rational. One gets that the image of $x$ in $R/zR$ is in $(x_2, \dots, x_d)^*$ in $R/zR$ because $c' \notin zR$ and $R/zR$ is a domain. So, the image of $x$ belongs to $(x_2, \dots, x_d)^* = (x_2, \dots, x_d)$ in the ring $R/zR$, which immediately gives that $x \in (z, x_2, \dots, x_d)$. □

We close this section with the statement of a result that describes the behavior of F-rationality under base change.

We recall here that if $R$ is local and Cohen-Macaulay, then $R$ is *F-injective* if every ideal $I$ generated by a system of parameters is Frobenius closed, see more details in Definition 3.8 and Theorem 3.9. We say that $K$-algebra $R$ is *geometrically F-rational*, respectively, *geometrically F-injective*, if $L \otimes_K R$ is F-rational, respectively, F-injective, for any finite field extension $K \subseteq L$.

THEOREM 2.16 (Aberbach-Enescu, see [**2**], Theorem 4.3). *Let $(R, \mathfrak{m}, K) \to (S, \mathfrak{n}, K)$ be a flat local map of excellent rings. Assume that $R$ is F-rational. Suppose that either*

(1) *there is a common (parameter) test element for $R$ and $S$ in $R^o$ and $S/\mathfrak{m}S$ is geometrically F-injective; or*
(2) *$S/\mathfrak{m}S$ is geometrically F-rational.*

*Then $S$ is F-rational.*

## 3. The Frobenius action on local cohomology

Let $(R, \mathfrak{m})$ be a reduced local Noetherian ring of dimension $d$ and let $x_1, \dots, x_d$ denote a system of parameters for $R$.

A *Frobenius action* on an Artinian $R$-module $M$ is an additive map $F_M : M \to M$ such that $F_M(rm) = r^p F_M(m)$ for all $r \in R, m \in M$. We will often drop the subscript "$-_M$" from our notation when there is no danger of confusion. The main example of Artinian $R$-modules that we will consider is that of local cohomology modules of $R$ with support in the maximal ideal $\mathfrak{m}$ and their submodules. Let $\underline{x} = x_1, \ldots, x_d$ be a system of parameters for $R$. The $i$th local cohomology module of $R$ equal the $i$th cohomology module of the Čech complex

$$0 \to R \to \oplus_{i=1}^d R_{x_i} \to \cdots \to \oplus_{i=1}^d R_{\mathbf{x}_{\hat{i}}} \to R_{\mathbf{x}} \to 0,$$

where $\mathbf{x}_{\hat{i}} = x_1 \cdots x_{i-1}x_{i+1} \cdots x_d$, and $\mathbf{x} = x_1 \cdots x_d$.

The Frobenius $F$ acts on $R$ and its localizations and hence induces an action to the cohomology modules of the Čech complex. We will denote this induced Frobenius action on $H^i_{\mathfrak{m}}(R)$ by $F$.

The particular case of $H^d_{\mathfrak{m}}(R)$ is very important, as this is the only nonzero local cohomology module of $R$ when $R$ is Cohen-Macaulay. An element of $\eta \in H^d_{\mathfrak{m}}(R)$ is denoted by $\eta = [\frac{z}{\mathbf{x}^s}]$ and the Frobenius action $F$ sends $\eta$ to $F(\eta) = [\frac{z^p}{\mathbf{x}^{ps}}]$.

The following alternate description of $H^d_{\mathfrak{m}}(R)$ will be useful later in the paper. Let $x_1, \ldots, x_d$ be a system of parameters in $R$ as before. The local cohomology module $H^d_{\mathfrak{m}}(R)$ can be obtained as a direct limit of $R/(x_1^t, \ldots, x_d^t)$ where the maps of the direct system are given by

$$R/(x_1^t, \ldots, x_d^t) \stackrel{\mathbf{x}^{l-t}}{\to} R/(x_1^l, \ldots, x_d^l),$$

where $l \geq t$ and $\mathbf{x} = x_1 \cdots x_d$.

With this description, we usually denote an element $\eta$ in $H^d_{\mathfrak{m}}(R)$ by $\eta = [z + (x_1^t, \ldots, x_d^t]$, for some $t$. One should remark that the maps in the direct system defining $H^d_{\mathfrak{m}}(R)$ are all injective when $R$ is Cohen-Macaulay.

We also recall that for a local ring $(R, \mathfrak{m}, K)$ , we will denote the injective hull of $K$ as an $R$-module by $E_R(K)$. For a Gorenstein ring $R$ of dimension $d$, there is an $R$-module isomorphism $E_R(K) \simeq H^d_{\mathfrak{m}}(R)$ (Lemma 3.5.4 in [**4**]).

The following result goes back to Smith's thesis and it is based upon a routine check of the definitions introduced above. We encourage the reader to establish it; for a proof see Lemma 10.3.20 in [**4**].

PROPOSITION 3.1. *Using the notation above*

(1)
$$\left[\frac{z}{\mathbf{x}^s}\right] = \left[\frac{z \cdot \mathbf{x}^n}{\mathbf{x}^{s+n}}\right].$$

(2)
$$\left[\frac{z}{\mathbf{x}^s}\right] = 0$$

*if and only if there exist* $n \geq 0$ *such that* $z \cdot \mathbf{x}^n \in (x_1^{s+n}, \ldots, x_d^{s+n})$;

(3) *if* $R$ *is Cohen-Macaulay, then*
$$\left[\frac{z}{\mathbf{x}^s}\right] = 0$$

*if and only if* $z \in (x_1^s, \ldots, x_d^s)$.

PROPOSITION 3.2. *Let* $(R, \mathfrak{m}, K)$ *be a local Cohen-Macaulay ring. Let* $J = (y_1, \ldots, y_d)$, $I = (x_1, \ldots, x_d)$ *be two ideals generated by systems of parameters and denote* $I_t = (x_1^t, \ldots, x_d^t)$, *for an integer* $t \geq 1$.*Then there exists* $t \geq 1$ *and an injective* $R$*-linear map*

$$0 \to R/J \to R/I_t.$$

PROOF. Because $R$ is Cohen-Macaulay, we have that $J_t, I_t$ are all generated by regular sequences in $R$.

Moreover, since $\operatorname{inj\,lim}_t R/J_t = H^d_{\mathfrak{m}}(R)$, we have that $R/J$ injects in $H^d_{\mathfrak{m}}(R)$. But $H^d_{\mathfrak{m}}(R) = \operatorname{inj\,lim}_t R/I_t$ and so there exists $t$ large enough such that $R/J$ injects into $R/I_t$. □

THEOREM 3.3 (Hochster-Huneke, see [**4**], Proposition 10.3.7). *Let $R$ be a Gorenstein ring. Then $R$ is F-rational if and only if $R$ is weakly F-regular.*

PROOF. For Gorenstein rings, both properties of weak F-regularity and F-rationality can be checked locally at maximal ideals. So, we can localize at such ideal and assume that $(R, \mathfrak{m}, K)$ is local. It remains to be shown now that if $R$ is local Gorenstein and F-rational then every ideal is tightly closed. Let $d = \dim(R)$. This statement can be further reduced to checking that every $\mathfrak{m}$-primary ideal is tightly closed (Proposition 10.1.12 (b) in [**4**]). So let us consider such an ideal $I$. Because $I$ is $\mathfrak{m}$-primary, we have that $R/I$ is Artinian and hence it can be embedded in a finite sum of copies of the injective hull $E_R(K) = H^d_{\mathfrak{m}}(R)$. But the local cohomology modules are direct limits of $R/(x_1^t, \ldots, x_d^t)$ for $t \geq 0$, where $x_1, \ldots, x_d$ form a system of parameters on $R$. So, each copy of $E_R(K)$ can be replaced by $R/(x_1^t, \ldots, x_d^t)$ for some large $t$. But 0 is tightly closed in $R/(x_1^t, \ldots, x_d^t)$ since $R$ is F-rational and hence 0 is tightly closed in their finite direct sum and hence in any submodule of theirs, and therefore 0 is tightly closed in $R/I$. But this implies that $I$ is tightly closed in $R$. □

The next two results are due to Vélez, see Lemma 1.3 in [**38**].

THEOREM 3.4. *Let $(R, \mathfrak{m}, K)$ be a local F-rational ring. Let $I = (x_1, \ldots, x_d)$ be an ideal generated by a system of parameters. Denote $I_t = (x_1^t, \ldots, x_d^t)$ for any nonnegative integer $t$ and fix $c \in R^o$. Then there exists $q_0$ such that for all $q \geq q_0$, and $t \geq 1$ the map*

$$R/I_t \to R^{1/q}/I_t R^{1/q}, \text{ sending } 1 \to c^{1/q}$$

*is injective.*

PROOF. Let $J_{t,q} = \{x \in R : cx^q \in I_t^{[q]}\}$. Clearly, $I_t \subseteq J_{t,q}$, for all $q$.

It is easy to see that $J_{t,q'} \subseteq J_{t,q}$ if $q' \geq q$: simply note that $cx^{q'} \in I_t^{[q']}$ implies that $(cx^q)^{q'/q} \in (I_i^{[q]})^{[q'/q]}$. However, this implies that $cx^q \in I_t^{[q]}$, because $q'/q$ is a power of $p$ and $R$ is F-injective being F-rational.

In fact, then there exists $q_0$ such that $J_{1,q_0} = J_{1,q} = I_t$, for all $q \geq q_0$. Indeed, $R/I$ is Artinian, and hence there exists $q_0$ such that $J_{1,q_0} = J_{1,q}$, for all $q \geq q_0$. Note that if $x \in J_{1,q_0} = J_{1,q}$, then $cx^q \in I^{[q]}$ and so $x \in I^* = I$.

Claim: for any $t \geq 1$, $I_t = J_{t,q}$ for all $q \geq q_0$. Let $\operatorname{Soc}(R/I)$ be generated by $v_1, \ldots, v_r$. Let $\mathbf{x} = x_1 \cdots x_d$. Then $\operatorname{Soc}(R/I_t)$ is generated by $\mathbf{x}^{t-1}v_1, \ldots, \mathbf{x}^{t-1}v_r$. If $J_{t,q} \not\subseteq I_t$, there exists $v \in R$ such that $\mathbf{x}^{t-1}v \in J_{t,q}$ while giving a nonzero element $\overline{\mathbf{x}^{t-1}v}$ in $\operatorname{Soc}(R/I_t)$. So, $\overline{v}$ is nonzero in $\operatorname{Soc}(R/I) \subseteq R/I$.

We have $c\mathbf{x}^{(t-1)q}v^q \in I_t^{[q]}$. But $R$ is Cohen-Macaulay and so $cv^q \in I_t^{[q]} : \mathbf{x}^{(t-1)q} \subseteq I^{[q]}$. So, $v \in J_{1,q} = I$, contradicting that $\overline{v}$ is nonzero in $R/I$.

Finally, it remains to note that the vanishing of the kernel of the map $R/I_t \to R^{1/q}/I_t R^{1/q}$, which sends $1 \to c^{1/q}$, follows from our claim because $c^{1/q}x \in /I_t R^{1/q}$ implies $x \in J_{t,q}$. □

THEOREM 3.5. *Let $(R, \mathfrak{m})$ be a local F-rational ring of dimension $d$. Let $c$ in $R$ nonzero. Let $I = (x_1, \ldots, x_d)$ be an ideal generated by a system of parameters. Then there exists $q_0$ such that for $q \geq q_0$ the map*

$$H^d_{\mathfrak{m}}(R) \xrightarrow{\cdot c^{1/q}} H^d_{\mathfrak{m}}(R^{1/q})$$

*is injective.*

PROOF. We let $I_t = (x_1^t, \ldots, x_d^t)$ for any nonnegative integer $t$ and $\mathbf{x} = x_1 \cdots x_d$. Then we have the following commutative diagram

$$\begin{array}{ccc} R/I & \xrightarrow{\cdot c^{1/q}} & R^{1/q}/IR^{1/q} \\ \mathbf{x}^{t-1}\downarrow & & \mathbf{x}^{t-1}\downarrow \\ R/I_t & \xrightarrow{\cdot c^{1/q}} & R^{1/q}/I_tR^{1/q} \\ \downarrow & & \downarrow \\ H^d_{\mathfrak{m}}(R) & \xrightarrow{\cdot c^{1/q}} & H^d_{\mathfrak{m}}(R^{1/q}) \end{array}$$

The vertical arrows are all injective because $R$ is Cohen-Macaulay and the first two horizontal arrows are injective in we choose $q_0$ as in Theorem 3.4 and let $q \geq q_0$. It follows that the last horizontal arrow is also injective, because local cohomology is the direct limit of the system constructed using the vertical arrows. □

COROLLARY 3.6. *Let $(R, \mathfrak{m})$ be a local F-rational ring. Then $R$ is strongly F-rational.*

PROOF. Let $c$ be a nonzero element in $R$. We need to show that if $I$ is a parameter ideal, then there exists $q_0$ such that for all $q \geq q_0$

$$R/I \to R^{1/q}/IR^{1/q}, \text{ sending } 1 \to c^{1/q}$$

is injective.

By Theorem 3.4, we know this statement is true if $I$ is generated by a full system of parameters for $R$. Let $x_1, \ldots, x_k$ be a sequence of parameters wtih $k < d = \dim(R)$ generating the ideal $I$, and complete the sequence to a full system of parameters $x_1, \ldots, x_k, x_{k+1}, \ldots, x_d$. We will write $I_n = (x_1, \ldots, x_k, x_{k+1}^n, \ldots, x_d^n)$, for all $n \geq 0$.

We choose $q_0$ such that

$$H^d_{\mathfrak{m}}(R) \xrightarrow{\cdot c^{1/q}} H^d_{\mathfrak{m}}(R^{1/q})$$

is injective, for $q \geq q_0$, a choice possible by Theorem 3.5.

In the following diagram, let $q \geq q_0$:

$$\begin{array}{ccc} R/I_n & \xrightarrow{\cdot c^{1/q}} & R^{1/q}/I_nR^{1/q} \\ \downarrow & & \downarrow \\ H^d_{\mathfrak{m}}(R) & \xrightarrow{\cdot c^{1/q}} & H^d_{\mathfrak{m}}(R^{1/q}) \end{array}$$

Observe that the vertical arrows are injective since $R$ is Cohen-Macaulay and the bottom horizontal arrow is also injective, as we already noticed above. This implies that the top horizontal arrow is injective. Therefore, $cx^q \in I_n^{[q]}$ implies $x \in I_n$.

Let $x$ in $R$ such that $cx^q \in I^{[q]}$. So, for all $n \geq 1$, $cx^q \in I_n^{[q]}$. By the considerations above we get $x \in I_n$ for all $n$ and so $x \in \cap_{n\geq 1} I_n = I$. □

If we apply the Frobenius functor (or the Peskine-Szpiro functor, as it is often called) to $H^d_{\mathfrak{m}}(R)$, we get

$$F^e(H^d_{\mathfrak{m}}(R)) = R^{(e)} \otimes_R H^d_{\mathfrak{m}}(R), \eta \mapsto 1 \otimes \eta = \eta^q.$$

Note that

$$F^e(H^d_{\mathfrak{m}}(R)) = R^{(e)} \otimes_R H^d_{\mathfrak{m}}(R) = \operatorname{inj}\lim_t R^{(e)} \otimes_R (R/(x_1^t, \ldots, x_d^t))$$
$$\simeq \operatorname{inj}\lim_t (R/(x_1^{qt}, \ldots, x_d^{qt}) \simeq H^d_{\mathfrak{m}}(R)$$

and $F^e(\eta) = \eta^q = [z^q + (x_1^{qt}, \ldots, x_d^{qt}]$, when $\eta = [z + (x_1^t, \ldots, x_d^t]$.

From the definition of tight closure, we see that an element $\eta = [\frac{z}{\mathbf{x}^s}] \in H^d_{\mathfrak{m}}(R)$ belongs to the tight closure of zero in $H^d_{\mathfrak{m}}(R)$, $0^*_{H^d_{\mathfrak{m}}(R)}$, if and only if there exists $c \in R^o$ such that $c \otimes \eta = 0$ in $F^e(H^d_{\mathfrak{m}}(R))$, for all large enough $q = p^e$. According to the discussion above, this is equivalent to $c\eta^q = 0$, for all large enough $q = p^e$.

In addition, $\eta$ belongs to $0^{*fg}_{H^d_{\mathfrak{m}}(R)}$ if and only if there exists a finitely generated submodule $M$ of $H^d_{\mathfrak{m}}(R)$ such that $\eta \in 0^*_M$. The submodule $0^{*fg}_{H^d_{\mathfrak{m}}(R)}$ is called the *finitistic tight closure of zero* in $H^d_{\mathfrak{m}}(R)$.

The following result is due to Smith, Proposition 3.1.1 in [**31**].

PROPOSITION 3.7.

$$0^*_{H^d_{\mathfrak{m}}(R)} = 0^{*fg}_{H^d_{\mathfrak{m}}(R)}.$$

DEFINITION 3.8. Let $(R, \mathfrak{m})$ be a local Noetherian ring of positive characteristic $p$, $p$ prime. Then $R$ is *F-injective* if $F$ acts injectively on $H^i_{\mathfrak{m}}(R)$ for all $i$.

Recall that, for an arbitrary ideal $I$,

$$I^F = \{x \in R : \text{ there exists } q = p^e \text{ such that } x^q \in I^{[q]}\}$$

is the Frobenius closure of $I$. The following characterization is due to Fedder and Watanabe. The reader can find an exploration of this result in the absence of the Cohen-Macaulay hypothesis in more recent work of Quy and Shimomoto [**30**].

THEOREM 3.9 (see Remark 1.9 and Theorem 2.2 in [**8**]). *Let $(R, \mathfrak{m}, K)$ be a Cohen-Macaulay ring. Then $R$ is F-injective if and only if $I = I^F$ for any ideal $I$ generated by a system of parameters.*

PROPOSITION 3.10. *Assume that $(R, \mathfrak{m})$ is local, equidimensional, and a homomorphic image of a Cohen-Macaulay ring.*

*Using the notation above,*

$$\{\eta = \left[\frac{z}{\mathbf{x}^s}\right] : \text{ where } z \in (x_1^s, \ldots, x_d^s)^*, \text{ for some } s\} \subseteq 0^*_{H^d_{\mathfrak{m}}(R)},$$

*with equality whenever $R$ admits a test element.*

PROOF. It is enough to show that $\eta = [\frac{z}{\mathbf{x}}] \in 0^*_{H^d_{\mathfrak{m}}(R)}$ if and only if $z \in (x_1, \ldots, x_d)^*$.

Say $z \in (x_1, \ldots, x_d)^*$ and so there exists $c \in R^o$ such that $cz^q \in (x_1^q, \ldots, x_d^q)$ for $q$ large enough. Therefore, $c\eta^q = 0$ in $H^d_{\mathfrak{m}}(R)$ and hence $\eta \in 0^*_{H^d_{\mathfrak{m}}(R)}$. This argument already shows the inclusion stated in the proposition.

For the reverse implication, note that

$$c\eta^q = 0$$

is equivalent to

$$cz^q\mathbf{x}^s \in (x_1^{q+s}, \dots, x_d^{q+s}) \text{ for some } s$$

This implies that $\eta \in 0^*_{H^d_{\mathfrak{m}}(R)}$ gives that, for a fixed large enough $q$, $cz^q\mathbf{x}^s \in (x_1^{q+s}, \dots, x_d^{q+s})$ for some $s$.

Now by the repeated application of colon-capturing, Theorem 1.3, we get $cz^q \in (x_1^q, \dots, x_d^q)^*$ for $q \gg 0$. If $R$ admits a test element, say $c' \in R^o$, then by multiplying with $c'$ we get $c'cz^q \in (x_1^q, \dots, x_d^q)$ for $q \gg 0$ and hence $z \in (x_1, \dots, x_d)^*$. □

Let $M$ be an $R$-module that admits a Frobenius action $F = F_M$, as defined at the beginning of this section. Let $N$ be a submodule of $M$. We say that $N$ is an *F-stable* submodule of $M$ if $F(N) \subseteq N$. In fact, let $R\{F\}$ be the noncommutative skew-polynomial ring over $R$ with indeterminate $F$ defined by the relations $Fr = r^pF$, for all $r \in R$. Then an $R\{F\}$-module $M$ can be simply regarded as an $R$-module $M$ with Frobenius action $F$, where the module structure on $M$ over $R\{F\}$ is defined by $F \cdot m = F(m)$, for all $m \in M$. Moreover, an F-stable submodule $M$ is a submodule of $M$ as an $R\{F\}$-module.

THEOREM 3.11 (Smith, see Theorem 3.1.6 in [**31**]). *Let $(R, \mathfrak{m}, K)$ be a local Noetherian ring of dimension d. Then $0^*_{H^d_{\mathfrak{m}}(R)}$ is an F-stable submodule of $H^d_{\mathfrak{m}}(R)$. Moreover, if $R$ is excellent and normal (or excellent and with completion a domain), then $0^*_{H^d_{\mathfrak{m}}(R)}$ is the unique maximal proper F-stable submodule of $H^d_{\mathfrak{m}}(R)$.*

PROOF. Let $\eta \in 0^*_{H^d_{\mathfrak{m}}(R)}$. We will show that $F(\eta) \in 0^*_{H^d_{\mathfrak{m}}(R)}$. So, there exists $c \in R^o$ such that $cF^e(\eta) = 0$, for all $e \gg 0$. But then $cF^e(F(\eta)) = 0$ for $e \gg 0$ and so $F(\eta) \in 0^*_{H^d_{\mathfrak{m}}(R)}$.

For the second part, we can complete and assume that $R$ is complete and a domain. We will use Matlis Duality.

Assume there is an F-stable submodule of $H^d_{\mathfrak{m}}(R)$ not contained in $0^*_{H^d_{\mathfrak{m}}(R)}$. Then there exists an element $\eta$ such that the $R$-submodule $N := \langle F^e(\eta) : e \in \mathbb{N}\rangle$ is not contained in $0^*_{H^d_{\mathfrak{m}}(R)}$. Consider the following short exact sequence of Artinian modules

$$0 \to N \to H^d_{\mathfrak{m}}(R) \to C \to 0.$$

Apply $(-)^\vee = \operatorname{Hom}_R(-, E)$, with $E = E_R(K)$ the injective hull of the residue field $K$ of $R$. We get

$$0 \to C^\vee \to H^d_{\mathfrak{m}}(R)^\vee \to N^\vee \to 0.$$

We know that $(H^d_{\mathfrak{m}}(R))^\vee = \omega_R$, where $\omega_R$ the canonical module of $R$. This is a torsion free finitely generated $R$-module of rank one, so $C^\vee$ is also a torsion free $R$-module. So either $Q(R) \otimes C^\vee = 0$ or $Q(R) \otimes_R N^\vee = 0$. Since $C^\vee$ is torsion free we get $Q(R) \otimes C^\vee \neq 0$ (otherwise $C^\vee = 0$ which implies $N = H^d_{\mathfrak{m}}(R)$) and so $Q(R) \otimes_R N^\vee = 0$. This last condition implies that there exists $c \in R$ such that $cN^\vee = 0$, because $N^\vee$ is a finitely generated $R$-module. But then $c(N^\vee)^\vee = 0$, and since $(N^\vee)^\vee \simeq N$ we get $cN = 0$. In particular, we obtain $c \cdot F^e(\eta) = 0$ for all $e \geq 0$. But $c \cdot F^e(\eta) = 0$, for all $e \geq 0$, implies that $\eta \in 0^*_{H^d_{\mathfrak{m}}(R)}$, which is a contradiction. □

THEOREM 3.12 (Smith, see Theorem 2.6 in [**34**]). *Let $R$ be an excellent Cohen-Macaulay local ring of dimension $d$. Then $R$ is F-rational if and only if there are no nonzero F-stable submodules in $H^d_{\mathfrak{m}}(R)$.*

PROOF. Assume that $R$ is F-rational, hence it is also normal by Theorem 2.3, and so $0^*_{H^d_{\mathfrak{m}}(R)}$ is a maximal F-stable submodule of $H^d_{\mathfrak{m}}(R)$, by Theorem 3.11. The F-rationality of $R$ implies that $0^*_{H^d_{\mathfrak{m}}(R)} = 0$ by Proposition 3.10 and so there are no nontrivial submodules of $H^d_{\mathfrak{m}}(R)$.

Conversely, if there no proper non-trivial F-stable submodules of $H^d_{\mathfrak{m}}(R)$ we get $0^*_{H^d_{\mathfrak{m}}(R)} = 0$. Let $x_1, \ldots, x_d$ be a system of parameters for $R$. Let $z \in (x_1, \ldots, x_d)^*$. Then $\eta = \left[\frac{z}{\mathbf{x}}\right] \in 0^*_{H^d_{\mathfrak{m}}(R)} = 0$. Then, when $R$ is Cohen-Macaulay, by Proposition 3.10, we obtain that $z \in (x_1, \ldots, x_d)$ and so $R$ is F-rational. □

DEFINITION 3.13. Let $R$ be a reduced Noetherian ring of characteristic $p$, $p$ prime.

(1) We define the *test ideal* of $R$ as
$$\tau(R) := \cap_{N \leq_{fg} M}(N : N^*_M),$$
where $N \leq_{fg} M$ indicates an inclusion of $R$-modules where $N$ and $M$ are finitely generated.
(2) We define the *big test ideal* of $R$ as
$$\tau_b(R) := \cap_{N \leq M}(N : N^*_M),$$
where $N \leq M$ indicates an inclusion of $R$-modules.
(3) We define the *parameter test ideal* as
$$\tau_{par} := \cap_{I \text{ parameter ideal}}(I : I^*).$$

From the definitions, $\tau_b(R) \subseteq \tau(R) \subseteq \tau_{par}$. Moreover, if $(R, \mathfrak{m}, K)$ is local then $\tau(R) = \mathrm{Ann}_R(0^{*fg}_E)$ according to Theorem 8.23 in [**14**] and similarly $\tau_b(R) = \mathrm{Ann}_R(0^*_E)$, where $E = E_R(K)$ the injective hull of the residue field of $R$. Also, $\tau_{par}(R) = \mathrm{Ann}_R(0^*_{H^d_{\mathfrak{m}}(R)}) = \mathrm{Ann}_R(0^{*fg}_{H^d_{\mathfrak{m}}(R)})$, according to Proposition 4.4 in [**33**]. Moreover, when $R$ is Gorenstein we have $\tau(R) = \tau_{par}(R)$ because $E \simeq H^d_{\mathfrak{m}}(R)$.

The following property of the parameter test ideal is due to K. E. Smith as well, see Theorem 5.2 in [**33**].

THEOREM 3.14. *(Smith) The parameter test ideal of a reduced complete local Cohen-Macaulay ring commutes with localization.*

## 4. Tight closure of parameter ideals

In this section, we let $R$ be a Noetherian ring of characteristic $p$ and want to describe how the tight closure of an ideal generated by parameters arises, in case of local ring or $\mathbb{N}$-graded over $R_0 = K$ a field of characteristic $p$. We structured this section based on the approach in [**22**], which contains more details for the interested reader.

Recall that for an arbitrary ideal $I$, we have already defined its Frobenius closure $I^F$ and its tight closure of $I^*$. The work of Huneke [**22**] has emphasized the relevance of several other operations on ideals to these closures. We will present this connection below.

DEFINITION 4.1. Let $I$ be an ideal generated by parameters $x_1, \ldots, x_m$.

(1) Then the *limit closure* of $I$ is

$$I^{lim} = \cup_{s\geq 1}(x_1^s, \ldots, x_m^s) :_R \mathbf{x}^{s-1},$$

where $\mathbf{x} = x_1 \cdots x_m$.

(2) Then the *germ closure* of $I$ is

$$I^{germ} = \sum_i (x_1, \ldots, x_{i-1}, x_{i+1}, \ldots, x_m)^*.$$

PROPOSITION 4.2 (per [**22**], page 211).

(1) *Let $(R, \mathfrak{m}, K)$ be a local Noetherian ring of dimension $d$. Assume that $R$ is equidimensional and a homomorphic image of a Cohen-Macaulay ring. Let $I$ be generated by a system of parameters. Then*

$$\overline{I^d} + I^{germ} + I^F + I^{lim} \subseteq I^*.$$

(2) *Assume that $R$ is a Noetherian $\mathbb{N}$-graded ring over $R_0 = K$ a field of characteristic $p$ and assume that $x_1, \ldots, x_d$ are a homogeneous system of parameters for $R$. Let $I = (x_1, \ldots, x_d)$ and $D = \sum_{i=1}^d deg(x_i)$. Then*

$$R_{\geq D} + I^{germ} + I^F + I^{lim} \subseteq I^*.$$

PROOF. Note that the colon-capturing theorem gives that $I^{lim} \subseteq I^*$. The only thing that needs to be remark is that, in the local case, $\overline{I^d} \subseteq I$ and, in the graded case, $R_{\geq D} \subseteq I^*$ by the Briançon-Skoda Theorem (as in [**14**], Theorem 5.4 for the local case, and [**16**], Theorem 7.7 and [**33**], Proposition 3.3 for the graded case). □

EXERCISE 7. Let $R$ be a local Cohen-Macaulay ring and $x_1, \ldots, x_d$ be a system of parameters that are test elements as well. Then

$$(x_1^t, \ldots, x_d^t)^* \subseteq (x_1^t, \ldots, x_d^t) + \mathbf{x}^{t-1} \cdot (x_1, \ldots, x_d)^*.$$

The following result of Huneke, see [**22**] Proposition 5.9, shows how sometimes in practice one can reduce the computation of tight closure to the computation of the Frobenius closure.

PROPOSITION 4.3. *Let $R$ be a Noetherian $\mathbb{N}$-graded ring over $R_0 = K$ a field of characteristic $p$ and assume that $x_1, \ldots, x_d$ are a homogeneous system of parameters for $R$. Let $I = (x_1, \ldots, x_d)$ and $D = \sum_{i=1}^d deg(x_i)$.*

*Assume that $R$ is Cohen-Macaulay and that $x_1, \ldots, x_d$ are test elements. Then*

$$(x_1, \ldots x_d)^* = (x_1, \ldots, x_d)^F + R_{\geq D}.$$

PROOF. We have already seen that

$$(x_1, \ldots, x_d)^F + R_{\geq D} \subseteq (x_1, \ldots x_d)^*.$$

Let $z \in (x_1, \ldots x_d)^*$. We can assume that $z$ is homogeneous of degree $n$. Then $z^q \in (x_1^q, \ldots, x_d^q)^*$. But $(x_1^q, \ldots, x_d^q)^* \subseteq (x_1^q, \ldots, x_d^q, \mathbf{x}^{q-1})$, where $\mathbf{x} = x_1 \cdots x_d$, by Exercise 7.

Now if $n < D$, then $nq < D(q-1)$ for large $q$ and hence by degree considerations

$$z^q \in (x_1^q, \ldots, x_d^q),$$

and so $z \in (x_1, \ldots, x_d)^F$. □

Recall that, if $R$ is a domain, we denote $Q(R)$ its fraction field and $R^+$ be the integral closure of $R$ in the algebraic closure of $Q(R)$. This ring is generally non-Noetherian. There is a fundamental result due to Hochster and Huneke in characteristic $p$, $p$ prime, specifically Theorem 5.15 in [**15**], that says the following.

THEOREM 4.4 (Hochster-Huneke). *Let $R$ be a local excellent domain of prime characteristic. Then $R^+$ is a balanced big Cohen-Macaulay $R$-algebra, that is every system of parameters in $R$ forms a regular sequence in $R^+$.*

In relation to F-rationality and, more specifically, tight closure of parameter ideals, we have the following celebrated result.

THEOREM 4.5 (Smith, [**32**]). *Let $(R, \mathfrak{m}, K)$ be locally excellent Noetherian domain of characteristic $p > 0$. Let $I$ be a parameter ideal. Then*

$$I^* = I^+ = IR^+ \cap R.$$

We will sketch below the proof strategy for Theorem 4.5. An additional definition is needed first.

DEFINITION 4.6. Let $(R, \mathfrak{m}, K)$ be a local Noetherian domain. We define the *plus closure* of 0 in $H^d_\mathfrak{m}(R)$ to equal the kernel of the map

$$H^d_\mathfrak{m}(R) \to H^d_\mathfrak{m}(R) \otimes_R R^+ \simeq H^d_\mathfrak{m}(R^+).$$

We will denote this plus closure of 0 by $0^+_{H^d_\mathfrak{m}(R)}$.

The following result is helpful in the proof of Theorem 4.5, since (3) will allow us to move from working with ideals to working on a statement in the highest local cohomology module.

PROPOSITION 4.7. *Let $(R, \mathfrak{m}, K)$ be an excellent, local, $d$-dimensional domain of prime characteristic $p > 0$. Let $x_1, \dots, x_d$ be a system of parameters for $R$ generating the ideal $I$. The following assertions hold.*

(1) *Let $\eta = [\frac{z}{\mathbf{x}}] \in H^d_\mathfrak{m}(R)$, where $\mathbf{x} = x_1 \cdots x_d$. We have*

$$\eta \in 0^+_{H^d_\mathfrak{m}(R)} \text{ if only if } z \in I^+.$$

(2)

$$0^+_{H^d_\mathfrak{m}(R)} \subseteq 0^*_{H^d_\mathfrak{m}(R)}$$

(3)

$$0^+_{H^d_\mathfrak{m}(R)} = 0^*_{H^d_\mathfrak{m}(R)}$$

*if and only if $J^+ = J^*$ for any ideal $J$ generated by a system of parameters.*

PROOF. (1) If $z \in I^+ = IR^+ \cap R$, then $z \in IR^+$ so by definition $\eta = [\frac{z}{\mathbf{x}}] \in 0^+_{H^d_\mathfrak{m}(R)} = Ker(H^d_\mathfrak{m}(R) \to H^d_\mathfrak{m}(R^+))$, because $\eta = 0$ in $H^d_\mathfrak{m}(R^+)$.

Conversely, let $\eta = [\frac{z}{\mathbf{x}}] \in 0^+_{H^d_\mathfrak{m}(R)}$, that is $\eta = 0$ in $H^d_\mathfrak{m}(R^+)$. Since $x_1, \dots, x_d$ form a regular sequence on $R^+$ per Theorem 4.4, the maps in the defining the local cohomology of $R^+$ via $x_1, \dots, x_d$ are injective and we get that $z \in IR^+$. But $z$ is already in $R$ so $z \in I^+$.

(2) Let $\eta = [\frac{z}{\mathbf{x}}] \in 0^+_{H^d_\mathfrak{m}(R)}$. Then, as above in (1), $z \in I^+ = IR^+ \cap R$. So, $z \in IS \cap R$, with $S$ a module-finite extension of $R$. Hence, $z \in I^*$. Therefore $\eta \in 0^*_{H^d_\mathfrak{m}(R)}$, by Proposition 3.10.

(3) Clearly, $J^* = J^+$ for any ideal generated by system of parameters implies that $0^*_{H^d_{\mathfrak{m}}(R)} = 0^+_{H^d_{\mathfrak{m}}(R)}$: indeed, let $\eta = [\frac{z}{\mathbf{x}}] \in 0^*_{H^d_{\mathfrak{m}}(R)}$, so since $R$ has a test element, we get that $z \in (x_1, \ldots, x_d)^*$ by Proposition 3.10. So, $z \in (x_1, \ldots, x_d)^+$, so $\eta \in 0^+_{H^d_{\mathfrak{m}}(R)}$.

Now, conversely, $\eta = [\frac{z}{\mathbf{x}}] \in 0^*_{H^d_{\mathfrak{m}}(R)}$ implies, by Proposition 3.10, that $z \in (x_1, \ldots, x_d)^*$. So, $z \in (x_1, \ldots, x_d)^+$. This further gives that $\eta \in 0^+_{H^d_{\mathfrak{m}}(R)}$.

□

The proof of Theorem 4.5 is based on the following important result, see Proposition 5.5.1 in [**31**].

THEOREM 4.8. *Let $(R, \mathfrak{m}, K)$ be an excellent, local, d-dimensional domain of prime characteristic $p > 0$ with $d > 1$. If*

$$\frac{0^*_{H^d_{\mathfrak{m}}(R)}}{0^+_{H^d_{\mathfrak{m}}(R)}}$$

*is of finite length, then it is zero.*

PROOF. Let $\eta \in 0^*_{H^d_{\mathfrak{m}}(R)}$ and $N_e = 0^+_{H^d_{\mathfrak{m}}(R)} + R\eta + \cdots + R\eta^q \subseteq 0^*_{H^d_{\mathfrak{m}}(R)}$. Due to the finite length assumption, we know that there exists some $e$ such that $N_e = N_{e+1} = \ldots$. The proof ends based on the following non-trivial claim:

If $N_e = N_{e+1}$ for some $e$, then $\eta \in 0^+_{H^d_{\mathfrak{m}}(R)}$.

This claim has a proof based on a so-called modified equational lemma, from work of Hochster, Huneke and Smith but we will not include it here. □

PROOF. (Sketch of proof of Theorem 4.5) The main idea of proof is to perform induction on the dimension. Assume that we have a counterexample, say $R$, of minimal dimension. By several technical results, one can assume that $R$ is local domain and complete and that the ideal $I$ for which $I^+ \neq I^*$ is generated by a full system of parameters $x_1, \ldots, x_d$. Also, one can assume $d \geq 3$ because for $d \leq 2$ the statement follows from more elementary considerations. We are sketching here the argument for $d = 1, 2$ and invite the reader to fill in the details or check Proposition 5.6.2 in [**31**]. For $d = 1$, one essentially needs to check the original statement for principal ideals $I$ but tight closure is the same as integral closure for principal ideals and $\overline{I} = IS \cap R \subseteq IS^+ \cap R = IR^+ \cap R = I^+$, while we already know that $I^+ \subseteq I^*$, where $S$ is the normalization of $R$ and $S^+ = R^+$. For $d = 2$, one reduces the statement to the case where $R$ is normal, hence an isolated singularity, and then one notices that $0^*_{H^d_{\mathfrak{m}}(R)}$ is of finite length and so $\frac{0^*_{H^d_{\mathfrak{m}}(R)}}{0^+_{H^d_{\mathfrak{m}}(R)}}$ is of finite length, and then Theorem 4.8 applies.

Assume that there exists $z \in I^* \setminus I^+$. We now first reduce to the case that $R$ is Gorenstein, and we will show this here: the Cohen Structure Theorem allows us to assume that $(R, \mathfrak{m}, K)$ is presented as a module-finite extension of $A = K[[x_1, \ldots, x_d]]$. There exists an integral dependence equation $F(Z)$ of $z$ over $A$ which gives a module finite extension

$$K[[x_1, \ldots, x_d]][Z]/(F(Z)) \to R.$$

Recall that $z \in I^*$, so, by the definition of tight closure, let $c \in R^o$ such that $cz^q \in (x_1^q, \dots, x_d^q)R$, for all large enough $q = p^e$. We can find an $A[z]$-linear map $\phi : R \to A[z]$ such that $\phi(c) = c' \neq 0$.

Note that

$$cz^q = a_1 x_1^q + \cdots + a_d x_d^q, \text{ where } a_i \in R, i = 1, \dots, d.$$

Apply $\phi$ and we get

$$c' z^q = \phi(a_1) x_1^q + \cdots + \phi(a_d) x_d^q$$

and so $z \in ((x_1, \dots, x_d)A[z])^*$. However, if $((x_1, \dots, x_d)A[z])^* = ((x_1, \dots, x_d)A[z])^+$, then automatically we get $z \in (x_1, \dots, x_d)^+$ in $R$ (note that $A[z]^+ = R^+$.) Therefore we can assume that $z \notin (x_1, \dots, x_d)A[z])^+$ and hence we can assume that $R$ is Gorenstein, since $A[z]$ is.

Now, it remains to show that

$$\frac{0^*_{H^d_{\mathfrak{m}}(R)}}{0^+_{H^d_{\mathfrak{m}}(R)}}$$

is of finite length and apply Theorem 4.8 and Proposition 4.7.

Let us denote $\tau_p(R) = \operatorname{Ann}(0^+_{H^d_{\mathfrak{m}}(R)})$. We take the Matlis dual of $\frac{0^*_{H^d_{\mathfrak{m}}(R)}}{0^+_{H^d_{\mathfrak{m}}(R)}}$ and get $\tau_p(R)/\tau(R)$. Denote $W = \tau_p(R)/\tau(R)$. By Matlis duality, it is enough to show that that this dual has finite length. Note that $R$ is Gorenstein and hence $\tau(R) = \tau_{par}(R)$ and it behaves well under localization at any prime ideal of $R$. Moreover, $\tau_p(R)R_P = \tau_p(R_P)$.

Note that $W$ is Noetherian, so it is enough to show that $W_P = 0$ for any prime ideal $P$ different the maximal ideal. By the assumption of minimal dimension, we have that $R_P$ is not a counterexample, and this implies that $\tau_p(R)R_P = \tau_p(R_P) = \tau(R_P) = \tau(R)R_P$. □

In fact, this result implies that in $R^+$ the ideals generated by a system of parameters in $R$ are tightly closed, suggesting that $R^+$ is, in some loose sense, F-rational, although not Noetherian.

Specifically, assume that $R$ is complete local domain and let $x_1, \dots, x_d$ be a system of parameters in $R^+$. Let $z \in ((x_1, \dots, x_d)R^+)^*$, so there exists nonzero $c$ such that $cz^q \in (x_1^q, \dots, x_d^q)R^+$, for all large $q$. We can find a module-finite ring that is a local complete extension of $R \subseteq S$ containing $c, z, x_1, \dots, x_d$. So, $cz^q \in S \cap (x_1^q, \dots, x_d^q)R^+ \subseteq ((x_1^q, \dots, x_d^q)S)^*$. But $S$ is complete domain and hence there exists a test element say $c'$ in $S$. Therefore we get $c'cz^q \in (x_1^q, \dots, x_d^q)S$. So, by applying Theorem 4.5, $z \in ((x_1, \dots, x_d)S)^* = ((x_1, \dots, x_d)S^+) \cap S = ((x_1, \dots, x_d)R^+) \cap S \subseteq (x_1, \dots, x_d)R^+$. In conclusion, $x_1, \dots, x_d$ generate a tightly closed ideal in $R^+$.

## 5. Connection to rational singularities

DEFINITION 5.1. Let $(R, \mathfrak{m}, K)$ be a normal local ring essentially of finite type over a field of characteristic zero. Let $f : Z \to \operatorname{Spec}(R) = X$ be a resolution of singularities of $X$. We say that the local ring $R$ has *rational singularities* if $R^j f_*(\mathcal{O}_Z) = 0$, for all $j > 0$. A ring $R$ has *rational singularities* if all its localizations $R_P$ at any prime ideal $P$ have rational singularities.

The definition is independent of the resolution of singularities $Z$. It is known that when $(R, \mathfrak{m}, K)$ is an isolated singularity then $R^j f_*(\mathcal{O}_Z) \simeq H^{j+1}_{\mathfrak{m}}(R)$, for $1 \leq j \leq \dim(R) - 2$. Since Spec$(R)$ is affine, $R^j f_*(\mathcal{O}_Z) = H^j(Z, \mathcal{O}_Z)$.

It is known that a rational singularity is Cohen-Macaulay and normal.

Let $R$ be a $\mathbb{N}$-graded ring with the largest homogeneous maximal ideal of $R$ denoted by $\mathfrak{m}$. The local cohomology modules with support in $\mathfrak{m}$ are hence graded themselves, with the grading induced from $R$.

DEFINITION 5.2. Let $R$ be an $\mathbb{N}$-graded ring with homogeneous maximal ideal $\mathfrak{m}$ and $R_0 = K$ a field. The *a-invariant* of $R$, denoted by $a(R)$, is the largest integer $n$ such that the $n$-graded piece of $H^d_{\mathfrak{m}}(R)$ is nonzero.

For example, when $R = K[X_1, \ldots, X_m]/(f_1, \ldots, f_n)$, where $f_i$ is a homogeneous polynomial of degree $d_i$, $i = 1, \ldots, n$ such that $f_1, \ldots, f_n$ form a regular sequence on $R$, then $a(R) = \sum_{i=1}^{n} deg(f_i) - \sum_{i=1}^{m} deg(X_i)$.

The $a$-invariant calculation is very helpful in giving examples of rational singularities due to a result due to Flenner in [**9**] and, independently, Watanabe in [**39**].

THEOREM 5.3 (Flenner-Watanabe; Example 3.9 in [**20**]). *Let $R$ be an $\mathbb{N}$-graded ring with homogeneous maximal ideal $\mathfrak{m}$ and $R_0 = K$ a field of characteristic zero. Then $R$ has rational singularities if and only if the following conditions hold:*

(1) *$R$ is Cohen-Macaulay and normal.*
(2) *$R_P$ has a rational singularity for all primes $P \neq \mathfrak{m}$;*
(3) *$a(R) < 0$.*

EXAMPLE 5.4. Let $R = K[X, Y, Z]/(X^2 + Y^3 + Z^5)$, where $K$ has characteristic zero. If we let $X$ have degree 15, $Y$ degree 10, and $Z$ degree 6, then the equation $X^2 + Y^3 + Z^5$ has degree 30 and the $a$-invariant of $R$ is $a(R) = 30 - (15 + 10 + 5) = -1 < 0$. It can be seen that $R$ is an isolated singularity and hence, by Theorem 5.3, $R$ will have rational singularities.

In fact, when $R$ is graded, we can see that if $\eta \in H^d_{\mathfrak{m}}(R)$ has nonnegative degree then it in fact belongs to $0^*_{H^d_{\mathfrak{m}}(R)}$. Indeed, let $x_1, \ldots, x_d$ be homogeneous parameters such that $deg(x_i) = d_i$ $i = 1, \ldots, d$. Let $\eta = [\frac{z}{\mathbf{x}}] \in H^d_{\mathfrak{m}}(R)$. The grading of $H^d_{\mathfrak{m}}(R)$ gives that the degree of $\eta$ equals $deg(z) - \sum_i deg(x_i)$. Let $D = \sum_i deg(x_i)$. Then the degree of $\eta$ is nonnegative and this implies that $z \in R_{\geq D}$. Now, by Proposition 4.2 (essentially the part due to the Briançon-Skoda theorem), we get $z \in (x_1, \ldots, x_d)^*$ and so $\eta \in 0^*_{H^d_{\mathfrak{m}}(R)}$.

So $a(R) < 0$ implies that there exist no elements of nonnegative degree in $H^d_{\mathfrak{m}}(R)$ and hence the nonzero part of $0^*_{H^d_{\mathfrak{m}}(R)}$ is concentrated in negative degree, if it exists.

A result of Hochster and Huneke from 1994 can be seen as the counterpart of the Flenner-Watanabe theorem stated above. It provides a practical way of checking for F-rationality, since F-injectivity can often be established much easier (for Gorenstein rings, one can employ the Fedder's criterion mentioned in Epstein's lectures).

THEOREM 5.5 (Hochster-Huneke, Theorem 7.2 in [**16**]). *Let $R$ be an $\mathbb{N}$-graded Cohen-Macaulay $K$-algebra, with $R_0 = K$. Assume that its homogeneous maximal ideal $\mathfrak{m}$ has the property that every element in $\mathfrak{m} \cap R^o$ has a power that is a test element. Then the following assertions are equivalent:*

(1) *every ideal generated by a part of a homogeneous system of parameters is tightly closed;*
(2) *there exists an ideal generated by a homogeneous system of parameters that is tightly closed;*
(3) $R_{\mathfrak{m}}$ *is F-rational;*
(4) $a(R) < 0$ *and there is an ideal generated by homogeneous system of parameters* $I$ *such that* $I = I^F$*;*
(5) $R$ *is F-rational.*

The following characterization of rational singularities in characteristic zero will be useful, see [**24**], pages 50-51.

THEOREM 5.6. *Let* $(R, \mathfrak{m}, K)$ *be a Cohen-Macaulay normal local ring essentially of finite type over a field of characteristic zero. Then* $X = \operatorname{Spec}(R)$ *has rational singularities if and only if* $\pi_*(\omega_Z) = \omega_X$ *for any desingularization* $Z$ *of* $X$*, where* $\omega_X$ *is the canonical sheaf of* $X$ *and* $\omega_Z$ *is the canonical sheaf of* $Z$*.*

Due to difficulties of obtaining a desingularization in prime characteristic, Lipman and Teissier defined, in [**26**], an alternate class of singularities, pseudo-rational singularities, that agree with rational singularities.

DEFINITION 5.7 (Lipman-Teissier, see [**26**], also [**34**], Definition 1.8). Let $(R, \mathfrak{m}, K)$ be local $d$-dimensional ring. A Cohen-Macaulay ring, normal, analytically unramified is called *pseudo-rational* if for any proper birational map $\pi : Z \to X = \operatorname{Spec}(R)$ with $Z$ normal and closed fiber $E = \pi^{-1}(\mathfrak{m})$, the canonical map

$$H^d_{\mathfrak{m}}(\pi_*\mathcal{O}_Z) \to H^d_E(\mathcal{O}_Z)$$

is injective.

They established the following theorem in [**26**].

THEOREM 5.8. *Let* $(R, \mathfrak{m}, K)$ *be a local* $d$*-dimensional ring that admits a canonical module. Then* $R$ *is pseudo-rational if and only if* $R$ *is Cohen-Macaulay, normal, analytically unramified and if for any proper birational map* $\pi : Z \to X = \operatorname{Spec}(R)$ *with* $Z$ *normal we have* $\pi_*(\omega_Z) = \omega_X$*, where* $\omega_Z$ *is the canonical sheaf of* $Z$ *and* $\omega_X$ *is the canonical sheaf of* $X$*.*

THEOREM 5.9 (Smith, Theorem 3.1 in [**34**]). *Let* $(R, \mathfrak{m}, K)$ *be an excellent local ring of prime characteristic* $p$*. If* $R$ *is F-rational, then* $R$ *is pseudo-rational.*

It is interesting to know how F-rationality (which is defined in positive characteristic) connects to rational singularities. For this purpose, we discuss briefly the notion of tight closure in characteristic zero and that of a characteristic zero ring of F-rational type.

Let $K$ be a field of characteristic zero and $R = K[X_1, \ldots, X_n]/(f_1, \ldots, f_m)$ be a finitely generated $K$-algebra. Consider an ideal $I$ of $R$ and an element $x$ in $R$. The ideal $I$ and $x$ can be described by providing a finite list of defining information, such as polynomials $g_j$ in $T = K[X_1, \ldots, X_n]$ describing the finitely many generators of $I$ and a polynomial $h$ also from $T$ mapping to $x$. One can collect all the coefficients of the polynomials $f_i$, $g_j$ and $h$ in a $\mathbb{Z}$-algebra $A$ and let $R_A = A[X_1, \ldots, X_d]/(f_1, \ldots, f_n)$ and then construct an ideal $I_A$, $h_A \in R_A$ such that when tensoring with $K$ over $A$ we obtain back $R$, $I$, and $h_A$ maps to $h$. By using the generic freenes criterion, we can invert an element of $A$ and assume that $R_A$ and $R_A/I_A$ are $A$-free in this construction.

Then we define $x$ to be in the *tight closure* of $I$ in $R$ if for all maximal ideals $\mathfrak{m}$ in a dense open subset of the maximal spectrum, $\mathrm{Max}(A)$, of $A$, the image of $x$ in $R(\mathfrak{m}) := K(\mathfrak{m}) \otimes_A R_A$, where $K(\mathfrak{m}) = A/\mathfrak{m}$ is in the tight closure of $IR(\mathfrak{m})$. Note that $A/\mathfrak{m}$ is a finite field and hence it is of prime positive characteristic.

Similarly, we say that $R$ has *F-rational type* if there exists a finitely generated $\mathbb{Z}$-algebra $A$ as above such that the ring $R(\mathfrak{m}) = K(\mathfrak{m}) \otimes_A R_A$ if F-rational for all maximal ideals $\mathfrak{m}$ in a dense open subset of $\mathrm{Max}(A)$.

K. E. Smith's result translates then to the following theorem which is one of the main results in [**34**], see Theorem 4.3.

THEOREM 5.10 (Smith). *Let $X$ be a scheme of finite type over a field of characteristic zero. If $X$ has F-rational type, then $X$ has rational singularities.*

N. Hara was able to provide a converse to this result in [**12**], using deep algebro-geometric techniques that go back to work of Deligne-Illusie on Akizuki-Kodaira-Nakano vanishing theorems in positive characteristic in addition to some work of Fedder and Watanabe. A similar result also obtained independently by Mehta and Srinivas in [**29**].

THEOREM 5.11 (Hara, Theorem 1.1 in [**12**]). *Let $R$ be a finitely generated algebra over a field $K$ of characteristic zero. If $X = \mathrm{Spec}(R)$ has at most rational singularities, then $R$ is of F-rational type.*

## 6. Extended list of exercises with select solutions by Kyle Maddox

EXERCISE 1. Let $R$ be a Noetherian ring, $M$ be a finitely generated $R$-module and $x_1, \ldots, x_n$ and $y_1, \ldots, y_n$ two regular sequences on $M$ that generate the ideals $I$ and, respectively, $J$, with $J \subseteq I$. Write $y_i = \sum_{j=1}^{n} a_{ij} x_j$, for $i = 1, \ldots, n$, with $a_{ij} \in R$ for all $j = 1, \ldots, n$. Denote $A = (a_{ij})_{1 \le i,j \le n}$ and $\Delta = \det(A)$. Then

(1) $\Delta I \subseteq J$.
(2) The map

$$M/IM \overset{\cdot \Delta}{\to} M/JM$$

is injective.

EXERCISE 2. Let $R$ be a Noetherian ring, $I$ an ideal of $R$ and $S$ a multiplicative closed set in $R$. [**4**, 10.3.9]

(1) Then there exists $s \in S$ such that

$$\bigcup_{w \in S} (I^m : w) = I^m : s^m$$

for all integers $m \ge 0$.
(2) If characteristic of $R$ is $p > 0$, with $p$ prime, and $I$ is generated by a regular sequence of length $n$, then for $s$ chosen as above in (1) we have

$$\bigcup_{w \in S} (I^{[q]} : w) = I^{[q]} : s^{(n+1)q}$$

for all integers $q = p^e$.

SOLUTION. Notice $\supset$ is free in both parts, we only need to prove $\subset$ in (1) or (2). Now, for an ideal $J \subset R$, write $J : S$ for the set $\cup_{w \in S} J : w$. Notice that $J : S$ is an ideal:

- $0 \in J \subset J : S$ since $1 \in S$,
- if $a, b \in J : S$ then $wa \in J$ and $w'b \in J$ for some $w, w' \in S$ which implies $ww' \in S$ – since $ww'(a+b) \in J$, we have $a + b \in J : S$,
- and if $r \in R$ and $a \in J : S$, $wa \in J$ so $rwa \in J$ as $J$ is an ideal – so $ra \in J : S$.

We will now show (a). Notice $m = 0$ is free so we assume $m \geq 1$. Recall the **associated graded ring** $\mathcal{G} = \operatorname{gr}_I(R)$ is a Noetherian $\mathbb{N}$-graded ring defined by $[\mathcal{G}]_n = I^n/I^{n+1}$ where $I^0 = R$. This is relevant since $(I^m : S)/I^m = [0 :_{\mathcal{G}} S]_{m-1}$, so we may study the effect of $I^m : S$ for all $m$ by studying the ideal $0 :_{\mathcal{G}} S$ of $\mathcal{G}$. As $\mathcal{G}$ is Noetherian, $0 :_{\mathcal{G}} S$ has finitely many associated primes. So, we can pick an element $s \in S$ for which $0 :_{\mathcal{G}} s$ is maximal among the elements of $S$, in particular, $0 :_{\mathcal{G}} (ws) \subset 0 :_{\mathcal{G}} s$ for all $w \in S$.

We want to show that if $wz \in I^m$ for some $w \in S$ then $s^m z \in I^m$. Since $R$ is local, we can pick a minimal $t$ so that $z \in I^t \setminus I^{t+1}$, and we will use induction on $t$ to show that $s^{m-t} z \in I^m$ which implies $s^m z \in I^m$ since $I^m$ is an ideal. We can clearly assume $t < m$ since if $t \geq m$, $z \in I^t \subset I^m : s^m$ holds trivially. Then $wz \in I^m \subset I^{t+1}$ shows that $swz \in I^{t+1}$ so $sz \in I^{t+1}$. By induction, $s^{m-t}z \in I^m$, so $s^m z \in I^m$. This shows $I^m : w \subset I^m : s^m$ for all $w \in S$, concluding the claim.

Turn to (b). Let $z \in I^{[p^e]} : w$ for some $w \in S$, $e \in \mathbb{N}$. We will show by induction on $t$ that $s^{p^e+t} z \in I^{[p^e]} + I^{p^e+t}$ for all $t$ – this will conclude the proof since $p^e + np^e = (n+1)p^e$ and a usual combinatorial argument shows that since $I$ is $n$-generated, $I^{(n+1)p^e} \subset I^{[p^e]}$ for any $e$. For convenience, write $J_t = I^{[p^e]} + I^{p^e+t}$.

If $t = 0$, $I^{[p^e]} + I^{p^e} = I^{p^e}$ and we know $s^{p^e} z \in I^{p^e}$ by part (a). As our induction hypothesis, suppose $s^{p^e+(t-1)} z \in J_{t-1}$ and we want to show $s^{p^e+t} z \in J_t$. We need to invoke some equations here, so let $I = (x_1, \cdots, x_n)$ to obtain:

$$s^{p^e+t-1} z = \sum_{i=1}^{n} r_i x_i^{p^e} + \sum_{|\mathbf{a}|=p^e+t-1} r_{\mathbf{a}} x_1^{a_1} \cdots x_n^{a_n}$$

where $\mathbf{a} \in \mathbb{N}^n$ is a multi-index and $|\mathbf{a}|$ is the sum of its entries. Furthermore, in the equation above, we can assume $a_i < p^e$ for each $i$, as if $a_i \geq p^e$ for any $i$, that term will live in $I^{[p^e]}$ and so we can move that piece to the left hand sum.

By assumption, $wz \in I^{[p^e]}$, so write $ws^{p^e+t-1} z = \sum_{i=1}^n y_i x_i^{p^e}$. Then we can use the equation above to get:

$$\sum_{i=1}^{n} y_i x_i^{p^e} = \sum_{i=1}^{n} (r_i w) x_i^{p^e} + \sum_{\substack{|\mathbf{a}|=p^e+t-1 \\ a_i < p^e \, \forall i}} (w r_{\mathbf{a}}) x_1^{a_1} \cdots x_n^{a_n}$$

$$\sum_{i=1}^{n} (y_i - r_i w) x_i^{p^e} = \sum_{\substack{|\mathbf{a}|=p^e+t-1 \\ a_i < p^e \, \forall i}} (w r_{\mathbf{a}}) x_1^{a_1} \cdots x_n^{a_n}$$

but since $\underline{x}$ forms a regular sequence, the right hand side is in $I^{[p^e]}$ requires $wr_{\mathbf{a}} \in I$. Thus we may use our trick one more time – $r_{\mathbf{a}} \in I : w \subset I : s$, so $sr_{\mathbf{a}} \in I$ by (a).

Then multiplying the first displayed equation above by $s$, we get:

$$s^{p^e+t}z = \sum_{i=1}^{n}(sr_i)x_i^{p^e} + \sum_{|\mathbf{a}|=p^e+t-1}(sr_{\mathbf{a}})x_1^{a_1}\cdots x_n^{a_n}$$

where the right hand summand is in $I^{p^e+t}$ since $sr_{\mathbf{a}} \in I$ for each $\mathbf{a}$ – this completes the induction and concludes the exercise. □

EXERCISE 3. Let $R$ be a Noetherian ring and $W$ a multiplicatively closed set in $R$. Then every element in $(W^{-1}R)^o$ can be written in the form $c/t$ with $c \in R^o$ and $t \in W$.

EXERCISE 4. Let $(R, \mathfrak{m}, K)$ be an excellent Noetherian local ring of characteristic $p$, $p$ prime. Let $I$ be an $\mathfrak{m}$-primary ideal. Then $I^*\hat{R} = (I\hat{R})^*$. [**4**, 10.3.18]

SOLUTION. First, suppose $R$ is reduced. Then, $R$ has a completely stable big test element $c \in R^\circ$. Since the natural map $R \to \widehat{R}$ is faithfully flat, we have $R^\circ \subset \widehat{R}^\circ$ and for $J = I\widehat{R}$, we have $I^*\widehat{R} \subset J^*$. We then have the inclusions $J \subset I^*\widehat{R} \subset J^*$ to which we may apply $\bullet^*$ to see $J^* \subset (I^*\widehat{R})^* \subset (J^*)^* = J^*$, i.e. $J^* = (I^*\widehat{R})^*$. So, to check that $J^* \subset I^*R$, it suffices to check that if $I$ is tightly closed, then so is $I\widehat{R} = J$.

We now use that there is a natural one-to-one correspondence between the $\mathfrak{m}$-primary ideals of $R$ and the $\widehat{\mathfrak{m}}$-primary ideals of $\widehat{R}$. This correspondence is given by ordinary extension-contraction, i.e. if $I$ is $\mathfrak{m}$-primary, then $I\widehat{R}$ is $\widehat{\mathfrak{m}}$-primary and if $J$ is an $\widehat{\mathfrak{m}}$-primary ideal then $J \cap R$ is an $\mathfrak{m}$-primary ideal.

If $I$ is tightly closed but $J$ is not, then $J^*$ is an $\widehat{\mathfrak{m}}$-primary ideal as $J$ is, so there is an $\mathfrak{m}$-primary ideal $I' \subset R$ with $I \subsetneq I'$. Since $c$ is a test element on $\widehat{R}$ we have $c(J^*)^{[p^e]} \subset J^{[p^e]}$ for all $e \geq 0$ which implies $c(I')^{[p^e]} \subset I^{[p^e]}$ for all $e \geq 0$ by contracting back to $R$. This shows $I' \subset I^* = I$, which gives $J = J^*$, completing the proof when $R$ is reduced.

When $R$ is not reduced, let $R' = R/\sqrt{0}$. Since $R'$ is reduced and excellent, we get a $c + \sqrt{0} \in R'^\circ$ which is a test element for $R'$. But if $e_0$ is smallest so that $\sqrt{0}^{[p^{e_0}]} = 0$ then $c$ is a weak test element for $R$ in the sense that we can repeat the proof above replacing "for all $e \geq 0$" with "for all $e \geq e_0$" and the conclusions still hold, so we are finished. □

EXERCISE 5. Let $I$ be an ideal in a Noetherian ring $R$. Then there exists $c \in R^o$ such that $c(I^*)^{[q]} \subseteq I^{[q]}$, for all $q$ large enough. (Hint: to solve the exercise, the reader should start by using the definition of tight closure for every element in a generating set of $I^*$.)

EXERCISE 6. Show that $R = K[X, Y, Z]/(X^2 + Y^3 + Z^5)$ is not F-rational if the characteristic of $K$ is $2, 3$ or $5$. In fact, show that $R$ is not even F-injective in these cases.

SOLUTION. Write $x, y, z$ for the images of $X, Y, Z$ in $R$. Notice first that $y, z$ forms a system of parameters of $R$ and $x \notin (y, z)$, but:

$$x^2 = -y^3 - z^5 = (-y)y^2 + (-z^3)z^2 \qquad \in (y^2, z^2)$$

$$x^3 = x(-y^3 - z^5) = (-x)y^3 + (-z^2)z^3 \qquad \in (y^3, z^3)$$

$$x^5 = x(-y^3 - z^5)^2 = x(y^6 + 2y^3z^5 + z^{10}) = (xy)y^5 + (2xy^3 + xz^5)z^5 \qquad \in (y^5, z^5)$$

independent of the characteristic of $K$. So if $\mathrm{char}(K) = 2, 3$, or $5$, one of the containments above gives a non-trivial Frobenius closure relation for the parameter ideal $(y, z)$. Hence, $(y, z)$ is not tightly closed and so $R$ is not $F$-rational. We can also see that $R$ is not $F$-injective here since the class $[x + (y, z)] \in H^2_{(x,y,z)}(R)$ is nonzero but vanishes under $F$. Alternatively, we can use Exercise 8 to see $R$ is not $F$-injective. □

EXERCISE 7. Let $R$ be a local Cohen-Macaulay ring and $x_1, \ldots, x_d$ be a system of parameters that are test elements as well. Then

$$(x_1^t, \ldots, x_d^t)^* \subseteq (x_1^t, \ldots, x_d^t) + \mathbf{x}^{t-1} \cdot (x_1, \ldots, x_d)^*.$$

EXERCISE 8. Let $(R, \mathfrak{m}, K)$ be a $d$-dimensional local Cohen-Macaulay ring. Then $R$ is F-injective if and only if there exists a system of parameters $x_1, \ldots, x_d$ generating an ideal that Frobenius-closed. [**6**, 1.4]

SOLUTION. We will show the following are equivalent when $R$ is local and Cohen-Macaulay.

(a) Every parameter ideal is Frobenius closed.
(b) There is a system of parameters $x_1, \cdots, x_d$ such that $(x_1^n, \cdots, x_d^n)$ is Frobenius closed for all $n \geq 1$.
(c) There is a single parameter ideal which is Frobenius closed.
(d) $R$ is $F$-injective.

Clearly (a)⇒(b)⇒(c). Now let $\underline{x} = x_1, \cdots, x_d$ be a system of parameters with $\mathfrak{q} = (\underline{x})$, where we write $\underline{x}^n = x_1^n, \cdots, x_d^n$ and $x = x_1 x_2 \cdots x_d$. Recall that the Frobenius action $F : H^d_{\mathfrak{m}}(R) \to H^d_{\mathfrak{m}}(R)$ is the natural map on the direct limit of the rows induced by the commutative diagram:

$$\begin{array}{ccc} R/(\underline{x}^n) & \xrightarrow{\cdot x} & R/(\underline{x}^{n+1}) \\ \downarrow{\scriptstyle f_R} & & \downarrow{\scriptstyle f_R} \\ R/(\underline{x}^{np}) & \xrightarrow{\cdot x^p} & R/(\underline{x}^{np+p}) \end{array}$$

where $f_R(z + (\underline{x}^n)) = z^p + (\underline{x}^{np})$ is the natural map $R/(\underline{x}^n) \to F^* R/(\underline{x}^n) \simeq R/(\underline{x}^{np})$.

Now suppose $\mathfrak{q} = \mathfrak{q}^F$. Since $R$ is Cohen-Macaulay and $\underline{x}$ is a system of parameters, we know the maps $\cdot x$ and $\cdot x^p$ in the diagram above are injective since $\underline{x}$ is a regular sequence. Let $S_n$ be the socle of $R/(\underline{x}^n)$, i.e. $S_n = (0 :_{R/(\underline{x}^n)} \mathfrak{m})$, which are $K$-vector spaces of the same dimension[1] for every $n$. Since $xS_n \subset S_{n+1}$ for every $n$, we have $\cdot x : S_n \to S_{n+1}$ is an injective linear transpformation of $K$-vector spaces of the same dimension, and is hence an isomorphism.

Now we let $[z + (\underline{x}^n)] \in H^d_{\mathfrak{m}}(R)$ vanish under $F$. We can assume $z \in S_n$, as $R/(\underline{x}^n)$ is Artinian and is thus an essential extension of its socle. Hence, the injection $(\underline{x}^n)^F/(\underline{x}^n) \hookrightarrow R/(\underline{x}^n)$ is nonzero if and only if it is nonzero after intersecting with $S_n$, so if there is any nontrivial Frobenius closure relation there must also be a nontrivial Frobenius closure relation for a socle element.

Now, if $F[z + (\underline{x}^n)] = [z^p + (\underline{x}^{np})] = 0$ then $z^p \in (\underline{x}^{np})$ since the direct limit system is injective. But then we can lift $z^p \in (\underline{x}^{np})$ to $z'^p \in (\underline{x}^p)$ where $z'$ is the preimage of $z$ under $\cdot x^{n-1}$ in $S_1$. This implies $z' \in (\underline{x})$ since $\mathfrak{q}^F = \mathfrak{q}$. So, $z \in (\underline{x}^n)$

[1]This is the fact that $\dim_K S_n = d$, but that they are of the same dimension is also easily derivable from Proposition 3.1

and $[z + (\underline{x}^n)] = 0$. Thus, $F : H^d_{\mathfrak{m}}(R) \to H^d_{\mathfrak{m}}(R)$ is injective, so $R$ is $F$-injective, showing (c)⇒(d).

Now suppose $R$ is $F$-injective, and let $\underline{x}$ be a system of parameters with $z \in (\underline{x})^F$, so there is an $e$ with $z^{p^e} \in (\underline{x}^{p^e})$. Then $0 = [z^{p^e} + (\underline{x}^{p^e})] = F^e[z + (\underline{x})]$, so $[z + (\underline{x})] = 0$ as $R$ is $F$-injective. This implies $z \in (\underline{x})$ since the direct limit system is injective. So, $(\underline{x}) = (\underline{x})^F$, showing (d)⇒(a). □

EXERCISE 9. Let $R$ be a local F-finite Cohen-Macaulay ring and $z$ a NZD on $R$ such that a power of $z$ is a parameter test element for $R$ and $R/zR$ is F-injective. Prove that $R$ is F-rational. [**8**, 2.13]

PROOF. Since $z$ is a regular element, it can be completed to a full system of parameters $\underline{x} = z, x_2, \cdots, x_d$, and we write $\mathfrak{q} = (z, \cdots, x_d)$ and $\mathfrak{q}' = (x_2, \cdots, x_d)$. To see that $R$ is $F$-rational, it suffices to show $\mathfrak{q}$ is tightly closed.

Let $w \in \mathfrak{q}^*$. We can assume for some $n \in \mathbb{N}$ and all $e \gg 0$ that $z^n w^{p^e} \in \mathfrak{q}^{[p^e]}$ since $z^n$ is a parameter test element. But then $w^{p^e} \in (z^{p^e-n}) + (\mathfrak{q}')^{[p^e]}$ since $\underline{x}$ is a system of parameters in a Cohen-Macaulay ring, and hence a regular sequence. This implies $w^{p^e} + zR \in (\mathfrak{q}')^{[p^e]} + zR$, so:

$$w + zR \in (\mathfrak{q}')^F + zR = \mathfrak{q}' + zR = \mathfrak{q} + zR$$

since $R/zR$ is $F$-injective and $\mathfrak{q}'$ is a parameter ideal in $R/zR$. Consequently, $w \in (z) + \mathfrak{q}' = \mathfrak{q}$, and so $\mathfrak{q}$ is tightly closed and $R$ is $F$-rational. □

EXERCISE 10. Let $R$ be a Noetherian domain. Then the plus closure of an ideal $I$ commutes with localization.

## Acknowledgments

The author thanks Kyle Maddox for his comments on these notes and the anonymous reviewer for valuable feedback that helped improved the exposition. In addition, the influence of the many writings on F-rationality and tight closure theory in general by M. Hochster, C. Huneke, and K. E. Smith in the preparation of these notes is acknowledged.

## References

[1] Ian M. Aberbach, *Tight closure in $F$-rational rings*, Nagoya Math. J. **135** (1994), 43–54, DOI 10.1017/S0027763000004943. MR1295816

[2] Ian M. Aberbach and Florian Enescu, *Test ideals and base change problems in tight closure theory*, Trans. Amer. Math. Soc. **355** (2003), no. 2, 619–636, DOI 10.1090/S0002-9947-02-03162-8. MR1932717

[3] Ian M. Aberbach, Melvin Hochster, and Craig Huneke, *Localization of tight closure and modules of finite phantom projective dimension*, J. Reine Angew. Math. **434** (1993), 67–114, DOI 10.1515/crll.1993.434.67. MR1195691

[4] Winfried Bruns and Jürgen Herzog, *Cohen-Macaulay rings*, Cambridge Studies in Advanced Mathematics, vol. 39, Cambridge University Press, Cambridge, 1993. MR1251956

[5] Florian Enescu, *On the behavior of F-rational rings under flat base change*, J. Algebra **233** (2000), no. 2, 543–566, DOI 10.1006/jabr.2000.8430. MR1793916

[6] Richard Fedder, *$F$-purity and rational singularity*, Trans. Amer. Math. Soc. **278** (1983), no. 2, 461–480, DOI 10.2307/1999165. MR701505

[7] Richard Fedder, *$F$-purity and rational singularity in graded complete intersection rings*, Trans. Amer. Math. Soc. **301** (1987), no. 1, 47–62, DOI 10.2307/2000327. MR879562

[8] Richard Fedder and Keiichi Watanabe, *A characterization of $F$-regularity in terms of $F$-purity*, Commutative algebra (Berkeley, CA, 1987), Math. Sci. Res. Inst. Publ., vol. 15, Springer, New York, 1989, pp. 227–245, DOI 10.1007/978-1-4612-3660-3_11. MR1015520

[9] Hubert Flenner, *Rationale quasihomogene Singularitäten* (German), Arch. Math. (Basel) **36** (1981), no. 1, 35–44, DOI 10.1007/BF01223666. MR612234

[10] Ofer Gabber, *Notes on some t-structures*, Geometric aspects of Dwork theory. Vol. I, II, Walter de Gruyter, Berlin, 2004, pp. 711–734. MR2099084

[11] A. Grothendieck, *Éléments de géométrie algébrique. IV. Étude locale des schémas et des morphismes de schémas. II* (French), Inst. Hautes Études Sci. Publ. Math. **24** (1965), 231. MR199181

[12] Nobuo Hara, *A characterization of rational singularities in terms of injectivity of Frobenius maps*, Amer. J. Math. **120** (1998), no. 5, 981–996. MR1646049

[13] Melvin Hochster, *Tight closure theory and characteristic p methods*, Trends in commutative algebra, Math. Sci. Res. Inst. Publ., vol. 51, Cambridge Univ. Press, Cambridge, 2004, pp. 181–210, DOI 10.1017/CBO9780511756382.007. With an appendix by Graham J. Leuschke. MR2132652

[14] Melvin Hochster and Craig Huneke, *Tight closure, invariant theory, and the Briançon-Skoda theorem*, J. Amer. Math. Soc. **3** (1990), no. 1, 31–116, DOI 10.2307/1990984. MR1017784

[15] Melvin Hochster and Craig Huneke, *Infinite integral extensions and big Cohen-Macaulay algebras*, Ann. of Math. (2) **135** (1992), no. 1, 53–89, DOI 10.2307/2946563. MR1147957

[16] Melvin Hochster and Craig Huneke, *Tight closure of parameter ideals and splitting in module-finite extensions*, J. Algebraic Geom. **3** (1994), no. 4, 599–670. MR1297848

[17] Melvin Hochster and Craig Huneke, *F-regularity, test elements, and smooth base change*, Trans. Amer. Math. Soc. **346** (1994), no. 1, 1–62, DOI 10.2307/2154942. MR1273534

[18] Melvin Hochster and Craig Huneke, *Comparison of symbolic and ordinary powers of ideals*, Invent. Math. **147** (2002), no. 2, 349–369, DOI 10.1007/s002220100176. MR1881923

[19] Melvin Hochster and Joel L. Roberts, *The purity of the Frobenius and local cohomology*, Advances in Math. **21** (1976), no. 2, 117–172, DOI 10.1016/0001-8708(76)90073-6. MR417172

[20] Craig Huneke, *Tight closure and its applications*, CBMS Regional Conference Series in Mathematics, vol. 88, Published for the Conference Board of the Mathematical Sciences, Washington, DC; by the American Mathematical Society, Providence, RI, 1996. With an appendix by Melvin Hochster, DOI 10.1016/0167-4889(95)00136-0. MR1377268

[21] Craig Huneke, *Tight closure and strong test ideals*, J. Pure Appl. Algebra **122** (1997), no. 3, 243–250, DOI 10.1016/S0022-4049(97)00053-4. MR1481089

[22] Craig Huneke, *Tight closure, parameter ideals, and geometry*, Six lectures on commutative algebra (Bellaterra, 1996), Progr. Math., vol. 166, Birkhäuser, Basel, 1998, pp. 187–239. MR1648666

[23] Craig Huneke and Karen E. Smith, *Tight closure and the Kodaira vanishing theorem*, J. Reine Angew. Math. **484** (1997), 127–152, DOI 10.1515/crll.1997.484.127. MR1437301

[24] G. Kempf, Finn Faye Knudsen, D. Mumford, and B. Saint-Donat, *Toroidal embeddings. I*, Lecture Notes in Mathematics, Vol. 339, Springer-Verlag, Berlin-New York, 1973. MR335518

[25] Ernst Kunz, *On Noetherian rings of characteristic p*, Amer. J. Math. **98** (1976), no. 4, 999–1013, DOI 10.2307/2374038. MR432625

[26] Joseph Lipman and Bernard Teissier, *Pseudorational local rings and a theorem of Briançon-Skoda about integral closures of ideals*, Michigan Math. J. **28** (1981), no. 1, 97–116. MR600418

[27] Hideyuki Matsumura, *Commutative algebra*, 2nd ed., Mathematics Lecture Note Series, vol. 56, Benjamin/Cummings Publishing Co., Inc., Reading, MA, 1980. MR575344

[28] Hideyuki Matsumura, *Commutative ring theory*, Cambridge Studies in Advanced Mathematics, vol. 8, Cambridge University Press, Cambridge, 1986. Translated from the Japanese by M. Reid. MR879273

[29] V. B. Mehta and V. Srinivas, *A characterization of rational singularities*, Asian J. Math. **1** (1997), no. 2, 249–271, DOI 10.4310/AJM.1997.v1.n2.a4. MR1491985

[30] Pham Hung Quy and Kazuma Shimomoto, *F-injectivity and Frobenius closure of ideals in Noetherian rings of characteristic $p > 0$*, Adv. Math. **313** (2017), 127–166, DOI 10.1016/j.aim.2017.04.002. MR3649223

[31] Karen Ellen Smith, *Tight closure of parameter ideals and F-rationality*, ProQuest LLC, Ann Arbor, MI, 1993. Thesis (Ph.D.)–University of Michigan. MR2689065

[32] K. E. Smith, *Tight closure of parameter ideals*, Invent. Math. **115** (1994), no. 1, 41–60, DOI 10.1007/BF01231753. MR1248078

[33] Karen E. Smith, *Test ideals in local rings*, Trans. Amer. Math. Soc. **347** (1995), no. 9, 3453–3472, DOI 10.2307/2155019. MR1311917

[34] Karen E. Smith, *F-rational rings have rational singularities*, Amer. J. Math. **119** (1997), no. 1, 159–180. MR1428062
[35] Karen E. Smith, *Tight closure and vanishing theorems*, School on Vanishing Theorems and Effective Results in Algebraic Geometry (Trieste, 2000), ICTP Lect. Notes, vol. 6, Abdus Salam Int. Cent. Theoret. Phys., Trieste, 2001, pp. 149–213. MR1919458
[36] *Stacks project*, `https://stacks.math.columbia.edu/`, 2024.
[37] Irena Swanson and Craig Huneke, *Integral closure of ideals, rings, and modules*, London Mathematical Society Lecture Note Series, vol. 336, Cambridge University Press, Cambridge, 2006. MR2266432
[38] Juan D. Vélez, *Openness of the F-rational locus and smooth base change*, J. Algebra **172** (1995), no. 2, 425–453, DOI 10.1016/S0021-8693(05)80010-9. MR1322412
[39] Keiichi Watanabe, *Rational singularities with $k^*$-action*, Commutative algebra (Trento, 1981), Heidelberger Taschenbücher, Dekker, New York, 1983, pp. 339–351. MR686954

Department of Mathematics and Statistics, Georgia State University, Atlanta Georgia 30303
*Email address*: `fenescu@gsu.edu`

Department of Mathematical Sciences, 850 West Dickson Street, Room 309, University of Arkansas, Fayetteville, Arkansas 72701
*Email address*: `kmaddox@uark.edu`

Contemporary Mathematics
Volume **831**, 2025
https://doi.org/10.1090/conm/831/16630

# Three lectures on test ideals

Kevin Tucker, with excercise solutions by Kenta Sato

## Lecture 1

Test ideals are at the heart of the celebrated theory of tight closure introduced by Hochster and Huneke. Consider a Noetherian ring $R$ of characteristic $p > 0$; we assume here that $R$ is a domain for simplicity. For an ideal $I \subseteq R$, the *tight closure* of $I$ is the ideal given by

$$I^* = \{x \in R \mid \text{there exists a non-zero } c \in R \text{ with } cx^{p^e} \in I^{[p^e]} \text{ for all } e > 0\}$$

where $I^{[p^e]} := F^e(I) \cdot R$ is the expansion of $I$ via the $e$-iterated (absolute) Frobenius endomorphism $F^e \colon R \longrightarrow R$ or $p^e$-th power map $r \mapsto r^{p^e}$. A priori, the non-zero element $c \in R$ witnessing an instance of tight closure may depend on the particular element $x \in I^*$. However, with mild conditions on $R$ satisfied in most arithmetic and geometric settings, Hochster and Huneke showed that there exist non-zero elements that witness all instances of tight closure. Such elements were coined test elements and the ideal of such (together with zero) as the test ideal.

DEFINITION 1.

$$\tau^{\mathrm{fin}}(R) := \{c \in R \mid cx^{p^e} \in I^{[p^e]} \text{ for all } I \subseteq R, x \in I^*, \text{ and all } e \geq 0\}$$

This ideal, which we shall refer to more precisely as the *finitistic (tight closure) test ideal* of $R$, is the first of many test ideals we shall come across in these notes.

Alternatively, given that all ideals in a regular ring are tightly closed, one is naturally led to consider ways in which tight closure might be used to study the singularities of $R$. It is straightforward to check that

$$\tau^{\mathrm{fin}}(R) = \bigcap_{I \subseteq R} (I :_R I^*) \tag{1.1}$$

is an equivalent formulation of the finitistic test ideal, so that $\tau^{\mathrm{fin}}(R)$ in a sense captures the severity of the singularities of $R$ by encoding the extent to which the ideals of $R$ fail to be tightly closed. We have that $\tau^{\mathrm{fin}}(R) = R$ if and only if $R$ is *weakly $F$-regular*, i.e., all of the ideals of $R$ are tightly closed. Here, the qualifier "weakly" is an avatar of one of the most vexing questions about tight closure that remains open to this day: does weak $F$-regularity pass to localizations? From the perspective of test ideals, one could hope the stronger statement that the

The author was partially supported by National Science Foundation Grant No. DMS-2200716 and No. DMS-2501904.

finitistic test ideal is compatible with localizations – say either at prime ideals (i.e., $\tau^{\text{fin}}(R_{\mathfrak{p}}) = \tau^{\text{fin}}(R)_{\mathfrak{p}}$ for $\mathfrak{p} \in \operatorname{Spec}(R)$) or at single elements (i.e., $\tau^{\text{fin}}(R[\frac{1}{f}]) = \tau^{\text{fin}}(R)[\frac{1}{f}]$ for $f \in R$). These too remain open, and it is easy to see from (1.1) where the difficulties lie. Not only does localization fail to commute with infinite intersections in general, tight closure fails to commute with localization at prime ideals as well [**BM10**].

The notion of "strong" $F$-regularity which we shall introduce here next, as well as the non-finitistic flavor of test ideal we shall introduce later on, arose from attempts to address these difficulties. Given an $R$-module $M$, let us write $F^e_* M$ for the $R$-module given by restriction of scalars for $F^e$. Thus, $F^e_* M = \{F^e_* m \mid m \in M\}$ is identified with $M$ as an Abelian group but with $F^e$-twisted $R$-multiplication given by $r(F^e_* m) = F^e_*(r^{p^e} m)$ for $r \in R$ and $m \in M$. Another perspective on the iterated Frobenius is as the map of $R$-modules $R \longrightarrow F^e_* R$ given by $r \mapsto F^e_*(r^{p^e})$. Alternatively, one can also think of the iterated Frobenius as the inclusion $R \subseteq R^{1/p^e}$, which identifies with $R^{1/p^e}$ as an $R$-module. We shall primarily rely on the notation $F^e_* R$ throughout.

We assume hereafter[1] the mild condition that $R$ is Noetherian and *$F$-finite*, i.e., $F_* R$ is a finitely generated $R$-module, from which it follows that $F^e_* R$ is a finitely generated $R$-module for all $e > 0$. This condition is satisfied, for example, when $R$ is essentially of finite type over a complete local ring with $F$-finite residue field – the reader is encouraged to think of the kinds of rings arising most naturally in geometric settings such as localizations of quotients of polynomial rings over perfect fields. $F$-finiteness is easily seen to pass to localization (as $W^{-1} F^e_* R = F^e_*(W^{-1} R)$ for $W \subseteq R$ a multiplicative set) and completion when $(R, \mathfrak{m})$ is local (as $\widehat{R} \otimes_R F^e_* R = \widehat{F^e_* R} = F^e_* \widehat{R}$). Moreover, Gabber showed [**Gab04**, Remark 13.6] that every Noetherian $F$-finite ring is a quotient of a finite-dimensional excellent regular ring. Among the most important consequences of $F$-finiteness for us here is that $\operatorname{Hom}_R(F^e_* R, R)$ is compatible with localization and completion (i.e., $W^{-1} \operatorname{Hom}_R(F^e_* R, R) = \operatorname{Hom}_{W^{-1}R}(F^e_* W^{-1}_R, W^{-1} R)$ for $W \subseteq R$ multiplicative, and $\widehat{R} \otimes_R \operatorname{Hom}_R(F^e_* R, R) = \operatorname{Hom}_{\widehat{R}}(F^e_* \widehat{R}, \widehat{R})$ for $(R, \mathfrak{m})$ local).

The ring $R$ is said to be *$F$-split* if the Frobenius map $R \longrightarrow F_* R$ splits as a map of $R$ modules, i.e., admits an $R$-module retraction. In other words, there should exist $\phi \in \operatorname{Hom}_R(F_* R, R)$ with $\phi(F_* 1) = 1$. Regular rings are always $F$-split; this can be seen easily using an important theorem of Kunz [**Kun69**] stating that $R$ is regular if and only if the Frobenius endomorphism is flat, i.e., $F_* R$ is a flat $R$-module. The composition of flat maps is flat, so the same holds for the iterated Frobenius as well. As we have assumed $R$ to be $F$-finite, it follows that $R$ is regular if and only if $F^e_* R$ is a (finitely-generated) projective (or equivalently locally free) $R$-module for all $e > 0$. If $G = R^{\oplus n}$ is any free module over a local ring $(R, \mathfrak{m})$ one sees that

$$\{g \in G \mid \text{ there exists } \phi \in \operatorname{Hom}_R(G, R) \text{ with } \phi(g) = 1\} = G \setminus \mathfrak{m} G$$

by considering the projection maps onto the various factors. Thus, for $(R, \mathfrak{m})$ regular and $r \in R$, we see that there exists $\phi \in \operatorname{Hom}_R(F^e_* R, R)$ with $\phi(F^e_* r) = 1$ precisely when $r \notin \mathfrak{m}^{[p^e]}$. As $1 \notin \mathfrak{m}^{[p]}$, we see regular local rings are $F$-split, and so

[1]Unless otherwise stated, our rings of interest will be assumed Noetherian and $F$-finite throughout these lectures. Occasionally, we reiterate the $F$-finite assumption to highlight instances where it has played an important role.

too are all regular rings as $F$-splitting can be checked locally (in the $F$-finite setting we are in). Nonetheless, the following much stronger condition holds as well.

DEFINITION 2. $R$ is said to be *strongly $F$-regular* if for all non-zero $d \in R$ there exists $\phi \in \mathrm{Hom}_R(F^e_* R, R)$ with $\phi(F^e_* d) = 1$.

LEMMA 3. *Regular* $\Longrightarrow$ *Strongly $F$-regular* $\Longrightarrow$ *All localizations of $R$ are weakly $F$-regular*

PROOF. Assume first that $R$ is regular and in particular $F$-split; take $\chi \in \mathrm{Hom}_R(F_* R, R)$ with $\chi(F_* 1) = 1$. Given an arbitrary (fixed) non-zero ring element $d \in R$, consider the sequence of ideals

$$I_e = \mathrm{im}\left(\mathrm{Hom}_R(F^e_* R, R) \xrightarrow{\phi \mapsto \phi(F^e_* d)} R\right).$$

For the first implication, we need to show $I_e = R$ for some $e \in \mathbb{Z}_{>0}$; let us first argue that $I_e \subseteq I_{e+1}$ for all $e \in \mathbb{Z}_{>0}$. If $x \in I_e$, then $x^p \in I_e$ as well, so there is some $\phi \in \mathrm{Hom}_R(F^e_* R, R)$ with $\phi(F^e_* d) = x^p$ by the definition of $I_e$. We then have that $\psi := (\chi \circ F_* \phi) \in \mathrm{Hom}_R(F^{e+1}_* R, R)$ and $\psi(F^{e+1}_* d) = \chi(F_*(\phi(F^e_* d))) = \chi(F_* x^p) = x \cdot \chi(F_* 1) = x$ so that $x \in I_{e+1}$. Thus, $I_e \subseteq I_{e+1}$ as claimed. Since $R$ is Noetherian, there is some $e_0 \in \mathbb{Z}_{>0}$ with $I_{e_0} = I_e$ for all $e \geq e_0$. To conclude, we verify that $I_{e_0} = R$. This equality can be checked locally, so we may further assume that $(R, \mathfrak{m})$ is a regular local ring. In this case, since $d \notin \bigcap_{e>0} \mathfrak{m}^{[p^e]} = 0$, taking $e \gg e_0$ gives that $d \notin \mathfrak{m}^{[p^e]}$ and hence $I_e = R$ as argued in the paragraph above. Thus, $I_{e_0} = I_e = R$ for $e \gg e_0$ as desired, giving that regular rings are strongly $F$-regular.

For the second implication, as strong $F$-regularity is preserved after localization, we need only show that strongly $F$-regular rings are weakly $F$-regular. In other words, given that $R$ is strongly $F$-regular, we must show all ideals are tightly closed. To that end, suppose $x \in I^*$ for some $I \subseteq R$. We have some non-zero $d \in R$ with $dx^{[p^e]} \in I^{[p^e]}$ for all $e > 0$. By the assumption that $R$ is strongly $F$-regular, we can find some $e > 0$ and $\phi \in \mathrm{Hom}_R(F^e_* R, R)$ with $\phi(F^e_* d) = 1$. Now observe that

$$x = \phi(F^e_*(dx^{p^e})) \in \phi(F^e_*(I^{[p^e]})) = \phi(I \cdot F^e_* R) = I \cdot \phi(F^e_* R) \subseteq I$$

to conclude that all ideals of $R$ are tightly closed. □

Central to the study of Frobenius splittings is the notion of a compatibly split ideal: an ideal $I$ is *$\phi$-compatible* provided that $\phi(F_* I) \subseteq I$, in which case $\phi$ descends to an $F$-splitting of the quotient $R/I$. More generally, we say that an ideal $I \subseteq R$ is uniformly $F$-compatible if it is compatible with all potential splittlings of all iterates of Frobenius. In other words:

DEFINITION 4. An ideal $I \subseteq R$ is *uniformly $F$-compatible* provided that

$$\sum_{e \geq 0} \sum_{\phi \in \mathrm{Hom}_R(F^e_* R, R)} \phi(F^e_* I) \subseteq I.$$

An alternative way to think of potential Frobenius splittings is as follows. A map $\phi \in \mathrm{Hom}_{\mathbb{Z}}(R, R)$ is said to be *$p^{-e}$-linear*[2] for some $e \in \mathbb{Z}_{\geq 0}$ if $r\phi(x) =$

[2]The term '$p^{-e}$-linear map' is meant to reflect that multiplication by elements of $R$ in a sense 'commutes' with $\phi$ after taking $p^e$-th roots. In contrast, one might also consider $\phi \in \mathrm{Hom}_{\mathbb{Z}}(R, R)$

$\phi(r^{p^e}x)$ for all $r, x \in R$; let $\mathscr{C}_e^R$ denote the set of all $p^{-e}$-linear maps. Function composition is compatible with such a structure in a natural way: if $\phi_1 \in \mathscr{C}_{e_1}^R$ and $\phi_2 \in \mathscr{C}_{e_2}^R$, then $\phi_1 \circ \phi_2 \in \mathscr{C}_{e_1+e_2}^R$. There is a straightforward identification of $\mathscr{C}_e^R$ with $\mathrm{Hom}_R(F_*^e R, R)$, where this composition structure is a bit cumbersome to exhibit. Specifically, if we have $\phi_1 \in \mathrm{Hom}_R(F_*^{e_1}R, R)$ and $\phi_2 \in \mathrm{Hom}_R(F_*^{e_2}R, R)$, their "composition" is the map

(4.1)
$$\phi_1 \cdot \phi_2 := \phi_1 \circ F_*^{e_1}\phi_2 = \left[ F_*^{e_1+e_2}R \xrightarrow{F_*^{e_1}\phi_2} F_*^{e_1}R \xrightarrow{\phi_1} R \right] \in \mathrm{Hom}_R(F_*^{e_1+e_2}R, R).$$

In particular, if $\phi \in \mathrm{Hom}_R(F_*^e R, R)$, it makes sense to talk about its powers (or iterates) $\phi^n \in \mathrm{Hom}_R(F_*^{ne}R, R)$. The *total Cartier algebra of* $R$ is the non-commutative graded ring $\mathscr{C}^R = \bigoplus_{e \geq 0} \mathscr{C}_e^R$ where the multiplicative structure is given by function composition. Note that while $\mathscr{C}_0^R = \mathrm{Hom}_R(R, R) = R$, $\mathscr{C}^R$ is not an $R$-algebra as $R$ is not in the center of $\mathscr{C}^R$: we have $r\phi = \phi r^{p^e}$ for $\phi \in \mathscr{C}_e^R$ and $r \in R$. Evaluation at a given element makes $R$ into a left $\mathscr{C}^R$-module, and we have that $I \subseteq R$ is uniformly $F$-compatible if and only if $I$ is preserved by this action (i.e., $\mathscr{C}^R(I) \subseteq I$).

EXERCISE 5. (Easy) What is the relationship between $(I^*)^{[p^e]}$ and $(I^{[p^e]})^*$? Use this to (1) verify that $\tau^{\mathrm{fin}}(R) = \bigcap_{I \subseteq R}(I :_R I^*)$ is an equivalent description of $\tau^{\mathrm{fin}}(R)$, and (2) show that $\tau^{\mathrm{fin}}(R)$ is uniformly $F$-compatible.

EXERCISE 6. If $I$ is a uniformly $F$-compatible ideal of $R$ and $W \subseteq R$ is a multiplicative set, show that $W^{-1}I$ is a uniformly $F$-compatible ideal of $W^{-1}R$. If $(R, \mathfrak{m})$ is local and $I$ is a uniformly $F$-compatible ideal of $R$, show that $I\widehat{R}$ is a uniformly $F$-compatible ideal of $\widehat{R}$.

As we shall see, uniformly $F$-compatible ideals arise in a number of different ways. For starters, given any ideal $I \subseteq R$ and $\phi \in \mathrm{Hom}_R(F_*^e R, R)$, one can check[3] easily that

$$\phi(F_*^e(I^{[p^e]} :_R (I^{[p^e]})^*)) \subseteq \phi(F_*^e(I^{[p^e]} :_R (I^*)^{[p^e]})) \subseteq (I :_R I^*)$$

and so in particular it follows that $\tau^{\mathrm{fin}}(R)$ is uniformly $F$-compatible. The crucial observation of Hochster and Huneke ultimately showing the existence of test elements for tight closure is the following.

THEOREM 7 ([**HH90**, Section 6] [**HH94**, Theorem 5.10]). *There is a non-zero element $c \in R$ contained in every non-zero uniformly $F$-compatible ideal. In particular, we have that $c \in \tau^{\mathrm{fin}}(R)$ is a test element.*

---

to be '$p^e$-linear' when $\phi(rx) = x^p\phi(x)$ for all $r, x \in R$. The iterated Frobenius $F^e$ is an example of a $p^e$-linear map, whence such maps are sometimes also referred to as 'Frobenius linear' instead. In contrast, $p^{-e}$-linear maps are sometimes referred to as 'Cartier linear' so as to reference the Cartier operator [**Car57**]. See [**Bli13**, Section 2] for additional exposition on $p^{-e}$-linear maps, or [**BS13**] for a survey on $p^{-e}$-linear maps from a geometric perspective.

[3]The first inclusion follows using Exercise 5. For the second, suppose that $x \in I^*$ and $y \in (I^{[p^e]} : ((I^*)^{[p^e]}))$. Then since $x^{p^e} \in ((I^*)^{[p^e]})$ we have $x\phi(F_*^e y) = \phi(F_*^e(x^{p^e}y)) \in \phi(F_*^e(I^{[p^e]})) = I \cdot \phi(F_*^e R) \subseteq I$.

PROOF. We can find a non-zero $x \in R$ so that $R[\frac{1}{x}]$ is regular,[4] so in particular $F$-split. Since the evaluation at one map

$$\operatorname{Hom}_R(F_*R, R) \xrightarrow{\psi \mapsto \psi(F_*1)} R$$

is surjective[5] after inverting $x$, it follows we can find $\phi \in \operatorname{Hom}_R(F_*R, R)$ with $\phi(F_*1) = x^N$ for some $N \gg 0$. Letting $y = x^N$, we then have that $y^2 = \phi(F_*y^p) \subseteq \phi(F_*(y^2R))$ and iterating[6] gives that

$$y^2R \subseteq \phi(F_*(y^2R)) \subseteq \phi^2(F^2_*(y^2R)) \subseteq \phi^3(F^3_*(y^2R)) \subseteq \cdots$$

or simply $y^2 \in \phi^e(F^e_*(y^2R))$ for all $e > 0$. Moreover, $y^3 \in \phi^e(F^e_*(y^{p^e+2}R))$ for all $e > 0$. We will show that $c = y^3 = x^{3N}$ is contained in every non-zero uniformly $F$-compatible ideal.

Suppose that $J \subseteq R$ is any non-zero uniformly compatible ideal, and pick non-zero $d \in J$. Checking locally, we deduce that

$$\operatorname{Hom}_{R[\frac{1}{y}]}(F^e_*R[\tfrac{1}{y}], R[\tfrac{1}{y}]) \xrightarrow{\psi \mapsto \psi(F^e_* \frac{d}{1})} R[\tfrac{1}{y}]$$

must be surjective for $e \gg 0$ as $R[\frac{1}{y}] = R[\frac{1}{x}]$ is strongly $F$-regular. In particular, since $J[\frac{1}{y}]$ is uniformly $F$-compatible in $R[\frac{1}{y}]$ and contains $\frac{d}{1}$, it follows that $J[\frac{1}{y}] = R[\frac{1}{y}]$ and so we must have $y^M \in J$ for some $M \gg 0$. Picking $e \gg 0$ so that $p^e + 2 \geq M$, it follows that

$$c = y^3 \in \phi^e(F^e_*y^{p^e+2}R) \subseteq \phi^e(F^e_*y^MR) \subseteq \phi^e(F^e_*J) \subseteq J$$

so that the desired conclusion now follows. ☐

One important and immediate consequence of the proof of the existence of test elements is the following:

DEFINITION 8. There exists a unique smallest non-zero and uniformly $F$-compatible ideal $\tau^{\text{non-fin}}(R)$ called the *non-finitistic (tight closure) test ideal.*

Indeed, if $c \in R$ is as in Theorem 7 and $J$ is a non-zero uniformly $F$-compatible ideal, we have not only that $c \in J$ but also $\phi(F^e_*c) \subseteq \phi(F^e_*J) \in J$ for any $e > 0$ and $p^{-e}$-linear map $\phi$. Thus we have that

$$\sum_{e \geq 0} \sum_{\phi \in \operatorname{Hom}_R(F^e_*R,R)} \phi(F^e_*c) \subseteq \sum_{e \geq 0} \sum_{\phi \in \operatorname{Hom}_R(F^e_*R,R)} \phi(F^e_*J) \subseteq J. \tag{8.1}$$

On the other hand, the sum on the left of (8.1) is non-zero (it contains $c$, for example) and uniformly $F$-compatible by design (again leveraging the "composition"

[4] If $K$ is the fraction field of $R$, $K \otimes_R F_*R = F_*K$ is a free $K$-module of finite rank (i.e., a finite dimensional $K$-vector space). As the locus in $\operatorname{Spec}(R)$ where any finitely-generated $R$-module is free is Zariski open, there must be a non-zero element $x \in R$ so that $R[\frac{1}{x}] \otimes_R F_*R = F_*R[\frac{1}{x}]$ too is a free $R[\frac{1}{x}]$-module. Appealing to Kunz's Theorem, we see that $R[\frac{1}{x}]$ is in fact a regular ring. More generally, Kunz was the first to observe that $F$-finite rings are excellent [**Kun76**, Theorem 2.5].

[5] Recall that, in general, $\operatorname{Hom}_R(\underline{\ \ },\underline{\ \ })$ is compatible with flat base-change provided the first entry is finitely-presented. Thus, as $F_*R$ is a finitely-generated $R$-module by assumption, we have $R[\frac{1}{x}] \otimes_R \operatorname{Hom}_R(F_*R, R) = \operatorname{Hom}_{R[\frac{1}{x}]}(F_*(R[\frac{1}{x}]), R[\frac{1}{x}])$.

[6] Here, recall that the "composition" of $\phi$ with itself is defined by $\phi^2 = \phi \cdot \phi := \phi \circ F_*\phi$ as in (4.1).

of $p^{-e}$-linear maps as in (4.1)). Thus the sum on the left of (8.1) *is* ths smallest non-zero uniformly $F$-compatible ideal of $R$, i.e., we have that

$$\tau^{\text{non-fin}}(R) = \sum_{e \geq 0} \sum_{\phi \in \operatorname{Hom}_R(F^e_* R, R)} \phi(F^e_* c)$$

and $\tau^{\text{non-fin}}(R) \subseteq J$ for any non-zero uniformly $F$-compatible ideal $J$. In particular, taking $J$ to be $\tau^{\text{fin}}(R)$ (which is uniformly $F$-compatible by Exercise 5), we may also conclude that $\tau^{\text{non-fin}}(R) \subseteq \tau^{\text{fin}}(R)$. One checks that $R$ is strongly $F$-regular if and only if $\tau^{\text{non-fin}}(R) = R$.

EXERCISE 9. Show that the non-finitistic (tight closure) test ideal is compatible with localization and completion. In other words, show that:

(1) for any multiplicative set $W \subseteq R$, we have $\tau^{\text{non-fin}}(W^{-1}R) = \tau^{\text{non-fin}}(R) \cdot W^{-1}R$;
(2) if $(R, \mathfrak{m})$ is local with $\mathfrak{m}$-adic completion $\widehat{R}$, then $\tau^{\text{non-fin}}(\widehat{R}) = \tau^{\text{non-fin}}(R) \cdot \widehat{R}$.

EXERCISE 10. Show that the following are equivalent:

(1) $R$ is strongly $F$-regular;
(2) for any non-zero $d \in R$, we have $\sum_{e \geq 0} \sum_{\phi \in \operatorname{Hom}_R(F^e_* R, R)} \phi(F^e_* d) = R$;
(3) $\tau^{\text{non-fin}}(R) = R$.

## Lecture 2

In addition to those descriptions given above, there is also an interpretation of strong $F$-regularity and $\tau^{\text{non-fin}}(R)$ in terms of tight closure for arbitrary submodules rather than just for ideals. If $N \subseteq M$, one writes

$$N^{[p^e]}_M := \operatorname{im}(N \otimes_R F^e_* R \longrightarrow M \otimes_R F^e_* R)$$

for $e \geq 0$, and the tight closure of $N$ in $M$ is the submodule given by

$$N^*_M = \{x \in M \mid \text{there exists a non-zero } c \in R \text{ with } x \otimes F^e_* c \in N^{[p^e]}_M \text{ for all } e > 0\}.$$

It can be shown that (see [**Hoc07**, pages 65, 151] and Exercise 11 below)

$$\tau^{\text{non-fin}}(R) = \bigcap_{N \subseteq M} (N :_R N^*_M) = \bigcap_{\substack{\mathfrak{m} \subseteq R \\ \text{maximal}}} \operatorname{Ann}_R(0^*_{E_R(R/\mathfrak{m})}) \tag{10.1}$$

where the first intersection ranges over arbitrary inclusions of $R$-modules $N \subseteq M$, and in the second intersection $E_R(R/\mathfrak{m})$ denotes the injective hull of the quotient field $R/\mathfrak{m}$ for a given maximal ideal $\mathfrak{m} \subseteq R$. Given what we already know about $\tau^{\text{non-fin}}(R)$, (10.1) essentially follows after showing $\tau^{\text{non-fin}}(R) = \operatorname{Ann}_R(0^*_{E_R(R/\mathfrak{m})})$ when $(R, \mathfrak{m})$ is a complete local domain – where it can deduced using Matlis duality. In contrast, one also has a similar interpretation for $\tau^{\text{fin}}(R)$, namely

$$\tau^{\text{fin}}(R) = \bigcap_{\substack{N \subseteq M \\ M/N \text{ finitely generated}}} (N :_R N^*_M) \tag{10.2}$$

where the intersection ranges only over the inclusions of $R$-modules $N \subseteq M$ so that $M$ is finitely generated over $N$ (see [**HH90**, Proposition 8.23] and Exercise 12 below). The terminology of the finitistic and non-finitistic (tight closure) test ideals arise via these descriptions.

EXERCISE 11 (Challenging). Suppose that $(R, \mathfrak{m})$ is a complete local domain and $E = E_R(R/\mathfrak{m})$ is the injective hull of the residue field. Show first that

$$0^*_E = \bigcap_{e \geq 0} \ker \left( E \xrightarrow{e \mapsto e \otimes F^e_* c} E \otimes_R F^e_* R \right)$$

where $c \in R$ is as in Theorem 7. Next, show using Matlis duality $(\_\_)^\vee = \operatorname{Hom}_R(\_\_, E)$ that

$$(0^*_E)^\vee = \frac{R}{\left( \sum_{e \geq 0} \sum_{\phi \in \operatorname{Hom}_R(F^e_* R, R)} \phi(F^e_* c) \right)}$$

and verify (10.1) for $R$.

EXERCISE 12 (Challenging). The goal of this exercise is to verify (10.2). One inclusion is clear; to see the reverse assume by way of contradiction that we have some $c \in \tau^{\text{fin}}(R)$, $R$-modules $N \subseteq M$ with $M/N$ finitely generated, and $x \in N^*_M$ with $cx \notin N$.

(a) Suppose that $V$ is a finitely generated $R$-module and $u \in V$ is non-zero. If $U \subseteq V$ is maximal with respect to $u \notin U$, show that: $V/U$ has finite length as an $R$-module, $\operatorname{Ass}(V/U) = \{\mathfrak{m}\}$ for some maximal ideal $\mathfrak{m}$ of $R$, and $(U :_V \mathfrak{m})/U$ is a one-dimensional vector space over the field $R/\mathfrak{m}$ spanned by the image of $u$ in $V/U$.

(b) Using (a) and passing to an appropriate quotient of $M$, show that there is a finitely length $R$-module $L$ and an element $y \in 0^*_L$ so that the following two conditions hold: (1) $\operatorname{Ass}_R(L) = \{\mathfrak{m}\}$ for some maximal ideal $\mathfrak{m}$ of $R$, and (2) $\operatorname{Ann}_L \mathfrak{m}$ is spanned by $cy \neq 0$.

(c) Recall that a local ring is said to be approximately Gorenstein if it has a sequence of irreducible ideals cofinite with the powers of its maximal ideal. Using that $R_\mathfrak{m}$ is approximately Gorenstein (as is every $F$-finite and hence excellent local domain), arrive at a contradiction by showing there is an $\mathfrak{m}$-primary irreducible ideal $I$ of $R$ and an element $z \in I^*$ where $cz \notin I$.

One of the key features of the Frobenius endomorphism in positive characteristic commutative algebra is that it is functorial: for any ring map $R \longrightarrow B$, the corresponding diagrams

$$\begin{array}{ccc} R & \longrightarrow & B \\ {\scriptstyle F_R}\uparrow & & \uparrow{\scriptstyle F_B} \\ R & \longrightarrow & B \end{array} \qquad\qquad \begin{array}{ccc} F^e_* R & \longrightarrow & F^e_* B \\ \uparrow & & \uparrow \\ R & \longrightarrow & B \end{array}$$

of maps of rings on the left or of maps of $R$-modules on the right both commute. As such, for any ideal $I \subseteq R$ and $\phi \in \operatorname{Hom}_R(F^e_* R, R)$, it follows easily that

$$\phi(F^e_*(I^{[p^e]} :_R (I^{[p^e]} B \cap R))) \subseteq (I :_R (IB \cap R))$$

and thus the ideal $\bigcap_{I \subseteq R}(I : IB \cap R)$ is is uniformly $F$-compatible. It is reasonable to think of this intersection as being a variant of test ideal for the operation on ideals of $R$ given by $I \mapsto IB \cap R$. Once again it is not clear from this perspective when to expect this intersection to be non-zero, i.e., whether "$B$ test elements" exist; similarly, it is unclear when to expect this infinite intersection to be compatible with localization or completion.

Recall that $B$ is said to be a solid $R$-algebra precisely when $\mathrm{Hom}_R(B, R) \neq 0$, in which case the trace of $B$ over $R$

$$\mathrm{tr}(B/R) := \mathrm{im}\left(\mathrm{Hom}_R(B, R) \xrightarrow{\phi \mapsto \phi(1)} R\right).$$

given by the image of the evaluation at one map is non-zero. Once more, note that $\mathrm{tr}(B/R)$ will not always be compatible with localization and completion, but is readily seen to be so provided $B$ is finitely presented over $R$. It is not difficult to show, essentially using the functoriality of Frobenius once more, that $\mathrm{tr}(B/R)$ is again uniformly $F$-compatible. One can also check $\mathrm{tr}(B/R) \subseteq (I : IB \cap R)$ for all $I \subseteq R$, and it follows

$$\tau^{\mathrm{non-fin}}(R) \subseteq \mathrm{tr}(B/R) \subseteq \bigcap_{I \subseteq R} (I :_R IB \cap R)$$

for all solid $R$-algebras $B$.

EXERCISE 13. Suppose that $(R, \mathfrak{m})$ is a complete local domain. Using once more that $R$ is approximately Gorenstein, show that $\mathrm{tr}(B/R) = \bigcap_{I \subseteq R}(I :_R IB \cap R)$.

EXERCISE 14. Suppose $B$ is an $R$-algebra, and $\phi \in \mathrm{Hom}_R(F^e_* B, R)$. Show that $\phi$ factors as

$$\begin{array}{ccc} F^e_* B & \xrightarrow{\tilde{\phi}} & \mathrm{Hom}_R(B, R) \\ & \searrow^{\phi} & \downarrow{\scriptstyle \psi \mapsto \psi(1)} \\ & & R \end{array}$$

where $\tilde{\phi}(F^e_* b) = \left(x \mapsto \phi(F^e_*(bx^{p^e}))\right)$ is in fact $B$-linear (and so also $R$-linear). Use this to show that $\mathrm{tr}(B/R)$ is uniformly $F$-compatible.

In the complete local setting, an amazing (and difficult) result of Hochster shows that a single $R$-algebra captures all of the tight closure relations on the ideals of $R$ at once and yields the following interpretation of $\tau^{\mathrm{fin}}(R)$.

THEOREM 15 (see [**DT23**, Proposition 1.0.3 and Corollary 3.3.5]). *If $(R, \mathfrak{m})$ is a complete local domain, there exists a (big balanced Cohen-Macaulay and hence solid) $R$-algebra $B$ so that $\tau^{\mathrm{fin}}(R) = \mathrm{tr}(B/R) = \bigcap_{I \subseteq R}(I : IB \cap R)$.*

In a different direction, building on arguments of Huneke and Lyubeznik [**HL07**], we have the following result globally.

THEOREM 16 ([**BST15**]). *If $R$ is normal $\mathbb{Q}$-Gorenstein,[7] there exists a (module) finite domain extension $R \subseteq S$ so that*

$$\tau^{\mathrm{non-fin}}(R) = \tau^{\mathrm{fin}}(R) = \mathrm{tr}(S/R).$$

*Moreover, if additionally $(R, \mathfrak{m})$ is complete local and $R^+ = \varinjlim_{\substack{R \subseteq S \\ \text{finite}}} S$ is the absolute integral closure of $R$, then*

$$\tau^{\mathrm{non-fin}}(R) = \tau^{\mathrm{fin}}(R) = \mathrm{tr}(R^+/R).$$

[7]A normal ring $R$ is called $\mathbb{Q}$-Gorenstein if the canonical module $\omega_R$, when viewed as a height-one fractional ideal, has a symbolic power $\omega_R^{(n)}$ that is locally free for some $n > 0$; see [**ST12**, Appendix A] for further details.

Over the years, a number of important additional flavors of test ideals have appeared with numerous applications. In the rest of these lectures, we will introduce some of these variants and discuss some important problems and properties. Many of these variants are motivated in some way or another via (finite) duality for the Frobenius endomorphism. For simplicity here, let us assume that $R$ is a Cohen-Macaulay domain. As $R$ is assumed to be Noetherian and $F$-finite, we have that $R$ is a quotient of an excellent $F$-finite regular domain $S$ by [**Gab04**, Remark 13.6] admitting a $p$-basis (over $S^p$); in other words, there are distinct elements $s_1, \ldots, s_n \in S$ so that the monomials $F^e_*(s_1^{a_1} \cdots s_n^{a_n})$ for $0 \le a_i < p^e$ are a free basis for $F^e_* S$ as an $S$-module. An example to keep in mind is where $S = k[x_1, \ldots, x_n]$ is a polynomial ring over a perfect field $k$ of characteristic $p > 0$ and $s_1 = x_1, \ldots, s_n = x_n$ are the variables. Via an explicit computation, it is not hard to show $\mathrm{Hom}_S(F^e_* S, S)$ is isomorphic to $F^e_* S$ as a $F^e_* S$-module (see Exercise 18 below). As a result, choosing the canonical module $\omega_R := \mathrm{Ext}_S^{\dim(S)-\dim(R)}(R, S)$, one has that $\mathrm{Hom}_R(F^e_* R, \omega_R)$ is identified via duality[8] with $\omega_{F^e_* R} = F^e_* \omega_R$. This yields a corresponding trace map

$$\Phi_e \colon F^e_* \omega_R \simeq \mathrm{Hom}_R(F^e_* R, \omega_R) \xrightarrow{\phi \mapsto \phi(F^e_* 1)} \omega_R$$

given by the evaluation at one map. One checks easily using duality that $\mathrm{Hom}_R(F^e_* \omega_R, \omega_R) \simeq F^e_* R$ and is generated by $\Phi_e$ as a module over $F^e_* R$. In other words, for every $\phi(\_\_) \in \mathrm{Hom}_R(F^e_* \omega_R, \omega_R)$ there is some $r_\phi \in R$ so that $\phi(\_\_) = \Phi_e(F^e_* r_\phi \cdot \_\_)$ is just a premultiple of $\Phi_e$.

EXERCISE 17. Show that $\Phi_{e+1} = \Phi_e \Phi_1$ up to premultiplication by a unit of $R$, and conclude $(\Phi_1)^e$ generates $\mathrm{Hom}_R(F^e_* \omega_R, \omega_R)$ over $F^e_* R$ for all $e > 0$. (Hint: Show that there is a factorization

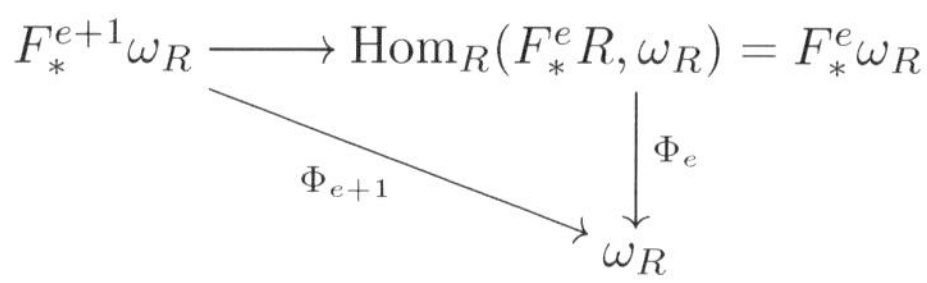


where the rightward arrow given by $F^{e+1}_* x \mapsto (F^e_* r \mapsto \Phi_{e+1}(F^{e+1}_* r^p x))$ is $F^e_* R$-linear and can be identified with a premultiple of $F^e_* \Phi_1$.)

EXERCISE 18. Suppose that $k$ is a perfect field of characteristic $p > 0$, and let $R = k[x_1, \ldots, x_d]$. Show that $F^e_* R$ is a free $R$-module with a free basis given by the set of monomials $\{F^e_* x_1^{a_1} \cdots x_n^{a_n}\}_{0 \le a_1, \ldots, a_n < p^e}$. Thus, the corresponding dual basis, i.e., the component functions determined by the individual monomials, give a set of $R$-module generators of $\mathrm{Hom}_R(F^e_* R, R)$. Let $\Phi_e$ be the dual basis element of $\mathrm{Hom}_R(F^e_* R, R)$ corresponding to the monomial $F^e_* x_1^{p^e-1} \cdots x_n^{p^e-1}$ where the exponents are largest. Check that $\Phi_e$ generates $\mathrm{Hom}_R(F^e_* R, R)$ as a $F^e_* R$-module, and that $\Phi_e = (\Phi_1)^e$. Said another way, after setting $\omega_R = R$ we can identify $\Phi_e$ with the trace map $\mathrm{Hom}_R(F^e_* \omega_R, \omega_R)$ coming from duality.

EXERCISE 19. Given an ideal $I \subseteq R$, recall that the $n$-th symbolic power of $I$ is defined to be the set of all elements $x \in R$ so that $x \in I^n R_{\mathfrak{p}}$ for all associated primes

[8] Here is a rough sketch of the argument. Writing $\pi \colon S \longrightarrow R$ for the quotient mapping, up to shifts we have $\pi^! S = \mathrm{Ext}_S^{\dim(S)-\dim(R)}(R, S)$. The identification $F^e_*(F^e)^! S = \mathrm{Hom}_S(F^e_* S, S) \cong F^e_* S$ is equivalent to $(F^e)^! S \cong S$. This gives that $(F^e)^! \omega_R = (F^e)^! \pi^! S \cong (\pi \circ F^e)^! S = (F^e \circ \pi)^! S \cong \pi^! (F^e)^! S \cong \pi^! S = \omega_R$, which is again equivalent to $\mathrm{Hom}_R(F^e_* R, R) \cong F^e_* \omega_R$.

$\mathfrak{p}$ of $I$. Suppose that $R$ is in addition normal and $I \subseteq R$ is an ideal of pure height one, i.e., so that every associated prime has height one. For any $\phi \in \mathrm{Hom}_R(F_*^e R, R)$, show that $\phi(F_*^e I^{(p^e)}) \subseteq I$. In particular, if $R$ is Cohen-Macaulay normal and $\omega_R \simeq J \subseteq R$ is a canonical ideal, conclude that $\phi(F_*^e J^{(p^e)}) \subseteq J$.

The *parameter test submodule* $\tau(\omega_R)$ of a (Noetherian $F$-finite) Cohen-Macaulay domain $R$ is the unique smallest non-zero submodule that is $\Phi_1$-compatible, i.e., so that $\Phi_1(F_*^e \tau(\omega_R)) \subseteq \omega_R$. As yet another consequence of the (proof of the) existence of test elements Theorem 7, let us show that this definition of $\tau(\omega_R)$ is well-posed. Note that since $\Phi_e = (\Phi_1)^e$ up to premultiplication by a unit and $\Phi_e$ generates $\mathrm{Hom}_R(F_*^e \omega_R, \omega_R)$, we have equivalently that $\tau(\omega_R)$ is the unique smallest non-zero submodule of $\omega_R$ that is $\phi$-compatible for all $\phi \in \mathrm{Hom}_R(F_*^e \omega_R, \omega_R)$ and all $e > 0$. If $N \subseteq \omega_R$ is non-zero and $\Phi_1$-compatible, we check $(N :_R \omega_R)$ is uniformly $F$-compatible. Supposing $\phi \in \mathrm{Hom}_R(F_*^e R, R)$ and $x \in (N :_R \omega_R)$, we want to show $\phi(F_*^e x)\eta \subseteq N$ for all $\eta \in \omega_R$. Consider the map $\phi_\eta \in \mathrm{Hom}_R(F_*^e R, \omega_R)$ given by $F_*^e r \mapsto \phi(F_*^e r)\eta$. We have $\phi_\eta \in \mathrm{Hom}_R(F_*^e R, \omega_R) \simeq F_*^e \omega_R$ and so it follows $F_*^e x \cdot \phi_\eta \in F_*^e N$ via this identification as $x \cdot \omega_R \subseteq N$. Using that $\Phi_e$ is identified with evaluation at $F_*^e 1$ and that $N$ is $\Phi_e$-compatible, this gives $\phi(F_*^e x) \cdot \eta = \Phi_e(F_*^e x \cdot \phi_\eta) \in \Phi_e(F_*^e N) \subseteq N$ as desired. Thus, it follows $\tau^{\mathrm{non-fin}}(R) \subseteq (N :_R \omega_R)$ so that $\tau^{\mathrm{non-fin}}(R) \cdot \omega_R$ is inside every non-zero $\Phi_1$-compatible submodule of $\omega_R$. In particular, $\tau(\omega_R)$ exists and moreover we have

$$\tau(\omega_R) = \sum_{e \geq 0} \Phi_e(F_*^e(\tau^{\mathrm{non-fin}}(R) \cdot \omega_R))$$

from which we see again $\tau(\omega_R)$ is compatible with localization and completion.

The ideal $\tau_{\mathrm{param}}(R) := (\tau(\omega_R) :_R \omega_R) \subseteq R$ is called the *parameter test ideal*, and when $(R, \mathfrak{m})$ is local we will now recover the result of Karen E. Smith [**Smi95**, Proposition 4.4] that

$$\tau_{\mathrm{param}}(R) = \mathrm{Ann}_R(0^*_{H^d_{\mathfrak{m}}(R)}) = \bigcap_{(\underline{x}) \subseteq R} ((\underline{x}) : (\underline{x})^*)$$

where the intersection varies over all (full) systems of parameters $\underline{x} = x_1, \ldots, x_d$ of $(R, \mathfrak{m})$. In order to show the first equality, we may assume $R$ is complete. As an Artinian or Noetherian $R$-module and its dual have the same annihilator in $R$, it suffices to observe that for any $c \in \tau^{\mathrm{non-fin}}(R)$ the Matlis dual of

$$0^*_{H^d_{\mathfrak{m}}(R)} = \bigcap_{e \geq 0} \ker\left(H^d_{\mathfrak{m}}(R) \xrightarrow{e \mapsto e \otimes F_*^e c} H^d_{\mathfrak{m}}(R) \otimes_R F_*^e R\right)$$

equals $\omega_R / \tau(\omega_R)$. For the second equality, recall that given a system of parameters $\underline{x} = x_1, \ldots, x_d$, we may view $H^d_{\mathfrak{m}} R = \varinjlim_t R/(x_1^t, \ldots, x_d^t)$ where the transition maps are given by multiplication by $x_1 x_2 \cdots x_d$. These maps are all injective given that we have assumed $R$ to be Cohen-Macaulay, so we may as well write $H^d_{\mathfrak{m}} R = \bigcup_t R/(x_1^t, \ldots, x_d^t)$. Let us show that $0^*_{H^d_{\mathfrak{m}}(R)} = \varinjlim_t (x_1^t, \ldots, x_d^t)^*/(x_1^t, \ldots, x_d^t) = \bigcup_t (x_1^t, \ldots, x_d^t)^*/(x_1^t, \ldots, x_d^t)$. We have that

$$H^d_{\mathfrak{m}}(R) \otimes_R F_*^e R = F_*^e H^d_{\mathfrak{m}}(R) = F_*^e \varinjlim_t R/(x_1^{tp^e}, \ldots, x_d^{tp^e})$$

where the transition maps are now given by multiplication by $F_*^e x_1^{p^e} x_2^{p^e} \cdots x_d^{p^e}$ and so are still injective. Thus, we see that if $\eta \in H^d_{\mathfrak{m}} R$ is represented by $y +$

$(x_1^t, \ldots, x_d^t) \in R/(x_1^t, \ldots, x_d^t)$, then $\eta \in 0^*_{H^d_{\mathfrak{m}}(R)}$ if and only if $cy^{p^e} \in (x_1^{tp^e}, \ldots, x_d^{tp^e})$ for all $e > 0$ (and $c \in \tau^{\text{non-fin}}(R)$), i.e., $y \in (x_1^t, \ldots, x_d^t)^*$ as claimed. The desired equality now follows easily as we see

$$\begin{aligned}\operatorname{Ann}_R(0^*_{H^d_{\mathfrak{m}}(R)}) &= \operatorname{Ann}_R\left(\bigcup_t (x_1^t, \ldots, x_d^t)^*/(x_1^t, \ldots, x_d^t)\right) \\ &= \bigcap_t ((x_1^t, \ldots, x_d^t) :_R (x_1^t, \ldots, x_d^t)^*)\end{aligned}$$

for any given fixed system of parameters $x_1, \ldots, x_d$.

In general, one says that $R$ is *$F$-rational* provided that the ideals generated by all systems of parameters (or equivalently any system of parameters by the exercise below) in all of the local rings of $R$ are tightly closed. From above, we see that a Cohen-Macaulay domain $R$ is $F$-rational if and only if $\tau_{\text{param}}(R) = R$ (as $\tau_{\text{param}}(R) = \operatorname{Ann}_R(0^*_{H^d_{\mathfrak{m}}(R)}) = \bigcap_{\substack{(\underline{x}) \subseteq R \\ \text{parameter}}} ((\underline{x}) : (\underline{x})^*))$ or equivalently $\tau(\omega_R) = \omega_R$ (as $\tau_{\text{param}}(R) = (\tau(\omega_R) :_R \omega_R)$).

EXERCISE 20. For a Cohen-Macaulay local ring $(R, \mathfrak{m})$, show that $R$ is $F$-rational if and only if $(x_1, \ldots, x_d)^* = (x_1, \ldots, x_d)$ for some system of parameters $x_1, \ldots, x_d$.

Note that in many ways the theory of $F$-rationality and parameter test submodules or ideals are better behaved: everything is compatible with localization and completion, and there is no need to distinguish between finitistic and non-finitistic constructions. Importantly, we also observe that in the Gorenstein case, $F$-rationality, weak $F$-regularity, and strong $F$-regularity all coincide, and we also have that the various notions of test ideals $\tau_{\text{param}}(R) = \tau^{\text{fin}}(R) = \tau^{\text{non-fin}}(R)$ all coincide. Indeed, these assertions follow from the local case, so consider a Gorenstein local ring $(R, \mathfrak{m})$. Using that $E_R(R/\mathfrak{m}) = H^d_{\mathfrak{m}}(R)$, it follows from what we have seen that

$$\begin{aligned}\tau_{\text{param}}(R) = \bigcap_{\substack{(\underline{x}) \subseteq R \\ \text{parameter}}} ((\underline{x}) : (\underline{x})^*) &\supseteq \bigcap_{I \subseteq R} (I :_R I^*) = \tau^{\text{fin}}(R) \supset \tau^{\text{non-fin}}(R) \\ &= \operatorname{Ann}_R(0^*_{E_R(R/\mathfrak{m})}) = \operatorname{Ann}_R(0^*_{H^d_{\mathfrak{m}}(R)}) = \tau_{\text{param}}(R)\end{aligned}$$

and so equality must hold throughout.

Our next variation of test ideals is that for a single non-zero $p^{-e}$-linear map.

DEFINITION-PROPOSITION 21 ([**HH90**, Section 6] [**Sch11**, Proposition 3.21]). For any non-zero $\phi \in \operatorname{Hom}_R(F^e_* R, R)$, there is a unique smallest non-zero ideal $\tau(\phi)$ that is $\phi$-compatible, i.e., so that $\phi(F^e_* \tau(\phi)) \subseteq \tau(\phi)$.

PROOF. To see that $\tau(\phi)$ is well-posed, we again need return to the proof of the existence of text elements and modify the argument. We can find non-zero $x \in R$ so that $R[\frac{1}{x}]$ is regular and the localization of $\phi$ generates $\operatorname{Hom}_{R[\frac{1}{x}]}(F^e_* R[\frac{1}{x}], R[\frac{1}{x}])$ as an $F^e_* R[\frac{1}{x}]$-module. We see that $\phi(F^e_* R[\frac{1}{x}]) = R[\frac{1}{x}]$, and so replacing $x$ by a power we may also assume $x \in \phi(F^e_* R)$. It follows $x^2 \in \phi(F^e_* x^2 R)$ and so $x^2 \in \phi^n(F^{en}_* x^2 R)$ and $x^3 \in \phi^n(F^{en}_* x^{2+p^{ne}} R)$ for all $n \geq 0$. Suppose that $J \subseteq R$ is any non-zero $\phi$-compatible ideal, and pick non-zero $d \in J$. Checking locally, we deduce that

$$\operatorname{Hom}_{R[\frac{1}{x}]}(F^{ne}_* R[\tfrac{1}{x}], R[\tfrac{1}{x}]) \xrightarrow{\psi \mapsto \psi(F^{ne}_* \frac{d}{1})} R[\tfrac{1}{x}]$$

must be surjective for $n \gg 0$. In particular, since the localization of $\phi^n$ also generates $\mathrm{Hom}_{R[\frac{1}{x}]}(F_*^{ne}R[\frac{1}{x}], R[\frac{1}{x}])$, we see that $\phi^n(J[\frac{1}{x}]) = R[\frac{1}{x}]$ and deduce that $J[\frac{1}{x}] = R[\frac{1}{x}]$. Thus, we have $x^{2+p^n e} \in J$ for all $e \gg 0$; since $x^3 \in \phi^n(F_*^{en}c^{2+p^{ne}}R)$ for all $n \geq 0$ and $J$ was assumed $\phi$-stable, we see $c = x^3 \in J$. Thus, it follows that

$$\tau(\phi) = \sum_{n \neq 0} \phi^n(F_*^{ne}cR)$$

is the smallest non-zero $\phi$-compatible ideal contained in all others. □

EXERCISE 22. Suppose $\phi \in \mathrm{Hom}_R(F_*^e R, R)$ is non-zero. Show that

(a) $\phi(F_*^e \tau(\phi)) = \tau(\phi)$;
(b) and $\tau(\phi) = \tau(\phi^n)$ for $n > 0$.

(Hint for (b): for $n = 2$, consider the ideal $\tau(\phi^2) \cap \phi(F_*^e \tau(\phi^2))$.)

A useful framework which allows for further variations on the notion of a test ideal is the use of certain subrings of the total Cartier algebra of $R$ – the non-commutative graded ring $\mathscr{C}^R = \bigoplus_{e \geq 0} \mathscr{C}_e^R$ introduced in the first lecture and where the multiplicative structure is given by function composition. A *Cartier subalgebra of* $R$ is a graded subring $\mathscr{D} \subseteq \mathscr{C}^R$ so that $\mathscr{D}_0 = \mathscr{C}_0^R = R$ and $\mathscr{D}_e \neq 0$ for some $e > 0$. The test ideal $\tau(\mathscr{D})$ is then defined as the unique smallest non-zero ideal compatible with all $\phi \in \mathscr{D}_e$ for all $e \geq 0$, i.e., so that $\mathscr{D}(\tau(\mathscr{D})) \subseteq \mathscr{D}$. In this case, it is easy to see that $\tau(\mathscr{D})$ is well-posed. If $\phi \in \mathscr{D}_e$ for some $e > 0$ is non-zero and $J \subseteq R$ is any non-zero ideal with $\mathscr{D}(J) \subseteq J$, we in particular must have $\phi(J) \subseteq J$ and so $0 \neq \tau(\phi) \subseteq \tau(\mathscr{D})$. It follows that $\tau(\mathscr{D}) = \mathscr{D}(\tau(\phi))$. In fact, while by no means obvious, we record the following result of Schwede.

THEOREM 23 ([**Sch11**, Theorem 5.1]). $\tau(\mathscr{D}) = \sum_{e>0} \sum_{\phi \in \mathscr{D}_e} \tau(\phi)$.

PROOF. Let $J = \sum_{e>0} \sum_{\phi \in \mathscr{D}_e} \tau(\phi) = \tau(\phi_1) + \cdots + \tau(\phi_r)$ for some $\phi_i \in \mathscr{D}_{e_i}$ and $i = 1, \ldots, r$ using that $R$ is Noetherian. It is clear that $J \subseteq \tau(\mathscr{D})$, and to show the reverse it suffices to check $\psi(J) \subseteq J$ for any given non-zero $\psi \in \mathscr{D}_e$. We can find a non-zero $c \in R$ so that:

- $R[\frac{1}{c}]$ is regular;
- $c \in \phi_i(R)$ for all $i$, and $c \in \psi(R)$;
- the localization of each $\phi_i$ generates $\mathrm{Hom}_{R[\frac{1}{c}]}(F_*^{e_i}R[\frac{1}{c}], R[\frac{1}{c}])$ over $F_*^{e_i}R[\frac{1}{c}]$ and that of $\psi_i$ generates $\mathrm{Hom}_{R[\frac{1}{c}]}(F_*^{e}R[\frac{1}{c}], R[\frac{1}{c}])$ over $F_*^{e}R[\frac{1}{c}]$.

One then checks easily that:

- $c^2 \in \phi_i(c^2R)$ and $c^2 \in \psi(c^2R)$, so also $c^2 \in \phi_i^\ell(c^2R)$ and $c^2 \in \psi \circ \phi^\ell(c^2R)$ for all $\ell \geq 0$;
- the localization of each $\phi_i^\ell$ generates $\mathrm{Hom}_{R[\frac{1}{c}]}(F_*^{\ell e_i}R[\frac{1}{c}], R[\frac{1}{c}])$ over $F_*^{\ell e_i}R[\frac{1}{c}]$ and that of $\psi \circ \phi_i^\ell$ generates $\mathrm{Hom}_{R[\frac{1}{c}]}(F_*^{e+\ell e_i}R[\frac{1}{c}], R[\frac{1}{c}])$ over $F_*^{e+\ell e_i}R[\frac{1}{c}]$ for all $\ell \geq 0$;
- $c^3 \in \tau(\phi_i)$ and $c^3 \in \tau(\psi \circ \phi^\ell)$ for all $i$ and all $\ell \geq 0$.

We then compute:

$$\begin{array}{rcl} \psi(J) & = & \sum_i \psi(\tau(\phi_i)) \\ & = & \sum_i \sum_{\ell \geq 0} \psi \circ \phi_i^\ell(c^3R) \\ & \subseteq & \sum_i \sum_{\ell \geq 0} \tau(\psi \circ \phi_i^\ell) \\ & \subseteq & J \end{array}$$

which completes the proof. □

In particular, this holds for $\mathscr{C}^R$, where we see that the non-finitistic tight closure test ideal

$$\tau^{\text{non-fin}}(R) = \sum_{e>0} \sum_{\phi \in \operatorname{Hom}_R(F^e_* R, R)} \tau(\phi)$$

can be described in terms of the test ideals of all potential splittings of iterates of Frobenius.

When $R$ is normal, an important duality argument can be used to give the so-called divisor correspondence for potential splittings of the iterates of Frobenius. Concretely, if $K_R$ is a canonical divisor of $R$, then duality gives an isomorphism

$$\operatorname{Hom}_R(F^e_* R, R) \simeq F^e_* R((1-p^e)K_R)$$

so that $\phi \in \operatorname{Hom}_R(F^e_* R, R)$ corresponds to a section $s_\phi \in R((1-p^e)K_R)$. Setting $\Delta_\phi := \frac{1}{p^e-1}\operatorname{div}(s_\phi) \sim_{\mathbb{Q}} -K_R$ gives a bijection [**Sch09**, Theorems 3.11, 3.13]

$$\left\{ \begin{matrix} \text{Effective } \mathbb{Q}\text{-divisors}\Delta \\ \text{on } \operatorname{Spec}(R) \text{ with} \\ (p^e-1)(K_R+\Delta) \sim 0 \end{matrix} \right\} \leftrightarrow \left\{ \begin{matrix} \phi \in \operatorname{Hom}_R(F^e_* R, R) \\ \text{modulo premultiplication by} \\ \text{units of } F^e_* R \end{matrix} \right\}.$$

This leads to the notion of a test ideals for divisor pairs $\tau(\Delta)$ which is particularly important in geometric settings. Letting $\mathscr{C}^\Delta_e = \{\phi \in \mathscr{C}^R_e \mid \Delta_\phi \geq \Delta\}$ and $\mathscr{C}^\Delta = \bigoplus_{e\geq 0} \mathscr{C}^\Delta_e$, one checks $\mathscr{C}^\Delta$ is a Cartier subalgebra on $R$ and sets $\tau(\Delta) = \tau(\mathscr{C}^\Delta)$. When $\Delta = \Delta_\phi$ for some $\phi \in \mathscr{C}^R_e$ and some $e > 0$, we have $\tau(\Delta_\phi) = \tau(\phi)$. According to Theorem 23, one also has

$$\tau(\Delta) = \sum_{e>0} \sum_{\substack{\phi \in \operatorname{Hom}_R(F^e_* R, R) \\ \Delta_\phi \geq \Delta}} \tau(\phi) = \sum_{\substack{\Delta' \geq \Delta,\ K_R+\Delta' \\ \mathbb{Q}\text{-Cartier of index prime to } p}} \tau(\Delta')$$

and so in particular

$$\tau^{\text{non-fin}}(R) = \sum_{\substack{\Delta \geq 0,\ K_R+\Delta \\ \mathbb{Q}\text{-Cartier of index prime to } p}} \tau(\Delta).$$

## Lecture 3

To see one use of the utility of the formalism of Cartier subalgebras, let us now describe the test ideals of so-called ideal pairs. If $\mathfrak{a} \subseteq R$ is non-zero and $t \in \mathbb{R}_{\geq 0}$, let

$$\mathscr{C}^{\mathfrak{a}^t}_e := \mathscr{C}^R \cdot \mathfrak{a}^{\lceil t(p^e-1)\rceil} = F^e_* \mathfrak{a}^{\lceil t(p^e-1)\rceil} \cdot \operatorname{Hom}_R(F^e_* R, R)$$

for $e \geq 0$ and set $\mathscr{C}^{\mathfrak{a}^t} = \bigoplus_{e\geq 0} \mathscr{C}^{\mathfrak{a}^t}$. In other words, $\phi \in \operatorname{Hom}_R(F^e_* R, R)$ is in $\mathscr{C}^{\mathfrak{a}^t}_e$ if and only if it can be written as a (finite) sum

$$\phi(\_\_) = \sum_j \phi_j(F^e_* a_j \cdot \_\_)$$

with $\phi_j \in \operatorname{Hom}_R(F^e_* R, R)$ and $a_j \in \mathfrak{a}^{\lceil t(p^e-1)\rceil}$. One checks that $\mathscr{C}^{\mathfrak{a}^t}$ is closed under composition using $p^{e_1}\lceil t(p^{e_2}-1)\rceil + \lceil t(p^{e_1}-1)\rceil \geq \lceil t(p^{e_1+e_2}-1)\rceil$ to show $\mathscr{C}^{\mathfrak{a}^t}_{e_1} \cdot \mathscr{C}^{\mathfrak{a}^t}_{e_2} \subseteq \mathscr{C}^{\mathfrak{a}^t}_{e_1+e_2}$.

DEFINITION 24 ([**HY03**] [**Sch08**]). If $\mathfrak{a} \subseteq R$ is non-zero and $t \in \mathbb{R}_{\geq 0}$, the $\mathfrak{a}^t$-*test ideal* is defined as $\tau(\mathfrak{a}^t) := \tau(\mathscr{C}^{\mathfrak{a}^t})$. In other words, $\tau(\mathfrak{a}^t)$ is the unique smallest non-zero ideal of $R$ so that

$$\sum_{e \geq 0} \sum_{\phi \in \operatorname{Hom}_R(F^e_* R, R)} \phi(F^e_* \mathfrak{a}^{\lceil t(p^e-1) \rceil} \tau(\mathfrak{a}^t)) \subseteq \tau(\mathfrak{a}^t).$$

The following alternative description of $\mathfrak{a}^t$-test ideals is very useful in practice.

LEMMA 25 ([**HT04**, Lemma 2.1]).

$$\begin{aligned} \tau(\mathfrak{a}^t) &= \textstyle\sum_{e \geq 0} \sum_{\phi \in \operatorname{Hom}_R(F^e_* R, R)} \phi(F^e_* \mathfrak{a}^{\lceil tp^e \rceil} \tau^{\text{non-fin}}(R)) \\ &= \textstyle\sum_{e \geq 0} \sum_{\phi \in \operatorname{Hom}_R(F^e_* R, R)} \phi(F^e_* \mathfrak{a}^{\lceil tp^e \rceil} c) \end{aligned}$$

for any non-zero $c \in \tau^{\text{non-fin}}(R)$.

Before diving into the proof, note in particular that this gives

$$\tau(\mathfrak{a}^t) = \sum_{e \geq 0} \sum_{\phi \in \operatorname{Hom}_R(F^e_* R, R)} \phi(F^e_* \mathfrak{a}^{\lceil tp^e \rceil}) \tag{25.1}$$

whenever $R$ is strongly $F$-regular.

PROOF. Set $J = \sum_{\phi \in \operatorname{Hom}_R(F^e_* R, R)} \phi(F^e_* \mathfrak{a}^{\lceil tp^e \rceil} \tau^{\text{non-fin}}(R))$. Using that

$$\mathscr{C}^R_{e_1} \mathfrak{a}^{\lceil t(p^{e_1}-1) \rceil} \mathscr{C}^R_{e_2} \mathfrak{a}^{\lceil tp^{e_2} \rceil} \subseteq \mathscr{C}^R_{e_1} \mathscr{C}^R_{e_2} \mathfrak{a}^{p^{e_2} \lceil t(p^{e_1}-1) \rceil + \lceil tp^{e_2} \rceil} \subseteq \mathscr{C}^R_{e_1+e_2} \mathfrak{a}^{\lceil tp^{e_1+e_2} \rceil}$$

it follows that $\mathscr{C}^{\mathfrak{a}^t}(J) \subseteq J$ and so we have $\tau(\mathfrak{a}^t) \subseteq J$. On the other hand, if $x \in \tau(\mathfrak{a}^t) \cap \tau^{\text{non-fin}}(R)$ is non-zero, we know

$$\tau^{\text{non-fin}}(R) = \sum_{e'} \mathscr{C}^R_{e'}(xR).$$

Acting on the left by $\mathscr{C}^R_e \mathfrak{a}^{\lceil tp^e \rceil}$ gives

$$\begin{aligned} \mathscr{C}^R_e \mathfrak{a}^{\lceil tp^e \rceil}(\tau^{\text{non-fin}}(R)) &= \textstyle\sum_{e'} \mathscr{C}^R_e \mathfrak{a}^{\lceil tp^e \rceil} \mathscr{C}^R_{e'}(xR) \\ &\subseteq \textstyle\sum_{e'} \mathscr{C}^R_{e+e'}(x \cdot \mathfrak{a}^{p^{e'} \lceil tp^e \rceil}) \\ &\subseteq \textstyle\sum_{e'} \mathscr{C}^R_{e+e'}(x \cdot \mathfrak{a}^{\lceil tp^{e+e'} \rceil}) \\ &\subseteq \tau(\mathfrak{a}^t) \end{aligned}$$

and summing over $e \geq 0$ gives $J = \tau(\mathfrak{a}^t)$. To complete the proof, note if $c \in \tau^{\text{non-fin}}(R)$ is non-zero, we can again leverage $\tau^{\text{non-fin}}(R) = \sum_{e'} \mathscr{C}^R_{e'}(cR)$ to get

$$\tau(\mathfrak{a}^t) = J = \sum_{e,e'} \mathscr{C}^R_e \mathscr{C}^R_{e'} \left( (\mathfrak{a}^{\lceil tp^e \rceil})^{[p^{e'}]} c \right) \subseteq \sum_e \mathscr{C}^R_e (\mathfrak{a}^{\lceil tp^e \rceil} c) \subseteq J = \tau(\mathfrak{a}^t)$$

where we see that equality must hold throughout. □

EXERCISE 26. Given non-zero ideals $\mathfrak{a}_1, \ldots, \mathfrak{a}_n \subseteq R$ and $t_1, \ldots, t_n \in \mathbb{R}_{\geq 0}$, explain how to define $\tau(\mathfrak{a}_1^{t_1} \cdots \mathfrak{a}_n^{t_n})$ and formulate and prove an analogue of Lemma 25 for $\tau(\mathfrak{a}_1^{t_1} \cdots \mathfrak{a}_n^{t_n})$. In the case that $\mathfrak{a} = \mathfrak{a}_1 = \cdots = \mathfrak{a}_n$ are all equal, show that $\tau(\mathfrak{a}^{t_1} \cdots \mathfrak{a}^{t_n}) = \tau(\mathfrak{a}^{t_1 + \cdots + t_n})$.

PROPOSITION 27 ([**HY03**] for (a) - (c); [**Tak06**, Theorem 2.7] for (d)). *Suppose that* $\mathfrak{a}, \mathfrak{b} \subseteq R$ *are non-zero ideals and* $s, t \in \mathbb{R}_{\geq 0}$.

(a) *If* $\mathfrak{a} \subseteq \mathfrak{b}$, *then* $\tau(\mathfrak{a}^t) \subseteq \tau(\mathfrak{b}^t)$ *with equality if* $\mathfrak{a}$ *is a reduction of* $\mathfrak{b}$.
(b) *We have* $\mathfrak{a}\tau(\mathfrak{b}^s) \subseteq \tau(\mathfrak{a}\mathfrak{b}^s)$ *with equality if* $\mathfrak{a}$ *is principal.*

(c) *(Skoda) If* $\mathfrak{a}$ *is generated by* $r$ *elements and* $t \geq r$*, then*

$$\tau(\mathfrak{a}^t\mathfrak{b}^s) = \mathfrak{a}\tau(\mathfrak{a}^{t-1}\mathfrak{b}^s).$$

(d) *(Subadditivity) If* $R$ *is regular, then* $\tau(\mathfrak{a}^t\mathfrak{b}^s) \subseteq \tau(\mathfrak{a}^t)\tau(\mathfrak{b}^s)$.

PROOF OF PROPOSITION 27 (a)–(c). For (a), note that if $\mathfrak{a} \subseteq \mathfrak{b}$, we have $\mathscr{C}^{\mathfrak{a}^t} \subseteq \mathscr{C}^{\mathfrak{b}^t}$. Thus, $\mathscr{C}^{\mathfrak{a}^t}(\tau(\mathfrak{b}^t)) \subseteq \mathscr{C}^{\mathfrak{b}^t}(\tau(\mathfrak{b}^t)) \subseteq \tau(\mathfrak{b}^t)$ so we must have $\tau(\mathfrak{a}^t) \subseteq \tau(\mathfrak{b}^t)$. If in addition $\mathfrak{a}$ is a reduction of $\mathfrak{b}$, there is some $\ell$ with $\mathfrak{b}^n = \mathfrak{b}^\ell\mathfrak{a}^{n-\ell}$ for all $n \geq \ell$ and so if $x \in \tau(\mathfrak{a}^t) \cap \mathfrak{a}^\ell$ is non-zero it follows that $x\mathfrak{b}^n \subseteq \mathfrak{a}^n$ for all $n \geq 0$. We then have

$$\begin{array}{rcl}\tau(\mathfrak{b}^t) & = & \sum_e \sum_{\phi \in \mathscr{C}_e^R} \phi(\mathfrak{b}^{\lceil t(p^e-1)\rceil}x^2) \\ & \subseteq & \sum_e \sum_{\phi \in \mathscr{C}_e^R} \phi(\mathfrak{a}^{\lceil t(p^e-1)\rceil}x) \\ & = & \tau(\mathfrak{a}^t)\end{array}$$

giving the reverse containment as well. For (b), choose non-zero $y \in \tau(\mathfrak{a}\mathfrak{b}^s) \subseteq \tau(\mathfrak{b}^s)$. Then compute

$$\begin{array}{rcl}\mathfrak{a}\tau(\mathfrak{b}^s) & = & \sum_e \sum_{\phi \in \mathscr{C}_e^R} \mathfrak{a}\phi(\mathfrak{b}^{\lceil s(p^e-1)\rceil}y) \\ & = & \sum_e \sum_{\phi \in \mathscr{C}_e^R} \phi(\mathfrak{a}^{[p^e]}\mathfrak{b}^{\lceil s(p^e-1)\rceil}y) \\ & \subseteq & \sum_e \sum_{\phi \in \mathscr{C}_e^R} \phi(\mathfrak{a}^{p^e}\mathfrak{b}^{\lceil s(p^e-1)\rceil}y) \\ & = & \tau(\mathfrak{a}\mathfrak{b}^s)\end{array}$$

and observe also that equality holds throughout if $\mathfrak{a}$ is principal as $\mathfrak{a}^{[p^e]} = \mathfrak{a}^{p^e}$ for all $e \geq 0$. Finally, for (c), if $\mathfrak{a}$ is generated by $r$ elements then one can check $\mathfrak{a}^n = \mathfrak{a}^{[p^e]}\mathfrak{a}^{n-p^e}$ for all $n \geq rp^e$ and $e \geq 0$. Using the analogue of Lemma 25, we have

$$\begin{array}{rcl}\tau(\mathfrak{a}^t\mathfrak{b}^s) & = & \sum_e \sum_{\phi \in \mathscr{C}_e^R} \phi(\mathfrak{a}^{\lceil tp^e\rceil}\mathfrak{b}^{\lceil sp^e\rceil}\tau^{\text{non-fin}}(R)) \\ & = & \sum_e \sum_{\phi \in \mathscr{C}_e^R} \phi(\mathfrak{a}^{[p^e]}\mathfrak{a}^{\lceil (t-1)p^e\rceil}\mathfrak{b}^{\lceil sp^e\rceil}\tau^{\text{non-fin}}(R)) \\ & = & \sum_e \sum_{\phi \in \mathscr{C}_e^R} \mathfrak{a}\phi(\mathfrak{a}^{\lceil (t-1)p^e\rceil}\mathfrak{b}^{\lceil sp^e\rceil}\tau^{\text{non-fin}}(R)) \\ & = & \mathfrak{a}\tau(\mathfrak{a}^{t-1}\mathfrak{b}^s)\end{array}$$

as claimed. □

THEOREM 28 ([**HH90**, Theorem 5.4]). *If* $R$ *is strongly* $F$*-regular and* $\mathfrak{a} \subseteq R$ *has a reduction generated by at most* $r$ *elements, then* $\overline{\mathfrak{a}^{r+n}} \subseteq \mathfrak{a}^{n+1}$ *for all* $n \geq 0$.

PROOF. Using properties (a), (b), and (c) from Proposition 27 we have

$$\overline{\mathfrak{a}^{r+n}} = \overline{\mathfrak{a}^{r+n}}\tau(R) \subseteq \tau(\overline{\mathfrak{a}^{r+n}}) = \tau(\mathfrak{a}^{r+n}) \subseteq \mathfrak{a}^n\tau(\mathfrak{a}^r) \subseteq \mathfrak{a}^{n+1}\tau(\mathfrak{a}^{r-1}) \subseteq \mathfrak{a}^{n+1}$$

where the first equality uses that $R = \tau^{\text{non-fin}}(R) = \tau(R)$ is strongly $F$-regular, the following containment uses (b), the following equality uses (a), the next equality uses (c) $n$-times, and the final inclusion is obvious. □

Note that subadditivity is quite a bit different from Proposition 27 (a) - (c) above, in that it is particular to regular rings (and essentially only those). We next wish to show the utility of the test ideals $\tau(\mathfrak{a}^t)$ by giving an application to symbolic powers of ideals in regular rings in Theorem 29 below. To that end, suppose first that $G$ is a finitely generated projective module over an arbitrary $R$. If $x \in G$ and $I \subseteq R$ is an ideal, it is easy to see that

$$x \in IG \quad \Leftrightarrow \quad \phi(x) \in I \text{ for all } \phi \in \operatorname{Hom}_R(G, R).$$

Indeed, the rightward implication is easy, and the leftward implication can be checked locally after which $G$ is then free and the conclusion follows by considering the projections onto each factor. For a regular ring $R$ where $F^e_*R$ is projective for all $e > 0$, this has the following consequence: for an ideal $I \subseteq R$ and $x \in R$,

$$x \in I^{[p^e]} \quad \Leftrightarrow \quad F^e_* x \in I(F^e_* R) \quad \Leftrightarrow \quad \phi(F^e_* x) \in I \text{ for all } \phi \in \mathrm{Hom}_R(F^e_* R, R).$$

PROOF OF PROPOSITION 27 (d) (SUBADDITIVITY). From (25.1), for each $e \geq 0$ we see that $\phi(\mathfrak{a}^{\lceil tp^e \rceil}) \subseteq \tau(\mathfrak{a}^t)$ for all $\phi \in \mathscr{C}^R_e$. Thus, it follows from the discussion immediately above that $\mathfrak{a}^{\lceil tp^e \rceil} \subseteq (\tau(\mathfrak{a}^t))^{[p^e]}$ for all $e \geq 0$. Thus, using (25.1) once again, we have

$$\begin{aligned} \tau(\mathfrak{a}^t \mathfrak{b}^s) &= \textstyle\sum_{e \geq 0} \sum_{\phi \in \mathscr{C}^R_e} \phi(\mathfrak{a}^{\lceil tp^e \rceil} \mathfrak{b}^{\lceil sp^e \rceil}) \\ &\subseteq \textstyle\sum_{e \geq 0} \sum_{\phi \in \mathscr{C}^R_e} \phi((\tau(\mathfrak{a}^t))^{[p^e]} \mathfrak{b}^{\lceil sp^e \rceil}) \\ &= \tau(\mathfrak{a}^t) \left( \textstyle\sum_{e \geq 0} \sum_{\phi \in \mathscr{C}^R_e} \phi(\mathfrak{b}^{\lceil sp^e \rceil}) \right) \\ &= \tau(\mathfrak{a}^t)\tau(\mathfrak{b}^s) \end{aligned}$$

as desired. □

THEOREM 29 ([**HH02**]). *If $R$ is regular and $\mathfrak{a} \subseteq R$ has a reduction generated by at most $r$ elements, then $\mathfrak{a}^{(rn)} \subseteq \mathfrak{a}^n$ for all integers $n > 0$.*

PROOF. We have that

$$\mathfrak{a}^{(rn)} \subseteq \tau(\mathfrak{a}^{(rn)}) = \tau\left( \underbrace{(\mathfrak{a}^{(rn)})^{1/n} \cdots (\mathfrak{a}^{(rn)})^{1/n}}_{n\text{-times}} \right) \subseteq \left( \tau((\mathfrak{a}^{(rn)})^{1/n}) \right)^n$$

using subadditivity $n$-times for the last inclusion. Thus, it suffices to check that $\tau((\mathfrak{a}^{(rn)})^{1/n}) \subseteq \mathfrak{a}$, which we can do after localizing at each associated prime $\mathfrak{p}$ of $\mathfrak{a}$. Say $\tilde{\mathfrak{a}} \subseteq \mathfrak{a}$ is a reduction generated by $r$ elements. We then have

$$\begin{aligned} \tau((\mathfrak{a}^{(rn)})^{1/n})R_{\mathfrak{p}} &= \tau((\mathfrak{a}^{(rn)}R_{\mathfrak{p}})^{1/n}) \\ &= \tau((\mathfrak{a}^{rn}R_{\mathfrak{p}})^{1/n}) \\ &= \tau(\mathfrak{a}^r R_{\mathfrak{p}}) \\ &= \tau(\tilde{\mathfrak{a}}^r R_{\mathfrak{p}}) \\ &= \tilde{\mathfrak{a}}\tau(\tilde{\mathfrak{a}}^{r-1}R_{\mathfrak{p}}) \\ &\subseteq \tilde{\mathfrak{a}}R_{\mathfrak{p}} \\ &\subseteq \mathfrak{a}R_{\mathfrak{p}} \end{aligned}$$

which concludes the proof. □

PROPOSITION 30 ([**MTW05**, Remark 2.12] [**BMS08**, Corollary 2.16] [**BSTZ10**, Lemma 3.23]). *Let $\mathfrak{a} \subseteq R$ be a non-zero ideal of $R$ and $t, t' \in \mathbb{R}_{\geq 0}$.*

(a) *If $t \leq t'$, then $\tau(\mathfrak{a}^{t'}) \subseteq \tau(\mathfrak{a}^t)$*
(b) *There is some $\epsilon > 0$ (depending on $t$) so that $\tau(\mathfrak{a}^{t'}) = \tau(\mathfrak{a}^t)$ for all $t' \in [t, t+\epsilon)$.*

PROOF. If $t \leq t'$, then $\mathfrak{a}^{\lceil t'p^e \rceil} \subseteq \mathfrak{a}^{\lceil tp^e \rceil}$ and hence (a) follows immediately from Lemma 25. For (b), note that we have

$$\tau(\mathfrak{a}^t) = \sum_{e=0}^{\infty} \mathscr{C}^R_e(\mathfrak{a}^{\lceil tp^e \rceil} \cdot \mathfrak{a}\tau^{\text{non-fin}}(R)) = \sum_{e=0}^{e_0} \mathscr{C}^R_e(\mathfrak{a}^{\lceil tp^e \rceil} \cdot \mathfrak{a}\tau^{\text{non-fin}}(R))$$

using Lemma 25 and for some $e_0$ as $R$ is Noetherian. Set $\epsilon = \frac{1}{p^{e_0}}$. For $t' \in [t, t+\epsilon)$, we have

$$\lceil t'p^e \rceil \leq \lceil (t+\epsilon)p^e \rceil \leq \lceil tp^e \rceil + 1$$

for all $0 \leq e \leq e_0$. Thus, it follows

$$\tau(\mathfrak{a}^{t'}) \subseteq \tau(\mathfrak{a}^t) = \sum_{e=0}^{e_0} \mathscr{C}_e^R(\mathfrak{a}^{\lceil tp^e \rceil + 1}\tau^{\text{non-fin}}(R)) \subseteq \sum_{e=0}^{e_0} \mathscr{C}_e^R(\mathfrak{a}^{\lceil t'p^e \rceil}\tau^{\text{non-fin}}(R)) = \tau(\mathfrak{a}^{t'})$$

so equality must hold throughout. □

Note in particular that, provided that $R$ is strongly $F$-regular, we see that $\tau(\mathfrak{a}^t) = R$ for all $0 < t \ll 1$. Thus, there must be a smallest value $\xi_1 > 0$ so that $\tau(\mathfrak{a}^{\xi_1}) \subsetneq R$. The real number $\xi_1$ is called the $F$-pure threshold of $\mathfrak{a}$. Similarly, there is a real number $\xi_2 > \xi_1$ so that $\tau(\mathfrak{a}^t) = \tau(\mathfrak{a}^{\xi_1})$ for all $t \in [\xi_1, \xi_2)$ and $\tau(\mathfrak{a}^{\xi_2}) \subsetneq \tau(\mathfrak{a}^{\xi_1})$. Continuing similarly, we arrive at a sequence of positive real numbers $\xi_1, \xi_2, \xi_3, \ldots$ called the jumping numbers of the ideals $\{\tau(\mathfrak{a}^t)\}_{t \in \mathbb{R}_{\geq 0}}$.

THEOREM 31 ([**BMS08, BSTZ10, ST14**]). *If $R$ is $\mathbb{Q}$-Gorenstein and $\mathfrak{a} \subseteq R$ is non-zero, then the jumping numbers of $\{\tau(\mathfrak{a}^t)\}_{t \in \mathbb{R}_{\geq 0}}$ are discrete and rational.*

PROOF OF THEOREM 31 WHEN $R = k[x_1, \ldots, x_n]$ WITH $k$ A PERFECT FIELD OF CHARACTERISTIC $p > 0$. Recall first the explicit structure of $\operatorname{Hom}_R(F^e_* R, R)$ described in Exercise 18. We have that $F^e_* R$ is a free $R$-module with a free basis given by the set of monomials $\{F^e_* x_1^{a_1} \cdots x_n^{a_n}\}_{0 \leq a_1, \ldots, a_n < p^e}$. The corresponding dual basis, i.e., the component functions determined by the individual monomials, give a set of $R$-module generators of $\operatorname{Hom}_R(F^e_* R, R)$. Moreover, if $\Phi_e$ is the dual basis element of $\operatorname{Hom}_R(F^e_* R, R)$ corresponding to the monomial $F^e_* x_1^{p^e-1} \cdots x_n^{p^e-1}$ where the exponents are largest, the entire dual basis is $\{\Phi_e(F^e_* x_1^{a_1} \cdots x_n^{a_n} \cdot \_\_)\}_{0 \leq a_1, \ldots, a_n < p^e}$ and in particular $\Phi_e$ generates $\operatorname{Hom}_R(F^e_* R, R)$ as a $F^e_* R$-module. Said another way, we can identify $\Phi_e$ with the trace map $\operatorname{Hom}_R(F^e_* \omega_R, \omega_R)$ coming from duality.

Using this explicit structure, let us first argue that if $I \subseteq R$ is any ideal generated by polynomials of degree at most $N$, then the ideal $\mathscr{C}_e^R(I)$ is generated by polynomials of degree at most $N/p^e$. Indeed, in order to see this we may assume $I = (f)$ is principal with generator $f \in R$ of degree $N$. Since the specified dual basis above gives a set of $R$-module generators for $\operatorname{Hom}_R(F^e_* R, R)$, the images of $f$ under the dual basis elements give a set of generators for the ideal $\mathscr{C}_e^R(I)$. Consider a particular dual basis element $\Phi_e(F^e_* x_1^{a_1} \cdots x_n^{a_n} \cdot \_\_)$ with $0 \leq a_1, \ldots, a_n < p^e$. For each term $m = \lambda x_1^{b_1} \cdots x_n^{b_n}$ appearing in $f$ where $\lambda \in k$ and necessarily $b_1 + \cdots + b_n \leq N$, we have that

$$\Phi_e(F^e_* x_1^{a_1} \cdots x_n^{a_n} \cdot m) = \begin{cases} \lambda^{1/p^e} \cdot \prod x_i^{\frac{a_i + b_i - p^e + 1}{p^e}} & a_i + b_i \equiv -1 \bmod p^e \\ 0 & \text{else} \end{cases}$$

which has degree bounded by

$$\sum_{i=1}^n \frac{a_i + b_i - p^e + 1}{p^e} \leq \sum_{i=1}^n \frac{b_i}{p^e} \leq \frac{N}{p^e}$$

and after summing up over all of the terms appearing in $f$ our assertion follows.

Proceeding now with the proof, let us first argue that the jumping numbers are discrete. In particular, if $\mathfrak{a}$ is generated by polynomials of degree at most $d$, we have that $\mathfrak{a}^{\lceil tp^e \rceil}$ is generated by polynomials of degree at most $d\lceil tp^e \rceil$, and so $\mathscr{C}_e^R(\mathfrak{a}^{\lceil tp^e \rceil})$ is generated by polynomials of degree at most $d\lceil tp^e \rceil / p^e < d(tp^e + 1)/p^e < dt + 1$. Summing up over $e$ and using (25.1), we see that $\tau(\mathfrak{a}^t)$ is generated by polynomials of degree at most $dt + 1$. In particular, given any fixed $T > 0$, we can conclude that there are only finitely many ideals $\tau(\mathfrak{a}^t)$ with $0 \leq t \leq T$ as each such $\tau(\mathfrak{a}^t)$ is determined by its intersection with the finite dimensional $k$ vector space of polynomials of degree at most $dT + 1$. Thus, we see that the jumping numbers are discrete.

The key to showing rationality is to show first that if $\xi$ is a jumping number, then $p\xi$ too is a jumping number. Using that $\Phi_1 \circ \Phi_e = \Phi_{e+1}$, we see that $\mathscr{C}_1^R \mathscr{C}_e^R = \mathscr{C}_{e+1}^R$. Thus, it follows that

$$\mathscr{C}_1^R(\tau(\mathfrak{a}^{pt})) = \mathscr{C}_1^R\left(\sum_e \mathscr{C}_e^R(\mathfrak{a}^{\lceil tp^{e+1}\rceil})\right) = \sum_e \mathscr{C}_{e+1}^R(\mathfrak{a}^{\lceil tp^{e+1}\rceil}) = \tau(\mathfrak{a}^t)$$

and by varying $t$ we see that in order for $\xi$ to be a jumping number we must have that $p\xi$ is a jumping number as well. Similarly, if $\mathfrak{a}$ is generated by $r$ elements and $t \geq r$, recall we have $\tau(\mathfrak{a}^t) = \mathfrak{a}\tau(\mathfrak{a}^{t-1})$ so that $\xi > r$ is a jumping number implies $\xi - 1$ is a jumping number. Iterating this, we see that $\{\xi\} + r$ must be a jumping number, where $\{\xi\} = \xi - \lfloor \xi \rfloor$ is the fractional part of $\xi$.

Finally, suppose $\xi$ is a jumping number. It follows that for all $e \gg 0$ we have that $\{p^e\xi\} + r$ is another jumping number and is less than $r + 1$. However, we already know there are only finitely many such jumping numbers by discreteness, so it follows we can find $e_2 > e_1 \gg 0$ with $\{p^{e_1}\xi\} + r = \{p^{e_2}\xi\} + r$. Equivalently, we have $p^{e_2}\xi - p^{e_1}\xi \in \mathbb{Z}$ and so $\xi \in \frac{1}{p^{e_2} - p^{e_1}}\mathbb{Z} \subseteq \mathbb{Q}$ is rational. □

EXERCISE 32. Suppose that $R$ is regular. If $f \in R$ is non-zero and $t = \frac{a}{p^b} \in \mathbb{R}_{\geq 0}$, show that

$$\tau(f^t) = \mathscr{C}_b^R(f^a R) = \operatorname{im}\left(\operatorname{Hom}_R(F_*^b R, R) \xrightarrow{\phi \mapsto \phi(F_*^b f^a)} R\right).$$

If $0 \neq \mathfrak{a} \subseteq R$ and $t = \frac{a}{p^b}$, show that

$$\mathscr{C}_{p^e b}^R(\mathfrak{a}^{p^e a}) \subseteq \mathscr{C}_{p^{e+1} b}^R(\mathfrak{a}^{p^{e+1} a}) \tag{32.1}$$

for all $e > 0$ and conclude $\tau(\mathfrak{a}^t) = \mathscr{C}_{p^e b}^R(\mathfrak{a}^{p^e a})$ for all $e \gg 0$. In fact, show $\tau(\mathfrak{a}^t) = \mathscr{C}_{p^e b}^R(\mathfrak{a}^{p^e a})$ for the smallest value of $e$ that achieves equality in (32.1).

In practice, much of the geometric intuition for working with test ideals stems from their somewhat mysterious connection to the mutliplier ideals that first appeared in complex birational algebraic geometry. To end these lectures, we highlight versions of two (hard) theorems which demonstrate this connection.

THEOREM 33 ([**BST15**]). *Suppose that $k$ is a field, $X = \mathbb{A}_k^n$, $t \geq 0$, and $\mathfrak{a} \subseteq \mathcal{O}_X = k[x_1, \ldots, x_n]$ a non-zero ideal sheaf. For all "sufficiently large" separable regular alterations $\pi \colon Y \longrightarrow X$ with $\mathfrak{a}\mathcal{O}_Y = \mathcal{O}_Y(-G)$, we have*

$$\operatorname{im}\left(\pi_* \mathcal{O}_Y(K_Y - \pi^* K_X - \lfloor tG \rfloor) \xrightarrow{\operatorname{Tr}_{K(Y)/K(X)}} \mathcal{O}_X\right)$$
$$= \begin{cases} \tau(\mathfrak{a}^t) & \text{the test ideal, if } k \text{ is a } (F\text{-finite}) \text{ field of characteristic } p > 0. \\ \mathcal{J}(\mathfrak{a}^t) & \text{the multiplier ideal, if } k \text{ is a field of characteristic } 0. \end{cases}$$

THEOREM 34 ([**Smi97**, **Har98**, **MS97**]). *Suppose that* $\mathfrak{a} \subseteq \mathbb{Z}[x_1, \ldots, x_n]$ *is a non-zero ideal and* $t \geq 0$. *Then*

$$\tau((\mathfrak{a} \bmod p)^t) = \mathcal{J}(\mathfrak{a}^t) \bmod p$$

*for all sufficently large primes* $p > 0$.

## SOLUTIONS TO SELECTED EXERCISES

by KENTA SATO

Throughout these exercises, assume that $R$ is an $F$-finite normal domain.

EXERCISE 5. *What is the relationship between* $(I^*)^{[p^e]}$ *and* $(I^{[p^e]})^*$*? Use this to*

(1) *verify that* $\tau^{\mathrm{fin}}(R) = \bigcap_{I \subseteq R}(I : I^*)$ *is an equivalent description of* $\tau^{\mathrm{fin}}(R)$,
(2) *and show that* $\tau^{\mathrm{fin}}(R)$ *is uniformly* $F$*-compatible.*

SOLUTION. We first prove the following claim: $(I^*)^{[p^e]} \subseteq (I^{[p^e]})^*$. Namely, take an element $x \in I^*$. We write $y := x^{p^e}$ and $J := I^{[p^e]}$. It is enough to show that $y \in J^*$. Since $x$ is an element of $I^*$, there exists an element $0 \neq c \in R$ such that for all $e' > 0$,

$$cx^{p^{e'}} \in I^{[p^{e'}]}.$$

Therefore, for an integer $e'' > 0$, we have

$$cy^{p^{e''}} = cx^{p^{e+e''}} \in I^{[p^{e+e''}]} = J^{[p^{e''}]}.$$

where we use Exercise.1 (1) of Epstein's lecture for the last equation. Therefore, we have $y \in J^*$ as desired, which proves the claim.

(1) We first recall that an element $c \in R$ is contained in $\tau^{\mathrm{fin}}(R)$ if and only if $cx^{p^e} \in I^{[p^e]}$ for all $I \subseteq R$, $x \in I^*$ and $e \geq 0$. In particular, considering $e = 0$, if $c$ is contained in $\tau^{\mathrm{fin}}(R)$, then we have $cx \in I$ for all $I \subseteq R$ and $x \in I^*$. Therefore,

$$\tau^{\mathrm{fin}}(R) \subseteq \bigcap_{I \subseteq R} (I : I^*).$$

On the other hand, take an element $c$ in the right hand side and we will show that $c$ is contained in $\tau^{\mathrm{fin}}(R)$. To do this, it is enough to show that

$$cx^{p^e} \in J^{[p^e]}$$

for all $J \subseteq R$, $x \in J^*$ and $e \geq 0$. But this inclusion holds since we have $x^{p^e} \in (J^*)^{[p^e]} \subseteq (J^{[p^e]})^*$ and $c$ is contained in $(J^{[p^e]} : (J^{[p^e]})^*$.

(2) Take an integer $e > 0$ and $\phi \in \mathrm{Hom}_R(F^e_* R, R)$. By (1), it is enough to show that for an ideal $I \subseteq R$, we have

$$\phi(F^e_*(\tau^{\mathrm{fin}}(R))) \subseteq (I : I^*).$$

Take an element $c \in \tau^{\mathrm{fin}}(R)$. It follows from (1) again that we have $c \in (I^{[p^e]} : (I^{[p^e]})^*$. Then we have

$$\begin{aligned}\phi(F^e_* c)I^* &= \phi(F^e_*(c(I^*)^{[p^e]}))\\ &\subseteq \phi(F^e_*(c(I^{[p^e]})^*))\\ &\subseteq \phi(F^e_* I^{[p^e]})\\ &= \phi(F^e_* R)I\\ &\subseteq I.\end{aligned}$$

EXERCISE 9. *Show that the non-finitistic (tight closure) test ideal is compatible with localization and completion. In other words, show that:*

(a) *for any multiplicative set $W \subseteq R$, we have $\tau^{\mathrm{non-fin}}(W^{-1}R) = \tau^{\mathrm{non-fin}}(R) \cdot W^{-1}R$;*
(b) *if $(R, \mathfrak{m})$ is local with $\mathfrak{m}$-adic completion $\widehat{R}$, then $\tau^{\mathrm{non-fin}}(\widehat{R}) = \tau^{\mathrm{non-fin}}(R) \cdot \widehat{R}$.*

SOLUTION. To prove (a), we first note that for every non-zero ideal $J$ of $W^{-1}R$, the contraction $J \cap R$ is non-zero. Therefore, there is a non-zero element $c \in R$ such that $c$ is contained in $\tau^{\mathrm{non-fin}}(W^{-1}R)$. After replacing $c$ by some multiplication, we may assume that $c$ is also contained in $\tau^{\mathrm{non-fin}}(R)$.

Then it follows from the equation after Definition. 8 in the lecture note that we have

$$\tau^{\mathrm{non-fin}}(R) = \Im\Big(\bigoplus_{e \geq 0} \mathrm{Hom}_R(F^e_* R, R) \xrightarrow{\phi \mapsto \phi(c)} R\Big).$$

By taking $\otimes_R W^{-1}R$, we have

$$\tau^{\mathrm{non-fin}}(R) \cdot W^{-1}R = \Im\Big(\bigoplus_{e \geq 0} (\mathrm{Hom}_R(F^e_* R, R) \otimes_R W^{-1}R) \xrightarrow{\phi \otimes a \mapsto a\phi(c)} W^{-1}R\Big).$$

Since $R$ is $F$-finite, we have

$$\mathrm{Hom}_R(F^e_* R, R) \otimes_R W^{-1}R \cong \mathrm{Hom}_{W^{-1}R}(F^e_*(W^{-1}R), W^{-1}R).$$

Therefore, one has

$$\tau^{\mathrm{non-fin}}(R) \cdot W^{-1}R = \Im\Big(\bigoplus_{e \geq 0} (\mathrm{Hom}_{W^{-1}R}(F^e_*(W^{-1}R), W^{-1}R) \xrightarrow{\phi \mapsto \phi(c)} W^{-1}R\Big),$$

where the right hand side coincides with $\tau^{\mathrm{non-fin}}(W^{-1}R)$. This proves (a).

The proof of (b) is similar.

EXERCISE 10. *Show that the following are equivalent:*

(a) *$R$ is strongly $F$-regular;*
(b) *for any $0 \neq d \in R$, we have $\sum_{e \geq 0} \sum_{\phi \in \mathrm{Hom}_R(F^e_* R, R)} \phi(F^e_* d) = R$;*
(c) *$\tau^{\mathrm{non-fin}}(R) = R$.*

SOLUTION. (a) $\Rightarrow$ (b) follows from the definition of strongly $F$-regularity.

To prove (b) $\Rightarrow$ (a), we first show that strongly $F$-regularity is a local property. Namely, we claim that if $R_{\mathscr{M}}$ is strongly $F$-regular for all maximal ideal $\mathscr{M} \subseteq R$, then $R$ is strongly $F$-regular. We first show that $R$ is $F$-split. To do this, it is enough to show that the natural morphism

$$\mathrm{Hom}_R(F_* R, R) \xrightarrow{\phi \mapsto \phi(1)} R$$

is surjective. Since surjectivity can be checked locally, it is enough to show that

$$\operatorname{Hom}_{R_{\mathscr{M}}}(F_*R_{\mathscr{M}}, R_{\mathscr{M}}) \xrightarrow{\phi \mapsto \phi(1)} R_{\mathscr{M}}$$

is surjective for all $\mathscr{M}$. But since $R_{\mathscr{M}}$ is strongly $F$-regular, it is $F$-split and this holds. Therefore, $R$ is $F$-split. Once we know $R$ is $F$-split, the assertion follows from the similar argument as in the proof of Lemma. 3 in the lecture note. This proves the claim.

By the claim, we may assume that $(R, \mathscr{M})$ is local. In this case, the condition (b) implies that there exists $e \geq 0$ and $\phi \in \operatorname{Hom}_R(F^e_*R, R)$ such that $u := \phi(F^e_*d)$ is not contained in $\mathscr{M}$, that is, a unit. Then $\psi := u^{-1}\phi \in \operatorname{Hom}_R(F^e_*R, R)$ satisfies $\psi(F^e_*d) = 1$, as desired.

(b) $\Rightarrow$ (c): Take an element $0 \neq c \in R$ as in Theorem. 7 in the lecture note. Then we have

$$\tau^{\text{non-fin}}(R) = \sum_{e \geq 0} \sum_{\phi \in \operatorname{Hom}_R(F^e_*R, R)} \phi(F^e_*c),$$

which coincides with $R$ by (b).

(c) $\Rightarrow$ (b): Take a non-zero element $d \in R$ and write

$$I := \sum_{e \geq 0} \sum_{\phi \in \operatorname{Hom}_R(F^e_*R, R)} \phi(F^e_*d) \subseteq R.$$

Since we can easily check that $I$ is a non-zero uniformly $F$-compatible ideal, we have

$$\tau^{\text{non-fin}}(R) \subseteq I,$$

which proves $I = R$.

EXERCISE 14. *Suppose $B$ is an $R$-algebra, and $\phi \in \operatorname{Hom}_R(F^e_*B, R)$. Show that $\phi$ factors as*

$$\begin{array}{ccc} F^e_*B & \xrightarrow{\tilde{\phi}} & \operatorname{Hom}_R(B, R) \\ & \underset{\phi}{\searrow} & \Big\downarrow {\scriptstyle \psi \mapsto \psi(1)} \\ & & R \end{array}$$

*where $\tilde{\phi}(F^e_*b) = \big(x \mapsto \phi(F^e_*bx^{p^e})\big)$ is in fact $B$-linear (and so also $R$-linear). Use this to show $\operatorname{tr}(B/R)$ is uniformly $F$-compatible.*

SOLUTION. (a) We first show that $\widetilde{\phi} : F^e_*B \longrightarrow \operatorname{Hom}_R(B, R)$ is B-linear. For $b, b' \in B$, we have

$$\begin{aligned} \widetilde{\phi}(F^e_*b + F^e_*b') = \widetilde{\phi}(F^e_*(b + b')) &= (x \mapsto \phi(F^e_*(b + b')x^{p^e})) \\ &= ((x \mapsto \phi(F^e_*bx^{p^e}) + \phi(F^e_*b'x^{p^e})) \\ &= \widetilde{\phi}(F^e_*b) + \widetilde{\phi}(F^e_*b'). \end{aligned}$$

Similarly, we have

$$\begin{aligned} \widetilde{\phi}(b(F^e_*b')) = \widetilde{\phi}(F^e_*(b^{p^e}b')) &= (x \mapsto \phi(F^e_*(b^{p^e}b')x^{p^e})) \\ &= ((x \mapsto \phi(F^e_*b'(bx)^{p^e}) = \widetilde{\phi}(F^e_*b')(bx)) \\ &= b \cdot \widetilde{\phi}(F^e_*b'). \end{aligned}$$

Therefore, $\widetilde{\phi}$ is $B$-linear.

We also note that the commutativity of the given diagram is obvious. In fact, for all $b \in B$, we have

$$\widetilde{\phi}(F^e_* b)(1) = \phi(F^e_* b),$$

which shows that the diagram

$$\begin{array}{ccc} F^e_* B & \xrightarrow{\tilde{\phi}} & \mathrm{Hom}_R(B, R) \\ & \searrow^{\phi} & \Big\downarrow {\scriptstyle \psi \mapsto \psi(1)} \\ & & R \end{array}$$

commutes.

(b) Take an integer $e > 0$, an $R$-linear map $\theta : F^e_* R \longrightarrow R$ and an $R$-linear map $f : B \longrightarrow R$. It is enough to show that

$$\theta(F^e_* \Im(f)) \subseteq \mathrm{tr}(B/R)$$

We write

$$\phi := \theta \circ F^e_* f : F^e_* B \longrightarrow R.$$

Since $\phi$ is an element in $\mathrm{Hom}_R(F^e_* B, R)$, it follows from (a) that we have

$$\phi = \psi \circ \widetilde{\phi}.$$

Therefore,

$$\theta(F^e_* \Im(f)) = \Im(\phi) \subseteq \Im(\psi) = \mathrm{tr}(B/R),$$

which completes the proof.

EXERCISE 18. *Suppose $k$ is a perfect field of characteristic $p > 0$, and let $R = k[x_1, \ldots, x_n]$. Show that $F^e_* R$ is a free $R$-module with a free basis given by the set $\{F^e_* x_1^{a_1} \cdots x_n^{a_n}\}_{0 \leq a_1, \ldots, a_n < p^e}$. Projection onto each of these factors, i.e. the dual basis, gives a set of $R$-module generators of $\mathrm{Hom}_R(F^e_* R, R)$. Let $\Phi_e$ be the dual basis element of $\mathrm{Hom}_R(F^e_* R, R)$ corresponding to projection onto the factor $F^e_* x_1^{p^e-1} \cdots x_n^{p^e-1}$ where the expontents are largest. Check that $\Phi_e$ generates $\mathrm{Hom}_R(F^e_* R, R)$ as a $F^e_* R$-module, and that $\Phi_e = (\Phi_1)^e$. Said another way, setting $\omega_R = R$ we can identify $\Phi_e$ with the trace map $\mathrm{Hom}_R(F^e_* \omega_R, \omega_R)$ coming from duality.*

SOLUTION. For an element $\mathbf{a} = (a_1, \ldots, a_n) \in (\mathbb{Z}_{\geq 0})^n$, we write

$$\mathbf{x}^{\mathbf{a}} := x_1^{a_1} \cdots x_n^{a_n} \in R.$$

We also define

$$I_e := \{\mathbf{a} = (a_1, \ldots, a_n) \in (\mathbb{Z}_{\geq 0})^n \mid 0 \leq a_1, a_2, \ldots, a_n < p^e\}.$$

We first show that $\{F^e_* \mathbf{x}^{\mathbf{a}}\}_{\mathbf{a} \in I_e}$ is a free basis of $F^e_* R$ over $R$. It is enough to show that the natural morphism

$$\phi : \bigoplus_{\mathbf{a} \in I_e} R \longrightarrow F^e_* R; \ ;(f_{\mathbf{a}})_{\mathbf{a} \in I_e} \mapsto \sum_{\mathbf{a}} f_{\mathbf{a}} F^e_* \mathbf{x}^{\mathbf{a}}$$

is bijective.

Since $\{\mathbf{x}^{\mathbf{b}}\}_{\mathbf{b} \in (\mathbb{Z}_{\geq 0})^n}$ is a free basis of $R$ over $k$, the natural morphism

$$f : \bigoplus_{\mathbf{b} \in (\mathbb{Z}_{\geq 0})^n} k \longrightarrow R \ ; \ (u_{\mathbf{b}})_{\mathbf{b} \in (\mathbb{Z}_{\geq 0})^n} \mapsto \sum_{\mathbf{b}} u_{\mathbf{b}} \mathbf{x}^{\mathbf{b}}$$

is bijective. On the other hand, since the map

$$I_e \times (\mathbb{Z}_{\geq 0})^n \longrightarrow (\mathbb{Z}_{\geq 0})^n \ ; \ (\mathbf{a}, \mathbf{b}) \mapsto \mathbf{a} + p^e\mathbf{b}$$

is bijective, we have a bijection

$$g : \bigoplus_{\mathbf{a}\in I_e} \bigoplus_{\mathbf{b}\in(\mathbb{Z}_{\geq 0})^n} k \longrightarrow \bigoplus_{\mathbf{c}\in(\mathbb{Z}_{\geq 0})^n} k \ ; \ (u_{\mathbf{a},\mathbf{b}})_{\mathbf{a},\mathbf{b}} \mapsto (u_{\mathbf{a},\mathbf{b}})_{\mathbf{a}+p^e\mathbf{b}}.$$

Hence, in order to prove that $\phi$ is bijective, it is enough to show that the composition

$$\psi : \bigoplus_{\mathbf{a}\in I_e} \bigoplus_{\mathbf{b}\in(\mathbb{Z}_{\geq 0})^n} k \xrightarrow{\oplus_{\mathbf{a}} f} \bigoplus_{\mathbf{a}\in I_e} R \xrightarrow{\phi} F^e_* R \xrightarrow{F^e_* r \mapsto r} R \xrightarrow{f^{-1}} \bigoplus_{\mathbf{c}\in(\mathbb{Z}_{\geq 0})^n} k \xrightarrow{g^{-1}} \bigoplus_{\mathbf{a}\in I_e} \bigoplus_{\mathbf{b}\in(\mathbb{Z}_{\geq 0})^n} k$$

is bijective. Since $\psi$ maps an element $(u_{\mathbf{a},\mathbf{b}})_{\mathbf{a},\mathbf{b}}$ to the element $(u^{p^e}_{\mathbf{a},\mathbf{b}})_{\mathbf{a},\mathbf{b}}$, the bijectivity follows from the assumption that $k$ is perfect.

We next show that $\Phi_e$ generates $\mathrm{Hom}_R(F^e_* R, R)$ as $F^e_* R$-module. Let $\{\phi_{\mathbf{a}} : F^e_* R \longrightarrow R\}_{\mathbf{a}}$ be the dual basis with respect to $\{F^e_* \mathbf{x}^{\mathbf{a}}\}_{\mathbf{a}}$. We write $\mathbf{a}_0 := (p^e - 1, \dots, p^e - 1) \in I_e$. For every $\mathbf{a} \in I_e$, noting that $\mathbf{a}_0 - \mathbf{a} \in I_e$, we have the equation

$$\mathbf{x}^{\mathbf{a}_0} = \mathbf{x}^{\mathbf{a}} \cdot \mathbf{x}^{\mathbf{a}_0 - \mathbf{a}}.$$

Then we can easily check the equation

$$\phi_{\mathbf{a}}(-) = \phi_{\mathbf{a}_0}(F^e_* \mathbf{x}^{\mathbf{a}_0 - \mathbf{a}} \cdot -),$$

and in particular, we have $\phi_{\mathbf{a}} \in F^e_* \cdot R\phi_{\mathbf{a}_0} = F^e_* R \cdot \Phi_e$. Therefore, we have

$$\mathrm{Hom}_R(F^e_* R, R) = \sum_{\mathbf{a}} R \cdot \phi_{\mathbf{a}} \subseteq F^e_* R \cdot \Phi_e.$$

We finally show the equation $\Phi_e = (\Phi_1)^e$. By the induction hypothesis, we may assume that $\Phi_{e-1} = (\Phi_1)^{e-1}$. Then, it is enough to show that $\Phi_e = \Phi_1 \circ \Phi_{e-1}$.

Since $F^e_* R$ is generated by $\{\mathbf{x}^{\mathbf{a}}\}_{\mathbf{a}}$, it is enough to show that for every $\mathbf{a} \in I_e$, we have

$$\Phi_1 \circ \Phi_{e-1}(F^e_* \mathbf{x}^{\mathbf{a}}) = \begin{cases} 0 & \text{if } \mathbf{a} \neq \mathbf{a}_0 \\ 1 & \text{if } \mathbf{a} = \mathbf{a}_0 \end{cases}$$

Let us consider the map

$$\theta : I_1 \times I_{e-1} \longrightarrow I_e \ ; \ (\mathbf{b}, \mathbf{c}) \mapsto p^{e-1}\mathbf{b} + \mathbf{c}.$$

Since this is bijective, there is a pair $(\mathbf{b}, \mathbf{c}) \in I_1 \times I_{e-1}$ such that $p^{e-1}\mathbf{b} + \mathbf{c} = \mathbf{a}$. We note that $\mathbf{a} = \mathbf{a}_0$ if and only if $\mathbf{b} = \mathbf{b}_0 := (p-1, \dots, p-1) \in I_1$ and $\mathbf{c} = \mathbf{c}_0 := (p^{e-1} - 1, \dots, p^{e-1} - 1) \in I_{e-1}$.

Then we have

$$\begin{aligned} \Phi_1 \circ \Phi_{e-1}(F^e_* \mathbf{x}^{\mathbf{a}}) &= \Phi_1 \circ \Phi_{e-1}(F^e_* \mathbf{x}^{\mathbf{p^{e-1}b+c}}) \\ &= \Phi_1 \circ \Phi_{e-1}(F^e_* (\mathbf{x}^{\mathbf{b}})^{p^{e-1}} \mathbf{x}^{\mathbf{c}}) \\ &= \Phi_1(F^1_* \mathbf{x}^{\mathbf{b}} \Phi_{e-1}(F^{e-1}_* \mathbf{x}^{\mathbf{c}})) \\ &= \begin{cases} 0 & \text{if } \mathbf{a} \neq \mathbf{a}_0; \\ 1 & \text{if } \mathbf{a} = \mathbf{a}_0. \end{cases} \end{aligned}$$

EXERCISE 21. *Suppose* $0 \neq \phi \in \mathrm{Hom}_R(F^e_* R, R)$. *Show that*

(a) $\phi(F^e_* \tau(\phi)) = \tau(\phi)$;
(b) *and* $\tau(\phi) = \tau(\phi^n)$ *for* $n > 0$.

*(Hint for (b): for $n = 2$, consider the ideal $\tau(\phi^2) \cap \phi(F^e_* \tau(\phi^2))$.)*

SOLUTION. (a) Since $\tau(\phi)$ is $\phi$-compatible, we have

$$\phi(F^e_* \tau(\phi)) \subseteq \tau(\phi).$$

For the converse inclusion, it is enough to show that $\phi(F^e_* \tau(\phi))$ is also $\phi$-compatible. Since we already know $\phi(F^e_* \tau(\phi)) \subseteq \tau(\phi)$, by taking $\phi$, we have

$$\phi(F^e_* \phi(F^e_* \tau(\phi))) \subseteq \phi(F^e_* \tau(\phi)),$$

which proves that $\phi(F^e_* \tau(\phi))$ is $\phi$-compatible.

(b) Since $\phi(F^e_* \tau(\phi)) = \tau(\phi)$, we have

$$\phi^n(F^{en}_* \tau(\phi)) = \phi^{n-1}(F^{e(n-1)}_* \tau(\phi) = \cdots = \tau(\phi).$$

Hence, $\tau(\phi)$ is $\phi^n$-compatible and we have

$$\tau(\phi^n) \subseteq \tau(\phi).$$

On the other hand, we write

$$J := \tau(\phi^n) \cap \phi(F^e_* \tau(\phi^n)) \cap \cdots \cap \phi^{n-1}(F^{e(n-1)}_* \tau(\phi^n)) \subseteq R.$$

Then we have

$$\begin{aligned}
\phi(F^e_* J) &= \phi(F^e_* \bigcap_{j=0}^{n-1} \phi^j(F^{ej}_* \tau(\phi^n))) \\
&\subseteq \bigcap_{j=0}^{n-1} \phi(F^e_* \phi^j(F^{ej}_* \tau(\phi^n))) \\
&= \bigcap_{j=1}^{n} \phi^j(F^{ej}_* \tau(\phi^n)) \\
&= (\bigcap_{j=1}^{n-1} \phi^j(F^{ej}_* \tau(\phi^n))) \cap \phi^n(F^{en}_* \tau(\phi^n)) \\
&= (\bigcap_{j=1}^{n-1} \phi^j(F^{ej}_* \tau(\phi^n))) \cap \tau(\phi^n) (\because \text{(a)}) \\
&= J.
\end{aligned}$$

Therefore, we have $\tau(\phi) \subseteq J \subseteq \tau(\phi^n)$.

EXERCISE 24. *Given non-zero ideals $\mathfrak{a}_1, \ldots, \mathfrak{a}_n \subseteq R$ and $t_1, \ldots, t_n \in \mathbb{R}_{\geq 0}$, explain how to define $\tau(\mathfrak{a}_1^{t_1} \cdots \mathfrak{a}_n^{t_n})$ and formulate and prove an analogue of Lemma 23 for $\tau(\mathfrak{a}_1^{t_1} \cdots \mathfrak{a}_n^{t_n})$. In the case that $\mathfrak{a} = \mathfrak{a}_1 = \cdots = \mathfrak{a}_n$ are all equal, show that $\tau(\mathfrak{a}^{t_1} \cdots \mathfrak{a}^{t_n}) = \tau(\mathfrak{a}^{t_1 + \cdots + t_n})$.*

SOLUTION. (a) Let $\mathscr{D}_e \subseteq \mathscr{C}_e$ be the submodule defined by

$$\begin{aligned}
\mathscr{D}_e &:= \operatorname{Hom}_R(F^e_* R, R) \cdot (\prod_{i=1}^{n} \mathfrak{a}_i^{\lceil t_i(p^e-1) \rceil}) \\
&= F^e_* (\prod_{i=1}^{n} \mathfrak{a}_i^{\lceil t_i(p^e-1) \rceil}) \cdot \operatorname{Hom}_R(F^e_* R, R)
\end{aligned}$$

and set $\mathscr{D} := \bigoplus_{e \geq 0} \mathscr{D}_e$. For every $i$ and $e_1, e_2 \geq 0$, since we have

$$p^{e_1} \lceil t_i(p^{e_2}-1) \rceil + \lceil t_i(p^{e_1}-1) \rceil \geq \lceil t_i(p^{e_1+e_2}-1) \rceil,$$

one has

$$\begin{aligned} (\mathfrak{a}_i^{\lceil t_i(p^{e_2}-1)\rceil})^{[p^{e_1}]} \mathfrak{a}_i^{\lceil t_i(p^{e_1}-1)\rceil} &\subseteq (\mathfrak{a}_i^{\lceil t_i(p^{e_2}-1)\rceil})^{p^{e_1}} \mathfrak{a}_i^{\lceil t_i(p^{e_1}-1)\rceil} \\ &= \mathfrak{a}_i^{p^{e_1}\lceil t_i(p^{e_2}-1)\rceil + \lceil t_i(p^{e_1}-1)\rceil} \\ &\subseteq \mathfrak{a}_i^{\lceil t_i(p^{e_1+e_2}-1)\rceil}. \end{aligned}$$

Therefore,

$$\begin{aligned} \mathscr{D}_{e_2}\mathscr{D}_{e_1} &= (\operatorname{Hom}_R(F_*^{e_2}R, R)\prod_{i=1}^n \mathfrak{a}_i^{\lceil t_i(p^{e_2}-1)\rceil}) \cdot (\operatorname{Hom}_R(F_*^{1_2}R, R)\prod_{i=1}^n \mathfrak{a}_i^{\lceil t_i(p^{e_1}-1)\rceil}) \\ &= \operatorname{Hom}_R(F_*^{e_2}R, R)(\prod_{i=1}^n \mathfrak{a}_i^{\lceil t_i(p^{e_2}-1)\rceil} \operatorname{Hom}_R(F_*^{e_1}R, R)) \prod_{i=1}^n \mathfrak{a}_i^{\lceil t_i(p^{e_1}-1)\rceil} \\ &= \operatorname{Hom}_R(F_*^{e_2}R, R)(\operatorname{Hom}_R(F_*^{e_1}R, R)(\prod_{i=1}^n \mathfrak{a}_i^{\lceil t_i(p^{e_2}-1)\rceil})^{[p^{e_1}]}) \prod_{i=1}^n \mathfrak{a}_i^{\lceil t_i(p^{e_1}-1)\rceil} \\ &\subseteq \mathscr{C}_{e_1+e_2} \prod_{i=1}^n \mathfrak{a}_i^{\lceil t_i(p^{e_1+e_2}-1)\rceil} \\ &= \mathscr{D}_{e_1+e_2}, \end{aligned}$$

which show that $\mathscr{D}$ is a Cartier subalgebra. Then we define the test ideal $\tau(\mathfrak{a}_1^{t_1} \cdots \mathfrak{a}_n^{t_n}) := \tau(\mathscr{D})$, that is, $\tau(\mathfrak{a}_1^{t_1} \cdots \mathfrak{a}_n^{t_n})$ is the unique smallest non-zero ideal of $R$ so that

$$\sum_{e \geq 0} \sum_{\phi \in \operatorname{Hom}_R(F_*^e R, R)} \phi(F_*^e \prod_{i=1}^n \mathfrak{a}_i^{\lceil t_i(p^e-1)\rceil} \tau(\mathfrak{a}_1^{t_1} \cdots \mathfrak{a}_n^{t_n})) \subseteq \tau(\mathfrak{a}_1^{t_1} \cdots \mathfrak{a}_n^{t_n}).$$

(b) We next prove an analogue of Lemma 23.

Claim. We have

$$\begin{aligned} \tau(\mathfrak{a}_1^{t_1} \cdots \mathfrak{a}_n^{t_n}) &= \sum_{e \geq 0} \sum_{\phi \in \operatorname{Hom}_R(F_*^e R, R)} \phi(F_*^e \prod_{i=1}^n \mathfrak{a}_i^{\lceil t_i p^e \rceil} \tau^{\text{non-fin}}(R)) \\ &= \sum_{e \geq 0} \sum_{\phi \in \operatorname{Hom}_R(F_*^e R, R)} \phi(F_*^e \prod_{i=1}^n \mathfrak{a}_i^{\lceil t_i p^e \rceil} c) \end{aligned}$$

for any $0 \neq c \in \tau^{\text{non-fin}}(R)$.

Proof of claim. Set

$$J := \sum_{e \geq 0} \sum_{\phi \in \operatorname{Hom}_R(F_*^e R, R)} \phi(F_*^e \prod_{i=1}^n \mathfrak{a}_i^{\lceil t_i p^e \rceil} \tau^{\text{non-fin}}(R)).$$

Using that

$$\begin{aligned}\mathscr{D}_{e_1}\mathscr{C}^R_{e_2}\prod_{i=1}^n \mathfrak{a}_i^{\lceil t_i p^{e_2}\rceil} &= \mathscr{C}^R_{e_1}\prod_{i=1}^n \mathfrak{a}_i^{\lceil t_i(p^{e_1}-1)\rceil}\mathscr{C}^R_{e_2}\prod_{i=1}^n \mathfrak{a}_i^{\lceil t_i p^{e_2}\rceil}\\ &= \mathscr{C}^R_{e_1}\mathscr{C}^R_{e_2}\prod_{i=1}^n (\mathfrak{a}_i^{\lceil t_i(p^{e_1}-1)\rceil})^{[p^{e_2}]}\mathfrak{a}_i^{\lceil t_i p^{e_2}\rceil}\\ &\subseteq \mathscr{C}^R_{e_1}\mathscr{C}^R_{e_2}\prod_{i=1}^n \mathfrak{a}_i^{p^{e_2}\lceil t_i(p^{e_1}-1)\rceil+\lceil t_i p^{e_2}\rceil}\\ &\subseteq \mathscr{C}^R_{e_1+e_2}\prod_{i=1}^n \mathfrak{a}_i^{\lceil t_i p^{e_1+e_2}\rceil},\end{aligned}$$

it follows that $\mathscr{D}(J)\subseteq J$ and so we have

$$\tau(\mathfrak{a}_1^{t_1}\cdots\mathfrak{a}_n^{t_n})\subseteq J.$$

On the other hand, if $0\neq x\in\tau(\mathfrak{a}_1^{t_1}\cdots\mathfrak{a}_n^{t_n})\cap\tau^{\text{non-fin}}(R)$, we know

$$\tau^{\text{non-fin}}(R)=\sum_{e'\geq 0}\mathscr{C}^R_{e'}(xR).$$

Acting on the left by $\mathscr{C}^R_e\prod_{i=1}^n\mathfrak{a}_i^{\lceil t_i p^e\rceil}$ gives

$$\begin{aligned}\mathscr{C}^R_e\prod_{i=1}^n\mathfrak{a}_i^{\lceil t_i p^e\rceil}(\tau^{\text{non-fin}}(R)) &= \sum_{e'}\mathscr{C}^R_e\prod_{i=1}^n\mathfrak{a}_i^{\lceil t_i p^e\rceil}\mathscr{C}^R_{e'}(xR)\\ &\subseteq \sum_{e'}\mathscr{C}^R_{e+e'}(x\prod_i(\mathfrak{a}_i^{\lceil t_i p^e\rceil})^{[p^{e'}]})\\ &\subseteq \sum_{e'}\mathscr{C}^R_{e+e'}(x\prod_i\mathfrak{a}_i^{p^{e'}\lceil t_i p^e\rceil})\\ &\subseteq \sum_{e'}\mathscr{C}^R_{e+e'}(x\prod_i\mathfrak{a}_i^{\lceil t_i p^{e+e'}\rceil})\\ &\subseteq \sum_{e'}\mathscr{D}_{e+e'}(x)\\ &\subseteq \tau(\mathfrak{a}_1^{t_1}\cdots\mathfrak{a}_n^{t_n}).\end{aligned}$$

Summing over $e\geq 0$ gives $J=\tau(\mathfrak{a}_1^{t_1}\cdots\mathfrak{a}_n^{t_n})$.

To complete the proof, note ig $0\neq c\in\tau^{\text{non-fin}}(R)$, we can again leverage $\tau^{\text{non-fin}}(R)=\sum_{e'}\mathscr{C}^R_{e'}(cR)$ to get

$$\begin{aligned}J &= \sum_{e,e'}\mathscr{C}_e\mathscr{C}_{e'}(\prod_i(\mathfrak{a}_i^{\lceil t_i p^e\rceil})^{[p^{e'}]}c)\\ &\subseteq \sum_{e,e'}\mathscr{C}_{e+e'}(\prod_i(\mathfrak{a}_i^{\lceil t_i p^{e+e'}\rceil})c)\\ &\subseteq \sum_{e'''}\mathscr{C}_{e''}(\prod_i(\mathfrak{a}_i^{\lceil t_i p^{e''}\rceil})\tau^{\text{non-fin}}(R))\\ &= J,\end{aligned}$$

where we see that equality must hold throughout. This finishes the proof of the claim.

(c) We show that $\tau(\mathfrak{a}^{t_1}\cdots\mathfrak{a}^{t_n}) = \tau(\mathfrak{a}^{t_1+t_2+\cdots+t_n})$. Set $t := t_1+\cdots+t_n$ and $m := \lceil t\rceil \in \mathbb{N}$. For every $e>0$, since we have

$$\lceil tp^e\rceil \leq \sum_{i=1}^{n}\lceil t_ip^e\rceil \leq \lceil tp^e\rceil + n$$

one has

$$\mathfrak{a}^{\lceil tp^e\rceil} \supseteq \prod_i \mathfrak{a}^{\lceil t_ip^e\rceil} \supseteq \mathfrak{a}^{\lceil tp^e\rceil}\mathfrak{a}^n.$$

Take an element $0\neq c\in \tau^{\text{non-fin}}(R)$ and $a\in\mathfrak{a}^n$. Then we have

$$\tau(\mathfrak{a}^t) = \sum_{e\geq 0}\mathscr{C}_e^R(\mathfrak{a}^{\lceil tp^e\rceil}c) \supseteq \sum_{e\geq 0}\mathscr{C}_e^R(\prod_i \mathfrak{a}^{\lceil t_ip^e\rceil}c) = \tau(\mathfrak{a}^{t_1}\cdots\mathfrak{a}^{t_n}).$$

On the other hand, since $ac$ is also contained in $\tau^{\text{non-fin}}(R)$, we conclude that

$$\begin{aligned}\tau(\mathfrak{a}^t) &= \sum_{e\geq 0}\mathscr{C}_e^R(\mathfrak{a}^{\lceil tp^e\rceil}ac)\\ &\subseteq \sum_{e\geq 0}\mathscr{C}_e^R(\mathfrak{a}^{\lceil tp^e\rceil}\mathfrak{a}^n c)\\ &\subseteq \sum_{e\geq 0}\mathscr{C}_e^R(\prod_i \mathfrak{a}^{\lceil t_ip^e\rceil}c)\\ &= \tau(\mathfrak{a}^{t_1}\cdots\mathfrak{a}^{t_n}),\end{aligned}$$

which completes the proof. □

## References


[Bli13] Manuel Blickle, *Test ideals via algebras of $p^{-e}$-linear maps*, J. Algebraic Geom. **22** (2013), no. 1, 49–83, DOI 10.1090/S1056-3911-2012-00576-1. MR2993047

[BMS08] Manuel Blickle, Mircea Mustaţă, and Karen E. Smith, *Discreteness and rationality of $F$-thresholds*, Michigan Math. J. **57** (2008), 43–61, DOI 10.1307/mmj/1220879396. Special volume in honor of Melvin Hochster. MR2492440

[BS13] Manuel Blickle and Karl Schwede, *$p^{-1}$-linear maps in algebra and geometry*, Commutative algebra, Springer, New York, 2013, pp. 123–205, DOI 10.1007/978-1-4614-5292-8_5. MR3051373

[BSTZ10] Manuel Blickle, Karl Schwede, Shunsuke Takagi, and Wenliang Zhang, *Discreteness and rationality of $F$-jumping numbers on singular varieties*, Math. Ann. **347** (2010), no. 4, 917–949, DOI 10.1007/s00208-009-0461-2. MR2658149

[BST15] Manuel Blickle, Karl Schwede, and Kevin Tucker, *$F$-singularities via alterations*, Amer. J. Math. **137** (2015), no. 1, 61–109, DOI 10.1353/ajm.2015.0000. MR3318087

[BM10] Holger Brenner and Paul Monsky, *Tight closure does not commute with localization*, Ann. of Math. (2) **171** (2010), no. 1, 571–588, DOI 10.4007/annals.2010.171.571. MR2630050

[Car57] Pierre Cartier, *Une nouvelle opération sur les formes différentielles* (French), C. R. Acad. Sci. Paris **244** (1957), 426–428. MR84497

[DT23] Rankeya Datta and Kevin Tucker, *Openness of splinter loci in prime characteristic*, J. Algebra **629** (2023), 307–357, DOI 10.1016/j.jalgebra.2023.03.025. MR4583731

[Gab04] Ofer Gabber, *Notes on some $t$-structures*, Geometric aspects of Dwork theory. Vol. I, II, Walter de Gruyter, Berlin, 2004, pp. 711–734. MR2099084

[Har98] Nobuo Hara, *A characterization of rational singularities in terms of injectivity of Frobenius maps*, Amer. J. Math. **120** (1998), no. 5, 981–996. MR1646049

[HT04] Nobuo Hara and Shunsuke Takagi, *On a generalization of test ideals*, Nagoya Math. J. **175** (2004), 59–74, DOI 10.1017/S0027763000008904. MR2085311

[HY03] Nobuo Hara and Ken-Ichi Yoshida, *A generalization of tight closure and multiplier ideals*, Trans. Amer. Math. Soc. **355** (2003), no. 8, 3143–3174, DOI 10.1090/S0002-9947-03-03285-9. MR1974679

[Hoc07] M. Hochster: *Foundations of tight closure theory*, lecture notes from a course taught on the University of Michigan Fall 2007 (2007), `https://dept.math.lsa.umich.edu/~hochster/711F07/fndtc.pdf`.

[HH90] Melvin Hochster and Craig Huneke, *Tight closure, invariant theory, and the Briançon-Skoda theorem*, J. Amer. Math. Soc. **3** (1990), no. 1, 31–116, DOI 10.2307/1990984. MR1017784

[HH94] Melvin Hochster and Craig Huneke, *$F$-regularity, test elements, and smooth base change*, Trans. Amer. Math. Soc. **346** (1994), no. 1, 1–62, DOI 10.2307/2154942. MR1273534

[HH02] Melvin Hochster and Craig Huneke, *Comparison of symbolic and ordinary powers of ideals*, Invent. Math. **147** (2002), no. 2, 349–369, DOI 10.1007/s002220100176. MR1881923

[HL07] Craig Huneke and Gennady Lyubeznik, *Absolute integral closure in positive characteristic*, Adv. Math. **210** (2007), no. 2, 498–504, DOI 10.1016/j.aim.2006.07.001. MR2303230

[Kun69] Ernst Kunz, *Characterizations of regular local rings of characteristic $p$*, Amer. J. Math. **91** (1969), 772–784, DOI 10.2307/2373351. MR252389

[Kun76] Ernst Kunz, *On Noetherian rings of characteristic $p$*, Amer. J. Math. **98** (1976), no. 4, 999–1013, DOI 10.2307/2374038. MR432625

[MS97] V. B. Mehta and V. Srinivas, *A characterization of rational singularities*, Asian J. Math. **1** (1997), no. 2, 249–271, DOI 10.4310/AJM.1997.v1.n2.a4. MR1491985

[MTW05] Mircea Mustaţă, Shunsuke Takagi, and Kei-ichi Watanabe, *$F$-thresholds and Bernstein-Sato polynomials*, European Congress of Mathematics, Eur. Math. Soc., Zürich, 2005, pp. 341–364. MR2185754

[Sch08] Karl Schwede, *Generalized test ideals, sharp $F$-purity, and sharp test elements*, Math. Res. Lett. **15** (2008), no. 6, 1251–1261, DOI 10.4310/MRL.2008.v15.n6.a14. MR2470398

[Sch09] Karl Schwede, *$F$-adjunction*, Algebra Number Theory **3** (2009), no. 8, 907–950, DOI 10.2140/ant.2009.3.907. MR2587408

[Sch11] Karl Schwede, *Test ideals in non-$\mathbb{Q}$-Gorenstein rings*, Trans. Amer. Math. Soc. **363** (2011), no. 11, 5925–5941, DOI 10.1090/S0002-9947-2011-05297-9. MR2817415

[ST12] Karl Schwede and Kevin Tucker, *A survey of test ideals*, Progress in commutative algebra 2, Walter de Gruyter, Berlin, 2012, pp. 39–99. MR2932591

[ST14] Karl Schwede and Kevin Tucker, *Test ideals of non-principal ideals: computations, jumping numbers, alterations and division theorems*, J. Math. Pures Appl. (9) **102** (2014), no. 5, 891–929, DOI 10.1016/j.matpur.2014.02.009. MR3271293

[Smi95] Karen E. Smith, *Test ideals in local rings*, Trans. Amer. Math. Soc. **347** (1995), no. 9, 3453–3472, DOI 10.2307/2155019. MR1311917

[Smi97] Karen E. Smith, *$F$-rational rings have rational singularities*, Amer. J. Math. **119** (1997), no. 1, 159–180. MR1428062

[Tak06] Shunsuke Takagi, *Formulas for multiplier ideals on singular varieties*, Amer. J. Math. **128** (2006), no. 6, 1345–1362. MR2275023

Department of Mathematics, University of Illinois at Chicago, Chicago, Illinois 60607

*Email address*: `kftucker@uic.edu`

Department of Mathematics Kyushu University, 744 Motooka, Nishiku, Fukuoka 819-0395, Japan

*Email address*: `ksato@math.kyushu-u.ac.jp`

Contemporary Mathematics
Volume **831**, 2025
https://doi.org/10.1090/conm/831/16631

# Hilbert-Kunz multiplicity

Vijayalaxmi Trivedi, with exercise solutions by Mandira Mondal

## 1. Introduction

The Hilbert-Kunz multiplicity $e_{HK}$ was introduced in a formal algebraic way by P. Monsky and soon caught the attention of commutative algebraists. Over the time HK multiplicity has turned out to be an important char $p$-invariant.

This expository article is very far from being exhaustive on the facts related to the invariant. Here we give a self contained proof of the existence of Hilbert-Kunz multiplicity and show that it is an additive function.

The article has drawn on material from the paper [**Mo**] of P. Monsky and also from the expository article of C. Huneke [**Hu**]. We state a few applications of the invariant in more detail, add some recent developments in the area and give some future directions.

## 2. Hilbert-Kunz multiplicity $e_{HK}$

Throughout this section $(R, \mathbf{m}, k)$ will denote a Noetherian local ring of prime characteristic $p > 0$ with maximal ideal $\mathbf{m}$ and residue field $k$. Also $R$ is of dimension $d \geq 1$.

Since the characteristic of $R$ is $p > 0$ the endomorphism map

$$F : R \longrightarrow R \quad \text{given by} \quad a \to a^p$$

is a ring homomorphism.

Further for any integer $e \geq 1$ we get the $e^{th}$-iterated Frobenius endomorphism $F^e : R \longrightarrow R$ where $a \to a^{p^e}$. This gives the ring $R$ a non canonical $R$-module structure, *i.e.*, via the map $F^e$. To emphasize this module structure on $R$, we would write the map $F^e : R \longrightarrow R$ as $F^e : R \longrightarrow F_*^e R$ and would denote the elements of $F_*^e R$ as $F_*^e r$ if $r \in R$. With this notation $F_*^e R$ is a module over $R$ with the module structure given as follows:

(1) If $F_*^e r_1, F_*^e r_2 \in F_*^e R$ then $F_*^e r_1 + F_*^e r_2 = F_*^e(r_1 + r_2)$ and
(2) if $r_1 \in R$ and $r_2 \in F_*^e R$ then $r_1 \cdot F_*^e r_2 = F_*^e(r_1^{p^e} r_2)$.

Similarly any $R$-module $M$ has another $R$-module structure, namely via the map $F^e$. Again we denote this module structure on $M$ by $F_*^e M$ and the element of $F_*^e M$ as $F_*^e m$, if $m \in M$.

With this notation $M$ and $F^e_*M$ both are the same as an additive group but the scalar multiplication on $F^e_*M$ is given as follows:

$$\text{if} \quad r \in R \quad \text{and} \quad m \in F^e_*M \quad \text{then} \quad r \cdot F^e_*m = F^e_*(r^{p^e} m).$$

REMARK 2.1. We note that $F^e_*M$ as $F^e_*R$-module is the same as $M$ as $R$-module, *i.e.*, for $F^e_*r \in F^e_*R$ and $F^e_*m \in F^e_*M$ we have $F^e_*r \cdot F^e_*m = F^e_*(rm)$. Hence we have a commutative diagram

$$\begin{array}{ccc} F^e_*R \times F^e_*M & \longrightarrow & F^e_*M \\ \| & & \| \\ R \times M & \longrightarrow & M, \end{array}$$

where we have identified the ring $F^e_*R$ with $R$ via the map $F^e_*r \to r$, similarly we have identified the $F^e_*M$ with $M$ via the map $F^e_*m \to m$.

Now if $I$ is an ideal of $R$ with generators $\{x_1, x_1, \ldots, x_n\}$ and $q = p^e$ then

$$I \cdot F^e_*R = \sum_i F^e_*(x_i^q R) \simeq \sum_i x_i^q R =: I^{[q]},$$

where the second identification is as mentioned in Remark 2.1.

E. Kunz has been one of the early ones to realize the importance of char $p$ methods in commutative algebra. We proceed by recalling two fundamental results of his (in [**K**]) in characteristic $p$.

THEOREM 1 (E. Kunz). *If $R$ is a Noetherian ring of characteristic $p > 0$ then it is regular if and only if the map $F^e : R \longrightarrow F^e_*R$ is flat for some $e > 0$ (equivalently for all $e > 0$).*

In particular if $R$ is $F$-finite, that is, if $F_*R$ is a finite $R$-module then the theorem of Kunz implies that the notion of $R$ being regular ring is same as $F^e_*R$ being a locally free $R$-module, for some $e$ (equivalently for all $e > 0$).

The following theorem of Kunz shows that the colengths of the Frobenius powers of the maximal ideal decide the regular property of the ring.

THEOREM 2 (E. Kunz). *Let $(R, \mathbf{m}, k)$ be a Noetherian local ring of dimension $d$ and characteristic $p > 0$. Then for $e \geq 1$ and $q = p^e$*

$$\ell_R(R/\mathbf{m}^{[q]}) \geq q^d.$$

*Moreover if the equality holds for some $e \geq 1$ then $R$ is regular.*

REMARK 2.2. To study these length functions we can assume without loss of generality that $(R, \mathbf{m})$ is complete and the residue field is algebraically closed.

For this we take the $\mathbf{m}$-adic completion $\widehat{R}$ of $R$. Now $\widehat{R}$ contains a coefficient field $k$. Then we take an algebraic closure $\bar{k}$ of the field $k$. Let $S = \widehat{R}\widehat{\otimes}_k\bar{k}$ then

(1) $S$ is a complete local ring with maximal ideal $\mathbf{m}S$ and
(2) $S$ is a faithfully flat extension of $R$ and
(3) the residue field of $S$ is $S/\mathbf{m}S = \bar{k}$, an algebraically closed field.

Now, if $I$ is an $\mathbf{m}$-primary ideal and $M$ is a finitely generated $R$-module then

$$\ell_R\Big(\frac{M}{I^{[q]}M}\Big) = \ell_S\Big(\frac{M}{I^{[q]}M} \otimes_R S\Big) = \ell_S\Big(\frac{M \otimes_R S}{I^{[q]}M \otimes_R S}\Big) = \ell_S\Big(\frac{M \otimes_R S}{I^{[q]}(M \otimes_R S)}\Big),$$

where the first and second equality follow from property (1) and (2) ively.

REMARK 2.3. Let $I$ and $M$ be as in Remark 2.2. Then to study the length function $\ell_R M/I^{[q]}M)$ we can assume that $(R, \mathbf{m})$ is a complete local ring where $R/\mathbf{m}$ is an algebraically closed field. In particular $R$ is $F$-finite and moreover, for any $e \geq 1$,

(2.1)
$$\ell_R\Big(\frac{F^e_* M}{I \cdot F^e_* M}\Big) = \ell_{F^e_* R}\Big(\frac{F^e_* M}{I \cdot F^e_* M}\Big) \cdot \ell_k\Big(\frac{F^e_* k}{k}\Big) = \ell_{F^e_* R}\Big(\frac{F^e_* M}{F^e_* I^{[q]} M}\Big) = \ell_R\Big(\frac{M}{I^{[q]} M}\Big).$$

NOTATION 2.4.

(1) $M$ is a finitely generated $R$-module and in that case $\mu_R(M)$ denotes the minimal number of generators of $M$.
(2) $I$ denotes an $\mathbf{m}$-primary ideal of $R$.
(3) $e(I, M) = \lim_{n \to \infty} \ell_R(M/I^n M) d!/n^d$ denotes the (Hilbert-Samuel) multiplicity of $M$ with respect to $I$.
(4)
$$\Lambda(R) = \{P \in \operatorname{Spec} R \mid \dim R/P = \dim R\}.$$
(5) If $f, g : \mathbb{N} \longrightarrow \mathbb{R}$ are two functions then we denote $f(n) = O(g(n))$ if there exists a positive constant $C$ such that $|f(n)| \leq C \cdot g(n)$, for all $n$.

Now we state and prove the following lemma which is the first assertion of Theorem 2 due to Kunz.

LEMMA 2.5. *Let $(R, \mathbf{m}, k)$ be a Noetherian local ring of dimension $d$ and characteristic $p > 0$. Then for $e \geq 1$ and $q = p^e$*
$$\ell_R(R/\mathbf{m}^{[q]}) \geq q^d.$$

PROOF. We can assume that $R$ is complete and has algebraically closed residue field $k$. Further we can assume $R$ is a domain by replacing $R$ by $R/P$, where $P \in \Lambda(R)$.

We choose (see Theorem 14.14 in [**Ma**]) a reduction ideal $(x_1, \ldots, x_d)$ of $\mathbf{m}$. Now consider the ring homomorphism $A = k[[X_1, \ldots, X_d]] \longrightarrow R$ mapping $X_i \to x_i$. Then, by Theorem 8.4 in [**Ma**], $R$ is a finitely generated $A$-module, and therefore the map $A \longrightarrow R$ is an injective map.

Consider the commutative diagram
$$\begin{array}{ccc} A & \longrightarrow & R \\ \downarrow & & \downarrow \\ F^e_* A & \longrightarrow & F^e_* R \end{array}$$

It is easy to check that $F^e_* A$ is a free $A$-module of rank $p^{ed} = q^d$, generated by $\{F^e_*(x_1^{i_1}), \cdots, F^e_*(x_d^{i_d}) \mid 0 \leq i_j < q = p^e\}$. We have
$$\operatorname{rank}_A F^e_* R = \big(\operatorname{rank}_R F^e_* R\big)\big(\operatorname{rank}_A R\big) = \big(\operatorname{rank}_{F^e_* A}(F^e_* R)\big)\big(\operatorname{rank}_A F^e_* A\big).$$
But $\operatorname{rank}_A R = \operatorname{rank}_{F^e_* A}(F^e_* R)$ which implies $\operatorname{rank}_R F^e_* R = \operatorname{rank}_A F^e_* A = q^d$.

If $T = R \setminus \{0\}$ then
$$\operatorname{rank}_R F^e_* R = \mu_{T^{-1}R}(T^{-1} R \otimes_R F^e_* R) \leq \mu_R(F^e_* R) = \ell_R\Big(\frac{F^e_* R}{\mathbf{m} \cdot F^e_* R}\Big) = \ell_R\left(R/\mathbf{m}^{[q]}\right).$$
$\square$

LEMMA 2.6. *For all $e \geq 0$ and $q = p^e$ we have*

$$\ell_R(M/I^{[q]}M) = O(q^{\dim M}).$$

PROOF. Let $t=\mu(I)$ then $I^{tq}\subset I^{[q]}$, for all $q$. Then $\ell_R(M/I^{[q]}M)\leq \ell_R(M/I^{tq}M)$. If $P(x) = c_0x^{d_1} + c_1x^{d_1-1} + \cdots + c_{d_1}$ is the Hilbert-Samuel polynomial of $M$ with respect to $I$, where $c_0 \neq 0$, then $d_1 = \dim M$ and for $q \gg 0$, $\ell_R(M/I^{tq}M) = P(tq)$. Now for

$$C = (d_1+1)t^{d_1} \cdot \max\{|c_i|\} \quad \text{we have} \quad \ell_R(M/I^{[q]}M) \leq C(q^{\dim M}).$$

□

LEMMA 2.7. *Let $M$ and $N$ be two finitely generated $R$-modules. Let $d = \dim R$. If $M_P \simeq N_P$, for each $P \in \Lambda(R)$ then for all $q = p^n$*

$$|\ell_R(M/I^{[q]}M) - \ell_R(N/I^{[q]}N)| = O(q^{d-1}).$$

PROOF. Let $S = R\setminus\bigcup_{P\in\Lambda(R)} P$. Then $S^{-1}R$ is an Artinian ring and $S^{-1}R \simeq \prod_{P\in\Lambda(R)} R_P$. This gives an isomorphism of $S^{-1}R$-modules

$$S^{-1}M = M\otimes_R S^{-1}R = \prod_{P\in\Lambda(R)} M_P \simeq \prod_{P\in\Lambda(R)} N_P = S^{-1}N, \tag{2.2}$$

well defined up to multiplication by a unit in $S^{-1}R$. Now the isomorphism

$$\mathrm{Hom}_{S^{-1}R}(S^{-1}M, S^{-1}N) \simeq S^{-1}\mathrm{Hom}_R(M,N)$$

implies that there is a map $\phi : M \longrightarrow N$ such that $S^{-1}\phi$ is a map as given in (2.2). In particular if $K = \operatorname{coker}\phi$ then $\dim K \leq d-1$.

The exact sequence

$$M/I^{[q]}M \longrightarrow N/I^{[q]}N \longrightarrow K/I^{[q]}K \longrightarrow 0$$

gives

$$\ell_R(N/I^{[q]}N) \leq \ell_R(M/I^{[q]}M) + \ell_R(K/I^{[q]}K) \leq \ell_R(M/I^{[q]}M) + O(q^{d-1}).$$

Now reversing the role of $M$ and $N$ we complete the proof of the lemma. □

LEMMA 2.8. *Let $0 \longrightarrow M' \longrightarrow M \longrightarrow M'' \longrightarrow 0$ be a short exact sequence of finitely generated $R$-modules and let $\dim A = d \geq 1$. Then*

$$\ell_R(M/I^{[q]}M) = \ell_R(M'/I^{[q]}M') + \ell_R(M''/I^{[q]}M'') + O(q^{d-1}).$$

PROOF. Again we can assume $R$ is complete and has an algebraically closed residue field.

CASE (a). $R$ is a reduced ring.

Let $\{P_1,\ldots,P_n\}$ be the set of minimal primes of $R$ and let $S = R\setminus(P_1\cup\cdots\cup P_n)$. Then $R_{P_i}$ is a field and therefore $(M'\oplus M'')_{P_i} \simeq M_{P_i}$ for each $i$. By Lemma 2.7

$$\begin{aligned}\ell_R\Big(\frac{M}{I^{[q]}M}\Big) &= \ell_R\Big(\frac{M'\oplus M''}{I^{[q]}(M'\oplus M'')}\Big) + O(q^{d-1})\\ &= \ell_R\Big(\frac{M'}{I^{[q]}M'}\Big) + \ell_R\Big(\frac{M''}{I^{[q]}M''}\Big) + O(q^{d-1}).\end{aligned}$$

CASE (b). $R$ is not a reduced ring. Let nilrad$(R)$ denote the nil radical of $R$. We choose $q_0 = p^{s_0}$ such that $(\mathrm{nilrad}(R))^{q_0} = 0$. Hence $F_*^{s_0}M$ is annihilated by nilrad$(R)$. Moreover

$$0 \longrightarrow N' = F_*^{s_0}M' \longrightarrow N = F_*^{s_0}M \longrightarrow N'' = F_*^{s_0}M'' \longrightarrow 0$$

is an exact sequence of $F_*^{s_0}R$-modules. Then it is an exact sequence of $R$-modules via the map $F_*^{s_0}$ and therefore of $R/\mathrm{nilrad}(R)$-modules. Now by Case (a)

$$\ell_R(N/I^{[q]}N) = \ell_R(N'/I^{[q]}N') + \ell_R(N''/I^{[q]}N'') + O(q^{d-1}).$$

By (2.1), for an $R$-module $N_0 = F_*^{s_0}M_0$

$$\ell_R\Big(\frac{N_0}{I^{[q]}N_0}\Big) = \ell_R\Big(\frac{F_*^sN_0}{I\cdot F_*^sN_0}\Big) = \ell_R\Big(\frac{F_*^{s+s_0}M_0}{I\cdot F_*^{s+s_0}M_0}\Big) = \ell_R\Big(\frac{M_0}{I^{[qq_0]}M_0}\Big).$$

This gives (taking $q > q_0$)

$$\ell_R(M/I^{[q]}M) = \ell_R(M'/I^{[q]}M') + \ell_R(M''/I^{[q]}M'') + O((q/q_0)^{d-1}).$$

But $O(q/q_0)^{d-1}) = O(q^{d-1})$ as $q_0$ is fixed. □

DEFINITION 2.9. We recall the following standard notations. If $\{x_n\}_n$ is a bounded sequence of elements in $\mathbb{R}$ then

$$\liminf_{n\to\infty} x_n = \lim_{n\to\infty} y_n \quad \text{where} \quad y_n = \inf\{x_m \mid m \geq n\}$$

and similarly

$$\limsup_{n\to\infty} x_n = \lim_{n\to\infty} z_n \quad \text{where} \quad z_n = \sup\{x_m \mid m \geq n\}.$$

Since $\{y_n\}_n$ and $\{z_n\}_n$ are both monotonic sequences, both limits do exist.

Exercise 2 (see the last section) proves that

$$e(I)/d! \leq \liminf_{q\to\infty} \ell_R(R/I^{[q]})/q^d \leq \limsup_{q\to\infty} \ell_R(R/I^{[q]})/q^d \leq e(I).$$

(These limits are well defined by Lemma 2.6).

LEMMA 2.10. *Let $(R, \mathbf{m}, k)$ be a complete local domain of dimension $d \geq 1$ such that $k$ is an algebraically closed field. Then there exists a constant $e_{HK}(R, I) > 0$ such that*

$$\ell_R(R/I^{[q]}) = e_{HK}(R, I)q^d + O(q^{d-1}).$$

PROOF. Let $r = \mathrm{rank}_R F_*R$, then for $S = R\setminus\{0\}$ we have an $S^{-1}R$-linear isomorphism

$$\widetilde{\phi} : \oplus^r S^{-1}R \longrightarrow S^{-1}R \otimes_R F_*R.$$

Since $F_*R$ is a finite $R$-module, the map $\widetilde{\phi}$ lifts to a map $\phi : \oplus^r R \longrightarrow F_*R$, after multiplying $\widetilde{\phi}$ by a unit in $S^{-1}R$. This gives us a short exact sequence of $R$-modules

$$0 \longrightarrow \oplus^r R \longrightarrow F_*R \longrightarrow \mathrm{coker}\,\phi \longrightarrow 0,$$

where $\dim \mathrm{coker}\,\phi \leq d-1$.

Therefore

$$O(q^{d-1}) = \Big|r\ell_R\Big(\frac{R}{I^{[q]}}\Big) - \ell_R\Big(\frac{F_*R}{I^{[q]}\cdot F_*R}\Big)\Big| = \Big|r\ell_R(R/I^{[q]}) - \ell_R(R/I^{[qp]})\Big|.$$

This gives

$$\Big|(r/(qp)^d)\ell_R(R/I^{[q]}) - (1/(qp)^d)\ell_R(R/I^{[qp]})\Big| = O(1/q).$$

Let $\{C_n\}_n$ be the sequence given by $C_n = (1/q^d)\ell_R\left(R/I^{[q]}\right)$, where $q = p^n$. Then we can write the above equality as

$$\left|(r/p^d)C_n - C_{n+1}\right| = O(1/q).$$

Therefore

$$\limsup_{n\to\infty} C_n = \limsup_{n\to\infty} C_{n+1} = \limsup_{n\to\infty}(r/p^d)C_n,$$

$$\liminf_{n\to\infty} C_n = \liminf_{n\to\infty} C_{n+1} = \liminf_{n\to\infty}(r/p^d)C_n.$$

Suppose $r/p^d > 1$ then

$$\limsup_{n\to\infty}(r/p^d)C_n = (r/p^d)\limsup_{n\to\infty} C_n > \limsup_{n\to\infty} C_n.$$

Suppose $r/p^d < 1$ then

$$\liminf_{n\to\infty}(r/p^d)C_n = (r/p^d)\liminf_{n\to\infty} C_n < \liminf_{n\to\infty} C_n.$$

Therefore $r/p^d = 1$ and $\{C_n\}_n$ is a Cauchy sequence converging to a constant $e_{HK}(R, I)$.

Now the assertion that $e_{HK}(R, I) > 0$ follows from Exercise (a). □

THEOREM 2.11. *Let $(R, \mathbf{m}, k)$ be a Noetherian local ring of dimension $d$. Let $M$ be a finitely generated $R$-module. Then there exists constant $e_{HK}(M, I) \in \mathbb{R}_{\geq 0}$ such that for all $q$*

$$\ell_R(M/I^{[q]}M) = e_{HK}(M, I)q^d + O(q^{d-1}). \tag{2.3}$$

*Further, the associativity formula holds:*

$$e_{HK}(M, I) = \sum_{P\in\Lambda(R)} e_{HK}(R/P, I)\ell_{R_P}(M_P).$$

PROOF. To prove the assertion (2.3), without loss of generality we can assume (from Remark 2.2) that $R$ is a complete local ring with algebraically closed residue field. Applying Lemma 2.8 to a prime filtration

$$0 = M_0 \subset M_1 \subset M_2 \subset \cdots \subset M_n = M,$$

where $M_{i+1}/M_i = R/P_i$, for $0 \leq i < n$ and where $P_i \in \operatorname{Spec} R$, we get

$$\ell_R(M/I^{[q]}M) = \sum_{0\leq i<n} \ell_R\Big(\frac{M_{i+1}/M_i}{I^{[q]}(M_{i+1}/M_i)}\Big) + O(q^{d-1})$$

$$= \sum_{\dim M_{i+1}/M_i = d} e_{HK}(M_{i+1}/M_i, I)q^d + O(q^{d-1}),$$

where the second equality follows from Lemma 2.10. Hence follows the first assertion of the theorem, that is, (2.3) holds.

Now consider a short exact sequence

$$0 \longrightarrow N \longrightarrow M \longrightarrow M/N \longrightarrow 0$$

of finitely generated $R$-modules. Then, by Lemma 2.8 it follows that for all $q$,

$$e_{HK}(M, I)q^d = e_{HK}(N, I)q^d + e_{HK}(M/N, I)q^d + O(q^{d-1})$$

and therefore $e_{HK}(M, I) = e_{HK}(N, I) + e_{HK}(M/N, I)$.

Now applying this additivity property of $e_{HK}$ to a prime filtration (over $R$, which may not be related to the earlier one over $\widehat{R}$)

$$0 = N_0 \subset N_1 \subset N_2 \subset \cdots \subset N_m = M,$$

of $M$ where $N_{i+1}/N_i = R/Q_i$, for $0 \leq i < m$ and some $Q_i \in \operatorname{Spec} R$, we get

$$\ell_R(M/I^{[q]}M) = \sum_{\dim N_{i+1}/N_i = d} e_{HK}(N_{i+1}/N_i, I)q^d + O(q^{d-1}).$$

Since for $Q \in \Lambda(R)$, the number of times $R/Q$ occurs as $N_{i+1}/N_i$ is equal to $\ell_{R_Q}(M_Q)$, the second assertion of the theorem follows. □

DEFINITION 2.12. We define the *Hilbert-Kunz (HK) multiplicity* of $M$ with respect to $I$ as the constant given by

$$e_{HK}(M, I) = \lim_{q \to \infty} \ell_R(M/I^{[q]}M)/q^d.$$

COROLLARY 2.13. *If $(R, \mathbf{m}, k)$ is a domain then*

$$e_{HK}(M, I) = e_{HK}(R, I)(\operatorname{rank} M).$$

REMARK 2.14. To study $e_{HK}(M, I)$ it is enough to know $e_{HK}(R/P, I)$, where $\dim R/P = \dim R$. Also $\dim M < \dim R$ if and only if $e_{HK}(M, I) = 0$.

COROLLARY 2.15. *If $(R, \mathbf{m}, k)$ is a ring of dimension 1 then $e_{HK}(R, I) = e(R, I)$.*

## 3. Some properties of $e_{HK}$

REMARK 3.1. We have

$$e_{HK}(R, I) = e_{HK}(\widehat{R}, I\widehat{R}) = \sum_{P \in \Lambda(\widehat{R})} e_{HK}(\widehat{R}/P, I\widehat{R}).$$

In particular the lower dimensional components of $R$ and $\widehat{R}$ do not contribute to $e_{HK}(R, I)$. Hence to study the implication of $e_{HK}$ on any property of $R$, we need to assume that $R$ is formally unmixed, that is, for every associated prime $P$ of $\widehat{R}$, we have $\dim \widehat{R}/P = \dim R$.

For example if $R = k[[x, y, z]]/(xy - xz)$ then by Theorem 2.11 $e_{HK}(R) = 1$ but $R$ is not regular.

DEFINITION 3.2. Let $R^o$ denote the complement of the union of all minimal primes of a ring $R$. An element $x \in R$ is the *tight closure* of an ideal $J$ of $R$, if there is $c \in R^o$ such that $cx^q \in I^{[q]}$, for $q = p^e \gg 0$.

We denote $I^* = \{x \in R \mid x \in \text{ tight closure of } I\}$. It is easy to check that $I^*$ is an ideal in $R$.

We have seen that analogous to multiplicity the HK multiplicity satisfies associativity formula. The following result shows that the HK multiplicity and the tight closures of the ideals have a relation very similar to the relation between multiplicity and the integral closures of the ideals.

THEOREM 3.3. *Let $(R, \mathbf{m}, k)$ be a Noetherian local ring of dimension $d$. Let $I$ be an $\mathbf{m}$-primary ideal and suppose $J \subseteq I \subseteq J^*$. Then $e_{HK}(R, I) = e_{HK}(R, J)$.*

*Conversely, if $R$ is formally unmixed and $J \subseteq I$, then*

$$e_{HK}(R, I) = e_{HK}(R, J) \implies I \subseteq J^*$$

*where for an ideal $I$, the ideal $I^*$ denotes the tight closure of $I$.*

PROOF. We prove the first assertion which is the easier part. If $J = \langle x_1, \ldots, x_n \rangle$ then $J^{[q]} = \langle x_1^q, \ldots, x_n^q \rangle$. We can find $c \in R^o$ such that $cJ^{[q]} \subset I^{[q]}$ for all $q$. Hence $J^{[q]}/I^{[q]}$ is a $R/(I^{[q]}, c)$-module. So there exists $m$ and a surjective map

$$\oplus^m R/(I^{[q]}, c) \longrightarrow J^{[q]}/I^{[q]}$$

of $R/(I^{[q]}, c)$-modules. Therefore Theorem 2.11 for $R/cR$-modules gives

$$\ell_R\left(J^{[q]}/I^{[q]}\right) \leq m\ell_R\left(R/(I^{[q]}, c)\right) = m\ell_{R/cR}\left(R/(I^{[q]}, c)\right) = O(q^{d-1})$$

and for the sequence of $R$-modules

$$0 \longrightarrow J^{[q]}/I^{[q]} \longrightarrow R/I^{[q]} \longrightarrow R/J^{[q]} \longrightarrow 0$$

Lemma 2.8 gives $e_{HK}(I) = e_{HK}(J)$.

The converse follows easily if we assume the following result due to Aberbach [**A**]: If $x \in R$ such that $x \notin I^*$ then there exists $k \geq 1$ such that for all $q$, we have $(I^{[q]} : x^q) \subset \mathbf{m}^{\lfloor q/k \rfloor}$. For this we consider the canonical exact sequence

$$0 \longrightarrow R/(I^{[q]} : x^q) \longrightarrow R/I^{[q]} \longrightarrow R/(I^{[q]}, x^q) \longrightarrow 0$$

which for $q \gg 0$ gives

$$\ell_R(R/I^{[q]}) - \ell_R(R/(I^{[q]}, x^q)) = \ell_R(R/(I^{[q]} : x^q)) \geq \ell_R(R/\mathbf{m}^{\lfloor q/k \rfloor}).$$

Note that the Hilbert-Samuel polynomial of $R$ with respect to $\mathbf{m}$ is

$$\ell_R(R/\mathbf{m}^n) = c_0 n^d + c_1 n^{d-1} + \cdots + c_d \quad \text{for} \quad n \gg 0.$$

Therefore for a choice of $q \gg 0$ such that

$$\lfloor q/k \rfloor \geq 2d \max\{|c_1|, \ldots, |c_d|\} \quad \text{and} \quad q \geq 2k^2$$

we have $\ell_R(R/\mathbf{m}^{\lfloor q/k \rfloor}) \geq (c_0 - 1/2)(q/2k)^d$. This implies $e_{HK}(R, I) > e_{HK}(R, I + xR)$. □

**3.1. Estimates on HK multiplicity.** Bound constraints on $e_{HK}$ reflect on the characteristic $p$-singularities. A detailed survey on these results can be found in [**Hu**]. Here we will discuss some more recent work on bound estimates. By the result (see Theorem 2) of Kunz we know that $e_{HK}$ is always bounded below by 1. Here we state a result due to Watanabe-Yoshida [?WY3] which characterizes rings achieving the minimum $e_{HK}(R, \mathbf{m})$. Again the statement is analogous to the fundamental result due to Nagata characterizing the rings with minimal multiplicity.

THEOREM 3.4. *Let $(R, \mathbf{m}, k)$ be a Noetherian local ring of dimension $d$ which is formally unmixed. Then*

$$e_{HK}(R) = 1 \iff R \text{ is regular.}$$

Examples (see Example 3.6) show that $e_{HK}$ need not be an integer and may depend on the characteristic of the ring. So the natural attempt is to give a characteristic-free lower bound on the HK multiplicity away from 1. The first such bound was proved by [**AE1**] which was improved later by [**CDHZ**] as follows.

*If $(R, \mathbf{m})$ is a non regular ring of dimension $d \geq 2$ then $e_{HK}(R) \geq 1 + (1/d!d^d)$.*

We recall the following conjecture of [**WY2**] along with a result of Gessel-Monsky [**GM**] which indicates that still this is not the best possible: First we recall a few notations.

Let $R_{p,d} = K[x_0, \ldots, x_d]/(x_0^2 + \cdots + x_d^2)$ and let $m_d$ denote the constants occurring as the coefficients of the following expression

$$\sec(x) + \tan(x) = 1 + \sum_{d=1}^{\infty} m_d x^d, \quad \text{where} \quad |x| < \pi/2,$$

where

$$m_3 = \frac{1}{3!}, \quad m_4 = \frac{5}{4!}, \quad m_5 = \frac{16}{5!}, \quad m_6 = \frac{61}{6!} \quad \text{etc.}$$

THEOREM ([**GM**]). $\lim_{p\to\infty} e_{HK}(R_{p,d}, \mathbf{m}) = 1 + m_d$.

CONJECTURE ([**WY2**]). *Let $p > 2$ be prime and $K = \bar{\mathbf{F}}_{\mathbf{p}}$.*

(a) *If $(A, \mathbf{m}_A, K)$ is a formally unmixed non regular local ring of dimension $d$. Then $e_{HK}(A, \mathbf{m}_A) \geq e_{HK}(R_{p,d}, \mathbf{m}) \geq 1 + m_d$.*
(b) *If $e_{HK}(A, \mathbf{m}_A) = e_{HK}(R_{p,d}, \mathbf{m})$ then the $\mathbf{m}$-adic completion $\widehat{A}$ of $A$ is isomorphic to $R_{p,d}$ as local rings.*

This conjecture has been proved for $d \leq 6$, see [**AE2**], [**WY2**]. Moreover the second inequality of the assertion (a) is proved recently in Theorem 4.3 of [**T7**] for general $d$ provided $p \geq d - 3$. Also the first inequality of the assertion (a) is proved in Theorem 4.6 of [**ES**] provided $(A, \mathbf{m}_A, K)$ is a complete intersection rings of dimension $d \geq 2$. In particular combining these two results we have the following.

THEOREM 3.5. *If $d \geq 4$ and $p \geq d-3$ and $(A, \mathbf{m}_A, K)$ is a complete intersection but non regular local ring of $d$. Then $e_{HK}(A, \mathbf{m}) \geq e_{HK}(R_{p,d}, \mathbf{m}) \geq 1 + m_d$.*

More recently [**JNSWY**] have revisited the conjecture and have conjectured a lower bound for the class of rings which are non regular and are not quadric hypersurfaces.

The following computation due to Han-Monsky of the Fermat quartic is very significant in the theory of HK multiplicity.

EXAMPLE 3.6. Let $R = k[[x, y, z]]/(x^4 + y^4 + z^4)$, where $\operatorname{char} k = p$. Then

$$\begin{aligned} e_{HK}(R, (x, y, z)) &= 3 + (1/p^2), \quad \text{if } p \equiv \pm 3 \pmod 8 \\ &= 3, \quad\quad\quad\quad\; \text{if } p \equiv \pm 1 \pmod 8. \end{aligned}$$

This example implies that

(1) $e_{HK}$ depends on the characteristic of the ring. Moreover it depends on the congruence class of $p$.
(2) Under any hyperplane section $e_{HK}$ can fail to remain same: Because if there is $x \in \mathbf{m}$ such that $e_{HK}(R/xR) = e_{HK}(R)$ then $e_{HK}(R) = e_{HK}(R/xR) = e(R/xR) \in \mathbb{N}$, whereas the above example implies otherwise.
(3) This example gives rise to an example of a vector bundle (see the next section) asserting that *reduction mod $p$* property is not open for the 'Frobenius semistability property' of $V$.

Failure of invariance under hyperplane section is one of the reasons which makes computations of $e_{HK}$ difficult. However, we will see some examples in the next section exhibiting that the HK multiplicity is a more refined invariant compared to the multiplicity.

**3.2. Behaviour of $e_{HK}$ in families of ideals.**

(1) Let $\phi : \operatorname{Spec} R \longrightarrow \mathbb{R}_{\geq 0}$ be the map given by $P \to e_{HK}(R_P, I_P)$. It was shown in [**S1**] that it is upper semi-continuous under some mild conditions on $R$.

(2) The growth of $e_{HK}(R, I^k)$ as $k \to \infty$ has been studied by [**Ha**] and [**WY1**]. In the graded situation in [**T6**] it was shown that

$$\lim_{k\to\infty} \frac{e_{HK}(M, I^k) - e_0(M, I^k)/d!}{k^{d-1}} = \frac{e_0(M,I)}{2(d-2)!} - \frac{E_1(M,I)}{(d-1)!} \geq \frac{e_0(M,I)}{2(d-1)\, p^{e_0(M,I)}},$$

where $e_0(M,I)$ and $e_1(M,I)$ are the coefficients of the Hilbert-Samuel polynomial of $M$ with respect to $I$ and $E_1(M,I) = \lim_{q\to\infty} e_1(M, I^{[q]})/q^d$ does exist. This formula was later generalised to local case in [**S2**].

(3) An affirmative answer to the following question will give a well defined notion of $e_{HK}$ in characteristic 0.

Open question. Let $(R, \mathbf{m}, k)$ be a Noetherian local ring where $k$ is a field of characteristic 0. Let $(R_p, \mathbf{m}_p, k_p)$ denote *reduction mod* $p$ of $(R, \mathbf{m}, k)$. Then

$$\text{does} \quad \lim_{p\to\infty} e_{HK}(R_p, \mathbf{m}_p) \quad \text{exist?}$$

An affirmative answer is obvious for toric rings, monomial rings (see [**C**], [**E**], [**W**]) where the $e_{HK}$ is independent of the characteristic. In the case of two dimensional standard graded rings [**T2**] or when $R$ is a diagonal hypersurface [**GM**] of any dimension the answer is affirmative though $e_{HK}(R_p, \mathbf{m}_p)$ depends on the $p$.

## 4. Some applications

**4.1. $e_{HK}$ and semistability of a vector bundle.** In the case $R$ is a graded ring the invariant $e_{HK}$ and the char $p$ behaviour of the associated *syzygy bundle* on $\operatorname{Proj} R$ are closely related. Here we elaborate when $R$ is two dimensional, equivalently when $\operatorname{Proj} R$ is a curve.

First we recall some relevant facts about vector bundles on a nonsingular projective curve. For more details on material related to semistable bundles reader can refer to [**HL**]. Given a vector bundle (locally free sheaf of $\mathcal{O}_X$-modules) $V$ on a projective curve $X$ is *semistable* if for any subbundle $W$ (locally free subsheaf) of $V$ we have $\mu(W) \leq \mu(V)$, where $\mu(E) = \deg(E)/\operatorname{rank}(E)$ denotes the $\mu$-slope of $E$.

A semistable bundle has cohomologically nice properties. We recall a fundamental result on vector bundles due to Harder-Narasimhan. Every vector bundle $V$ has a (unique) filtration

$$0 \subset E_1 \subset \cdots \subset E_n \subset V$$

of subbundles such that

(1) each $E_{i+1}/E_i$ is semistable and
(2) $\mu(E_1) > \mu(E_2/E_1) > \cdots > \mu(V/E_n)$.

This filtration is now called the Harder-Narasimhan (HN) filtration of $V$.

It is known that semistability property is an open property for *reduction mod* $p$: Suppose $V$ is a vector bundle on a nonsingular curve $X$ which is defined over a field of characteristic 0 and if $V_p$ denotes the *reduction mod* $p$ of $V$, then $V$ semistable implies that $V_p$ is semistable for $p \gg 0$.

It is known that if $V$ is a semistable vector bundle on $X$ and $\pi : Y \longrightarrow X$ is a a finite map of nonsingular projective curves then $\pi^*V$ is a semistable vector bundle on $Y$ where by a finite map means that if Spec $A$ is an affine open set of $X$ then $\pi^{-1}(\operatorname{Spec} A) = \operatorname{Spec} B$ is affine open set of $Y$ and the induced homomorphism $A \longrightarrow B$ is finite. If $X_p$ is *reduction mod* $p$ of $X$ then the Frobenius $F : X_p \longrightarrow X_p$ is a finite map, It is natural to ask that if $V$ is semistable then is $F^*V_p$ is semistable for $p \gg 0$?

The following relation between HK multiplicity and the semistability behaviour of the relevant syzygy bundle gives a negative answer.

Let $R$ be a standard graded domain of dimension two over a perfect field $k$. Then $C = \operatorname{Proj} R$ is an irreducible projective curve. Let $X$ be the normalization of $C$ and let $h_1, \ldots, h_s$ be a set of homogeneous generators of $\mathbf{m}$. Then we have a canonical short exact sequence of sheaves of $\mathcal{O}_X$-modules:

$$0 \longrightarrow V \longrightarrow \oplus^s \mathcal{O}_X \longrightarrow \mathcal{O}_X(1) \longrightarrow 0.$$

where $V$ is a vector bundle on $X$ of rank $s - 1$.

Note that for a given vector bundle $V$ *syzygy bundle* (associated to the above data) on a nonsingular projective curve we can associate a *strong HN data* $\big((a_1(V), a_2(V), \ldots, ), \quad (r_1(V), r_2(V), \ldots, )\big)$ as follows. Given a vector bundle $V$ there is $s_0 \gg 0$, see Theorem 2.7 in [**L**] due to Langer, such that $F^{*s_0}(V)$ has a *strong HN filtration*, that is, if

$$0 \subset V_1 \subset V_2 \subset \cdots \subset V_m \subset V_{m+1} = F^{*s_0}V$$

is the HN filtration of $F^{*s_0}V$ then for $s \geq s_0$ the filtration

$$0 \subset F^{*s-s_0}V_1 \subset F^{*s-s_0}V_2 \subset \cdots \subset F^{*s-s_0}V_m \subset F^{*s-s_0}V_{m+1} = F^{*s}V$$

is the HN filtration of $F^{*s}V$. We define $a_i(V) = \mu(V_{i+1}/V_i)/p^{s_0}$ and $r_i(V) = \operatorname{rank}(V_{i+1}/V_i)$.

THEOREM 4.1 ([**B**], [**T1**]). *With the notations as above we have*

$$e_{HK}(R) = \frac{\deg(X)}{2}\Big(\sum_{i=0}^{m}(r_i(V)a_i(V)^2 - s)\Big).$$

THEOREM 4.2 ([**T1**]). *Let $R = k[X, Y, Z]/(h)$ be a graded domain, where $h$ is a homogeneous polynomial of degree $d \geq 3$. Then*

$$e_{HK}(R) = (3d/4) + (l^2/4dp^{2s}),$$

*where $l \leq d(d-3)$ and $s$ are nonnegative integers with the following description: For $s_1 < s$ the bundle $F^{s_1*}V$ is semistable and therefore $0 \subset F^{s_1*}V$ is its HN filtration. Further $F^{s*}V$ has strong HN filtration and is given by*

$$0 \subset \mathcal{L} \subset F^{s*}V \quad \text{such that} \quad \mu(\mathcal{L}) = (-dp^s + l)/2 \quad \text{and} \quad \mu(F^{*s}V/\mathcal{L}) = (-dp^s - l)/2.$$

In particular this bound on $l$ (which was obtained using a result in [**SB**]) gives a dictionary between $e_{HK}$ and the integers $s$ and $l$ provided $p > d(d-3)$.

Applying this to the Example 3.6, for $\operatorname{char} p \geq 5$, we get

(1) If $p \equiv \pm 1 \pmod 8$ then $F^{s*}V$ is semistable for all $s \geq 0$.
(2) If $p \equiv \pm 3 \pmod 8$ then $F^*V$ has the HN filtration $\mathcal{L} \subset F^*V$ where $\mu(\mathcal{L}) = \mu(F^*V) + 2$. In particular $F^*V$ is not semistable.

Building on the idea of 'taxicab distance' due to Han-Monsky, one can compute (see [**T3**]) $e_{HK}(R)$ for any explicit trinomial plane curve $R = k[X,Y,Z]/h$ and therefore exhibit interesting and varied examples of the Frobenius semistability behaviour of the syzygy bundle $V$. In fact under *reduction mod $p$*, the Frobenius semistability behaviour is a function of the congruence class of $p$ modulo an integer determined explicitly in terms of the exponents of the monomials appearing in $h$.

**4.2. Tiling of a convex polytope and asymptotic growth of $e_{HK}(R, \mathbf{m}^k)$.** We know that (see [**E**] and [**W**]) that $e_{HK}$ of a toric ring does not depend on the characteristic of the ring. However $e_{HK}$ reflects on some combinatorial aspect of the toric rings.

DEFINITION 4.3. A rational convex polytope $P$ in $\mathbb{R}^{d-1}$ *tiles* the space $\mathbb{R}^{d-1}$ with respect to the lattice $\mathbb{Z}^{d-1}$ if there exists $\lambda \in \mathbb{R}_{>0}$ such that

(1) $\bigcup_{z\in\mathbb{Z}^{d-1}}(\lambda P + z) = \mathbb{R}^{d-1}$, and
(2) $\dim\big((\lambda P + z) \cap (\lambda P + z')\big) < d-1$ if $z \neq z'$.

For the basic facts which are used here about toric varieties one can refer to [**F**].

Let $(X, D)$ be a toric pair of dimension $d-1 \geq 1$ and where $X$ is a projective toric variety and $D$ is a torus invariant (very ample, Cartier) divisor. Let $R = \oplus_{m\in\mathbb{N}} H^0(X, \mathcal{O}(mD))$ (see Section 3.4 of [**F**]) denote the associated homogeneous coordinate ring with the graded maximal ideal $\mathbf{m}$. It is well known that a toric pair $(X, D)$ corresponds to a *very ample* convex integral polytope $P_{X,D}$ in $\mathbb{R}^{d-1}$.

THEOREM 4.4 ([**MT**]). *For a given toric pair $(X, D)$ of dimension $d-1$, where $d \geq 2$ and for associated ring $(R, \mathbf{m})$ as above we have*

$$\lim_{k\to\infty} \frac{e_{HK}(\mathbf{m}^k) - e(\mathbf{m}^k)/d!}{k^{d-1}} \geq \left[\frac{d-1}{d}\right]\left[\frac{e(\mathbf{m})}{(d-1)!}\right]^{\frac{d-2}{d-1}}$$

*and the equality holds if and only if the associated polytope $P_{X,D}$ tiles the space $\mathbb{R}^{d-1}$ with respect to the lattice $\mathbb{Z}^{d-1}$.*

In other words the normalized asymptotic growth of $e_{HK}(R, \mathbf{m}^k)$ is the slowest iff $P_{X,D}$ tiles $\mathbb{R}^{d-1}$ with respect to $\mathbb{Z}^{d-1}$.

Further we note that such a tiling property of any rational polytope $P$ in $\mathbb{R}^{d-1}$ can be formulated in terms of the Hilbert-Kunz multiplicity as follows: We choose $m_1 \geq 0$ such that $m_1 P$ is an integral polytope. Then by Corollary 2.2.18 in [**CLS**] the polytope $mP$, with $m = (d-2)m_1$, is a very ample integral convex polytope. In particular, there is a toric pair $(X, D)$ such that the associated polytope $P_{X,D} = mP$.

## 5. HK density function

Here we discuss a relatively new char $p$ invariant which was introduced to have a better understanding of the HK multiplicity.

DEFINITION 5.1. A *graded pair* $(R, I)$ means $R$ is a standard graded ring over a perfect field of characteristic $p > 0$ and $I$ is homogeneous ideal of finite colength. We also assume that $\dim R = d \geq 2$.

For a graded pair $(R, I)$ consider a sequence $\{f_n(R,I) : \mathbb{R}_{\geq 0} \longrightarrow \mathbb{R}_{\geq 0}\}_{n\in\mathbb{N}}$ of functions given by

$$f_n(R,I)(x) = \frac{\ell_R(R/I^{[q]})_{\lfloor xq \rfloor}}{q^{d-1}}, \quad \text{where} \quad q = p^n.$$

Then we define the *HK density (HKd) function* $f_{R,I} : \mathbb{R}_{\geq 0} \longrightarrow \mathbb{R}_{\geq 0}$ as $f_{R,I}(x) = \lim_{n\to\infty} f_n(R,I)(x)$. The following theorem asserts that this is indeed a well defined map.

THEOREM 5.2 ([**T4**]). *For a graded pair $(R, I)$ the sequence $\{f_n(R,I)\}_{n\in\mathbb{N}}$ converges uniformly to the continuous compactly supported function $f_{R,I}$ and*

$$\int_0^\infty f_{R,I}(x)dx = e_{HK}(R,I).$$

PROPERTIES.

(1) *(Associativity) The associativity formula holds for the HK density function.*
(2) *(Tight closure) If $R$ is an equidimensional ring and $(R, I)$ and $(R, J)$ two graded pairs with $I \subseteq J$ are two graded ideals of $R$ then*

$$f_{R,I} \cong f_{R,J} \quad \textit{if and only if} \quad J \subseteq I^*,$$

*where $I^*$ denotes the* tight closure *of $I$ in $R$.*
(3) *(Multiplicative property): If $(R, I)$ and $(S, J)$ are two graded pairs and $H_R(x) = e(R)x^{d-1}(d-1)!$, where $e(R)$ denotes the multiplicity of $R$ with respect to its irrelevant maximal ideal* $\mathbf{m}$ *and $d = \dim R$, then the Segre product $(R\#S, I\#J)$ satisfies*

$$H_{R\#S}(x) - f_{R\#S,I\#J}(x) = \big(H_R(x) - f_{R,I}(x)\big)\big(H_S(x) - f_{S,J}(x)\big).$$

Recall that the set of compactly supported continuous functions are in bijective correspondence with the set of their Fourier transforms which are holomorphic functions. Further if $\widehat{f}_{R,I}$ denotes the Fourier transform of $f_{R,I}$ then $\widehat{f}_{R,I}(0) = e_{HK}(R,I)$. This suggests possible applications of harmonic analysis in the study of HK multiplicities.

Two applications of HK density function we have already stated:

(1) To give a lower bound on the HK multiplicity of quadratic hypersurfaces as in the earlier mentioned conjecture of [**WY2**] (see Theorem 4.3 of [**T7**]) and
(2) relating tiling property a convex polytope with the growth of $e_{HK}(\mathbf{m}^k)$ (see Theorem 6.3 of [**MT**]).

Now we give a third application of HK density function.

Since $f_{R,I}$ is compactly supported function we can define the notion of maximal support $\alpha(R,I) = \sup\{x \mid f_{R,I}(x) \neq 0\}$. This relates to another char $p$ invariant, namely $F$-threshold $c^I(\mathbf{m})$ of $\mathbf{m}$ with respect to $I$, which is defined (see [**DsNbP**]) as

$$c^I(\mathbf{m}) = \lim_{q\to\infty} \frac{\min\{r \mid \mathbf{m}^{r+1} \subseteq I^{[q]}\}}{q}.$$

THEOREM 5.3 ([?TW]). *Let $(R, I)$ be a graded pair, where $R$ is a domain such that its normalization $S$ is strongly $F$-regular on its punctured spectrum. Then $\alpha(R,I) = c^I(\mathbf{m})$.*

*In particular the equality* $\alpha(R, I) = c^I(\mathbf{m})$ *holds for any two dimensional standard graded pair* $(R, I)$, *where* $R$ *is a domain.*

Note if $\dim R = 2$ then we can describe the function $f_{R,I}$ and relate $c^I(\mathbf{m})$ to the strong HN data of the syzygy bundle as described in Section 4. Using this relation and an example due to Gieseker [**G**] of a family of vector bundles with certain Frobenius semistability property on a nonsingular projective curve, one gives an affirmative answer to a question posed by Mustaţă-Takagi-Watanabe ([**MTW**], 2005) as follows.

THEOREM 5.4 ([**T5**]). *Given a prime* $p$ *and an integer* $g > 1$, *there exists a two dimensional standard graded normal* $\mathbb{Q}$*-Gorenstein domain* $R$ *with the graded maximal ideal* $\mathbf{m}$ *such that the set of* $F$*-thresholds of* $\mathbf{m}$ *has accumulation points, where* $\operatorname{Proj} R = X$ *is a nonsingular projective curve of genus* $g$ *over a field of char* $p$.

OPEN QUESTION. Let $(R, I)$ be a standard graded pair and $\mathbf{m}$ be the graded maximal ideal of $R$. Then, is $\alpha(R, I) = c^I(\mathbf{m})$?

## 6. Exercises, with solutions by Mandira Mondal

EXERCISE 1. Show that if $J \subset R$ is an ideal generated by a system of parameters then $\lim_{q\to\infty} \ell_R(R/J^{[q]})/q^d$ exists and is equal to $e(R, J)$.

SOLUTION. If $J$ is generated by a system of parameters, say $x_1, \ldots, x_d$ then by Lech's lemma

$$e(R, J) = \lim_{n\to\infty} \ell_R(R/x_1^n, \ldots, x_d^n)/n^d.$$

In particular

$$\lim_{q\to\infty} \ell_R(R/J^{[q]})/q^d = \lim_{q\to\infty} \ell_R(R/x_1^q, \ldots, x_d^q)/q^d = e(R, J).$$

EXERCISE 2. Show that

$$e(I)/d! \leq \liminf \ell_R(R/I^{[q]})/q^d \leq \limsup \ell_R(R/I^{[q]})/q^d \leq e(I).$$

SOLUTION. Without loss of generality, we can assume, $R$ is a complete local ring and the residue field $R/\mathbf{m}$ is algebraically closed.

We have $I^{[q]} \subseteq I^q$ for all $q = p^n$ with $n \in \mathbb{N}$. Hence $\ell_R(R/I^q) \leq \ell_R(R/I^{[q]})$. Multiplying by $1/q^d$ and taking limit as $n \to \infty$, we have

$$e(R, I)/d! = \lim_{q\to\infty} \ell_R(R/I^q)/q^d \leq \liminf_{q\to\infty} \ell_R(R/I^{[q]})/q^d.$$

Let $J$ be a reduction ideal of $I$ which is generated by system of parameters, i.e., $J \subseteq I$ and there exists $r > 0$ such that $JI^r = I^{r+1}$ (see Theorem 14.14 in [**Ma**]). Therefore $I^n = J^{n-r}I^r$ for all $n \geq r + 1$. This gives

$$\ell_R(R/J^n) \geq \ell_R(R/I^n) \geq \ell_R(R/J^{n-r}), \quad \text{for all} \quad n \gg 0.$$

Multiplying by $d!/n^d$ and taking limit as $n \to \infty$, we get $e(R, J) = e(R, I)$. Now we have

$$e(R, I) = e(R, J) = \lim_{q\to\infty} \ell_R(R/J^{[q]})/q^d \geq \limsup_{q\to\infty} \ell_R(R/I^{[q]})/q^d,$$

where the second equality follows by Exercise 1.

EXERCISE 3. Let $(R, \mathbf{m}, k)$ be a regular local ring, where $I$ is $\mathbf{m}$-primary then $e_{HK}(R, I) = \ell_R(R/I)$.

SOLUTION. Let $F : R \longrightarrow R$ be the Frobenius homomorphism and $e \in \mathbb{N}$. As discussed in Remark 2.3, we can assume that $(R, \mathbf{m})$ is complete local where $R/\mathbf{m}$ is an algebraically closed field. In particular $R = k[[x_1, \ldots, x_d]]$ a power series ring. Since $R$ is a direct summand of $F_*^e R$ as $R$ modules the map $F^e : R \longrightarrow F_*^e R$ is a flat map.

Note that the exactness of the sequence of $R$-modules $0 \longrightarrow I^{[q]} \longrightarrow R \longrightarrow R/I^{[q]} \longrightarrow 0$ is equivalent to the exactness of the sequence of $F_*^e R$-modules $0 \longrightarrow F_*^e I^{[q]} \longrightarrow F_*^e R \longrightarrow F_*^e(R/I^{[q]}) \longrightarrow 0$.

Therefore

$$\ell_{F_*^e R}\big(F_*^e(R/I^{[q]})\big) = \ell_{F_*^e R}\big(F_*^e R/F_*^e I^{[q]}\big) = \ell_{F_*^e R}\left(F_*^e R/I \cdot F_*^e R\right).$$

This implies

$$\ell_R\left(\frac{R}{I^{[q]}}\right) = \ell_{F_*^e R}\left(F_*^e\big(\frac{R}{I^{[q]}}\big)\right) = \ell_{F_*^e R}\left(\frac{F_*^e R}{I \cdot F_*^e I}\right) = \ell_{F_*^e R}\left(\frac{R}{I} \otimes_R F_*^e R\right).$$

Now we have

$$\ell_{F_*^e R}\left(\frac{R}{I} \otimes_R F_*^e R\right) = \ell_R\left(\frac{R}{I}\right)\ell_{F_*^e R}\left(\frac{F_*^e R}{\mathbf{m} F_*^e R}\right) = \ell_R\left(\frac{R}{I}\right)\ell_R\left(\frac{R}{\mathbf{m}^{[q]}}\right), \tag{6.1}$$

where the first equality holds as the map $F_*^e : R \longrightarrow F_*^e R$ is flat.

Now $R/\mathbf{m}^{[q]}$ is a $k$-vector space of dimension $q^d$ with a basis $\{x_1^{i_1} \cdots x_d^{i_d} \mid 0 \leq i_l \leq q-1\}$ and which gives $\ell_R(R/\mathbf{m}^{[q]}) = q^d$. Therefore $\ell_R(R/I^{[q]}) = \ell_R(R/I)q^d$ and hence $e_{HK}(R, I) = \ell_R(R/I)$.

EXERCISE 4. Let If $R = k[[X_1^{i_1} \cdots X_d^{i_d} \mid i_1, \ldots, i_d \geq 0, \sum_j i_j = r]]$ denote the $r$-th Veronese subring of $S = k[[X_1, \ldots, X_d]]$. Then show that

$$e_{HK}(R) = \frac{1}{r}\binom{d+r-1}{r}.$$

SOLUTION. For $0 \leq i < r$, set $S_{i,r}$ be the $R$-module $S_{i,r} := \bigoplus_{k=0}^{\infty} S_{kr+i}$. Since each $S_{i,r}$ is a finitely generated $R$-module and $S = \oplus_{i=0}^{r-1} S_{i,r}$, we have

$$\operatorname{rank}_R(S) = \sum_{i=0}^{r-1} \operatorname{rank}_R(S_{i,r}) = r,$$

where the last equality because $S_{i,r}$ is isomorphic to a nonzero submodule $X_1^{r-i} S_{i,r}$ of $R$ and therefore is of rank $= 1$.

By associativity formula,

$$e_{HK}(R) = \frac{1}{r} e_{HK}(S, \mathbf{m}) = \frac{1}{r} e_{HK}(S, \mathbf{m}S) = \frac{1}{r} \ell_S(S/\mathbf{m}S),$$

where $e_{HK}(S, \mathbf{m})$ is the HK multiplicity of $S$ as a $R$-module with respect to the ideal $\mathbf{m}$ and $e_{HK}(S, \mathbf{m}S)$ is the HK multiplicity of the ring $S$ with respect to the ideal $\mathbf{m}S$, whereas the third equality follows by Exercise 3.

Now the assertion follows as

$$\begin{aligned}\ell_S(S/\mathbf{m}S) = \dim_k S/\mathbf{m}S &= \# \quad \text{monomials of degree} \quad < r \quad \text{in } d \text{ variables} \\ &= 1 + \binom{d}{1} + \binom{d+1}{2} + \cdots + \binom{d+r-2}{r-1} = \binom{d+r-1}{r-1}.\end{aligned}$$

EXERCISE 5. Let $S = k[X_1, \ldots, X_d]$ and let $G$ be a finite group of automorphisms of the $k$-algebra $S$ and let $R = S^G$ be the invariant ring with a maximal ideal $\mathbf{m}$. Then

$$e_{HK}(R) = \ell_S(S/\mathbf{m}S)/|G|.$$

SOLUTION. For $\alpha \in S$ let

$$F_\alpha(X) = \prod_{g \in G}(X - g \cdot \alpha).$$

We note that $F_\alpha(X) \in S^G[X]$ such that $F_\alpha(\alpha) = 0$ and all the roots belong to $S$.

This implies $S^G \longrightarrow S$ is a finite extension of rings and $K = T^{-1}S^G \longrightarrow L = T^{-1}S$ is a finite extension of fields, where $T = S^G \setminus \{0\}$. Further $G$ is a finite group of automorphism of $L$ over $K$.

We show that $K = L^G$. We only need to show that $L^G \subset K$. Let $x \in L^G$ then $x = a/b$, where $a, b \in S$. We can rewrite

$$x = a_1/b_1 \quad \text{where} \quad a_1 = a \prod_{g \in G\setminus\{e\}} g \cdot b \quad \text{and} \quad b_1 = \prod_{g \in G} g \cdot b.$$

Then $b_1 \in S^G$ implies $a_1 \in S^G$ and therefore $x \in K$.

Hence $L/K = L/L^G$ is a finite Galois extension of degree $|G|$. This gives $\operatorname{rank}_R S = |G|$. Now using the associativity formula for the first equality, we have

$$e_{HK}(R) = (1/\operatorname{rank}_R(S))e_{HK}(S, \mathbf{m}) = (1/|G|)\ell_S(S/\mathbf{m}S).$$

EXERCISE 6. Show that $e_{HK}\left(k[[x,y]]/(x^2 - y^3)\right) = 2$.

SOLUTION. Note that $R \simeq k[[t^2, t^3]] \hookrightarrow k[[t]] =: S$. Since $R$ and $S$ have same fraction field, $\operatorname{rank}_R S = 1$. Let us denote the maximal ideal of $R$ by $\mathbf{m}$. Under the isomorphism, $\mathbf{m} = (t^2, t^3)$. As before $e_{HK}(R) = e_{HK}(S, \mathbf{m}S)/(\operatorname{rank}_R S) = \ell_S(S/\mathbf{m}S) = 2$.

EXERCISE 7. What is $e_{HK}\left(k[[x,y,z]]/(xy^3 - xz^3)\right)$?

SOLUTION. Without loss of generality, we can assume $k$ is algebraically closed. Write $R = k[[x,y,z]]/(xy^3 - xz^3)$. First we assume $p = 3$. By associativity of $e_{HK}$, we have

$$e_{HK}(R) = e_{HK}(R/(x)R) + 3e_{HK}(R/(y-z)R) = 1 + 3 = 4.$$

Now suppose $p \neq 3$. Then $xy^3 - xz^3 = x(y-z)(y+\alpha z)(y+\beta z)$ for some $\alpha, \beta \in k$ with $\alpha \neq \beta$ and $\alpha, \beta \notin \{-1, 1\} \subset k$. Again, by associativity of $e_{HK}$, we get

$$e_{HK}(R) = e_{HK}\Big(\frac{R}{xR}\Big) + e_{HK}\Big(\frac{R}{(y-z)R}\Big) + e_{HK}\Big(\frac{R}{(y+\alpha z)R}\Big) + e_{HK}\Big(\frac{R}{(y+\beta z)R}\Big) = 4.$$

for $n \gg 0$

EXERCISE 8. Let $(R, \mathbf{m}, k)$ be a Noetherian local ring of characteristic $p > 0$. Suppose for all $e \geq 1$ and $q = p^e$, $\ell_R(R/\mathbf{m}^{[q]}) = q^{\dim R}$. Prove that $F^e : R \longrightarrow F_*^e R$ is a free $R$-module of rank $p^{ed}$.

SOLUTION. Write $d = \dim R$. Without loss of generality we can assume $R$ is complete and $R/\mathbf{m} = k$ is algebraically closed. Let $P$ be a minimal prime ideal of

$R$ with $\dim R/P = d$. Then $\ell_R(R/P + \mathbf{m}^{[q]}) \geq q^d$, by Lemma 2.5. But we have, for all $q \geq 1$,

$$q^d = \ell_R(R/\mathbf{m}^{[q]}) \geq \ell_R(R/P + \mathbf{m}^{[q]}) \geq q^d.$$

Hence $\ell_R(\frac{P+\mathbf{m}^{[q]}}{\mathbf{m}^{[q]}}) = 0$, i.e., $P + \mathbf{m}^{[q]} = \mathbf{m}^{[q]}$ for all $q$. Hence $P \subseteq \bigcap_{q\geq 1} \mathbf{m}^{[q]} = 0$. Therefore $R$ is a domain.

Let $F^e : R \longrightarrow F^{e*}R$ be the $e^{th}$-iterated Frobenius map. Using the same argument as in the proof of Lemma 2.5 we conclude $\operatorname{rank}_R(F^e_* R) = q^d$. Now

$$\mu_R(F^e_* R) = \ell_R(F^e_* R/\mathbf{m}F^e_* R) = \ell_R(R/\mathbf{m}^{[q]}) = q^d,$$

where the second equality holds because of Remark 2.3. Therefore, we have $F^e_* R$ is a free $R$-module of rank $= p^{ed}$.

## References


[A] Ian M. Aberbach, *Extension of weakly and strongly F-regular rings by flat maps*, J. Algebra **241** (2001), no. 2, 799–807, DOI 10.1006/jabr.2001.8785. MR1843326

[AE1] Ian M. Aberbach and Florian Enescu, *Lower bounds for Hilbert-Kunz multiplicities in local rings of fixed dimension*, Michigan Math. J. **57** (2008), 1–16, DOI 10.1307/mmj/1220879393. Special volume in honor of Melvin Hochster. MR2492437

[AE2] Ian M. Aberbach and Florian Enescu, *New estimates of Hilbert-Kunz multiplicities for local rings of fixed dimension*, Nagoya Math. J. **212** (2013), 59–85, DOI 10.1215/00277630-2335204. MR3290680

[B] Holger Brenner, *The rationality of the Hilbert-Kunz multiplicity in graded dimension two*, Math. Ann. **334** (2006), no. 1, 91–110, DOI 10.1007/s00208-005-0703-x. MR2208950

[C] Aldo Conca, *Hilbert-Kunz function of monomial ideals and binomial hypersurfaces*, Manuscripta Math. **90** (1996), no. 3, 287–300, DOI 10.1007/BF02568307. MR1397658

[CDHZ] Olgur Celikbas, Hailong Dao, Craig Huneke, and Yi Zhang, *Bounds on the Hilbert-Kunz multiplicity*, Nagoya Math. J. **205** (2012), 149–165, DOI 10.1215/00277630-1543805. MR2891167

[CLS] David A. Cox, John B. Little, and Henry K. Schenck, *Toric varieties*, Graduate Studies in Mathematics, vol. 124, American Mathematical Society, Providence, RI, 2011, DOI 10.1090/gsm/124. MR2810322

[DsNbP] Alessandro De Stefani, Luis Núñez-Betancourt, and Felipe Pérez, *On the existence of F-thresholds and related limits*, Trans. Amer. Math. Soc. **370** (2018), no. 9, 6629–6650, DOI 10.1090/tran/7176. MR3814343

[E] Kazufumi Eto, *Multiplicity and Hilbert-Kunz multiplicity of monoid rings*, Tokyo J. Math. **25** (2002), no. 2, 241–245, DOI 10.3836/tjm/1244208851. MR1948662

[ES] Florian Enescu and Kazuma Shimomoto, *On the upper semi-continuity of the Hilbert-Kunz multiplicity*, J. Algebra **285** (2005), no. 1, 222–237, DOI 10.1016/j.jalgebra.2004.11.014. MR2119113

[F] William Fulton, *Introduction to toric varieties*, Annals of Mathematics Studies, vol. 131, Princeton University Press, Princeton, NJ, 1993. The William H. Roever Lectures in Geometry, DOI 10.1515/9781400882526. MR1234037

[JNSWY] Jack Jeffries, Yusuke Nakajima, Ilya Smirnov, Kei-ichi Watanabe, and Ken-Ichi Yoshida, *Lower bounds on Hilbert-Kunz multiplicities and maximal F-signatures*, Math. Proc. Cambridge Philos. Soc. **174** (2023), no. 2, 247–271, DOI 10.1017/S0305004122000238. MR4545206

[Ha] Douglas Hanes, *Notes on the Hilbert-Kunz function*, J. Algebra **265** (2003), no. 2, 619–630, DOI 10.1016/S0021-8693(03)00233-3. MR1987020

[Hu] Craig Huneke, *Hilbert-Kunz multiplicity and the F-signature*, Commutative algebra, Springer, New York, 2013, pp. 485–525, DOI 10.1007/978-1-4614-5292-8_15. MR3051383

[HM] C. Han and P. Monsky, *Some surprising Hilbert-Kunz functions*, Math. Z. **214** (1993), no. 1, 119–135, DOI 10.1007/BF02572395. MR1234602

[HL] Daniel Huybrechts and Manfred Lehn, *The geometry of moduli spaces of sheaves*, 2nd ed., Cambridge Mathematical Library, Cambridge University Press, Cambridge, 2010, DOI 10.1017/CBO9780511711985. MR2665168

[K] Ernst Kunz, *Characterizations of regular local rings of characteristic p*, Amer. J. Math. **91** (1969), 772–784, DOI 10.2307/2373351. MR252389

[Ma] Hideyuki Matsumura, *Commutative ring theory*, 2nd ed., Cambridge Studies in Advanced Mathematics, vol. 8, Cambridge University Press, Cambridge, 1989. Translated from the Japanese by M. Reid. MR1011461

[Mo] P. Monsky, *The Hilbert-Kunz function*, Math. Ann. **263** (1983), no. 1, 43–49, DOI 10.1007/BF01457082. MR697329

[GM] I. Gessel and P. Monsky, *The limit as $p \to \infty$ of the Hilbert-Kunz multiplicity of $\sum x_i^{d_i}$*, Preprint, `arXiv:1007.2004` [math.AC], 2010.

[G] David Gieseker, *Stable vector bundles and the Frobenius morphism*, Ann. Sci. École Norm. Sup. (4) **6** (1973), 95–101. MR325616

[MT] Mandira Mondal and V. Trivedi, *Hilbert-Kunz density function and asymptotic Hilbert-Kunz multiplicity for projective toric varieties*, J. Algebra **520** (2019), 479–516, DOI 10.1016/j.jalgebra.2018.10.038. MR3885209

[MTW] Mircea Mustaţă, Shunsuke Takagi, and Kei-ichi Watanabe, *F-thresholds and Bernstein-Sato polynomials*, European Congress of Mathematics, Eur. Math. Soc., Zürich, 2005, pp. 341–364. MR2185754

[L] Adrian Langer, *Semistable sheaves in positive characteristic*, Ann. of Math. (2) **159** (2004), no. 1, 251–276, DOI 10.4007/annals.2004.159.251. MR2051393

[SB] N. I. Shepherd-Barron, *Semi-stability and reduction mod p*, Topology **37** (1998), no. 3, 659–664, DOI 10.1016/S0040-9383(97)00038-4. MR1604907

[S1] Ilya Smirnov, *Upper semi-continuity of the Hilbert-Kunz multiplicity*, Compos. Math. **152** (2016), no. 3, 477–488, DOI 10.1112/S0010437X15007800. MR3477638

[S2] Ilya Smirnov, *Hilbert-Kunz multiplicity of the powers of an ideal*, Proc. Amer. Math. Soc. **147** (2019), no. 8, 3331–3338, DOI 10.1090/proc/14513. MR3981111

[T1] V. Trivedi, *Semistability and Hilbert-Kunz multiplicities for curves*, J. Algebra **284** (2005), no. 2, 627–644, DOI 10.1016/j.jalgebra.2004.10.016. MR2114572

[T2] V. Trivedi, *Hilbert-Kunz multiplicity and reduction mod p*, Nagoya Math. J. **185** (2007), 123–141, DOI 10.1017/S0027763000025770. MR2301461

[T3] V. Trivedi, *Arithmetic behaviour of Frobenius semistability of syzygy bundles for plane trinomial curves*, Bull. Sci. Math. **141** (2017), no. 7, 703–726, DOI 10.1016/j.bulsci.2017.07.008. MR3710674

[T4] V. Trivedi, *Hilbert-Kunz density function and Hilbert-Kunz multiplicity*, Trans. Amer. Math. Soc. **370** (2018), no. 12, 8403–8428, DOI 10.1090/tran/7268. MR3864381

[T5] Vijaylaxmi Trivedi, *Nondiscreteness of F-thresholds*, Math. Res. Lett. **27** (2020), no. 6, 1885–1895, DOI 10.4310/MRL.2020.v27.n6.a13. MR4216609

[T6] V. Trivedi, *Asymptotic Hilbert-Kunz multiplicity*, J. Algebra **492** (2017), 498–523, DOI 10.1016/j.jalgebra.2017.09.019. MR3709163

[T7] Vijaylaxmi Trivedi, *The Hilbert-Kunz density functions of quadric hypersurfaces*, Adv. Math. **430** (2023), Paper No. 109207, 63, DOI 10.1016/j.aim.2023.109207. MR4617943

[TW1] Vijaylaxmi Trivedi and Kei-Ichi Watanabe, *Hilbert-Kunz density functions and F-thresholds*, J. Algebra **567** (2021), 533–563, DOI 10.1016/j.jalgebra.2020.09.025. MR4163076

[TW2] Vijaylaxmi Trivedi and Kei-Ichi Watanabe, *Hilbert-Kunz density function for graded domains*, J. Pure Appl. Algebra **226** (2022), no. 2, Paper No. 106835, 28, DOI 10.1016/j.jpaa.2021.106835. MR4287492

[W] K.I. Watanabe, *Hilbert-Kunz multiplicity of toric rings*, Proc. Inst. Nat. Sci, (Nihon Univ.) 35 (2000), 173-177.

[WY1] Kei-Ichi Watanabe and Ken-Ichi Yoshida, *Hilbert-Kunz multiplicity of two-dimensional local rings*, Nagoya Math. J. **162** (2001), 87–110, DOI 10.1017/S0027763000007819. MR1836134

[WY2] Kei-ichi Watanabe and Ken-ichi Yoshida, *Hilbert-Kunz multiplicity of three-dimensional local rings*, Nagoya Math. J. **177** (2005), 47–75, DOI 10.1017/S0027763000009053. MR2124547

School of Mathematics, Tata Institute of Fundamental Research, Homi Bhabha Road, Mumbai-400005, India
*Email address*: vija@math.tifr.res.in

University of Hyderabad, Hyderabad, Telangana 500046, India

Contemporary Mathematics
Volume **831**, 2025
https://doi.org/10.1090/conm/831/16632

# Briançon-Skoda Theorems in positive characteristic

Ian M. Aberbach, with solutions by Suprajo Das

## 1. Lecture 1—History, Background, and the easy characteristic-$p$ proof

We will use the same conventions and notation as established in earlier lectures. In particular, $(R, \mathfrak{m}, k)$ will denote a local (which includes Noetherian) ring with residue field $k = R/\mathfrak{m}$, $R^\circ$ is the complement of the union of the minimal primes of $R$, and when $R$ has positive prime characteristic $p$, then $I^*$ is the tight closure of $I$, $e$ is a (varying) non-negative integer and $q$ is the (varying) corresponding power of $p$, i.e., $q = p^e$.

**1.1. A little bit of history.** Let $O_d = \mathbb{C}\{z_1, \ldots, z_d\}$, the ring of convergent power series over the complex numbers (this is a regular local ring of dimension $d$), and let $f \in O_d$ be a non-unit. It is then known that $f$ is in the integral closure of the ideal $(\partial f/\partial z_1, \ldots, \partial f/\partial z_d)$. In particular, for some $k > 0$, $f^k \in (\partial f/\partial z_1, \ldots, \partial f/\partial z_d)$.

Mather raised the following question (pre 1971):

Question 1. *Is there a bound on the minimum value of $k$ above that is independent of $f$?*

This question was answered, using entirely analytic techniques, by Briançon and Skoda in 1974:

Theorem 1.1 ([**SB74**]). *Let $O_d$ be as above. Suppose that $I \subseteq O_d$ can be generated by $\ell$ elements. Then for all $w \geq 0$,*

$$\overline{I^{\ell+w}} \subseteq I^{w+1}.$$

*In particular, $f^d \in (\partial f/\partial z_1, \ldots, \partial f/\partial z_d)$.*

In 1979, Mel Hochster noted that this result is entirely algebraic in its statement and ought to be true for any regular local ring. He then challenged the commutative algebra community to provide a purely algebraic proof!

Within the next year or so, Lipman and Sathaye, and Lipman and Teissier provided proofs.

Lipman and Sataye proved results that imply the following theorem (we will review analytic spread and reduction below):

Theorem 1.2 ([**LS81**]). *Let $(R, \mathfrak{m})$ be a regular local ring. If $I \subseteq R$ is an ideal of analytic spread $\ell$, and $J \subseteq I$ is a reduction, then for all $w \geq 0$*

$$\overline{I^{\ell+w}} \subseteq J^{w+1}.$$

In the case that the ring is not regular, Lipman and Teissier obtained results for pseudo-rational singularities (see notes from Enescu's lectures for the definition):

THEOREM 1.3 ([**LT81**]). *Let $(R, \mathfrak{m})$ be a ring with pseudo-rational singularities and dimension $d$. For any ideal $I \subseteq R$ and all $w \geq 0$,*

$$\overline{I^{d+w}} \subseteq I^{w+1}.$$

Note that this second result is not quite as powerful as the first result, since it is often the case that non-$\mathfrak{m}$-primary ideals have analytic spread strictly smaller than $d = \dim(R)$. Lipman and Teissier do prove that when an ideal in a pseudo-rational ring has a reduction generated by a regular sequence, then we may replace $d$ by $\ell$, the analytic spread (here, equal to the number of elements in the regular sequence). In lecture 2 we will show how to prove the full Briançon-Skoda theorem for all ideals in an F-rational ring of prime characteristic!

The above results, while done using algebraic, not analytic, techniques, are still quite difficult to achieve. Tight closure offers an easily understandable proof in characteristic $p$, and this proof gives results with almost no assumptions on the ring (of course, the conclusions above are true only up to tight closure).

**1.2. Review of reductions and analytic spread.** The following material can be found in the paper of Northcott and Rees ([**NR54**]), in section 4.5 of [**BH98**], or in the book [**HS06**].

DEFINITION 1.4. Let $R$ be a Noetherian ring and let $I \subseteq R$ be an ideal. The *integral closure* of $I$ in $R$, $\overline{I}$, consists of all $x \in R$ satisfying any of the following equivalent conditions:

(1) There exists $n > 0$ such that $x^n \in I(I, x)^{n-1}$.
(2) There exists $c \in R^\circ$ such that for all $n > 0$, $cx^n \in I^n$.
(3) There exists $c \in R^\circ$ such that for all $n \gg 0$, $cx^n \in I^n$.
(4) There exists $c \in R^\circ$ and a sequence $\{n_j\}_{j=1}^{\infty}$ such that $\lim_{j\to\infty} n_j = \infty$ and $cx^{n_j} \in I^{n_j}$.

DEFINITION 1.5. Let $R$ be a Noetherian ring and let $I \subseteq R$ be an ideal.

(1) An ideal $J \subseteq I$ such that for some $r > 0$, $I^{r+1} = JI^r$ is called a *reduction* of $I$.
(2) The smallest $r$ above which works is called the *reduction number of $J$ with respect to $I$*, denoted $r_J(I)$.
(3) Assume that $(R, \mathfrak{m}, k)$ is local.
    (a) The $k$-algebra $\mathcal{F}(I) = \oplus_{n=0}^{\infty} I^n/\mathfrak{m}I^n$ is the *fiber cone* of $I$.
    (b) $\ell(I) = \dim(\mathcal{F}(I))$ is the *analytic spread* of $I$.

Here are some fundamental results:

THEOREM 1.6. *Let $(R, \mathfrak{m}, k)$ be a local ring of dimension $d$, $I \subseteq R$ an ideal of analytic spread $\ell$, and $J \subseteq I$ a reduction. For convenience, we will assume that $R$ is a domain, although much of what is below is true much more generally.*

(1) $\operatorname{ht}(I) \leq \ell \leq d$.
(2) $\overline{J} = \overline{I}$ *and $J$ is a reduction of $\overline{I}$.*
(3) *If $k$ is infinite then there always exist minimal (with respect to inclusion) reductions, and each such reduction has $\mu(J) = \ell$.*

Since it may be of use below, we recall that the proof of Theorem 1.6(3) is that in the fiber cone $\mathcal{F}(I)$ we may pick a Noether normalization over $k$ consisting of degree-one elements (using that $k$ is infinite). Any lift of these elements back to $I$ gives a minimal reduction.

REMARK 1.7. Containment in both integral closure and tight closure may be checked modulo each minimal prime in $R$. Thus, in most of our work below, we will assume immediately that we are in the domain case.

EXERCISE 1. Let $J \subseteq I$ be a reduction.

(1) For all $n$, $J^n$ is a reduction of $I^n$.
(2) Let $J = (a_1, \ldots, a_\ell)$. Show that $(a_1^n, \ldots, a_\ell^n)$ is a reduction of $I^n$.
(3) If $r_J(I) = r$, then for all $n > r$, $I^n = J^{n-r} I^r$.

**1.3. The Briançon-Skoda Theorem in characteristic $p$.** One of the first uses of tight closure was to give a greatly simplified and vastly generalized version of the Briançon-Skoda theorem! (See section 5 of [**HH90**].)

THEOREM 1.8. *Let $R$ be a Noetherian ring of characteristic $p$, $I \subseteq R$, and $J \subseteq I$ a reduction generated by $\ell$ elements. Then for all $w \geq 0$,*

$$\overline{I^{\ell+w}} \subseteq (J^{w+1})^*.$$

*In particular, if $(R, \mathfrak{m})$ is local and $\ell = \ell(I)$, then $\overline{I^{\ell+w}} \subseteq (I^{w+1})^*$.*

A fundamental step in many of our proofs is the following exercise.

EXERCISE 2. Let $R$ be a Noetherian domain, and $J \subseteq I$ a reduction. Then there is a $c \in R^\circ$ such that for all $n > 0$,

$$c(\overline{I})^n \subseteq J^n.$$

PROOF. We will prove the case that $w = 0$ and leave the general case as an exercise.

By Remark 1.7, we may assume that $R$ is a domain. Let $J = (a_1, \ldots, a_\ell)$.

Let $x \in \overline{I^\ell}$. Choose $c \in R^\circ$ as in Exercise 2 for the ideal $I^\ell$. Then for all $q$,

$$cx^q \in c(\overline{I^\ell})^q \subseteq J^{\ell q} = (a_1, \ldots, a_\ell)^{\ell q}.$$

A typical element in $J^{\ell q}$ is an $R$-linear sum of monomials of degree $\ell q$ in the $\ell$ elements of $J$. For such a monomial, we must have an exponent of some $a_i$ to be at least $q$, or else the monomial has degree at most $(q-1)\ell < q\ell$, a contradiction. In other words, for all $q$,

$$cx^q \in (a_1^q \ldots, a_\ell^q) = J^{[q]}.$$

Hence $\overline{I^\ell} \subseteq J^*$. □

EXERCISE 3. Expand the proof above to the case $w > 0$.

EXERCISE 4. Theorem 1.8 is true when $J$ is any reduction. If $R$ has infinite residue field we can simply pick a minimal reduction (which has $\ell(I)$ minimal generators) inside of $J$. This exercise outlines the technique of passing to a ring with infinite residue field when the residue field of $R$ is finite.

Let $(R, \mathfrak{m}, k)$ be a local ring, where $k$ is not necessarily infinite. Let $t$ be a variable over $R$ and set $R_1 = (R[t])_{\mathfrak{m}R[t]}$. Let $I \subseteq R$ be an ideal of analytic spread $\ell$.

(1) Show that $R_1$ is a faithfully flat extension of $R$ and $\ell(IR_1) = \ell(I)$.
(2) Show that if $J \subseteq I$ is a reduction then $JR_1$ is a reduction of $IR_1$.
(3) Assume $R$ is positive characteristic. Show that the tight closure Briançon-Skoda Theorem holds for any reduction $J \subseteq I$.

COROLLARY 1.9. *Let $R$ be a weakly F-regular ring. If $I$ has an $\ell$-generated reduction $J$, then for all $w \geq 0$, $\overline{I^{\ell+w}} \subseteq J^{w+1}$.*

Thus, for instance, every characteristic $p$ ring that is a summand of a regular ring (so weakly F-regular) satisfies the full Briançon-Skoda Theorem.

EXERCISE 5. Theorem 3 in [**McA80**] (see also Theorem 2.1 in [**CHS10**]) can be used to show that if $(R, \mathfrak{m})$ is an excellent local domain and $I \subseteq R$ is an ideal with $\ell = \ell(I) < \dim(R)$, then for all $n$, $\overline{I^n} : \mathfrak{m}^\infty = \overline{I^n}$ (here $A : J^\infty = \cup_k(A : J^k)$). Use this result and Theorem 1.2 to prove:

Let $(R, \mathfrak{m})$ be an excellent local regular ring. Let $I$ be an ideal of $R$ of analytic spread $\ell < \dim R$. then for all $n$, $I^{n+\ell} : \mathfrak{m}^\infty \subseteq I^{n+1}$.

DISCUSSION 1.10. Let us look a bit more carefully at the proof of Theorem 1.8. The proof uses that

$$(a_1, \dots, a_\ell)^{\ell q} \subseteq (a_1, \dots, a_\ell)^{[q]}. \tag{1.1}$$

But Equation (1.1) omits a lot of data. Our justification was that we could consider monomials in $a_1, \dots, a_\ell$ of degree $\ell \cdot q$. Indeed, each such monomial must be divisible by $a_i^q$ for some $i$, but the coefficient of $a_i^q$ is still in quite a high power of $(a_1, \dots, a_\ell)$. What we really have is

$$(a_1, \dots, a_\ell)^{\ell q} = (a_1, \dots, a_\ell)^{[q]}(a_1, \dots, a_\ell)^{(\ell-1)q} \tag{1.2}$$

(the above two equations are for the case $w = 0$ — they can be adjusted accordingly when $w > 0$).

Equation (1.2) raises the question of whether we can gain any information in the Briançon-Skoda Theorem by using the additional information.

## 2. Lecture 2—Briançon-Skoda theorems with coefficients, and in F-rational rings

**2.1. A first Briançon-Skoda theorem with coefficients.** We would like to find some way to exploit the additional information in Equation (1.2). We will start with a result from [**AH93**].

DEFINITION 2.1. Let $I \subseteq R$ be an ideal.

(1) Let $I^{\text{un}}$ be the intersection of the minimal primary components of $I$. Warning: this is not a widely used or standardized term.
(2) Let the *big height* of $I$ be $\text{bight}(I) = \max\{\text{ht}(P) : P \text{ is minimal over } I\}$.

We can now prove our first theorem with coefficients.

THEOREM 2.2 ([**AH93**, Theorem 3.3]). *Let $(R, \mathfrak{m})$ be a regular local ring containing a field. Let $I$ be an ideal of analytic spread $\ell$ and $h = \text{bight}(I)$. If $J \subseteq I$ is a reduction then for all $w \geq 0$,*

$$\overline{I^{\ell+w}} \subseteq J^{w+1}\left(J^{\ell-h}\right)^{\text{un}}.$$

The proof when $R/\mathfrak{m}$ is equicharacteristic 0 is via reduction to characteristic $p$. We will sketch the proof in the characteristic $p$ case when $w=0$. For convenience we will assume that $R/\mathfrak{m}$ is infinite. Recall that when $R$ is regular, the Frobenius endomorphism is a faithfully flat map. In particular, we always have $\operatorname{Ass}(R/I^{[q]}) = \operatorname{Ass}(R/I)$ when $R$ is regular.

We encapsulate the main idea as a lemma:

LEMMA 2.3. *Let $R$, $I$, $J$, $\ell$, and $h$ be as above. Then there exists $f \neq 0$ such that for all $q \gg 0$,*

$$fJ^{(\ell-1)q} \subseteq ((J^{\ell-h})^{\mathrm{un}})^{[q]}.$$

PROOF. $J$ and $I$ are equal up to radical, so have the same bigheight.

For any ideal $A$, the minimal primes of any power of $A$ are the same as the minimal primes of $A$. So, by the observation about associated primes of $q^{th}$ powers in regular rings, it suffices to show that the desired containment holds after localizing.

Let $P$ be a minimal prime of $J$. Then $JR_P$ has a minimal reduction generated by $\ell(JR_P) \leq \operatorname{ht}(PR_P) \leq \operatorname{bight}(J) = h$ elements. Call this reduction $L_P$. Pick $c_P$ as in Exercise 2 for $L_P \subseteq JR_P$ (we can clear denominators and assume that $c_P \in R$). By the same computations as in the proof of Theorem 1.8 and Exercise 3 (using $w = \ell$, $\ell(L_P) \leq h$) we get

$$c_P(JR_P)^{(\ell-1)q} \subseteq ((L_P)^{(\ell-1-h+1)})^{[q]} = (L_P^{\ell-h})^{[q]} \subseteq (JR_P^{\ell-h})^{[q]}.$$

If we now let $f$ be the product of the $c_P$ such that $P \in \min(R/J)$, we obtain the desired result. □

PROOF OF 2.2 WHEN $w = 0$. We can assume that $J$ is a minimal reduction. Let $c \neq 0$ be as in Exercise 2 for $I^\ell$ and $f \neq 0$ as in Lemma 2.3. Suppose that $x \in \overline{I^\ell}$. Then for all $q$,

$$cfx^q \in fJ^{\ell q} \subseteq fJ^{[q]}J^{(\ell-1)q} \subseteq J^{[q]}((J^{\ell-h})^{\mathrm{un}})^{[q]} = (J(J^{\ell-h})^{\mathrm{un}})^{[q]}.$$

Hence $x \in (J(J^{\ell-h})^{\mathrm{un}})^* = (J(J^{\ell-h})^{\mathrm{un}})$. □

Note that Theorem 2.2 cannot be extended to non-regular weakly F-regular rings via the same argument. The argument depends strongly on the flatness of Frobenius in regular rings. However, we will see in Lecture 3 that there are results to be obtained using different methods!

**2.2. An application.** Recall that if $R$ is a ring and $I \subseteq R$, then we can form the *Rees ring of $R$ with respect to $I$*,

$$R[It] = R \oplus It \oplus I^2t^2 \oplus \cdots,$$

which is a subring of the polynomial ring over $R$ in the variable $t$, and also the *associated graded ring of $R$ with respect to $I$*,

$$\operatorname{gr}_I R = R/I \oplus I/I^2 \oplus I^2/I^3 \oplus \cdots.$$

Note that $\operatorname{gr}_I R \cong R[It]/IR[It]$. These rings are intimately involved in understanding reductions and integral closure, as well as the geometric properties of $V(I) \subseteq \operatorname{Spec}(R)$. Therefore understanding the properties of these two rings (e.g., Cohen-Macaulayness) and their interrelationships is of interest.

For instance, Huneke in [**Hun82**] proved

THEOREM 2.4. *Let $(R, \mathfrak{m})$ be a Cohen-Macaulay local ring and $I$ an ideal of positive height. If $R[It]$ is CM, then so is $\operatorname{gr}_I R$.*

The converse is false. In order to get some form of converse, one needs some information about the (local) reduction numbers of $I$. For instance, Goto and Shimoda proved in [**GS82**], that if $(R, \mathfrak{m})$ is CM, $I$ is $\mathfrak{m}$-primary, and $I$ has a reduction $J$ such that $r_J(I) < d$, then the CMness of $\operatorname{gr}_I R$ implies Cohen-Macaulayness of $R[It]$. Trung and Ikeda, in [**TI89**] give a very general if and only if statement (with no assumption that $R$ is CM), that involves $a$-invariants.

One of the motivations behind the paper [**AH93**] was the realization that Theorem 2.2, along with the assumption that $\operatorname{gr}_I R$ is CM, was sometimes enough information to get a converse of Huneke's statement. For instance (see Theorem 5.2 of [**AH93**]):

THEOREM 2.5. *Let $(R, \mathfrak{m})$ be a regular local ring with infinite residue field. Suppose $I$ is an ideal of height $h$ and analytic spread $\ell = h+1$, and is generically a complete intersection (i.e., generated locally at each minimal prime by a regular sequence). Suppose that $\operatorname{gr}_I R$ is CM. Then $r(I) < \ell$ and $R[It]$ is CM.*

While we will omit any proof here, the main ideas will come up in the proof that F-rational rings satisfy the Briançon-Skoda theorem.

Lipman showed the full converse of Huneke's theorem (for all regular local rings) via his idea of adjoints of ideals—see [**Lip94**]. Additional versions of Briançon-Skoda theorems with coefficients, and applications to Rees rings are in [**AHT95**]. See also, [**JK95**] for an independent (and nicer) proof of much of the material in [**AHT95**].

**2.3. F-rational rings.** Recall that $(R, \mathfrak{m})$ is *F-rational* if every ideal generated by a system of parameters (sop) is tightly closed. Under mild conditions (e.g., complete), this implies that $R$ is a normal, CM, domain. Our goal is to outline the ideas behind the proof of [**AH01a**], Theorem 3.6:

THEOREM 2.6. *Let $(R, \mathfrak{m})$ be an F-rational local ring of positive prime characteristic. Let $I \subseteq R$ be an ideal of analytic spread $\ell$, and let $J \subseteq I$ be a reduction. Then for all $w \geq 0$,*

$$\overline{I^{\ell+w}} \subseteq J^{w+1}.$$

This is a *better* result than Lipman and Teissier's result, as it applies to all ideals in $R$, although it does not apply outside of positive characteristic. Standard methods of reduction to characteristic $p$ do give a version in characteristic 0:

THEOREM 2.7. *Let $R$ be an algebra of finite type over a field of characteristic 0 having rational singularities. Let $I$, $J$, $\ell$, be as above. Then for all $w \geq 0$,*

$$\overline{I^{\ell+w}} \subseteq J^{w+1}.$$

As usual, we will concentrate on the case $w = 0$. Also, in order to elucidate the ideas of the proof, we will take the case that $\ell(I) = \operatorname{ht}(I) + 1$.

A key fact in the results in [**AH93, AHT95, AH01a**] is that when $(R, \mathfrak{m}, k)$ has infinite residue field, then not only does an ideal $I$ have a minimal reduction with $\ell(I)$ generators, but, in fact, taking $\ell(I)$ "general" $k$-linear combinations of the generators of $I$ will have extremely nice properties locally. Specifically,

LEMMA 2.8. *Let $(R, \mathfrak{m}, k)$ be a local ring with $k$ infinite. Suppose that $I \subseteq R$ is an ideal of analytic spread $\ell$, and $J \subseteq I$ is a minimal reduction (so $\mu(J) = \ell$). Then there exists a generating set $a_1, \ldots, a_\ell$ for $J$ such that*

(1) *If* $P \supseteq I$ *is prime of height* $i \le \ell$ *then* $(a_1, \dots, a_i)R_P$ *is a reduction of* $IR_P$ *(not necessarily minimal).*
(2) *For* $n \gg 0$, $\operatorname{ht}\big(((a_1, \dots, a_i)I^n : I^{n+1}) + I\big) \ge i+1$.
(3) *If* $c_i \equiv a_i \mod I^2$, *then (1) and (2) hold with* $c_i$ *replacing* $a_i$.

EXERCISE 6. Show that property (2) is a consequence of property (1) in Lemma 2.8.

In general, when $\ell(I) > \operatorname{ht}(I)$, we are unlikely to be able to find any system of parameters containing a reduction of $I$. Since only parameter ideals in an F-rational ring are known to be tightly closed, we need to develop a different approach. The method utilized to prove Theorem 2.6 is to find a sequence of parameter ideals $\{B_n\}_{n\to\infty}$, such that our reduction $J$ is congruent to $B_n \mod \mathfrak{m}^n$, and then apply the Krull Intersection Theorem.

Our main tool is Proposition 3.2 from [**AH01a**]. We give a simplified version of the proposition here:

PROPOSITION 2.9. *Let* $(R, \mathfrak{m})$ *be an equidimensional, catenary local ring with infinite residue field. Let* $I \subseteq R$ *be an ideal of height* $h$ *and analytic spread* $\ell$. *Let* $J \subseteq I$ *be a minimal reduction, with basic generating set* $(a'_1, \dots, a'_\ell)$ *as in Lemma 2.8. Fix* $n > 0$ *and* $w \ge 0$.

*There exist elements* $a_1, \dots, a_\ell$ *and* $t_{h+1}, \dots, t_\ell$ *such that (we let* $t_1 = \cdots = t_h = 0$ *for convenience) setting* $J_i = (a_1, \dots, a_i)$ *and* $b_i = a_i + t_i$ *for* $1 \le i \le t$:

(1) $a_i \equiv a'_i \mod I^2$ *for* $1 \le i \le \ell$.
(2) $t_i \in \mathfrak{m}^n$ *for* $i \ge h+1$.
(3) $b_1, \dots, b_\ell$ *is part of a system of parameters for* $R$.
(4) *The images of* $t_{h+1}, \dots, t_\ell$ *are part of an sop for* $R/I$.
(5) *For* $M \gg 0$ *and* $h \le i \le \ell - 1$, $t_{i+1} \in (J_i^k I^M : I^{M+k})$ *for* $0 \le k \le w + \ell$.

If we now set $B_n = (b_1, \dots, b_\ell)$, as in Proposition 2.9, we can prove that $\overline{I^\ell} \subseteq (B_n)^*$, and when $R$ is F-rational, $B_n^* = B_n$. As $n \to \infty$, these ideals are better and better approximations of $J$.

PROOF OF THEOREM 2.6 IN THE CASE $\ell(I) = \operatorname{ht}(I) + 1$ AND $w = 0$. Fix $n > 0$. Pick $J = (a_1, \dots, a_{h+1})$ and $t_{h+1}$ as in Proposition 2.9. Suppose $M$ is large enough to get $t_{h+1}I^{M+k} \subseteq J_h^k I^M$ for $0 \le k \le \ell$. We claim that this implies that

$$t_{h+1}^{n'} I^{M+n'k} \subseteq J_h^{n'k} I^M$$

for all $n' \ge 1$. The case $n' = 1$ is given. Suppose we know the containment for a given $n'$. Multiply the above equation by $t_{h+1}I^k$. We get

$$\begin{aligned}(t_{h+1}I^k)(t_{h+1}^{n'} I^{M+n'k}) &\subseteq (t_{h+1}I^k) J_h^{n'k} I^M = J_h^{n'k} t_{h+1} I^{M+k} \\ &\subseteq J_h^{n'k} J_h^k I^M = J_h^{(n'+1)k} I^M.\end{aligned}$$

Let $x \in \overline{I^\ell}$. Pick $c$ as in Exercise 2. Pick $c' \in I^M - \{0\}$ with $M$ large enough that $t_{h+1}$ satisfies part (5) in Proposition 2.9. We have

$$cc'x^q \in c'J^{q\ell} \subseteq c'(a_1^q, \dots, a_h^q, a_{h+1}^q)J^{(h+1-1)q}.$$

Write

$$cc'x^q = c'(r_1 a_1^q + \cdots + r_h a_h^q + r_{h+1} a_{h+1}^q), \tag{2.1}$$

where each $r_i \in J^{hq}$. All we really care about is the last coefficient, $r_{h+1}$. When multiplied by $t_{h+1}^q$ we have

$$t_{h+1}^q(c'r_{h+1}) \in t_{h+1}^q I^{hq+M} \subseteq (J_h^{hq}) \subseteq J_h^{[q]}.$$

Now add and subtract $t_{h+1}^q(c'r_{h+1})$ to Equation (2.1) and use that $b_{h+1} = a_{h+1} + t_{h+1}$ to get

$$cc'x^q \in (a_1^q, \ldots, a_h^q, (a_{h+1} + t_{h+1})^q) = B_n^{[q]}$$

for all $q$. Thus $x \in B_n^* = B_n$, since $R$ is F-rational and $B_n$ is part of a system of parameters.

Since $n$ was arbitrary, we have $x \in \cap_n B_n \subseteq \cap_n(J + \mathfrak{m}^n) = J$, by the Krull Intersection Theorem. □

## 3. Lecture 3—Coefficient ideals, cancellation results

**3.1. Coefficients in F-rational rings.** Having generalized the Briançon-Skoda Theorem to F-rational rings, can we improve theorems such as 2.2 to F-rational rings? The answer is yes. We mention two related results here, without proof. These are [**AH01b**], Theorem 3.6 and [**AH11**], Theorem 3.8, respectively. The former result was the initial theorem showing that a version of Briançon-Skoda with coefficients was possible in F-rational rings. The latter result showed that, in fact, most of the hypotheses in the first result were unneccessary.

THEOREM 3.1. *Let $(R, \mathfrak{m})$ be an excellent F-rational local ring, $J \subseteq I$ a reduction, $\ell = \ell(I) > h = \mathrm{ht}(I)$. Assume that $R/I$ is equidimensional, $I$ is generically of reduction number at most one, and that $R/I^{\mathrm{un}}$ satisfies Serre's depth condition $S_{\ell-h+1}$. Then*

$$\overline{I^\ell} \subseteq JI^{\mathrm{un}}.$$

THEOREM 3.2. *Let $(R, \mathfrak{m})$ be an F-rational CM local ring. Let $J \subseteq I$ be a reduction and $\ell(I) > \mathrm{ht}(I)$. For all $w \geq 0$,*

$$\overline{I^{\ell(I)+w}} \subseteq J^{w+1}I^{\mathrm{un}}.$$

**3.2. Coefficient ideals.** So far, when getting information on coefficients in the Briançon-Skoda theorem, we have exploited the difference between $\ell(I)$ and $\mathrm{ht}(I)$. But what kind of coefficients can we get when $\ell(I) = \mathrm{ht}(I)$, for example, when $I$ is $\mathfrak{m}$-primary (so $\ell(I) = \mathrm{ht}(I) = \dim(R)$)?

DEFINITION 3.3. Let $R$ be any Noetherian ring, and $J \subseteq I$ any two ideals. The *coefficient ideal* of $I$ relative to $J$, $\mathfrak{a}(I, J)$, is the largest ideal, $\mathfrak{b} \subseteq R$ such that $I\mathfrak{b} = J\mathfrak{b}$. We will write $\mathfrak{a}$ when $I$ and $J$ are understood.

EXERCISE 7. Let $R$ be Noetherian (not necessarily characteristic $p$ or local), and let $J \subseteq I$ be ideals.

(1) Show $\mathfrak{a} = \mathfrak{a}(I, J)$ is always defined.
(2) If $\mathfrak{a}$ contains a NZD then $\overline{J} = \overline{I}$, and hence $J$ is a reduction of $I$. In this case there exists $n \geq 0$ such that $I^n \subseteq \mathfrak{a}$.

EXERCISE 8. Let $(R, \mathfrak{m}, k)$ be local, $J$ a reduction of an ideal $I$, $r = r_J(I)$. Set $\mathfrak{a}_1 = (J : I)$ and $\mathfrak{a}_{n+1} = (J\mathfrak{a}_n : I)$ for $n \geq 1$.

(1) Show: $I^r \subseteq \mathfrak{a}$, $\mathfrak{a} \subseteq \mathfrak{a}_n$ and $\mathfrak{a}_{n+1} \subseteq \mathfrak{a}_n$ for all $n$.
(2) If $I$ is $\mathfrak{m}$-primary then the descending sequence stabilizes at $\mathfrak{a}$.

THEOREM 3.4. *Let $(R, \mathfrak{m}, k)$ be a regular local ring containing a field and having infinite residue field. Let $I$ be an $\mathfrak{m}$-primary ideal and $J$ a minimal reduction of $I$. For all $w \geq -1$,*

$$\overline{I^{d+w}} \subseteq J^{w+1}\mathfrak{a}(I, J).$$

We will prove the $w = -1$ and 0, characteristic $p$ cases. The equal characteristic 0 case follows from reduction to characteristic $p$.

PROOF. Most of the work goes into proving the case $w = -1$, i.e., that

$$\overline{I^{d-1}} \subseteq \mathfrak{a}(I, J).$$

We show that for all $n$, $\overline{I^{(d-1)q}} \subseteq (\mathfrak{a}_n)^{[q]}$. Then by Exercise 8(2), we know that $\overline{I^{(d-1)q}} \subseteq \mathfrak{a}(I, J)^{[q]}$ (the latter is $\mathfrak{a}_n$ for $n$ large enough).

We proceed by induction. We will use repeatedly that for any ideals $A$ and $B$ in a *regular* ring, $(A^{[q]} : B^{[q]}) = (A : B)^{[q]}$ (in general, we only have $\supseteq$).

For the case $n = 1$ we must show that $\overline{I^{(d-1)q}} \subseteq (J : I)^{[q]}$, i.e., that $I^{[q]}\overline{I^{(d-1)q}} \subseteq J^{[q]}$, since $(J : I)^{[q]} = (J^{[q]} : I^{[q]})$.

Let $x \in \overline{I^{(d-1)q}}$, $r \in I^{[q]}$ and choose $c \neq 0$ as in Exercise 2 for $I^{dq}$, so that for all $q'$,

$$c(rx)^{q'} \in J^{dqq'} \subseteq J^{[qq']} = (J^{[q]})^{[q']}.$$

This shows that $I^{[q]}\overline{I^{(d-1)q}} \subseteq (J^{[q]})^* = J^{[q]}$ (again, using that $R$ is regular, so weakly F-regular).

Assume now that the claim has been shown for $n - 1$. Then for $r \in I^{[q]}$,

$$c(rx)^{q'} \subseteq J^{dqq'} \subseteq J^{[qq']}J^{(d-1)qq'} \subseteq J^{[qq']}I^{(d-1)qq'} \subseteq (J\mathfrak{a}_{n-1})^{[qq']}$$

for all $q'$. Hence $\overline{I^{(d-1)q}} \subseteq ((J\mathfrak{a}_{n-1})^{[q]} : I^{[q]})^* = ((J\mathfrak{a}_{n-1} : I)^{[q]})^* = ((\mathfrak{a}_n)^{[q]})^* = \mathfrak{a}_n^{[q]}$.

This completes the proof of the case $w = -1$.

For $w = 0$, let $x \in \overline{I^d}$. After choosing $c$ as in Exercise 2 for $I^d$,

$$cx^q \in J^{dq} = J^{[q]}J^{(d-1)q} \subseteq J^{[q]}\mathfrak{a}^{[q]}$$

for all $q$. Hence $\overline{I^d} \subseteq (J\mathfrak{a})^* = J\mathfrak{a}$. □

Notice that the proof uses that $I$ is $\mathfrak{m}$-primary very strongly. Outside of that case, it is not at all clear that the descending sequence of ideals, $\{\mathfrak{a}_n\}$, will stabilize. One situation where we can overcome this obstacle is when there is enough "completeness" to invoke Chevalley's theorem (which says that if $(R, \mathfrak{m})$ is a complete local ring and $\{J_n\}_n$ is a decreasing sequence of ideals with $\cap_n J_n = 0$, then for all $t > 0$ there exists $n_t$ such that $J_{n_t} \subseteq \mathfrak{m}^t$). In [**AH11**] we were able to prove:

THEOREM 3.5. *Let $(R, \mathfrak{m})$ be a complete regular local ring containing a field. For any ideal $I$ and reduction $J$ of $I$, and for all $w \geq -1$,*

$$\overline{I^{\ell(I)+w}} \subseteq J^{w+1}\mathfrak{a}(I, J).$$

One avenue that I am not aware of being explored is trying to get *both* coefficient ideals (of the sort $\mathfrak{a}(I, J)$) and coefficients as in 2.2 when $\ell(I) > \mathrm{ht}(I)$. For instance, if $R$ is a complete regular local ring containing a field, $J \subseteq I$ is a minimal reduction, and $\ell(I) \geq \mathrm{ht}(I) + 1$, then is it possible that $\overline{I^{\ell+w}} \subseteq J^{w+1}I^{\mathrm{un}}\mathfrak{a}(I, J)$?

**3.3. Cancellation results.** In its most basic form, the Briançon-Skoda theorem says that when $J$ is a reduction of $I$ then $\overline{I^{\ell(I)}} \subseteq J$. But often, a much smaller power of the integral closure is already in $J$. This raises the

QUESTION 2. *Let* $(R, \mathfrak{m})$ *be local and* $J \subseteq I$ *a reduction. Under what circumstances is there an integer* $n < \ell(I)$ *such that* $\overline{I^n} \subseteq J$*?*

Based on the ideas behind the cancellation theorem of [**Hun00**], we were able to prove (see [**AH01a**], Theorem 4.1):

THEOREM 3.6. *Let* $(R, \mathfrak{m})$ *be an F-rational Gorenstein local ring of dimension* $d$ *and positive characteristic. Suppose that* $I$ *is an ideal of height* $h$ *and analytic spread* $\ell > h$ *with* $R/I$ *Cohen-Macaulay. Then, for any reduction* $J$ *of* $I$*,*

$$\overline{I^{\ell-1}} \subseteq J.$$

In Theorem 4.2 of [**AH11**] we generalize this slightly (to a condition weaker than $R/I$ being Cohen-Macaulay).

**3.4. Uniform Brianąn-Skoda results.** One of the most interesting facets of Noetherian rings is the surprisingly strong uniformity results that often hold. One example that you have seen in earlier lectures is the existence of test elements under mild conditions (e.g., excellent reduced local). The test ideal uniformly annihilates $I^*/I$ independently of $I$.

Huneke, in [**Hun92**], shows that for many rings there is a uniform version of the Artin-Rees theorem. In particular, under reasonably mild conditions on the ring $R$ (see Theorem 3.7 below for the hypotheses), given finitely generated modules $N \subseteq M$, there is a $k = k(N, M)$, such that for all ideals $I \subseteq R$ and for all $n \geq k$,

$$I^n M \cap N \subseteq I^{n-k} N$$

(there is no uniform version of the full Artin-Rees theorem where there is an equality). As a consequence of this theorem, Huneke proved

THEOREM 3.7 ([**Hun92**, Theorem 4.13]). *Let* $R$ *be a Noetherian reduced ring satisfying one of the following conditions*

(1) *$R$ is essentially of finite type over an excellent Noetherian local ring.*
(2) *$R$ is characteristic $p > 0$ prime and is $F$-finite ($R^{1/p}$ is module finite over $R$).*
(3) *$R$ is essentially of finite type over $\mathbb{Z}$.*

*There exists an integer $k > 0$ such that for all ideals $I \subseteq R$ and all $n \geq k$,*

$$\overline{I^n} \subseteq I^{n-k}.$$

## Solutions to exercises

by Suprajo Das

SOLUTION TO EXERCISE 1. *Let* $R$ *be a Noetherian ring and* $J \subseteq I$ *be a reduction.*

(1) *For all* $n$*,* $J^n$ *is a reduction of* $I^n$*.*
(2) *Let* $J = (a_1, \ldots, a_l)$*. Show that* $(a_1^n, \ldots, a_l^n)$ *is a reduction of* $I^n$*.*
(3) *If* $r_J(I) = r$*, then for all* $n > r$*,* $I^n = J^{n-r} I^r$*.*

PROOF. (1) Since $J \subseteq I$ is a reduction there exists an integer $r > 0$ such that $I^{r+1} = JI^r$. For any integer $n > 0$ we get that

$$\left(I^{r+1}\right)^n = (JI^r)^n \implies (I^n)^{r+1} = J^n (I^n)^r.$$

Thus $J^n$ is a reduction of $I^n$.
(2) It follows from Discussion 1.10 that

$$J^{n(l+1)} = (a_1^n, \dots, a_l^n)\, J^{nl} \implies (J^n)^{l+1} = (a_1^n, \dots, a_l^n)\, (J^n)^l.$$

This shows that $(a_1^n, \dots, a_l^n)$ is a reduction of $J^n$. We know from part (1) that $J^n$ is a reduction of $I^n$. By transitivity $(a_1^n, \dots, a_l^n)$ is a reduction of $I^n$.
(3) We shall prove $I^n = J^{n-r}I^r$ by induction on $n$. Since $r_J(I) = r$ we have that $I^{r+1} = JI^r$. By induction hypothesis we may assume that $I^n = J^{n-r}I^r$. Now

$$I^{n+1} = II^n = I\left(J^{n-r}I^r\right) = J^{n-r}I^{r+1} = J^{n-r}(JI^r) = J^{n-r+1}I^r. \qquad \square$$

SOLUTION TO EXERCISE 2. *Let $R$ be a Noetherian domain and $J \subseteq I$ a reduction. Then there is a $c \in R^\circ$ such that for all $n > 0$,*

$$c\,(\bar{I})^n \subseteq J^n.$$

PROOF. We have that $\bar{J} = \bar{I}$ as $J$ is a reduction of $I$, see Theorem 1.6. It now suffices to assume that $J = I$. By Exercise 1 part (3) there exists an integer $r > 0$ such that

$$I^r\left(\bar{I}\right)^n \subseteq \left(\bar{I}\right)^r\left(\bar{I}\right)^n = \left(\bar{I}\right)^{n+r} = I^n\left(\bar{I}\right)^r \subseteq I^n$$

for all integers $n > 0$. Now we may pick $c$ to be any non-zero element in $I^r$. $\square$

SOLUTION TO EXERCISE 3. *Expand the proof of Theorem 1.8 given in the lecture notes to the case $w > 0$.*

PROOF. By Remark 1.7 we may assume that $R$ is a domain. Let $J = (a_1, \dots, a_\ell)$ and let $w > 0$ be an integer. Let $x \in \overline{I^{\ell+w}}$. Exercise 1 part (1) tells us that $J^{\ell+w}$ is a reduction of $I^{l+w}$. By Exercise 2 there exists $c \in R^\circ$ such that $c\left(\overline{I^{\ell+w}}\right)^n \subseteq J^{(\ell+w)n}$ for all $n \in \mathbb{N}$. In other words, for all $q = p^e$,

$$cx^q \in c\left(\overline{I^{\ell+w}}\right)^q \subseteq J^{(\ell+w)q}.$$

Repeatedly applying Discussion 1.10 give us that

$$J^{(\ell+w)q} = J^{[q]}J^{(\ell+w-1)q} = \cdots = \left(J^{[q]}\right)^{w+1} J^{(l-1)q} \subseteq \left(J^{[q]}\right)^{w+1} = \left(J^{w+1}\right)^{[q]}$$

for all $q = p^e$. The last equality $\left(J^{[q]}\right)^{w+1} = \left(J^{w+1}\right)^{[q]}$ is a consequence of Exercise 1, part (6) in Assignment 1. Hence $x \in \left(J^{w+1}\right)^*$. $\square$

SOLUTION TO EXERCISE 4. *Let $(R, \mathfrak{m}, k)$ be a local ring, where $k$ is not necessarily infinite. Let $t$ be a variable over $R$ and set $R_1 = R[t]_{\mathfrak{m}R[t]}$. Let $I \subseteq R$ be an ideal of analytic spread $\ell$.*

(1) *Show that $R_1$ is a faithfully flat extension of $R$ and $\ell(IR_1) = \ell(I)$.*
(2) *Show that if $J \subseteq I$ is a reduction then $JR_1$ is a reduction of $IR_1$.*
(3) *Assume that $R$ is of positive characteristic. Show that the tight closure Briançon-Skoda theorem holds for any reduction $J \subseteq I$.*

Proof. (1) The composite map $R \to R[t] \to R[t]_{\mathfrak{m}R[t]}$ is flat since it is a composition of two flat maps and hence it is faithfully flat since it is also a local homomorphism. Further observe that

$$\mathcal{F}(I) = \bigoplus_{n\in\mathbb{N}} \frac{I^n}{\mathfrak{m}I^n} \hookrightarrow \bigoplus_{n\in\mathbb{N}} \frac{I^nR_1}{\mathfrak{m}I^nR_1} = \mathcal{F}(IR_1) \cong \mathcal{F}(I)\otimes_R R_1$$

is also a faithfully flat extension and therefore

$$\ell(IR_1) = \dim \mathcal{F}(IR_1) = \dim \mathcal{F}(I) + \dim R_1/\mathfrak{m}R_1 = \dim \mathcal{F}(I) = \ell(I).$$

(2) If $J \subseteq I$ is a reduction then $I^{r+1} = JI^r$ for some integer $r > 0$. Therefore $I^{r+1}R_1 = JR_1I^rR_1$ and so $JR_1 \subseteq IR_1$ is also a reduction.
(3) Suppose that $(R, \mathfrak{m})$ is a local ring of characteristic $p > 0$ and $J \subseteq I$ is any reduction. Then $JR_1 \subseteq IR_1$ is also a reduction and since $R_1$ has infinite residue field, there exists a minimal reduction $J'$ contained in $JR_1$ and generated by $\ell(IR_1) = \ell(I) =: \ell$ elements, see Theorem 1.6. Note that $J'^{\ell+w} \subseteq \overline{I^{\ell+w}R_1}$ is a reduction and there exists an integer $r > 0$ such that

$$I^{(\ell+w)r}R_1\left(\overline{I^{\ell+w}R_1}\right)^n \subseteq \left(\overline{I^{\ell+w}R_1}\right)^{n+r} = J'^{(\ell+w)n}\left(\overline{I^{\ell+w}R_1}\right)^r \subseteq J'^{(\ell+w)n}$$

for all integers $n > 0$. Pick a non-zero element $c$ in $I^{(\ell+w)r}$ and let $x \in \overline{I^{\ell+w}}$ be an arbitrary element. Then

$$cx^q \in c\left(\overline{I^{\ell+w}}\right)^q \subseteq J'^{(\ell+w)q} \subseteq \left(J'^{w+1}\right)^{[q]} \subseteq \left(J^{w+1}\right)^{[q]} R_1 \cap R = \left(J^{w+1}\right)^{[q]}$$

for all $q = p^e$. The last equality $\left(J^{w+1}\right)^{[q]} R_1 \cap R = \left(J^{w+1}\right)^{[q]}$ is a consequence of faithful flatness. Thus $x \in \left(J^{w+1}\right)^*$ as $c \in R^\circ$. □

Solution to Exercise 5. [**CHS10**, Theorem 2.1] *can be used to show that if* $(R, \mathfrak{m})$ *is an excellent local domain and* $I \subseteq R$ *an ideal of analytic spread* $\ell < \dim R$, *then for all* $n \in \mathbb{N}$, $\overline{I^n} : \mathfrak{m}^\infty = \overline{I^n}$ *(here* $A : J^\infty = \bigcup_{k\geq 0}(A : J^k)$*). Use this result and Theorem* 1.2 *to prove the following:*

*Let* $(R, \mathfrak{m})$ *be an excellent local regular ring. Let* $I$ *be an ideal in* $R$ *of analytic spread* $\ell < \dim R$. *Then for all* $n \in \mathbb{N}$,

$$I^{\ell+n} : \mathfrak{m}^\infty \subseteq I^{n+1}.$$

Proof. From Theorem 1.2 and [**CHS10**, Theorem 2.1] we get

$$I^{\ell+n} : \mathfrak{m}^\infty \subseteq \overline{I^{\ell+n}} : \mathfrak{m}^\infty = \overline{I^{\ell+n}} \subseteq I^{n+1}$$

for all $n \in \mathbb{N}$. □

Solution to Exercise 6. *Show that property* (2) *is a consequence of property* (1) *in Lemma* 2.8 *of the notes.*

Proof. It should be first noted that $\left\{(a_1, \dots, a_i)I^n : I^{n+1}\right\}_{n=1}^\infty$ is an increasing sequence of ideals in $R$ and hence must stabilize as $R$ is Noetherian. In particular, there exists an integer $N > 0$ such that

$$\left((a_1, \dots, a_i)I^N : I^{N+1}\right) + I = \bigcup_{n=1}^{\infty} \left(\left((a_1, \dots, a_i)I^n : I^{n+1}\right) + I\right).$$

Let $P$ be a prime ideal in $R$ of height $i$ and containing $I$. By property (1) in Lemma 2.8 of the notes we know that $(a_1, \ldots, a_i)R_P$ is a reduction of $IR_P$. So there exists an integer $r > 0$ (depending on $P$) such that

$$I^{r+1}R_P = (a_1, \ldots, a_i)I^r R_P.$$

Let $n$ be an integer such that $n > r$ and observe that

$$\begin{aligned}\left(((a_1, \ldots, a_i)I^n : I^{n+1}) + I\right) R_P &= \left((a_1, \ldots, a_i)I^n R_P : I^{n+1}R_P\right) + IR_P \\ &= \left((a_1, \ldots, a_i)I^r I^{n-r} R_P : I^{n+1}R_P\right) + IR_P \\ &= \left(I^{r+1}I^{n-r}R_P : I^{n+1}R_P\right) + IR_P \\ &= R_P.\end{aligned}$$

This means that

$$((a_1, \ldots, a_i)I^N : I^{N+1}) + I = \bigcup_{n=1}^{\infty} \left(((a_1, \ldots, a_i)I^n : I^{n+1}) + I\right) \not\subset P.$$

In other words, $\operatorname{ht}\left(((a_1, \ldots, a_i)I^n : I^{n+1}) + I\right) \geq i + 1$ for all integers $n \geq N$. □

Solution to Exercise 7. *Let $R$ be a Noetherian (not necessarily characteristic $p$ or local) ring and let $J \subseteq I$ be ideals.*

(1) *Show that $\mathfrak{a}(I, J)$ is always defined.*
(2) *If $\mathfrak{a}(I, J)$ contains a non-zero divisor then $\bar{I} = \bar{J}$ and hence $J$ is a reduction of $I$. In this case there exists $n \geq 0$ such that $I^n \subseteq \mathfrak{a}(I, J)$.*

Proof. (1) Let $\Lambda$ be the collection of all ideals $\mathfrak{b}$ in $R$ such that $J\mathfrak{b} = I\mathfrak{b}$. The set $\Lambda$ is partially ordered by inclusion. The set $\Lambda$ is non-empty since $(0) \in \Lambda$. Let $\mathfrak{a} = \sum_{\mathfrak{b} \in \Lambda} \mathfrak{b}$ which is an ideal and is the smallest ideal containing $\bigcup_{\mathfrak{b} \in \Lambda} \mathfrak{b}$. By construction $J\mathfrak{a} = I\mathfrak{a}$ and $\mathfrak{a}$ is the largest element of $\Lambda$ with respect to inclusion.

(2) Let $x$ be an arbitrary element of $I$ and we may write $\mathfrak{a} = (a_1, \ldots, a_t)$. The equality $J\mathfrak{a} = I\mathfrak{a}$ allows us to write for each $i = 1, \ldots, t$,

$$xa_i = \sum_{j=1}^{t} b_{ij}a_j$$

where $b_{ij} \in J$. Let $B$ be the $t \times t$-matrix defined by $B_{ij} = \delta_{ij}x - b_{ij}$ where $\delta_{ij}$ is the Kronecker delta function. Let $a$ be the column vector $(a_1, \ldots, a_t)^T$. By construction $Ba = 0$ and therefore $\det(B)a_i = 0$ for all $i = 1, \ldots, t$. In other words, $\det(B)\mathfrak{a} = 0$ but the ideal $\mathfrak{a}$ contains a non-zero divisor. This forces $\det(B) = 0$ and an expansion of $\det(B) = 0$ yields an equation of integral dependence of $x$ over $J$. Thus $\bar{J} = \bar{I}$.

*Alternative solution for $\overline{I} = \overline{J}$:* Note that $\overline{J} \subseteq \overline{I}$ as $J \subseteq I$. Suppose that $x \in \overline{I}$ and so there exists $c \in R^\circ$ such that $cx^n \in I^n$ for all $n > 0$. Let $y \in \mathfrak{a}(I, J) =: \mathfrak{a}$ be a non-zero divisor. Then

$$cyx^n \in I^n\mathfrak{a} = J^n\mathfrak{a} \subseteq J^n$$

for all $n > 0$. Since $y$ is a non-zero divisor, $y \in R^\circ$ and hence $cy \in R^\circ$. By Definition 1.4 we have that $x \in \overline{J}$ and thus $\overline{I} = \overline{J}$.

For any $x \in I$ we have $x \in \overline{J}$ and thus there exists $n > 0$ such that $x^n \in J(J,x)^{n-1}$. In other words, $(J,x)^n = J(J,x)^{n-1}$ and so $J \subseteq (J,x)$ is a reduction. Let $I = (a_1, \ldots, a_t)$ and consider the chain

$$J \subseteq (J, a_1) \subseteq \cdots \subseteq (J, a_1, \ldots, a_t) = I$$

where $(J, a_1, \ldots, a_{i-1}) \subseteq (J, a_1, \ldots, a_i)$ is a reduction for all $i = 1, \ldots, t$. By transitivity $J \subseteq I$ is reduction. So there exists an integer $r \geq 0$ such that $I^{r+1} = II^r = JI^r$ and from the maximality of $\mathfrak{a}(I,J)$ it follows that $I^r \subseteq \mathfrak{a}(I,J)$. □

SOLUTION TO EXERCISE 8. *Let $(R, \mathfrak{m})$ be a local ring, $J \subseteq I$ a reduction and $r := r_J(I)$. Set $\mathfrak{a} := \mathfrak{a}(I,J)$, $\mathfrak{a}_1 = (J\colon I)$ and inductively define $\mathfrak{a}_{n+1} = (J\mathfrak{a}_n\colon I)$ for all $n \geq 1$.*

(1) *Show that $I^r \subseteq \mathfrak{a}$, $\mathfrak{a} \subseteq \mathfrak{a}_n$ and $\mathfrak{a}_{n+1} \subseteq \mathfrak{a}_n$ for all $n \geq 1$.*
(2) *If $I$ is $\mathfrak{m}$-primary then $\mathfrak{a}_n = \mathfrak{a}$ for all $n \gg 0$.*

PROOF. (1) We have $I^{r+1} = II^r = JI^r$ and from the maximality of $\mathfrak{a}$ it follows that $I^r \subseteq \mathfrak{a}$. We shall prove $\mathfrak{a} \subseteq \mathfrak{a}_n$ by induction on $n$. For $n = 1$ observe that

$$I\mathfrak{a} = J\mathfrak{a} \implies \mathfrak{a} \subseteq J\mathfrak{a}\colon I \subseteq J\colon I = \mathfrak{a}_1.$$

By induction hypothesis we assume that $\mathfrak{a} \subseteq \mathfrak{a}_n$ for some $n \geq 1$. Then

$$I\mathfrak{a} = J\mathfrak{a} \implies \mathfrak{a} \subseteq J\mathfrak{a}\colon I \subseteq J\mathfrak{a}_n\colon I = \mathfrak{a}_{n+1}.$$

We shall prove $\mathfrak{a}_{n+1} \subseteq \mathfrak{a}_n$ by induction on $n$. For $n = 1$ observe that

$$\mathfrak{a}_2 = J\mathfrak{a}_1\colon I \subseteq J\colon I = \mathfrak{a}_1.$$

By induction hypothesis we assume that $\mathfrak{a}_{n+1} \subseteq \mathfrak{a}_n$ for some $n \geq 1$. Then

$$\mathfrak{a}_{n+2} = J\mathfrak{a}_{n+1}\colon I \subseteq J\mathfrak{a}_n\colon I = \mathfrak{a}_{n+1}.$$

(2) It follows from the previous part that $I^r \subseteq \mathfrak{a}$ and so $\mathfrak{a}$ is also $\mathfrak{m}$-primary. Now observe that $R/\mathfrak{a}$ is an Artinian local ring and

$$\cdots \mathfrak{a}_3/\mathfrak{a} \subseteq \mathfrak{a}_2/\mathfrak{a} \subseteq \mathfrak{a}_1/\mathfrak{a}$$

is a descending chain of ideals in $R/\mathfrak{a}$. Therefore this chain must stabilize. In other words, there exists an integer $N > 0$ such that $\mathfrak{a}_{n+1} = \mathfrak{a}_n$ for all $n \geq N$. Let $n \geq N$ and note that

$$\mathfrak{a}_n = \mathfrak{a}_{n+1} = J\mathfrak{a}_n\colon I \implies I\mathfrak{a}_n = I\,(J\mathfrak{a}_n\colon I) \subseteq J\mathfrak{a}_n \implies I\mathfrak{a}_n = J\mathfrak{a}_n.$$

The maximality of $\mathfrak{a}$ forces $\mathfrak{a}_n \subseteq \mathfrak{a}$ and therefore $\mathfrak{a}_n = \mathfrak{a}$ for all $n \geq N$. □

## Acknowledgments

The author thanks Suprajo Das for his comments and suggestions during the preparation of this manuscript.

## References

[AH93] Ian M. Aberbach and Craig Huneke, *An improved Briançon-Skoda theorem with applications to the Cohen-Macaulayness of Rees algebras*, Math. Ann. **297** (1993), no. 2, 343–369, DOI 10.1007/BF01459507. MR1241812

[AH01a] Ian M. Aberbach and Craig Huneke, *F-rational rings and the integral closures of ideals*, Michigan Math. J. **49** (2001), no. 1, 3–11, DOI 10.1307/mmj/1008719031. MR1827071

[AH01b] Ian M. Aberbach and Craig Huneke, *F-rational rings and the integral closures of ideals. II*, Ideal theoretic methods in commutative algebra (Columbia, MO, 1999), Lecture Notes in Pure and Appl. Math., vol. 220, Dekker, New York, 2001, pp. 1–12. MR1836587

[AH11] Ian M. Aberbach and Aline Hosry, *A less restrictive Briançon-Skoda theorem with coefficients*, J. Algebra **345** (2011), 72–80, DOI 10.1016/j.jalgebra.2011.08.016. MR2842054

[AHT95] Ian M. Aberbach, Craig Huneke, and Ngô Việt Trung, *Reduction numbers, Briançon-Skoda theorems and the depth of Rees rings*, Compositio Math. **97** (1995), no. 3, 403–434. MR1353282

[BH98] Winfried Bruns and Jürgen Herzog, *Cohen-Macaulay rings*, Cambridge Studies in Advanced Mathematics, vol. 39, Cambridge University Press, Cambridge, 1993. MR1251956

[CHS10] Steven Dale Cutkosky, Jürgen Herzog, and Hema Srinivasan, *Asymptotic growth of algebras associated to powers of ideals*, Math. Proc. Cambridge Philos. Soc. **148** (2010), no. 1, 55–72, DOI 10.1017/S0305004109990144. MR2575372

[GS82] Shiro Goto and Yasuhiro Shimoda, *On the Rees algebras of Cohen-Macaulay local rings*, Commutative algebra (Fairfax, Va., 1979), Lect. Notes Pure Appl. Math., vol. 68, Dekker, New York, 1982, pp. 201–231. MR655805

[HH90] Melvin Hochster and Craig Huneke, *Tight closure, invariant theory, and the Briançon-Skoda theorem*, J. Amer. Math. Soc. **3** (1990), no. 1, 31–116, DOI 10.2307/1990984. MR1017784

[HS06] Craig Huneke and Irena Swanson, *Integral closure of ideals, rings, and modules*, London Mathematical Society Lecture Note Series, vol. 336, Cambridge University Press, Cambridge, 2006. MR2266432

[Hun82] Craig Huneke, *On the associated graded ring of an ideal*, Illinois J. Math. **26** (1982), no. 1, 121–137. MR638557

[Hun92] Craig Huneke, *Uniform bounds in Noetherian rings*, Invent. Math. **107** (1992), no. 1, 203–223, DOI 10.1007/BF01231887. MR1135470

[Hun00] Craig Huneke, *A cancellation theorem for ideals*, J. Pure Appl. Algebra **152** (2000), no. 1-3, 123–132, DOI 10.1016/S0022-4049(99)00148-6. Commutative algebra, homological algebra and representation theory (Catania/Genoa/Rome, 1998). MR1783992

[JK95] Bernard Johnston and Daniel Katz, *Castelnuovo regularity and graded rings associated to an ideal*, Proc. Amer. Math. Soc. **123** (1995), no. 3, 727–734, DOI 10.2307/2160792. MR1231300

[Lip94] Joseph Lipman, *Adjoints of ideals in regular local rings*, Math. Res. Lett. **1** (1994), no. 6, 739–755, DOI 10.4310/MRL.1994.v1.n6.a10. With an appendix by Steven Dale Cutkosky. MR1306018

[LS81] Joseph Lipman and Avinash Sathaye, *Jacobian ideals and a theorem of Briançon-Skoda*, Michigan Math. J. **28** (1981), no. 2, 199–222. MR616270

[LT81] Joseph Lipman and Bernard Teissier, *Pseudorational local rings and a theorem of Briançon-Skoda about integral closures of ideals*, Michigan Math. J. **28** (1981), no. 1, 97–116. MR600418

[McA80] Stephen McAdam, *Asymptotic prime divisors and analytic spreads*, Proc. Amer. Math. Soc. **80** (1980), no. 4, 555–559, DOI 10.2307/2043422. MR587926

[NR54] D. G. Northcott and D. Rees, *Reductions of ideals in local rings*, Proc. Cambridge Philos. Soc. **50** (1954), 145–158, DOI 10.1017/s0305004100029194. MR59889

[SB74] Henri Skoda and Joël Briançon. *Sur la clôture intégrale d' un idéal de germes de fonctions holomorphes en un point de Cn.* Académie des Sciences, 1974.

[TI89] Ngô Việt Trung and Shin Ikeda, *When is the Rees algebra Cohen-Macaulay?*, Comm. Algebra **17** (1989), no. 12, 2893–2922, DOI 10.1080/00927878908823885. MR1030601

Department of Mathematics, University of Missouri, Columbia, Missouri 65211
*Email address*: aberbachi@missouri.edu

Indian Institute of Technology Bombay, Mumbai, Maharashtra 400076, India

Contemporary Mathematics
Volume **831**, 2025
https://doi.org/10.1090/conm/831/16633

# On the applications of Big Cohen-Macaulay modules and algebras

Thomas Polstra, with exercise solutions by Mitra Koley

Throughout these notes, $R$ denotes a commutative Noetherian ring. We do not make assumptions on the characteristic of $R$ and will specify the characteristic whenever necessary. These notes are part of the Graduate Course on Tight Closure of Ideals and its Applications, an ICTP virtual meeting that took place during the Summer of 2022.

## 1. Introduction

Big Cohen-Macaulay modules and algebras over Noetherian rings are remarkably well-behaved, they serve as useful tools to study homological behavior of Noetherian rings, they can be used to study the singularities of a ring, and it is known that every complete local domain admits plentiful big Cohen-Macaulay modules and algebras. The existence of big Cohen-Macaulay modules and algebras imply most of the *homological conjectures*, e.g. the Monomial Conjecture, the Direct Summand Conjecture, and the Vanishing Map of Tor Conjecture.

Weakly functorial big Cohen-Macaulay algebras seemingly promote a robust singularity theory in all characteristics, see [**MS21**] and [**BMP$^+$20**]. Indeed, without referencing the characteristic, the notions of $F$-rationality and $F$-regularity can be defined in terms of the homological behavior big Cohen-Macaulay algebras $B$. For example, by work of Smith, [**Smi97**], under mild hypotheses, a local ring $(R, \mathfrak{m}, k)$ of Krull dimension $d$ and of prime characteristic $p > 0$ is $F$-rational if and only if for all big Cohen-Macaulay $R$-algebras $R \to B$ the induced map of local cohomology modules $H^d_\mathfrak{m}(R) \to H^d_\mathfrak{m}(B)$ is injective.

Hochster proved that every domain, which contains field, must admit a big Cohen-Macaulay module, [**Hoc75**]. Hochster and Huneke improved this line of research by showing that every equicharacteristic local domain admits weakly functorial big Cohen-Macaulay algebras, [**HH92, HH95**]. Hochster and Huneke proved the existence of weakly functorial Big Cohen-Macaulay algebras for rings of positive characteristic in [**HH92**]. It is then through a process of reduction to positive characteristic, Hochster and Huneke achieve the same result for rings that are of equicharacteristic 0, [**HH95**]. Hochster and Huneke proved that the absolute integral closure of a complete local domain is big Cohen-Macaulay if $R$ is positive

The author was supported in part by NSF Grant DMS #2101890 and a grant from the Simons Foundation, Grant Number 814268, MSRI.

characteristic. Very recently, Bhatt has shown that if $R$ is a complete local domain of mixed characteristic $(0, p)$, then the $p$-adic completion of the absolute integral closure of $R$ is a big Cohen-Macaulay algebra, [**Bha20**]. If $R$ is of equicharacteristic 0 and of dimension at least 3, then the absolute integral closure of $R$ is not a big Cohen-Macaulay algebra, see Exercise 5.

The greatest tool at a our disposal will be André's theorem on the existence of weakly functorial big Cohen-Macaulay algebras, [**And20**]. André's methodology heavily leans upon the theory of perfectoid spaces developed by Scholze, [**Sch16**].

THEOREM 1 (Weakly functorial big Cohen-Macaulay algebras, [**And20**]). Every complete local domain admits a big Cohen-Macaulay algebra. Even further, let $R \to S$ be a map of complete local domains and $R \to B_R$ a big Cohen-Macaulay $R$-algebra. Then there exists a big Cohen-Macaulay $S$-algebra $S \to B_S$ and a map $B_R \to B_S$ so that the following diagram commutes:

$$\begin{array}{ccc} B_R & \longrightarrow & B_S \\ \uparrow & & \uparrow \\ R & \longrightarrow & S \end{array}$$

## 2. Big Cohen-Macaulay modules

Let $(R, \mathfrak{m}, k)$ be a Noetherian local ring of Krull dimension $d$, $x_1, \ldots, x_d$ a system of parameters on $R$, and $M$ an $R$-module, not necessarily finitely generated.

DEFINITION 2.1. The module $M$ is a *big Cohen-Macaulay module* with respect to $x_1, \ldots, x_d$ if $x_1, \ldots, x_d$ is a regular sequence on $M$, i.e.,

(a) $(x_1, \ldots, x_d)M \neq M$;
(b) $(x_1, \ldots, x_i)M :_M x_{i+1} = (x_1, \ldots, x_i)M$.

The module $M$ is a *balanced big Cohen-Macaulay module*, or simply a *big Cohen-Macaulay module*, if every system of parameters on $R$ is a regular sequence on $M$. If $M$ is finitely generated and if $M$ is big Cohen-Macaulay with respect to a system of parameters then it is a big Cohen-Macaulay module. This phenomenon fails if $M$ is not finitely generated, see Exercise 1. An $R$-algebra $R \to S$ is called a *big Cohen-Macaulay algebra* if $S$ is a big Cohen-Macaulay $R$-module.

EXERCISE 1. Let $k$ be a field and $R$ the 2-dimensional regular local ring $k[x, y]_{(x,y)}$. Let $E$ be the fraction field of of $R/(y)$ and let $M = R \oplus E$. Show that $x, y$ is a regular sequence on $M$ but $y, x$ is not. Therefore $M$ is a big Cohen-Macaulay module with respect to $x, y$, but $M$ is not a balanced big Cohen-Macaulay module.

SOLUTION. Let $m = (r, e) \in M$ be such that $xm = 0$. We must show that $r = 0$ and $e = 0$. The element $x$ is a nonzerodivisor on $R$ therefore $r = 0$. The map $E \xrightarrow{\cdot x} E$ is an automorphism. Therefore if $xe = 0$ then $e = 0$. Notice further that $M/xM \cong R/xR$ and $y$ is regular on $R/xR$. Therefore $x, y$ is a regular sequence on $M$.

Any element of the form $(0, e) \in M$ is annihilated by $y$. Therefore $M$ is not $y$-regular and so $y, x$ is not a regular sequence on $M$. □

**2.1. Koszul complexes and Koszul homology.** Let $R$ be a Noetherian ring and $x \in R$. The Koszul complex on $x$ is

$$K_\bullet(x;R) : 0 \to \underset{1}{R} \xrightarrow{x} \underset{0}{R} \to 0;$$

the numbers in the complex indicate the degree of the complex.

Given a sequence of elements $x_1, \ldots, x_d$ the Koszul complex on $x_1, \ldots, x_d$ is

$$K_\bullet(\underline{x};R) = K_\bullet(x_1;R) \otimes_R \cdots \otimes_R K_\bullet(x_d;R).$$

If $M$ is an $R$-module then $K_\bullet(\underline{x};M) = K_\bullet(\underline{x};R) \otimes_R M$ is the Koszul complex on $M$ with respect to $\underline{x}$. For each natural number $i$ let $H_i(\underline{x};M)$ denote the $i$th homology group of $K_\bullet(\underline{x};M)$. For example, if $x, y$ is a sequence of elements then, up to a choice of sign, we can express the Koszul complex on $x, y$ as

$$K_\bullet(x,y;R) : 0 \to \underset{2}{R} \xrightarrow{\begin{bmatrix} -y \\ x \end{bmatrix}} \underset{1}{R^{\oplus 2}} \xrightarrow{\begin{bmatrix} x & y \end{bmatrix}} \underset{0}{R} \to 0.$$

Similarly, if $x, y, z$ is a sequence of 3-elements then, up to a choice of sign, we can express the Koszul complex on $x, y, z$ as

$$K_\bullet(x,y,z;R) : 0 \to \underset{3}{R} \xrightarrow{\begin{bmatrix} z \\ y \\ x \end{bmatrix}} \underset{2}{R^{\oplus 3}} \xrightarrow{\begin{bmatrix} -y & z & 0 \\ x & 0 & -z \\ 0 & -x & y \end{bmatrix}} \underset{1}{R^{\oplus 3}} \xrightarrow{\begin{bmatrix} x & y & z \end{bmatrix}} \underset{0}{R} \to 0.$$

Koszul homology groups detect if a module is a big Cohen-Macaulay module.

PROPOSITION 2.2. *Let $(R, \mathfrak{m}, k)$ be a local Noetherian ring of Krull dimension $d$ and $M$ an $R$-module. Then $M$ is a big Cohen-Macaulay module if and only if for all parameter sequences $\underline{x} = x_1, \ldots, x_h$ of $R$ and for all $i \geq 1$, $H_i(\underline{x};M) = 0$ and $(\underline{x})M \neq M$.*

PROOF. First suppose that $M$ is a big Cohen-Macaulay $R$-module. The Koszul complex on a parameter element $x_1$ with respect to the module $M$ is

$$K_\bullet(x;M) : 0 \to M \xrightarrow{\cdot x_1} M \to 0.$$

Therefore $H_1(x_1;M) = \ker(M \xrightarrow{\cdot x_1} M)$, which is 0 as $x_1$ is a regular sequence on $M$. Suppose that $x_1, \ldots, x_h$ is a parameter sequence of $R$ and let $K_\bullet(x_1, \ldots, x_{h-1};M)'$ denote a shift of the Koszul complex $K_\bullet(x_1, \ldots, x_{h-1};M)$ up by a degree. We assume by induction that all higher Koszul homology groups of $K_\bullet(x_1, \ldots, x_{h-1};M)$ are 0. By [**Mat89**, Theorem 16.4], there is an exact sequence of complexes

(2.2.1)

$$0 \to K_\bullet(x_1, \ldots, x_{h-1};M) \to K_\bullet(x_1, \ldots, x_{h-1}, x_h;M) \to K_\bullet(x_1, \ldots, x_{h-1};M)' \to 0$$

and the connecting map $\delta$ of Koszul homology groups

$$H_i(K_\bullet(x_1, \ldots, x_{h-1};M)') \cong H_{i-1}(x_1, \ldots, x_{h-1};M) \xrightarrow{\delta} H_{i-1}(x_1, \ldots, x_{h-1};M)$$

is multiplication by $x_h$. For $i \geq 2$ there are exact sequences

$$H_i(x_1, \ldots, x_{h-1};M) \to H_i(x_1, \ldots, x_{h-1}, x_h;M) \to H_{i-1}(x_1, \ldots, x_{h-1};M).$$

The left-most and right-most Koszul homology modules are 0 by induction. Therefore $H_i(x_1,\ldots,x_{h-1},x_h;M)=0$ for all $i\geq 2$. In low degrees, there is a left-exact sequence

$$0\to H_1(x_1,\ldots,x_{h-1},x_h;M)\to H_0(x_1,\ldots,x_{h-1};M)\xrightarrow{\cdot x_h} H_0(x_1,\ldots,x_{h-1};M).$$

There is an isomorphism $H_0(x_1,\ldots,x_{h-1};M)\cong M/(x_1,\ldots,x_{h-1})M$ and multiplication by $x_h$ on this module is injective the assumption $M$ is a big Cohen-Macaulay module. Therefore the remaining Koszul homology group $H_1(x_1,\ldots,x_{h-1},x_h;M)$ $=0$.

For the converse assume that for all parameter sequences $\underline{x}=x_1,\ldots,x_h$ of $R$ and for all $i\geq 1$, $H_i(\underline{x};M)=0$ and $(\underline{x})M\neq M$. Let $x_1,\ldots,x_d$ be a full system of parameters. We use the exact sequence of complexes in (2.2.1) for each of the partial parameter sequences $x_1,\ldots,x_h$ with $h\leq d-1$ and our assumption of higher Koszul homology groups vanishing to produce short exact sequences

$$0\to H_0(x_1,\ldots,x_{h-1};M)\xrightarrow{\cdot x_h} H_0(x_1,\ldots,x_{h-1};M)\to H_0(x_1,\ldots,x_h;M)\to 0.$$

Then $H_0(x_1,\ldots,x_{h-1};M)\cong M/(x_1,\ldots,x_{h-1})M$ and $x_h$ is a regular on $M/(x_1,\ldots,x_{h-1})M$, i.e., $x_1,\ldots,x_d$ is a regular sequence on $M$. □

**2.2. Local cohomology.** Let $(R,\mathfrak{m},k)$ be a local ring. The $i$th local cohomology functor with support in the maximal ideal, denoted $H^i_{\mathfrak{m}}(-)$, is the $i$th derived right functor of the $\mathfrak{m}$-torsion functor $M\mapsto H^0_{\mathfrak{m}}(M):=\{m\in M\mid \exists t\in\mathbb{N}:\mathfrak{m}^t m=0\}$. If $x_1,\ldots,x_d$ is a system of parameters of $R$ and $M$ an $R$-module then

$$H^d_{\mathfrak{m}}(M)\cong\varinjlim\left(\frac{M}{(x_1^t,\ldots,x_d^t)M}\xrightarrow{\cdot x_1\cdots x_d}\frac{M}{(x_1^{t+1},\ldots,x_d^{t+1})M}\right).$$

More generally, by [**BH93**, Theorem 3.5.6], local cohomology groups are isomorphic to a direct limit of Koszul homology groups

$$H^i_{\mathfrak{m}}(M)\cong\varinjlim\left(H_{d-i}(x_1^t,\ldots,x_d^t;M)\right).$$

In particular, if $M$ is a big Cohen-Macaulay $R$-module, then all lower local cohomology groups of $M$ with support in the maximal ideal $\mathfrak{m}$ vanish by Proposition 2.2.

Proposition 2.3. *Let $(R,\mathfrak{m},k)$ be a complete local Noetherian ring of Krull dimension $d$ and $M\neq 0$ an $R$-module. If $M$ is a big Cohen-Macaulay then $H^i_{\mathfrak{m}}(M)=0$ for all $i<d$. Moreover, if $M$ is a big Cohen-Macaulay module and $x_1,\ldots,x_d$ is a system of parameters then the maps in the direct limit system*

$$H^d_{\mathfrak{m}}(M)\cong\varinjlim\left(\frac{M}{(x_1^t,\ldots,x_d^t)M}\xrightarrow{\cdot x_1\cdots x_d}\frac{M}{(x_1^{t+1},\ldots,x_d^{t+1})M}\right).$$

*are injective.*

Proof. Vanishing of the lower local cohomology modules is the content of the discussion which precedes the statement of the proposition. Injectivity of the maps in the direct limit system is an application of Exercise 2. □

Exercise 2. Let $(R,\mathfrak{m},k)$ be a Noetherian local ring, $x_1,\ldots,x_d$ a system of parameters, and $M$ a big Cohen-Macaulay module. Show that for all natural numbers $n_1,\ldots,n_d\in\mathbb{N}$ that

$$(x_1,x_2,\ldots,x_d)M=(x_1^{n_1+1},x_2^{n_2+1},\ldots,x_d^{n_d+1})M:_M x_1^{n_1}x_2^{n_2}\cdots x_d^{n_d}.$$

SOLUTION. If $m \in (x_1, x_2, \ldots, x_d)M$ then

$$x_1^{n_1} x_2^{n_2} \cdots x_d^{n_d} m \in x_1^{n_1} x_2^{n_2} \cdots x_d^{n_d}(x_1, \ldots, x_d)M \subseteq (x_1^{n_1+1}, x_2^{n_2+1}, \ldots, x_d^{n_d+1})M$$

and therefore $m \in (x_1^{n_1+1}, x_2^{n_2+1}, \ldots, x_d^{n_d+1})M :_M x_1^{n_1} x_2^{n_2} \cdots x_d^{n_d}$.

Conversely, if $m \in (x_1^{n_1+1}, x_2^{n_2+1}, \ldots, x_d^{n_d+1})M :_M x_1^{n_1} x_2^{n_2} \cdots x_d^{n_d}$ then

$$x_1^{n_1} x_2^{n_2} \cdots x_d^{n_d} m \in (x_1^{n_1+1}, x_2^{n_2+1}, \ldots, x_d^{n_d+1})M.$$

Therefore there exists an element $m' \in M$ such that

$$x_1^{n_1} x_2^{n_2} \cdots x_d^{n_d} m - x_d^{n_d+1} m' \in (x_1^{n_1+1}, x_2^{n_2+1}, \ldots, x_{d-1}^{n_{d-1}+1})M.$$

Equivalently,

$$x_1^{n_1} x_2^{n_2} \cdots x_{d-1}^{n_{d-1}} m - x_d m' \in (x_1^{n_1+1}, x_2^{n_2+1}, \ldots, x_{d-1}^{n_{d-1}+1})M :_M x_d^{n_d}$$

The module $M$ is balanced big Cohen-Macaulay and therefore

$$(x_1^{n_1+1}, x_2^{n_2+1}, \ldots, x_{d-1}^{n_{d-1}+1})M :_M x_d^{n_d} = (x_1^{n_1+1}, x_2^{n_2+1}, \ldots, x_{d-1}^{n_{d-1}+1})M.$$

Hence

$$x_1^{n_1} x_2^{n_2} \cdots x_{d-1}^{n_{d-1}} m \in (x_1^{n_1+1}, x_2^{n_2+1}, \ldots, x_{d-1}^{n_{d-1}+1}, x_d)M.$$

The desired result follows by induction. □

EXERCISE 3. Let $(R, \mathfrak{m}, k)$ be a Cohen-Macaulay local ring and $M$ an $R$-module, not necessarily finitely generated. Show that if $M$ is faithfully flat then $M$ is a big Cohen-Macaulay module.

SOLUTION. Let $\underline{x}$ be a system of parameters of $R$. Because $R$ is Cohen-Macaulay there are short exact sequences

$$0 \to \frac{R}{(x_1, \ldots, x_{i-1})R} \xrightarrow{\cdot x_i} \frac{R}{(x_1, \ldots, x_{i-1})R} \to \frac{R}{(x_1, \ldots, x_{i-1}, x_i)R} \to 0.$$

The module $M$ is flat, therefore we can apply $- \otimes_R M$ and produce short exact sequences

$$0 \to \frac{M}{(x_1, \ldots, x_{i-1})M} \xrightarrow{\cdot x_i} \frac{M}{(x_1, \ldots, x_{i-1})R} \to \frac{M}{(x_1, \ldots, x_{i-1}, x_i)M} \to 0.$$

Hence $M$ is a big Cohen-Macaulay $R$-module. □

The converse statement of Exercise 3 is true if $R$ is a regular local ring. Recall that a module $M$ over a local ring $(R, \mathfrak{m}, k)$ is faithfully flat if and only if for all finitely generated $R$-modules $N$ and all $i \geq 1$, $\mathrm{Tor}_i(M, N) = 0$.

EXERCISE 4. Let $(R, \mathfrak{m}, k)$ be a regular local ring of Krull dimension $d$. Show that a module $M$ is a balanced big Cohen-Macaulay module if and only if $M$ is faithfully flat as an $R$-module by proving the following:

(a) If $N$ is a finitely generated $R$-module then $\mathrm{Tor}_{d+1}(M, N) = 0$, therefore by induction, we may assume that if $i \geq 1$ that $\mathrm{Tor}_{i+1}(M, N) = 0$ for all finitely generated $R$-modules $N$.
(b) Show that $\mathrm{Tor}_i(M, N) = 0$ for all finitely generated $R$-modules $N$ if and only if $\mathrm{Tor}_i(M, R/P) = 0$ for all $P \in \mathrm{Spec}(R)$;
(c) Show that $\mathrm{Tor}_i(M, R/\mathfrak{m}) = 0$ for all $i \geq 1$;
(d) Show that if $P$ is a prime ideal of height $h \leq d - 1$ then there exists a parameter sequence $\underline{x} = x_1, \ldots, x_h \in P$ and an element $d \in \mathfrak{m} \setminus P$ such that $P = (\underline{x}) :_R d$;

(e) Conclude that $\operatorname{Tor}_i(M, R/P) = 0$. (Hint: By part ($d$) there is a short exact sequence $0 \to R/P \xrightarrow{\cdot d} R/(\underline{x}) \to R/(\underline{x}, d) \to 0$.)

SOLUTION. (a) The ring $R$ is regular therefore every finitely generated module $N$ has resolution not exceeding $d$ and so $\operatorname{Tor}_{d+1}(M, N) = 0$ for every finitely generated $R$-module $N$.

(b) Every finitely generated module $N$ admits a filtration

$$0 = N_0 \subseteq N_1 \subseteq \cdots \subseteq N_t = N$$

so that $N_i/N_{i-1} \cong R/P_i$ for some $P_i \in \operatorname{Spec}(R)$. It follows that there are exact sequences

$$\operatorname{Tor}_j(M, N_{i-1}) \to \operatorname{Tor}_j(M, N_i) \to \operatorname{Tor}_j(M, R/P_i)$$

from which the solution follows.

(c) The ideal $\mathfrak{m}$ is generated by a regular system of parameters $\underline{x} = x_1, \ldots, x_d$ and so $\operatorname{Tor}_i(M, k) \cong H_i(\underline{x}; M) = 0$ if $i \geq 1$ as $M$ is assumed to be a big Cohen-Macaulay module.

(d) The localized ring $R_P$ is a regular local ring of Krull dimension $h$. Therefore there exists elements $x_1, \ldots, x_h$ so that $PR_P = (x_1, \ldots, x_h)R_P$. Therefore one of two things can happen: either $(x_1, \ldots, x_h) = P$ or $(x_1, \ldots, x_h) = P \cap Q$ where $\sqrt{Q} = P_2 \cap \cdots \cap P_t$ is an intersection of height $h$ primes so that $P \notin \{P_2, \ldots, P_t\}$. In the former case we can take $d$ to be any element of $\mathfrak{m}$ that avoids $P$. In the latter case, $d \in P_2 \cap \cdots P_t$ that avoids $P$ has the desired property.

(e) There are short exact sequences

$$0 \to R/P \xrightarrow{\cdot d} R/(\underline{x}) \to R/(\underline{x}, d) \to 0$$

and so there are exact sequences

$$\operatorname{Tor}_{i+1}(M, R/(\underline{x}, d)) \to \operatorname{Tor}_i(M, R/P) \to \operatorname{Tor}_i(M, R/(\underline{x})).$$

We have that $\operatorname{Tor}_{i+1}(M, R/(\underline{x}, d)) = 0$ by induction and

$$\operatorname{Tor}_i(M, R/(\underline{x})) \cong H_i(\underline{x}; M) = 0$$

by Proposition 2.2. □

**2.3. Absolute integral closure.** Let $R$ be a Noetherian domain, $K$ the fraction field of $R$, and $\overline{K}$ a choice of algebraic closure of $K$.

DEFINITION 2.4. The *absolute integral closure* of $R$, denoted by $R^+$, is the integral closure of $R$ in $\overline{K}$.

The following two theorems on the homological behavior of the absolute integral closure of a complete local domain are of upmost importance:

THEOREM 2.5 ([**HH92**], c.f. [**HL07**]). *Let $(R, \mathfrak{m}, k)$ be a complete local domain of prime characteristic $p > 0$. Then $R^+$ is a big Cohen-Macaulay algebra.*

THEOREM 2.6 ([**Bha20**]). *Let $(R, \mathfrak{m}, k)$ be a complete local domain of mixed characteristic $p$. Then the $p$-adic completion of $R^+$ is a big Cohen-Macaulay algebra.*

The algebra $R^+$ is big Cohen-Macaulay in prime characteristic and therefore must be of use to study the singularities of $R$. Smith proved that under mild hypotheses if $x_1, \ldots, x_d$ is a parameter sequence of a prime characteristic ring $R$ then $(x_1, \ldots, x_d)^* = (x_1, \ldots, x_d)R^+ \cap R$, [**Smi94**, Theorem 5.1]. It is **not** the case that $I^* = IR^+ \cap R$ for every ideal $I \subseteq R$. If such a property were always true then it would follow that the computation of tight closure commutes with a localization, which it does not, [**BM10**]. What is not clear though is the following:

QUESTION 2.7 (Open problem). Let $(R, \mathfrak{m}, k)$ be an excellent local domain of prime characteristic $p > 0$. Suppose that for every ideal $I \subseteq R$ we have that $IR^+ \cap R = I$. Then is it necessarily the case that $I^* = I$ for every ideal $I$, i.e., is $R$ weakly $F$-regular?

A ring map $R \to S$ is a *finite extension* if $R \to S$ is an injective map and $S$ is module-finite as an $R$-module. A ring $R$ is called a *splinter* if $R$ has the property that $IR^+ \cap R = I$ for every ideal $I \subseteq R$. If $R \to S$ is module finite inclusion of rings, then $S$ being algebraic over $R$ implies that $S$ can be identified up to isomorphism as a subring of $R^+$. In particular, if $R$ is splinter then $IS \cap R = I$. The absolute integral closure $R^+$ is a direct limit of finite extensions of $R$. It follows that if $R$ has the property that for every finite extension $R \to S$ and every ideal $I \subseteq R$ that $IS \cap R = I$ then $R$ is a splinter, in particular the property of being splinter localizes. The above question asks if every splinter local ring of prime characteristic is weakly $F$-regular. An affirmative answer to the above question would be a celebrated theorem as it would imply that the property of weak $F$-regularity localizes and thus we can remove the adjective "weak" from the terminology. Question 2.7 has been answered affirmatively for rings whose anticanonical algebras are Noetherian, see [**CEMS18**]. To date, Question 2.7 is settled for rings of Krull dimension at most 2 and the Krull dimension 3 case remains open.

The term splinter is derived from the notion of a map of $R$-modules $\varphi : N \to M$ being *split*; there exists an $R$-linear map $\psi : M \to N$ such that $\psi \circ \varphi : N \to N$ is the identity map on $N$. If $(R, \mathfrak{m}, k)$ is a complete local domain, $M$ a finitely generated $R$-module, and $R \to M$ is an $R$-linear map, then $R \to M$ splits if and only if for every ideal $I \subseteq R$, $IM \cap R = I$, [**Hoc77**, Theorem 1.7].

**2.4. Absolute integral closure in equicharacteristic 0.** The purpose of this section is to demonstrate that the absolute integral closure of an equicharacteristic 0 ring is not a big Cohen-Macaulay algebra whenever $R$ is of dimension at least 3.

LEMMA 2.8. *Let $R$ be a Noetherian normal domain of equicharacteristic* 0*. Let $R \to S$ be a finite extension, then $R \to S$ splits.*

PROOF. We may replace $S$ by $S/P$ for some $P \in \operatorname{Spec}(S)$ that contracts to the 0-prime of $R$. We therefore assume that $R \to S$ is a finite extension of domains. Let $K$ and $L$ denote the fractions fields of $R$ and $S$ respectively. Let $T : L \to K$ denote the field trace map, i.e., for each element $s \in L$ let $T(s)$ denote the trace of the $K$-linear map $L \xrightarrow{\cdot s} L$. The trace of an element $s \in L$ is a coefficient of the minimal polynomial of $s$ over $K$. If $s \in S$ then the coefficients of the minimal polynomial of $s$ over $K$ are integral over $R$ and therefore belong to $R$. In conclusion, if $s \in S$ then $T(s) \in R$ and the field trace map defines an $R$-linear map $T : S \to R$.

If $r \in R$ then $T(r) = rT(1) = r[L : K]$. The integer $[L : K]$ is invertible in $R$ as $\mathbb{Q} \subseteq R$. Therefore $\frac{1}{[L:K]}T : S \to R$ is an $R$-linear map that splits $R \to S$. □

REMARK 2.9. If $R$ is a normal domain of prime characteristic $p$ and $R \to S$ is a finite extension of domains then the proof of Lemma 2.8 shows that $R \to S$ splits whenever $p$ does not divide the degree of the induced maps of fraction fields of $R$ and $S$.

COROLLARY 2.10. *Let $R$ be a Noetherian normal domain of equicharacteristic 0 and $I \subseteq R$ an ideal. Then $IR^+ \cap R = I$.*

PROOF. The algebra $R^+$ is direct limit of finite extensions of $R$,

$$R^+ = \bigcup_{R \to S \text{ finite}} S.$$

Therefore $IR^+ \cap R = \bigcup IS \cap R = I$ by Lemma 2.8. □

PROPOSITION 2.11. *If $(R, \mathfrak{m}, k)$ is a normal domain of equicharacteristic 0 that is of dimension at least 3 and of depth 2, then $R^+$ is not a big Cohen-Macaulay algebra.*

PROOF. Suppose by way of contradiction that $R^+$ is a big Cohen-Macaulay algebra. Let $x, y, z$ be a parameter sequence of elements of $R$ such that $x, y$ is a regular sequence on $R$ but $x, y, z$ is not. In particular, $(x, y) :_R z \neq (x, y)R$. Observe that

$$((x, y) :_R z)R^+ \subseteq (x, y)R^+ :_{R^+} z = (x, y)R^+ \subseteq ((x, y) :_R z)R^+$$

with the equality being derived from the assumption that $R^+$ is a big Cohen-Macaulay algebra. By Corollary 2.10 we then have that

$$((x, y) :_R z) = ((x, y) :_R z)R^+ \cap R = (x, y)R^+ \cap R = (x, y),$$

a contradiction. □

EXERCISE 5. Let $(R, \mathfrak{m}, k)$ be a complete local equicharacteristic 0 domain of Krull dimension at least 3. Show that $R^+$ is not a big Cohen-Macaulay algebra.

SOLUTION. Suppose that $x_1, \dots, x_d$ is a system of parameters of $R$. The ring $R$ is module-finite over a power series ring $A = k[[x_1, x_2, x_3, \dots, x_d]]$ and therefore $A^+ = R^+$. Let $S$ be the normalization of $R$. Then $S$ is a complete normal local domain of Krull dimension $d$ so that $S$ is finite over $A$ and $S$ has depth 2. Therefore $A^+ \cong S^+$. One can then follow the method of Proposition 2.11 to show that $R^+$ is not big Cohen-Macaulay. □

## 3. Applications of big Cohen-Macaulay modules and algebras

Every complete local domain admits a big Cohen-Macaulay algebra, [**HH95**, **Bha20**], and therefore every complete local domain admits a big Cohen-Macaulay module. We use the existence of Big Cohen-Macaulay modules to prove the Monomial Conjecture and the Direct Summand Conjecture. We will then resort to the stronger result on the existence of weakly functorial big Cohen-Macaulay algebras to prove the Vanishing Maps of Tor Conjecture and that rings that are direct summands of regular rings are Cohen-Macaulay.

**3.1. The monomial conjecture.** Recall that the Direct Summand Conjecture asserts that every regular ring is a splinter, i.e. it splits out of every finite extension. To the novice, it may not be apparent that the Direct Summand Conjecture is an easy consequence of the Monomial Conjecture, Theorem 3.1 below.

THEOREM 3.1 (Monomial conjecture). *Let $(R, \mathfrak{m}, k)$ be a local ring of Krull dimension $d$. Let $x_1, \ldots, x_d$ be a system of parameters of $R$. Then for all natural numbers $n \in \mathbb{N}$;*

$$(x_1 \cdots x_d)^n \notin (x_1^{n+1}, \ldots, x_d^{n+1}).$$

PROOF. By standard reductions, we may assume that $R$ is a complete domain. Suppose by way of contradiction that $(x_1 \ldots x_d)^n \in (x_1^{n+1}, \ldots, x_d^{n+1})$. Then the map

$$\frac{R}{(x_1, \ldots, x_d)} \xrightarrow{\cdot (x_1 \cdots x_d)^n} \frac{R}{(x_1^{n+1}, \ldots, x_d^{n+1})}$$

is the 0-map. Let $M$ be a big Cohen-Macaulay module. Then we tensor the above map with $M$ and find that

$$\frac{M}{(x_1, \ldots, x_d)M} \xrightarrow{\cdot (x_1 \cdots x_d)^n} \frac{M}{(x_1^{n+1}, \ldots, x_d^{n+1})M}$$

is the 0-map. Equivalently, for every element $m \in M$

$$(x_1 \cdots x_d)^n m \in (x_1^{n+1}, x_2^{n+1}, \ldots, x_d^{n+1})M.$$

Therefore

$$M = (x_1^{n+1}, x_2^{n+1}, \ldots, x_d^{n+1})M :_M (x_1 \cdots x_d)^n.$$

By Exercise 2 we know that

$$(x_1^{n+1}, x_2^{n+1}, \ldots, x_d^{n+1})M :_M (x_1 \cdots x_d)^n = (x_1, \ldots, x_d)M,$$

a contradiction since $M \neq (x_1, \ldots, x_d)M$ by assumption. □

**3.2. Split maps and injective hulls.** Before establishing the Direct Summand Conjecture via the Monomial Conjecture, we first explore basic criteria to determine if a map $R \to M$ to a finitely generated $R$-module $M$ splits.

LEMMA 3.2. *Let $R$ be a Noetherian ring, $M$ a finitely generated $R$-module, and $R \to M$ an $R$-linear map. Then the following are equivalent:*

(a) *$R \to M$ splits;*
(b) *$\operatorname{Hom}_R(M, R) \to \operatorname{Hom}_R(R, R) \cong R$ is onto;*
(c) *$R_\mathfrak{m} \to M_\mathfrak{m}$ splits for every maximal ideal $\mathfrak{m}$;*
(d) *$R_P \to M_P$ splits for every prime ideal $P$.*

PROOF. The property of a map of finitely generated modules being surjective is a well-known property that may be checked locally. Moreover, if $M$ and $N$ are finitely generated $R$-modules, then $\operatorname{Hom}_R(M, N)_P \cong \operatorname{Hom}_{R_P}(M_P, N_P)$. Therefore it suffices to prove that $(a)$ and $(b)$ are equivalent. If $R \to M$ splits then there exists an $R$-linear map $M \to R$ such that $R \to M \to R$ is the identity map on $R$. We therefore apply $\operatorname{Hom}_R(-, R)$ and find that $\operatorname{Hom}_R(R, R) \to \operatorname{Hom}_R(M, R) \to \operatorname{Hom}_R(R, R)$ is the identity map on $\operatorname{Hom}_R(R, R)$. Therefore $\operatorname{Hom}_R(M, R) \to \operatorname{Hom}_R(R, R) \cong R$ is onto.

Conversely, suppose that $\operatorname{Hom}_R(M, R) \to \operatorname{Hom}_R(R, R)$ is onto. If $\varphi \in \operatorname{Hom}_R(M, R)$ is the pre-image of $1 \in R$ then $R \to M \xrightarrow{\varphi} R$ is the identity map on $R$. □

Exercise 6 (Easy). Let $R$ be a Noetherian ring and $M_1 \to M_2$ a map of $R$-modules. Show that if $M_1 \to M_2$ splits then for every $R$-module $N$ and every $i \in \mathbb{N}$ the map $\mathrm{Tor}_i^R(M_1, N) \to \mathrm{Tor}_i^R(M_2, N)$ is injective.

Solution. If $M_1 \to M_2$ splits then there exists an $R$-linear map $M_2 \to M_1$ so that the composition $M_1 \to M_2 \to M_1$ is the identity map on $M_1$. Therefore $\mathrm{Tor}_i^R(M_1, N) \to \mathrm{Tor}_i^R(M_2, N) \to \mathrm{Tor}_i^R(M_1, N)$ is the identity map on $\mathrm{Tor}_i^R(M_1, N)$. In conclusion, $\mathrm{Tor}_i^R(M_1, N) \to \mathrm{Tor}_i^R(M_2, N)$ is injective. □

Lemma 3.3. *Let $(R, \mathfrak{m}, k)$ be a Noetherian local ring, $M$ a finitely generated $R$-module, and $R \to M$ an $R$-linear map. Then the following are equivalent:*

(a) *$R \to M$ splits*

(b) *$\widehat{R} \to \widehat{M}$ splits where $\widehat{N}$ denotes the completion of an $R$-module $N$ with respect to the maximal ideal.*

Proof. The map $R \to M$ splits if and only if the induced map $\mathrm{Hom}_R(M, R) \to \mathrm{Hom}_R(R, R) \cong R$ is surjective. Equivalently,

$$\mathrm{Hom}_R(M, R) \otimes_R \widehat{R} \to \widehat{R}$$

is surjective. But $M$ is finitely generated and therefore

$$\mathrm{Hom}_R(M, R) \otimes_R \widehat{R} \cong \mathrm{Hom}_{\widehat{R}}(\widehat{M}, \widehat{R}),$$

equivalently $\widehat{R} \to \widehat{M}$ is split. □

Suppose $(R, \mathfrak{m}, k)$ is a Noetherian local ring, $M$ a finitely generated $R$-module, and $R \to M$ an $R$-linear map. Then the injective hull of the residue field of $R$ plays an important role in determining if the map $R \to M$ splits. The injective hull of the residue field of $R$, denoted by $E_R(k)$, is the module, unique up to isomorphism, satisfying the following properties:

(a) There is an essential inclusion $k \subseteq E_R(k)$, i.e., if $0 \neq N \subseteq E_R(k)$ is a nonzero submodule of $E_R(k)$, then $k \subseteq N$.

(b) $E_R(k)$ is an injective $R$-module.

If $I \subseteq R$ is an ideal and $E$ an injective $R$-module then $(0 :_E I)$ is an injective $R/I$-module. Therefore if $E_R(k)$ is the injective hull of the residue field of $k$ then $(0 :_{E_R(k)} I)$ is the injective hull of the residue field of $R/I$ by the defining properties of $E_{R/I}(k)$.

If $(R, \mathfrak{m}, k)$ is complete, then the Matlis duality functor $(-)^\vee := \mathrm{Hom}_R(-, E_R(k))$ provides an anti-equivalence between Noetherian and Artinian $R$-modules and satisfies the following properties, [**Mat89**, Theorem 18.6]:

(a) $M \to M^{\vee\vee}$ is an isomorphism whenever $M$ is either Artinian or Noetherian;

(b) If $M$ is Noetherian then $M^\vee$ is Artinian;

(c) If $M$ is Artinian then $M^\vee$ is Noetherian;

(d) A map of Artinian (or Noetherian) modules $N \to M$ is injective (respectively surjective) if and only if $M^\vee \to N^\vee$ is surjective (respectively injective).

Lemma 3.4. *Let $(R, \mathfrak{m}, k)$ be a Noetherian local ring, $M$ a finitely generated $R$-module, and $R \to M$ an $R$-linear map. Let $E_R(k)$ be the injective hull of the residue field of $R$. Then the following are equivalent:*

(a) *$R \to M$ splits;*
(b) *$E_R(k) \to M \otimes_R E_R(k)$ is injective.*

PROOF. By Lemma 3.3, the map $R \to M$ splits if and only if $\widehat{R} \to \widehat{M}$ splits. The injective hull $E_R(k)$ is a module over the completion $\widehat{R}$. In particular,

$$M \otimes_R E_R(k) \cong M \otimes_R (\widehat{R} \otimes_{\widehat{R}} E_R(k)) \cong \widehat{M} \otimes_{\widehat{R}} E_R(k).$$

Therefore we may replace $R$ and $M$ by $\widehat{R}$ and $\widehat{M}$ respectively and assume that $R$ is a complete local ring.

Let $E = E_R(k)$. If $R \to M$ splits then $E \cong R \otimes_R E \to M \otimes_R E$ is injective. Conversely, suppose that $E \to M \otimes_R E$ is injective. Equivalently, by Matlis duality,

$$\mathrm{Hom}_R(M \otimes_R E, E) \to \mathrm{Hom}_R(E, E)$$

is onto. There is an isomorphism $\mathrm{Hom}_R(E, E) \cong R$, therefore by Tensor-Hom adjunction

$$\mathrm{Hom}_R(M, R) \to R$$

is an onto map. Equivalently, $R \to M$ splits. $\square$

A 0-dimensional local ring $(R, \mathfrak{m}, k)$ is said to be *Gorenstein* if the *socle*, $(0 :_R \mathfrak{m})$, is a 1-dimensional $k$-vector space. A local Cohen-Macaulay ring $(R, \mathfrak{m}, k)$ of positive dimension $d$ is said to be Gorenstein if for some system of parameters $y_1, \dots, y_d$ of $R$, $R/(y_1, \dots, y_d)$ is a 0-dimensional Gorenstein ring.

THEOREM 3.5. *Let $(R, \mathfrak{m}, k)$ be a Cohen-Macaulay local ring of Krull dimension $d$. If $d = 0$ then $R$ is Gorenstein if and only if $R$ can be realized as an essential extension of $k$ if and only if $H^0_{\mathfrak{m}}(R) = R \cong E_R(k)$. If $d > 0$ and $x_1, \dots, x_d$ a system of parameters of $R$. Then $R$ is Gorenstein if and only if*

$$E_R(k) \cong H^d_{\mathfrak{m}}(R) \cong \varinjlim_t \left( \frac{R}{(x_1^t, \dots, x_d^t)} \xrightarrow{x_1 \cdots x_d} \frac{R}{(x_1^{t+1}, \dots, x_d^{t+1})} \right).$$

PROOF. Suppose that $R$ is 0-dimensional. Then $R$ is a complete local ring and $H^0_{\mathfrak{m}}(R) = R$ as some power of the maximal ideal is the 0-ideal.

Suppose $(R, \mathfrak{m}, k)$ is a 0-dimensional Gorenstein ring. We will show that $R$ is an essential extension of $k \cong (0 :_R \mathfrak{m}) \subseteq R$. Let $r \in R$ be a non-zero element. If $\mathfrak{m} \not\subseteq (0 :_R r)$ then there exists an element $s_1 \in \mathfrak{m}$ such that $s_1 r \neq 0$. Then $(0 :_R r) \subseteq (0 :_R s_1 r)$. If $\mathfrak{m} \not\subseteq (0 :_R s_1 r)$ then there exists an element $s_2 \in \mathfrak{m}$ such that $s_2 s_1 r \neq 0$. Let $r_1 = s_1 r$ and $r_2 = s_2 s_1 r$. Then $(0 :_R r) \subseteq (0 :_R r_1) \subseteq (0 :_R r_2)$. Some power of the maximal ideal is the 0-ideal and so the we can not find an infinite sequence of multiples of $r$ by elements in higher powers of the maximal ideal that produces a nonzero element. Therefore at some point we will find a multiple of $r$, $r_n$, such that $r_n \neq 0$ and $\mathfrak{m} \subseteq (0 :_R r_n)$. In particular, $0 \neq r_n \in (0 :_R \mathfrak{m})$ and $k \cong (0 :_R \mathfrak{m}) \subseteq R$ is indeed an essential extension.

Now suppose that the 0-dimensional ring $R$ can be realized as an essential extension of $k$. We first show that $k \subseteq E_R(k)$ extends to an inclusion $R \subseteq E_R(k)$. There exists a commutative diagram with horizontal arrow being an essential extension:

$$\begin{array}{ccc} k & \xrightarrow{\subseteq} & R \\ {\scriptstyle\subseteq}\big\downarrow & & \\ E_R(k) & & \end{array}$$

The injective hull $E_R(k)$ is an injective $R$-module and therefore there exists $\psi : R \to E_R(k)$ such that the following diagram commutes:

$$\begin{array}{ccc} k & \xrightarrow{\subseteq} & R \\ {\scriptstyle\subseteq}\downarrow & \swarrow{\scriptstyle\psi} & \\ E_R(k) & & \end{array}$$

If $r \in R$ then $k \subseteq R$ is an essential extension. Therefore there exists $a \in R$ such that $0 \neq ar \in k \subseteq E_R(k)$. Hence $0 \neq \psi(ar) = a\psi(r)$, $\psi(r) \neq 0$, and $\psi : R \to E_R(k)$ is an injective map. The Matlis dual of the injective map $\psi$ is an onto map

$$\operatorname{Hom}_R(E_R(k), E_R(k)) \cong \widehat{R} \cong R \xrightarrow{\operatorname{Hom}_R(\psi, E_R(k))} \operatorname{Hom}_R(R, E_R(k)) \cong E_R(k).$$

Therefore $E_R(k) \cong R/I$ for some ideal $I \subseteq R$. In particular, $(0 :_R E_R(k)) = I$. If $r \in R$ is a non-zero element then the multiplication map $R \xrightarrow{\cdot r} R$ is a non-zero map. The Matlis dual of the multiplication map is the multiplication map $E_R(k) \xrightarrow{\cdot r} E_R(k)$. The multiplication by $r$ on $E_R(k)$ must be a non-zero map, else we apply Matlis duality again and find that the original multiplication map by $r$ on $R$ is the zero map, a contradiction. Hence $(0 :_R E_R(k)) = 0$ and $I = 0$, i.e., $E_R(k) \cong R$.

Consider the socle $(0 :_R \mathfrak{m}) \subseteq R \cong E_R(k)$ and identify $k \cong R \cdot u \subseteq R$ as an essential extension. Then $u \in (0 :_R \mathfrak{m})$. Let $u' \in (0 :_R \mathfrak{m})$. Then $R \cdot u \subseteq R$ is essential and therefore there exists an element $\alpha \in R$ such that $\alpha u' = u$. The element $u'$ is annihilated by $\mathfrak{m}$. Therefore $\alpha$ is a unit of $R$ and hence $(0 :_R \mathfrak{m})$ is a 1-dimensional $k$-vector space, i.e., $R$ is Gorenstein.

Now suppose that $R$ has positive Krull dimension $d > 0$. Suppose first that $R$ is Gorenstein. If $x_1, \ldots, x_d$ is any system of parameters of $R$ then

$$H^d_{\mathfrak{m}}(R) \cong \varinjlim_t \left( \frac{R}{(x_1^t, \ldots, x_d^t)} \xrightarrow{x_1 \cdots x_d} \frac{R}{(x_1^{t+1}, \ldots, x_d^{t+1})} \right).$$

For any choice of system of parameters $x_1, \ldots, x_d$, the maps in the above direct limit system are injective by Exercise 2. We are assuming $R$ is Gorenstein. We therefore can choose our system of parameters $x_1, \ldots, x_d$ so that $R/(x_1, \ldots, x_d)$ is a 0-dimensional Gorenstein ring. The 0-dimensional quotient $R/(x_1, \ldots, x_d)$ is an essential extension of $k$, therefore the 0-dimensional quotients $R/(x_1^{t+1}, \ldots, x_d^{t+1})$, which contain $R/(x_1, \ldots, x_d)$ as a submodule, are essential extensions of $k$. Therefore each quotient $R/(x_1^{t+1}, \ldots, x_d^{t+1})$ is a 0-dimensional Gorenstein ring and the image of $u$ in $R/(x_1^{t+1}, \ldots, x_d^{t+1})$, specifically $x_1^t \cdots x_d^t u$, principally generates the socle of $R/(x_1^{t+1}, \ldots, x_d^{t+1})$.

For each $t \in \mathbb{N}$ let $I_t = (x_1^t, \ldots, x_d^t)$ and let $E = E_R(k)$. For each integer $t$ let $E_t = (0 :_E I_t)$. Then $E_t \cong R/I_t$ and $E = \bigcup E_t$. Therefore there exists injective maps $\beta_t : R/I_t \to R/I_{t+1}$ so that $E \cong \varinjlim \left( R/I_t \xrightarrow{\beta_t} R/I_{t+1} \right)$. We are tasked with showing that we can take $\beta_t$ to be multiplication by $x_1 \cdots x_d$.

The map $\beta_t : R/I_t \to R/I_{t+1}$ is multiplication by an element of $(I_{t+1} :_R I_t)$. Let $u_t = (x_1 \cdots x_d)^t u$ so that $u_t$ is a choice of socle generator of $R/I_t$. Since $\beta_t$ is injective we have that $\beta_t(u_t) = \varepsilon_{t+1} u_{t+1}$ for some unit $\varepsilon_{t+1}$. It is then simple to show that $\beta_t = \cdot\varepsilon_{t+1}x_1\cdots x_d$ and that

$$\varinjlim\left(R/I_t \xrightarrow{\cdot x_1\cdots x_d} R/I_{t+1}\right) \cong \varinjlim\left(R/I_t \xrightarrow{\beta_t} R/I_{t+1}\right)$$

as needed.

Now suppose that $R$ is a Cohen-Macaulay ring of dimension $d > 0$, $x_1, \ldots, x_d$ is a system of parameters, and

$$E_R(k) \cong H^d_{\mathfrak{m}}(R) \cong \varinjlim\left(\frac{R}{(x_1^t, \ldots, x_d^t)} \xrightarrow{\cdot x_1\cdots x_d} \frac{R}{(x_1^{t+1}, \ldots, x_d^{t+1})}\right).$$

The maps in the direct limit system are injective and $E_R(k)$ is an essential extension of the residue field $k$. Therefore $R/(x_1, \ldots, x_d)$ as a nonzero submodule of $H^d_{\mathfrak{m}}(R)$ is an essential extension of $k$. Therefore the 0-dimensional ring $R/(x_1, \ldots, x_d)$ is the injective hull of its residue field and therefore $R$ is Gorenstein. □

Let $(R, \mathfrak{m}, k)$ be a Cohen-Macaulay local ring. A *canonical module* of $R$ is a finitely generated maximal Cohen-Macaulay module $\omega_R$ defined by the property that $H^d_{\mathfrak{m}}(\omega_R) \cong E_R(k)$. If $R$ is Gorenstein, then $R$ serves as its own canonical module by Theorem 3.5. Conversely, if $R$ is Cohen-Macaulay and $R$ serves as a canonical module then $R$ is Gorenstein by Theorem 3.5.

EXERCISE 7. Let $(R, \mathfrak{m}, k)$ be a Cohen-Macaulay local ring. Suppose that $R$ admits a canonical ideal $J \subsetneq R$, i.e., $J$ is a proper ideal of $R$ and $J \cong \omega_R$ is a canonical module. Let $x_1 \in J$ be a nonzerodivisor of $R$ and $x_1, x_2, \ldots, x_d$ a system of parameters of $R$. For each integer $t$ let $I_t = (x_1^{t-1}J, x_2^t, \ldots, x_d^t)$. Show that

$$E_R(k) \cong \varinjlim\left(\frac{R}{I_t} \xrightarrow{\cdot x_1\cdots x_d} \frac{R}{I_{t+1}}\right).$$

SOLUTION. Consider the short exact sequence $0 \to J \to R \to R/J \to 0$. As $R$-modules, $J$ and $R$ are maximal Cohen-Macaulay $R$-modules. Examining the functorial long exact sequence of local cohomology modules with support in the maximal ideal, we find that $H^i_{\mathfrak{m}}(R/J) = 0$ for all $i \leq d-2$ and there is a short exact sequence

$$0 \to H^{d-1}_{\mathfrak{m}}(R/J) \to H^d_{\mathfrak{m}}(J) \to H^d_{\mathfrak{m}}(R) \to 0.$$

Therefore $R/J$ is a $d-1$-dimensional Cohen-Macaulay local ring. Let $x_2, \ldots, x_d \in R$ be a system of parameters of $R$. Then

$$H^{d-1}_{\mathfrak{m}}(R/J) \cong \varinjlim\left(\frac{R}{(J, x_2^t, \ldots, x_d^t)} \xrightarrow{\cdot x_2\cdots x_d} \frac{R}{(J, x_2^{t+1}, \ldots, x_d^{t+1})}\right)$$

and the maps in the direct limit system are injective as $R/J$ is Cohen-Macaulay. Therefore there are inclusions

$$\frac{R}{(J, x_2, \ldots, x_d)} \subseteq H^{d-1}_{\mathfrak{m}}(R/J) \subseteq H^d_{\mathfrak{m}}(J) \cong E_R(k).$$

As a consequence, the 0-dimensional ring $R/(J, x_2, \ldots, x_d)$ is an essential extension of $k$ and therefore a 0-dimensional Gorenstein ring. In particular, the quotient $R/J$ is a Gorenstein ring.

As modules, the ideals $x_1^{t-1}J$ are isomorphic to the canonical ideal $J$ and therefore also serve as canonical ideals. Therefore the quotients $R/I_t$ are 0-dimensional Gorenstein rings by the above. The solution to the exercise now continues as in the proof of Theorem 3.5. □

THEOREM 3.6 (Direct summand conjecture). *Let $A$ be a regular ring and $A \to R$ a finite extension. Then $A \to R$ splits as an $A$-linear map.*

PROOF. By Lemma 3.3 we may assume that $(A, \mathfrak{m}, k)$ is a complete regular local ring. If $A$ is 0-dimensional, then $A$ is a field and every non-zero element of $R$ is part of an $A$-basis. In particular, $1 \in R$ is part of an $A$-basis and there exists an $A$-linear map $R \to A$ sending the identity element of $R$ to the identity element of $A$. If $A$ has positive dimension then $A$ admits a system of parameters $x_1, \dots, x_d$ such that $A/(x_1, \dots, x_d)$ is a field. Therefore $A$ is Gorenstein. By Theorem 3.5,

$$E_A(k) \cong \varinjlim \left( \frac{A}{(x_1^t, \dots, x_d^t)A} \xrightarrow{\cdot x_1 \cdots x_d} \frac{A}{(x_1^{t+1}, \dots, x_d^{t+1})A} \right).$$

By Lemma 3.4, the map $A \to R$ splits if and only if

$$E_A(k) \to R \otimes_A E_A(k) \cong \varinjlim \left( \frac{R}{(x_1^t, \dots, x_d^t)R} \xrightarrow{\cdot x_1 \cdots x_d} \frac{R}{(x_1^{t+1}, \dots, x_d^{t+1})R} \right)$$

is injective. The element $[1 + (x_1, \dots, x_d)A]$ of $E_R(k)$ principally generates the socle $(0 :_{E_R(k)} \mathfrak{m})$. Therefore $A \to R$ does not split if and only if $[1 + (x_1, \dots, x_d)R]$ represents the 0-element of

$$\varinjlim \left( \frac{R}{(x_1^t, \dots, x_d^t)R} \xrightarrow{\cdot x_1 \cdots x_d} \frac{R}{(x_1^{t+1}, \dots, x_d^{t+1})R} \right).$$

Equivalently, there exists an integer $n$ so that $(x_1 \cdots x_d)^n + (x_1^{n+1}, \dots, x_d^{n+1})$ is the 0-element of $R/(x_1^{n+1}, \dots, x_d^{n+1})R$. This is to assert that

$$(x_1 \cdots x_d)^n \in (x_1^{n+1}, \dots, x_d^{n+1})R,$$

a containment that cannot occur by the Monomial Conjecture, Theorem 3.1. □

## 4. Vanishing maps of Tor conjecture and direct summands of regular rings are Cohen-Macaulay

THEOREM 4.1. *Let $A$ be a regular local ring and $A \to R$ a finite extension. Let $R \to S$ be a ring map to a regular ring $S$. Then for every $A$-module $M$ and for every integer $i \geq 1$;*

$$\operatorname{Tor}_i^A(R, M) \to \operatorname{Tor}_i^A(S, M)$$

*is the 0-map.*

PROOF. By a direct limit argument, we assume without loss of generality that $M$ is finitely generated. Vanishing of the map $\operatorname{Tor}_i^A(R, M) \to \operatorname{Tor}_i^A(S, M)$ is local on $S$. We can therefore localize at a prime ideal $P$ of $S$ and replace $A$ and $R$ be their respective localizations of the contracted prime $A \cap P$ of $A$. Even further, we can replace $A, R$, and $S$ by their respective completions with respect to the maximal ideal of $A$. The ring $R$ will then be a complete semi-local ring, i.e., $R$ is a semi-local ring and localization of $R$ at a maximal ideal is a complete local ring. The maximal ideals of $R$ are necessarily of common height, namely $\dim(A)$, as $A \to R$ is finite. The kernel of the map $R \to S$ contains a minimal prime $P \in \operatorname{Spec}(R)$. The ring $A$ is a regular local ring and $A \to R$ is a finite extension. Therefore $A \to R/P$ remains a finite extension and there are maps $A \to R \to R/P \to S$. Hence there are maps of Tor-modules

$$\operatorname{Tor}_i^A(R, M) \to \operatorname{Tor}_i^A(R/P, M) \to \operatorname{Tor}_i^A(S, M).$$

If the second map is the 0-map, then the composition of the maps is the 0-map. We may therefore assume that $R$ is a complete semi-local domain. By Theorem 1, for each of the maximal ideals $\mathfrak{m}$ of $R$ there are big Cohen-Macaulay algebras $R_{\mathfrak{m}} \to B_{R_{\mathfrak{m}}}$ and $S \to B_S$ so that the following diagram commutes:

$$\begin{array}{ccc} B_{R_{\mathfrak{m}}} & \longrightarrow & B_S \\ \uparrow & & \uparrow \\ R_{\mathfrak{m}} & \longrightarrow & S \\ \uparrow & & \uparrow \\ R & \longrightarrow & S \end{array}$$

The map $A \to R$ is finite, therefore $B_{R_{\mathfrak{m}}}$ is a big Cohen-Macaulay $A$-algebra. Hence $B_{R_{\mathfrak{m}}}$ is a flat $A$-module by Exercise 4 and $\operatorname{Tor}_i^A(B_{R_{\mathfrak{m}}}, M) = 0$ for all $i \geq 1$. The ring $S$ is regular and so $S \to B_S$ is a faithfully flat map of $A$-algebras.

We claim that $\operatorname{Tor}_i^A(S, M) \to \operatorname{Tor}_i^A(B_S, M)$ is injective for all $i \geq 0$. The module $M$ is a finitely generated $A$-module. We can surject a finitely generated free $A$-module onto $M$ and consider a short exact sequence of the following form:

$$0 \to \Omega_1 \to A^{\oplus n_0} \to M \to 0$$

Consider the induced commutative diagram Tor-modules derived from $S \to B_S$:

$$\begin{array}{ccccccccccc} 0 & \longrightarrow & \operatorname{Tor}_1^A(S, M) & \longrightarrow & S \otimes_A \Omega_1 & \longrightarrow & S^{\oplus n_0} & \longrightarrow & S \otimes_A M & \longrightarrow & 0 \\ & & \downarrow & & \downarrow & & \downarrow & & \downarrow & & \\ 0 & \longrightarrow & \operatorname{Tor}_1^A(B_S, M) & \longrightarrow & B_S \otimes_A \Omega_1 & \longrightarrow & B_S^{\oplus n_0} & \longrightarrow & B_S \otimes_A M & \longrightarrow & 0 \end{array}$$

The rows of the above commutative diagram are exact and the three right-most vertical arrows are injective by flatness of $S \to B_S$. A simple diagram chase verifies that the map of $\operatorname{Tor}_1^A$-modules is also injective.

As for the higher Tor-modules, the short exact sequence $0 \to \Omega_1 \to A^{\oplus n_0} \to M \to 0$ and the map $S \to B_S$ yield the following commutative diagrams whose horizontal arrows are isomorphisms for $i \geq 2$:

$$\begin{array}{ccc} \operatorname{Tor}_i^A(S, M) & \xrightarrow{\cong} & \operatorname{Tor}_{i-1}^A(S, \Omega_1) \\ \downarrow & & \downarrow \\ \operatorname{Tor}_i^A(B_S, M) & \xrightarrow{\cong} & \operatorname{Tor}_{i-1}^A(B_S, \Omega_1) \end{array}$$

If $i = 2$ then the induced map of $\operatorname{Tor}_2^A$-modules is injective by repeating the above argument with respect to $\Omega_1$ to show that the map $\operatorname{Tor}_1^A(S, \Omega_1) \to \operatorname{Tor}_1^A(B_S, \Omega_1)$ injective. By induction, $\operatorname{Tor}_i^A(S, M) \to \operatorname{Tor}_i^A(B_S, M)$ is injective for all $i \geq 0$.

In conclusion, for each $i \geq 1$ there are commutative diagrams

$$\begin{array}{ccc} 0 = \operatorname{Tor}_i^A(B_{R_{\mathfrak{m}}}, M) & \longrightarrow & \operatorname{Tor}_i^A(B_S, M) \\ \uparrow & & \uparrow {\scriptstyle 1\text{-}1} \\ \operatorname{Tor}_i^A(R_{\mathfrak{m}}, M) & \longrightarrow & \operatorname{Tor}_i^A(S, M) \end{array}$$

and the properties discussed above paired with a simple diagram chase imply $\mathrm{Tor}_i^A(R_\mathfrak{m}, M) \to \mathrm{Tor}_i^A(S, M)$ is the 0-map. Therefore $\mathrm{Tor}_i^A(R_\mathfrak{m}, M) \to \mathrm{Tor}_i^A(S, M)$ is the 0-map for every maximal ideal $\mathfrak{m}$ of $R$. This allows us to conclude $\mathrm{Tor}_i^A(R, M) \to \mathrm{Tor}_i^A(S, M)$ is the 0-map as desired. □

THEOREM 4.2. *Let $R \to S$ be a split map of Noetherian ring and suppose that $S$ is regular. Then $R$ is Cohen-Macaulay.*

PROOF. By standard reductions, we can assume that $R$ is a complete local domain and that $R \to S$ is an extension of rings such that $S$ is a regular domain. By Cohen's Structure Theorem the ring $R$ is a finite extension of a regular local ring $A$. Let $x_1, \ldots, x_d$ generate the maximal ideal of $A$. By Theorem 4.1, for each integer $i \geq 1$, the maps

$$\mathrm{Tor}_i^A(R, A/(x_1, \ldots, x_d)) \to \mathrm{Tor}_i^A(S, A/(x_1, \ldots, x_d))$$

are the 0-map for all $i \geq 1$. But, the above maps are injective since $R \to S$ splits, see Exercise 6. Therefore $\mathrm{Tor}_i^A(R, A/(x_1, \ldots, x_d)) \cong H_i(x_1, \ldots, x_d; R) = 0$ for all $i \geq 1$. Equivalently, $R$ is a Cohen-Macaulay ring. □

## 5. BCM-rational singularities

**5.1. Big Cohen-Macaulay rational singularities.** Recall that a local ring $(R, \mathfrak{m}, k)$ is called *$F$-rational* if every parameter ideal of $R$ is tightly closed. Suppose that $R$ is of Krull dimension $d$. Then the Frobenius map $F : R \to R$ admits a Frobenius action $F : H^d_\mathfrak{m}(R) \to H^d_\mathfrak{m}(R)$. More generally, if $R \to B$ is an algebra, the Frobenius map $F : B \to B$ induces a Frobenius map on $H^d_\mathfrak{m}(B) \to H^d_\mathfrak{m}(B)$. In particular, we make $H^d_\mathfrak{m}(R)$ an $R\{F\}$-module by letting $F$ act as the Frobenius. Recall that Smith proved that $F$-rationality is characterized by simplicity of $H^d_\mathfrak{m}(R)$ as an $R\{F\}$-module.

THEOREM 5.1 ([**Smi97**, Theorem 2.6]). *Let $(R, \mathfrak{m}, k)$ be a complete local domain of prime characteristic $p > 0$ and of Krull dimension $d$. Then $0^*_{H^d_\mathfrak{m}(R)}$ is the largest Frobenius stable submodule of $H^d_\mathfrak{m}(R)$ and $R$ is $F$-rational if and only if $R$ is Cohen-Macaulay and $0 = 0^*_{H^d_\mathfrak{m}(R)}$. Equivalently, $R$ is $F$-rational if and only if $R$ is Cohen-Macaulay and $H^d_\mathfrak{m}(R)$ is simple as an $R\{F\}$-module.*

THEOREM 5.2 ([**Smi94**, Theorem 5.1]). *Let $(R, \mathfrak{m}, k)$ be a complete local domain of prime characteristic $p > 0$ and Krull dimension $d$. Let $x_1, \ldots, x_d$ be a system of parameters. Then $(x_1, \ldots, x_d)^* = (x_1, \ldots, x_d)R^+ \cap R$. Consequently,*

$$0^*_{H^d_\mathfrak{m}(R)} \cong \varinjlim \frac{(x_1^t, \ldots, x_d^t)^*}{(x_1^t, \ldots, x_d^t)} = \ker(H^d_\mathfrak{m}(R) \to H^d_\mathfrak{m}(R^+)).$$

PROOF. We do not prove here that $(x_1, \ldots, x_d)^* = (x_1, \ldots, x_d)R^+ \cap R$ for each system of parameters. We only use this to point out the latter isomorphism.

Observe that the kernel of $R/(x_1^t, \ldots, x_d^t)R \to R^+/(x_1^t, \ldots, x_d^t)R^+$ is

$$(x_1^t, \ldots, x_d^t)R^+ \cap R/(x_1^t, \ldots, x_d^t) = (x_1^t, \ldots, x_d^t)^*/(x_1^t, \ldots, x_d^t)$$

by Theorem 5.1. Even further, the maps of the direct limit system

$$H^d_\mathfrak{m}(R^+) \cong \varinjlim \left(R^+/(x_1^t, \ldots, x_d^t)R^+\right)$$

are injective as $R^+$ is a big Cohen-Macaulay algebra, see Proposition 2.3. Therefore the kernel of $H^d_\mathfrak{m}(R) \to H^d_\mathfrak{m}(R^+)$ is $\varinjlim \frac{(x_1^t, \ldots, x_d^t)^*}{(x_1^t, \ldots, x_d^t)}$ as claimed. □

EXERCISE 8. Let $(R, \mathfrak{m}, k)$ be a complete local domain of prime characteristic $p > 0$ and suppose that $R \to B$ is an algebra. Show that $\ker(H^d_{\mathfrak{m}}(R) \to H^d_{\mathfrak{m}}(B)) \subseteq 0^*_{H^d_{\mathfrak{m}}(R)}$.

SOLUTION. By Theorem 5.1 it suffices to show that $\ker(H^d_{\mathfrak{m}}(R) \to H^d_{\mathfrak{m}}(B))$ is a Frobenius stable submodule. There exists a commutative diagram

$$\begin{array}{ccc} H^d_{\mathfrak{m}}(R) & \longrightarrow & H^d_{\mathfrak{m}}(B) \\ {\scriptstyle F}\uparrow & & {\scriptstyle F}\uparrow \\ H^d_{\mathfrak{m}}(R) & \longrightarrow & H^d_{\mathfrak{m}}(B) \end{array}$$

If $\eta$ belongs to lower left copy of $H^d_{\mathfrak{m}}(R)$ and is mapped to 0 in $H^d_{\mathfrak{m}}(B)$ then clearly $F(\eta)$ in the upper left copy of $H^d_{\mathfrak{m}}(R)$ will be mapped to 0 in $H^d_{\mathfrak{m}}(B)$ by a diagram chase. Hence the kernel is indeed a Frobenius stable submodule. □

EXERCISE 9. Let $(R, \mathfrak{m}, k)$ be a complete local domain and $R \to B$ a big Cohen-Macaulay algebra. Show that $H^d_{\mathfrak{m}}(R) \to H^d_{\mathfrak{m}}(B)$ is nonzero.

SOLUTION. Let $x_1, \ldots, x_d$ be a system of parameters so that $H^d_{\mathfrak{m}}(R) \cong \varinjlim (R/(x_1^t, \ldots, x_d^t))$. Then the image of $[1 + (x_1, \ldots, x_d)]$ in $H^d_{\mathfrak{m}}(B)$ will be nonzero. Else, $[1+(x_1, \ldots, x_d)B]$ represents the 0-element of $H^d_{\mathfrak{m}}(B) \cong \varinjlim (B/(x_1^t, \ldots, x_d^t)B)$. The module $B$ is big Cohen-Macaulay, therefore the maps of the direct limit system are injective and so $1 \in (x_1, \ldots, x_d)B$, a contradiction as $(x_1, \ldots, x_d)B \subsetneq B$. □

THEOREM 5.3. *Let $(R, \mathfrak{m}, k)$ be a complete local domain of prime characteristic $p > 0$ and of Krull dimension $d$. Then the following are equivalent:*

(a) *$R$ is $F$-rational;*
(b) *$R$ is Cohen-Macaulay and for each big Cohen-Macaulay algebra $R \to B$, the induced map of local cohomology modules $H^d_{\mathfrak{m}}(R) \to H^d_{\mathfrak{m}}(B)$ is injective.*

PROOF. Suppose that $R$ is $F$-rational, then $R$ is Cohen-Macaulay, [**HH94**, Theorem 4.2(c)]. Let $R \to B$ be a big Cohen-Macaulay algebra. Then $\ker(H^d_{\mathfrak{m}}(R) \to H^d_{\mathfrak{m}}(B))$ is an $F$-stable submodule of $H^d_{\mathfrak{m}}(R)$ by Exercise 8 and is therefore either 0 or $H^d_{\mathfrak{m}}(R)$. By Exercise 9 the kernel is 0.

Conversely, suppose that $R$ is Cohen-Macaulay and that for every big Cohen-Macaulay algebra $R \to B$ the induced map of local cohomology modules $H^d_{\mathfrak{m}}(R) \to H^d_{\mathfrak{m}}(B)$ is injective. Let $B = R^+$. Then $H^d_{\mathfrak{m}}(R) \to H^d_{\mathfrak{m}}(R^+)$ is injective by assumption and therefore by Theorem 5.2, $0^*_{H^d_{\mathfrak{m}}(R)} = 0$. We are assuming $R$ is Cohen-Macaulay and so $R$ is $F$-rational by Theorem 5.1. □

Theorem 5.3 inspires the following definition of BCM-rational singularities.

DEFINITION 5.4 ([**MS21**]). Let $(R, \mathfrak{m}, k)$ be a complete Cohen-Macaulay local domain of Krull dimension $d$ and $R \to B$ a big Cohen-Macaulay $R$-algebra. We say that $R$ is $\mathrm{BCM}_B$-rational if $H^d_{\mathfrak{m}}(R) \to H^d_{\mathfrak{m}}(B)$ is injective. We say that $R$ is BCM-rational if $R$ is $\mathrm{BCM}_B$-rational for all big Cohen-Macaulay algebras $R \to B$.

EXERCISE 10 (BCM-rational implies normal). Let $(R, \mathfrak{m}, k)$ be a complete local domain and suppose that $R$ is BCM-rational.

(a) Let $R \to R^N$ be the normalization of $R$. Show that $H^d_{\mathfrak{m}}(R) \to H^d_{\mathfrak{m}}(R^N)$ is injective.

(b) Suppose that $x, y \in R$, $y/x \in R^N$, complete $x$ to a full system of parameters $x, x_2, \ldots, x_d$ of $R$, and identify $H^d_{\mathfrak{m}}(R) \cong \varinjlim (R/(x_1^t, \ldots, x_d^t))$. Use injectivity of $H^d_{\mathfrak{m}}(R) \to H^d_{\mathfrak{m}}(R^N)$ to conclude that $y \in \bigcap_{t \in \mathbb{N}}(x, x_2^t, \ldots, x_d^t) = (x)$. Therefore $y/x \in R$ and so $R = R^N$.

SOLUTION. (a) The normalization of a complete local domain is again a complete local domain. The map $R \to R^N$ is finite and so any big Cohen-Macaulay $R^N$-algebra is a big Cohen-Macaulay $R$-algebra. Let $R \to R^N \to C$ be such an algebra. By assumption the composition of $H^d_{\mathfrak{m}}(R) \to H^d_{\mathfrak{m}}(R^N) \to H^d_{\mathfrak{m}}(C)$ is injective and therefore $H^d_{\mathfrak{m}}(R) \to H^d_{\mathfrak{m}}(R^N)$ is injective.

(b) The element $y/x \in R^N$ therefore $y \in xR^N$. Hence the element $[y + (x, x_2^t, \ldots, x_d^t)] \in H^d_{\mathfrak{m}}(R)$ is mapped to 0 in $H^d_{\mathfrak{m}}(R^N)$. Therefore $[y + (x, x_2^t, \ldots, x_d^t)]$ represents the 0-element of $H^d_{\mathfrak{m}}(R)$ by part (a) and so $y \in (x, x_2^t, \ldots, x_d^t)$ for every $t \in \mathbb{N}$. □

In prime characteristic, the notion of $F$-rationality *deforms*: That is if $x \in R$ is a nonzerodivisor such that $R/xR$ is $F$-rational then $R$ is $F$-rational, [**HH94**, Theorem 4.2(h)]. The notion of BCM-rational singularities also deform.

EXERCISE 11 (Deformation of BCM-rational singularities). Let $(R, \mathfrak{m}, k)$ be a complete local Cohen-Macaulay domain and $x \in R$ a nonzero divisor.

(a) Let $R \to B$ be a big Cohen-Macaulay algebra. Observe that $R/xR \to B/xB$ is a big Cohen-Macaulay $R/xR$-algebra and show that there exists a commutative diagram of short exact sequences of local cohomology modules

$$\begin{array}{ccccccccc} 0 & \longrightarrow & H^{d-1}_{\mathfrak{m}}(B/xB) & \longrightarrow & H^d_{\mathfrak{m}}(B) & \xrightarrow{\cdot x} & H^d_{\mathfrak{m}}(B) & \longrightarrow & 0 \\ & & \uparrow & & \uparrow & & \uparrow & & \\ 0 & \longrightarrow & H^{d-1}_{\mathfrak{m}}(R/xR) & \longrightarrow & H^d_{\mathfrak{m}}(R) & \xrightarrow{\cdot x} & H^d_{\mathfrak{m}}(R) & \longrightarrow & 0 \end{array}$$

(b) Show that if $R/xR$ is BCM-rational then so to is $R$.

SOLUTION. (a) There is a commutative diagram of short exact sequences

$$\begin{array}{ccccccccc} 0 & \longrightarrow & B & \xrightarrow{\cdot x} & B & \longrightarrow & B/xB & \longrightarrow & 0 \\ & & \uparrow & & \uparrow & & \uparrow & & \\ 0 & \longrightarrow & R & \xrightarrow{\cdot x} & R & \longrightarrow & R/xR & \longrightarrow & 0 \end{array}$$

The desired diagram is obtained by examining the induced diagram of local cohomology modules with support in $\mathfrak{m}$.

(b) Let $R \to B$ be a big Cohen-Macaulay $R$-algebra. We must show that $H^d_{\mathfrak{m}}(R) \to H^d_{\mathfrak{m}}(B)$ is injective. By part (a) there is a commutative diagram

$$\begin{array}{ccccccccc} 0 & \longrightarrow & H^{d-1}_{\mathfrak{m}}(B/xB) & \longrightarrow & H^d_{\mathfrak{m}}(B) & \xrightarrow{\cdot x} & H^d_{\mathfrak{m}}(B) & \longrightarrow & 0 \\ & & f_1 \uparrow & & f_2 \uparrow & & \uparrow & & \\ 0 & \longrightarrow & H^{d-1}_{\mathfrak{m}}(R/xR) & \longrightarrow & H^d_{\mathfrak{m}}(R) & \xrightarrow{\cdot x} & H^d_{\mathfrak{m}}(R) & \longrightarrow & 0 \end{array}$$

We are assuming that $R/xR$ is BCM-rational, therefore $f_1$ is an injective map. Suppose that $\eta \in H^d_{\mathfrak{m}}(R)$ and belongs to the kernel of $f_2$. Every element of $H^d_{\mathfrak{m}}(R)$ is annihilated by a power of $x$. Therefore, if $\eta$ is nonzero, we can replace $\eta$ by $x^{t-1}\eta \neq 0$ for some $t \geq 1$ and assume that $x\eta = 0$, i.e., $\eta$ is the image of an element of belonging to $H^{d-1}_{\mathfrak{m}}(R/xR)$. A simple diagram chase verifies that $\eta = 0$, a contradiction. □

## Acknowledgments

The authors express their gratitude to the referee for reviewing these notes and offering numerous suggestions that have enhanced the content and presentation of the materials.

## References

[And20] Yves André, *Weak functoriality of Cohen-Macaulay algebras*, J. Amer. Math. Soc. **33** (2020), no. 2, 363–380, DOI 10.1090/jams/937. MR4073864

[BH93] Winfried Bruns and Jürgen Herzog, *Cohen-Macaulay rings*, Cambridge Studies in Advanced Mathematics, vol. 39, Cambridge University Press, Cambridge, 1993. MR1251956

[Bha20] Bhargav Bhatt, *Cohen-Macaulayness of absolute integral closures*, arXiv e-prints, arXiv:2008.08070, August 2020.

[BM10] Holger Brenner and Paul Monsky, *Tight closure does not commute with localization*, Ann. of Math. (2) **171** (2010), no. 1, 571–588, DOI 10.4007/annals.2010.171.571. MR2630050

[BMP+20] Bhargav Bhatt, Linquan Ma, Zsolt Patakfalvi, Karl Schwede, Kevin Tucker, Joe Waldron, and Jakub Witaszek, *Globally +-regular varieties and the minimal model program for threefolds in mixed characteristic*, Publ. Math. Inst. Hautes Études Sci. **138** (2023), 69–227, DOI 10.1007/s10240-023-00140-8. MR4666931

[CEMS18] Alberto Chiecchio, Florian Enescu, Lance Edward Miller, and Karl Schwede, *Test ideals in rings with finitely generated anti-canonical algebras*, J. Inst. Math. Jussieu **17** (2018), no. 1, 171–206, DOI 10.1017/S1474748015000456. MR3742559

[HH92] Melvin Hochster and Craig Huneke, *Infinite integral extensions and big Cohen-Macaulay algebras*, Ann. of Math. (2) **135** (1992), no. 1, 53–89, DOI 10.2307/2946563. MR1147957

[HH94] Melvin Hochster and Craig Huneke, *F-regularity, test elements, and smooth base change*, Trans. Amer. Math. Soc. **346** (1994), no. 1, 1–62, DOI 10.2307/2154942. MR1273534

[HH95] Melvin Hochster and Craig Huneke, *Applications of the existence of big Cohen-Macaulay algebras*, Adv. Math. **113** (1995), no. 1, 45–117, DOI 10.1006/aima.1995.1035. MR1332808

[HL07] Craig Huneke and Gennady Lyubeznik, *Absolute integral closure in positive characteristic*, Adv. Math. **210** (2007), no. 2, 498–504, DOI 10.1016/j.aim.2006.07.001. MR2303230

[Hoc75] Melvin Hochster, *Topics in the homological theory of modules over commutative rings*, Conference Board of the Mathematical Sciences Regional Conference Series in Mathematics, No. 24, Published for the Conference Board of the Mathematical Sciences, Washington, DC; by American Mathematical Society, Providence, RI, 1975. Expository lectures from the CBMS Regional Conference held at the University of Nebraska, Lincoln, Neb., June 24–28, 1974. MR371879

[Hoc77] Melvin Hochster, *Cyclic purity versus purity in excellent Noetherian rings*, Trans. Amer. Math. Soc. **231** (1977), no. 2, 463–488, DOI 10.2307/1997914. MR463152

[Mat89] Hideyuki Matsumura, *Commutative ring theory*, 2nd ed., Cambridge Studies in Advanced Mathematics, vol. 8, Cambridge University Press, Cambridge, 1989. Translated from the Japanese by M. Reid. MR1011461

[MS21] Linquan Ma and Karl Schwede, *Singularities in mixed characteristic via perfectoid big Cohen-Macaulay algebras*, Duke Math. J. **170** (2021), no. 13, 2815–2890, DOI 10.1215/00127094-2020-0082. MR4312190

[Sch16] Peter Scholze, *Perfectoid Shimura varieties*, Jpn. J. Math. **11** (2016), no. 1, 15–32, DOI 10.1007/s11537-016-1484-6. MR3510678

[Smi94] K. E. Smith, *Tight closure of parameter ideals*, Invent. Math. **115** (1994), no. 1, 41–60, DOI 10.1007/BF01231753. MR1248078

[Smi97] Karen E. Smith, *F-rational rings have rational singularities*, Amer. J. Math. **119** (1997), no. 1, 159–180. MR1428062

Department of Mathematics, University of Alabama, Tuscaloosa, Alabama 35487
*Email address*: tmpolstra@ua.edu

Indian Institute of Science Education and Research, Thiruvananthapuram, India

Contemporary Mathematics
Volume **831**, 2025
https://doi.org/10.1090/conm/831/16634

# Uniform Artin Rees results

Irena Swanson, with exercise solutions by Kriti Goel

The Artin–Rees Lemma is a very important result, and perhaps we should correspondingly call it a theorem, but for historical (and perhaps affectionate) reasons it remains a lemma. In these lectures we trace its history, its uses, and focus mostly on the uniform version, as developed by Craig Huneke in [**13**].

## 1. The classical Artin–Rees Lemma

Its classical version is as follows:

LEMMA 1.1. *Let $R$ be a Noetherian ring, $I$ an ideal in $R$ and $N \subseteq M$ finitely generated $R$-modules. Then there exists an integer $k$ such that for all integers $n \geq k$,*

$$I^n M \cap N = I^{n-k}(I^k M \cap N).$$

It first appeared in print in David Rees's 1956 paper [**20**] as Lemma 1 for the case $M = R$. Rees needed this lemma for a new proof of the Krull Principal Ideal Theorem. According to the footnote on page 145 in David Eisenbud's book [**6**], the name Artin–Rees Lemma seems to have been coined by Masayoshi Nagata, because Emil Artin lectured on the general case of the lemma in 1955. Hideyuki Matsumura's book [**16**] has on page 59 that the lemma was proved independently by Artin and Rees.

Artin–Rees Lemma has been applied in many contexts, including in the following:

(1) To compare $I$-adic topologies on submodules.
(2) To prove the Krull Intersection Theorem.
(3) To prove the Krull Principal Ideal Theorem.
(4) To prove inclusion properties of completions.
(5) To establish the Hilbert-Samuel function/polynomials and hence multiplicities, especially additivity of multiplicity on short exact sequences.
(6) To prove the existence of superficial elements and sequences.
(7) To prove Rees-Izumi's Theorem for comparing valuations.
(8) To determine finite generation of ideal transform rings.

Most applications do not use the full power of the lemma, but only use that there exists an integer $k$, depending on $I$, $M$ and $N$, such that for all integers $n \geq k$,

$$I^n M \cap N \subseteq I^{n-k} N.$$

This form is usually called the **weak Artin–Rees Lemma**, and the version in Lemma 1.1 is called the **strong Artin–Rees Lemma**, for distinguishing between the two.

It is not true that we can choose an integer $k$ that works for all $M$, $N$, either for the weak or the strong Artin–Rees Lemma. Namely, if for some ideal $I$, $I^n M \cap N \subseteq I^{n-k} N$ for all finitely generated $R$-modules $N \subseteq M$ and all integers $n \geq k$, then in particular $I^{\max\{m,n\}} = I^n R \cap I^m \subseteq I^{n-k+m}$, which cannot be when both $n$ and $m$ are strictly greater than $k$ and $I$ is not nilpotent.

Thus any uniform statements of the Artin–Rees Lemma of necessity restrict $M$ and $N$. They do allow $I$ to vary, as we see in the next section.

## 2. First statements of the uniform Artin–Rees Lemma

A first uniform Artin–Rees Lemma statement appeared in 1979 in a strong form as a problem on page 159 of the paper [**7**] of David Eisenbud and Mel Hochster; they proposed such a result to lead to yet another proof of a sharpening of Zariski's Main Lemma on holomorphic functions from paper [**25**]. (This paper of Zariski gives another example in which the Main Lemma as opposed to a Main Theorem seems to be playing a large role.) Eisenbud and Hochster asked the question in case the ring is affine and $I$ varies only over the maximal ideals.

The first proof of a uniform Artin–Rees result, in a weak form, is due to Liam O'Carroll in his 1987 paper [**17**], using a prime filtration of $M/N$ to reduce the problem to submodules being prime ideals of the ring $R$, and then using much machinery to prove the weak uniform Artin–Rees lemma in that case. Subsequently, Andrew Duncan and Liam O'Carroll proved affirmatively the Eisenbud–Hochster's strong uniform Artin–Rees problem in their 1989 paper [**5**]:

THEOREM 2.1 (Duncan, O'Carroll [**5**]). *Let $R$ be an excellent (or more generally, a J-2) ring. Then for any finitely generated $R$-modules $N \subseteq M$ there exists a positive integer $k$ such that for all positive integers $n \geq k$ and all maximal ideals $\mathfrak{m}$ of $R$,*

$$\mathfrak{m}^n M \cap N = \mathfrak{m}^{n-k}(\mathfrak{m}^k M \cap N).$$

O'Carroll proved in [**18**] that for any Noetherian ring $R$ and any finitely generated $R$-modules $N \subseteq M$ there exists a positive integer $k$ such that for all integers $n \geq k$ and for all principal ideals $I$,

$$I^n M \cap N = I^{n-k}(I^k M \cap N).$$

In the analytic context, with $I$ varying over maximal ideals, the first weak versions of the uniform Artin–Rees-type statement were given by Abdelhafed El Khadiri and Jean-Claude Tougeron in their 1985 paper [**8**], and by Edward Bierstone and Pierre Milman in their 1987 paper [**2**, Theorem 7.4].

## 3. Huneke's version of the uniform Artin–Rees Lemma

The following theorem of Huneke's is the focus of these lectures:

THEOREM 3.1. *[Uniform Artin–Rees Lemma,* [**13**, Theorem 4.12]*] Let $R$ be a Noetherian ring. Let $N \subseteq M$ be two finitely generated $R$-modules. If $R$ satisfies at least one of the conditions below, then there exists an integer $k$ such that for all ideals $I$ of $R$, and for all integers $n > k$*

$$I^n M \cap N \subseteq I^{n-k} N.$$

(i) *$R$ is essentially of finite type over a Noetherian local ring.*
(ii) *$R$ is a ring of characteristic $p$, and $R$ is module-finite over $R^p$.*
(iii) *$R$ is essentially of finite type over $\mathbb{Z}$.*

With the interpretation that any non-positive power of an ideal equals the whole ring, in the sequel we omit specifying $n > k$.

The rings in (ii) and (iii) are excellent (in a technical sense not defined here). The proof of (i) will proceed via many reductions, including a reduction to an excellent ring.

**Step 1: Reduction to faithfully flat extensions.**

Being able to extend the ring to a better, fuller ring, often gives access to more machinery. We will use this in the sequel.

LEMMA 3.2. *Let $R$ be a Noetherian ring and $S$ a faithfully flat extension. Let $N \subseteq M$ be finitely generated $R$ modules, and $k$ a positive integer such that for all integers $n$, $I^n MS \cap NS \subseteq I^{n-k} NS$. Then for all integers $n$, $I^n M \cap N \subseteq I^{n-k} N$.*

PROOF. This proof is a straightforward exercise. □

**Step 2: Reduction to the case M = R, N is a prime ideal.**

The reductions in this part are the content of [**13**, Proposition 2.2].

LEMMA 3.3. *Let $R$ be a Noetherian ring and $N \subseteq M$ finitely generated $R$-modules. Then there exist modules $L_0 = N \subseteq L_1 \subseteq L_2 \subseteq \cdots \subseteq L_s = M$ such that for all $i = 1, \ldots, s$, $L_i/L_{i-1}$ is isomorphic to $R/P_i$ for some prime ideal $P_i$ in $R$.*

PROOF. Use the Noetherian property. First pick an associated prime $P_1$ of $M/N$. Thus $P_1 = N :_R x$ for some $x \in M$. Set $L_1 = N + xR$. Then $L_1/N \cong R/(N :_R x) = R/P_1$. If $L_1 = M$, we are done, otherwise by Noetherian induction there exists a filtration for $L_1 \subseteq M$. □

LEMMA 3.4. *Let $R$ be a ring and $N \subseteq L \subseteq M$ be inclusions of $R$-modules. Suppose that there exist positive integers $k, l$ such that for all ideals $I$ and all integers $n$,*

$$I^n M \cap L \subseteq I^{n-k} L \quad \text{and} \quad I^n L \cap N \subseteq I^{n-l} N.$$

*Then for all positive integers $n$,*

$$I^n M \cap N \subseteq I^{n-k-l} N.$$

PROOF. This is a very easy exercise. □

LEMMA 3.5. *Let $R$ be a ring, $N \subseteq M$ be an inclusion of $R$-modules, and let $P$ be an ideal in $R$ such that $M/N \cong R/P$. Suppose that there exists a positive integer $k$ such that for all ideals $I$ and all integers $n \geq k$,*

$$I^n R \cap P \subseteq I^{n-k} P.$$

*Then for all ideals $I$ and all integers $n \geq k$,*

$$I^n M \cap N \subseteq I^{n-k} N.$$

PROOF. (This proof is as in [**13**, Proposition 2.2] or as in [**17**, Theorem].) We can write $M$ as $N + xR$ with $N :_R x = P$. First let $y \in I^n xR \cap N$. Write $y = zx$ for some $z \in I^n$. Then $z \in N :_R x = P$, and so $z \in I^n \cap P \subseteq I^{n-k} P$. It follows that $y \in I^{n-k} Px \subseteq I^{n-k} N$. Thus

$$I^n M \cap N = I^n (N + xR) \cap N = I^n N + I^n xR \cap N \subseteq I^{n-k} N. \qquad \square$$

> It thus suffices to prove that for every prime ideal $P$ in relevant $R$ there exists a positive integer $k$ such that for all ideals $I$ and all integers $n$,
> $$I^n \cap P \subseteq I^{n-k}P.$$

**Step 3: Reduction to using only ideals primary to maximal ideals.** This reduction is the content of [**13**, Proposition 2.3].

LEMMA 3.6. *Let $R$ be a Noetherian ring, let $N \subseteq M$ be $R$-modules, and let $k$ be a positive integer such that for all zero-dimensional primary ideals $I$ and for all integers $n \geq k$, $I^nM \cap N \subseteq I^{n-k}N$. Then for all ideals $I$ and for all integers $n \geq k$, $I^nM \cap N \subseteq I^{n-k}N$.*

PROOF. Let $I$ be an ideal in $R$ and $\mathfrak{m}$ a maximal ideal. Then for all positive integers $l$, $I + \mathfrak{m}^l$ is either $\mathfrak{m}$-primary or it is the unit ring. In either case, by assumption or trivially, $I^nM \cap N \subseteq (I+\mathfrak{m}^l)^nM \cap N \subseteq (I+\mathfrak{m}^l)^{n-k}N \subseteq I^{n-k}N + \mathfrak{m}^lN$. Thus
$$I^nM \cap N \subseteq \bigcap_{l,\mathfrak{m}} (I^{n-k}N + \mathfrak{m}^lN) = I^{n-k}N. \qquad \square$$

> It thus suffices to prove that for every prime ideal $P$ in relevant $R$ there exists a positive integer $k$ such that for all zero-dimensional primary ideals $I$ and all integers $n$,
> $$I^n \cap P \subseteq I^{n-k}P.$$

LEMMA 3.7. *Let $R$ be a Noetherian ring, $P$ a prime ideal, $\mathfrak{m}$ a maximal ideal, and $I$ an $\mathfrak{m}$-primary ideal. Then for any integers $n, k$,*
$$I^n \cap P \subseteq I^{n-k}P \textit{ if and only if } I^nR_{\mathfrak{m}} \cap PR_{\mathfrak{m}} \subseteq I^{n-k}PR_{\mathfrak{m}}.$$

PROOF. Certainly the global inclusion implies the local inclusion. Now suppose that $I^nR_{\mathfrak{m}} \cap PR_{\mathfrak{m}} \subseteq I^{n-k}PR_{\mathfrak{m}}$. If $\mathfrak{n}$ is any maximal ideal of $R$ other than $\mathfrak{m}$, then $IR_{\mathfrak{n}} = R_{\mathfrak{n}}$, so the localized equality at $\mathfrak{n}$ holds trivially. Thus $I^n \cap P \subseteq I^{n-k}P$ after localization at any maximal ideal of $R$, hence it holds globally. $\square$

> It thus suffices to prove that for every prime ideal $P$ in a Noetherian ring $R$ there exists a positive integer $k$ such that for all maximal ideals $\mathfrak{m}$ (containing $P$), all $\mathfrak{m}$-primary ideals $I$, and all integers $n$,
> $$I^nR_{\mathfrak{m}} \cap PR_{\mathfrak{m}} \subseteq I^{n-k}PR_{\mathfrak{m}}.$$

Let $X$ be an indeterminate over $R$. Then $R[X]_{\mathfrak{m}R[X]}$ is a faithfully flat extension of $R_{\mathfrak{m}}$, and by Lemma 3.7, it suffices to prove the uniform Artin–Rees Lemma in this extension. The advantage of this new ring is that its residue field is infinite.

> It thus suffices to prove that for every prime ideal $P$ in a Noetherian ring $R$ there exists a positive integer $k$ such that for all prime ideals $\mathfrak{m}$ with an infinite residue field, for all $\mathfrak{m}$-primary ideals $I$, and for all integers $n$,
> $$I^nR_{\mathfrak{m}} \cap PR_{\mathfrak{m}} \subseteq I^{n-k}PR_{\mathfrak{m}}.$$

**Step 4: Reduction to minimal reductions, or at least an attempt at it.**

It is sometimes the case that ideals generated by a system of parameters are easier to handle. In this step we try to replace an arbitrary $\mathfrak{m}$-primary ideal by a reduction; in the next step we will use this in case of minimal reductions in Lemma 5.3.

LEMMA 3.8 (This is Huneke [**13**, Lemma 3.1]). *Let $I$, $J$ and $P$ be ideals in a ring $R$ with $J \subseteq I$, and let $k, h$ be non-negative integers such that for all $n$,*

(i) $J^n \cap P \subseteq J^{n-k}P$, *and*
(ii) $I^n \subseteq J^{n-h} + P$.

*Then $I^n \cap P \subseteq I^{n-k-h-1}P$ for all $n$.*

PROOF. We first prove that $I^n \subseteq J^{n-h} + I^{n-k-h-1}P$ for all $n$. This is assumption (ii) if $n \leq k+h+1$. Suppose that we know the conclusion for some $n \geq k+h+1$. Then by multiplying through by $I$ and by using assumption (ii) we have that

$$\begin{aligned} I^{n+1} &\subseteq \left(IJ^{n-h} + I^{n-k-h}P\right) \cap \left(J^{n+1-h} + P\right) \\ &= IJ^{n-h} \cap \left(J^{n+1-h} + P\right) + I^{n-k-h}P \\ &\subseteq J^{n-h} \cap \left(J^{n+1-h} + P\right) + I^{n-k-h}P \\ &\subseteq J^{n-h} \cap P + J^{n+1-h} + I^{n-k-h}P \\ &\subseteq J^{n-h-k}P + J^{n+1-h} + I^{n-k-h}P \\ &= J^{n+1-h} + I^{n-k-h}P. \end{aligned}$$

This proves the claim by induction. It follows that

$$\begin{aligned} I^n \cap P &\subseteq (J^{n-h} + I^{n-k-h-1}P) \cap P \\ &= J^{n-h} \cap P + I^{n-k-h-1}P \\ &\subseteq J^{n-h-k-1}P + I^{n-k-h-1}P \\ &= I^{n-k-h-1}P. \end{aligned}$$

□

Condition (ii) in Lemma 3.8 says that $J(R/P)$ is a reduction of $I(R/P)$. So the point of this lemma is to use the Artin–Rees Lemma established for a reduction to prove it for the original ideal.

One weakness of the lemma above is that the integer $h$ in (ii) depends on $I$ and on $J$, which is not promising if we would like a uniform $h$. We handle this obstacle by making the following wish-definition:

DEFINITION 3.9 (This is [**13**, Definition 2.8]). Let $R$ be a Noetherian ring. For any positive integer $t$ set

$$T_t(R) = \bigcap_{n,I} (I^{n-t} : \overline{I^n}) \quad \text{and} \quad T(R) = \bigcup_t T_t(R).$$

Suppose that we can prove that $T(R/P)$ is not the zero ideal for $R/P$ in Lemma 3.8. Then there exists a non-zero element $c$ in $T_t(R/P)$ for some positive integer $t$. This means that for all positive integers $n$, for all $\mathfrak{m}$-primary ideals $I$, and for any ideal $J \subseteq I$ that is a reduction of $I$ modulo $P$,

$$cI^n(R/P) \subseteq c\,\overline{(I(R/P))^n} = c\,\overline{(J(R/P))^n} \subseteq J^{n-t}(R/P).$$

Thus $cI^n \subseteq J^{n-t}+P$. If we can take $c$ to be 1, then we get uniform (ii) in Lemma 3.8 for all ideals $I$ and $J$ there. In general, more effort is needed; keep reading.

## 4. When is T(R) non-zero

We have made some reductions towards the proof of the weak version of the Uniform Artin–Rees Lemma, but we have not yet completed it. We need some very important ingredients from integral and tight closure.

For $T(R)$ to contain a nonzero-divisor, it is necessary that the ring be reduced, because the integral closure of an ideal always contains the nilradical, but high powers of an ideal do not.

THEOREM 4.1 (This is [**13**, Proposition 4.7]). *Let $R$ be a reduced Noetherian ring and of characteristic $p$. Then $T(R)$ is non-zero in any of the following cases:*

(1) *$R$ has a test element and has finite Krull dimension.*
(2) *$R$ is finitely generated over an excellent local ring.*
(3) *$R/^{1/p}$ is module-finite over a domain $R$.*

PROOF. Clearly the rings in (2) have finite Krull dimension, and the rings in (3) have finite Krull dimension by [**14**, Proposition 1.1]. Assumptions in (2) and (3) imply that the ring has a test element, so the proof of the theorem follows if we can prove (1). The proof on the existence of test elements for rings in (2) takes up much of the 14-page Section 6 in [**12**], so it is beyond the scope of these notes. A proof on the existence of test elements for rings in (3) has already appeared in Kevin Tucker's lectures (Theorem 7), but I present another proof of the existence of test elements here for completeness. This proof is taken from [**10**, 3.4 Theorem]. Since $R^{1/p}$ is finitely generated over the domain $R$, there exists a non-zero element $c$ in $R$ such that $(R^{1/p})_c = (R_c)^{1/p}$ is free over $R_c$. Thus for every non-zero element $d$ there exists an $R_c$-module homomorphism $R_c^{1/p} \to R_c$ that takes $d^{1/p}$ to 1. This means that for every non-zero $d \in R$ there exists an $R$-module homomorphism $f_d : R^{1/p} \to R$ that takes $d^{1/p} \in R^{1/p}$ to $c^{N_d}$ for some non-negative integer $N_d$. By multiplying by a power of $c$ we may assume that all $N_d$ are powers of $p$. By replacing $c$ by $c^{N_1}$ we may assume that $f_1$ takes $u$ to $c$. We now claim that for all powers $q$ of $p$ there exists an $R$-module homomorphism $g_q : R^{1/q} \to R$ that takes 1 to $c^2$. We can take $g_1 = cf_1$. If we have $g_q$, we then set $g_{qp}$ to be the composition

$$R^{1/pq} \xrightarrow{g_1^{1/p}} R^{1/p} \xrightarrow{c^{(p-2)/p}} R^{1/p} \xrightarrow{f_1} R,$$

which takes $1 \mapsto c^{2/p} \mapsto c \mapsto f_1(c) = cf_1(1) = c^2$. Now let $x \in I^*$. Then there exists a non-zero $a \in R$ such that $ax^q \in I^{[q]}$ for all sufficiently large powers $q$ of $p$. We will show that $c^3x^q \in I^{[q]}$ for all sufficiently large powers $q$ of $p$.

Recall that $f_a$ takes $a^{1/p} \in R^{1/p}$ to $c^{N_a}$ in $R$. Thus $f_a^{1/pqN_a} : R^{1/pqN_a} \to R^{1/qN_a}$ takes $a^{1/pqN_a}$ to $c^{1/q}$. Then $ax^{pqN_a} \in I^{[pqN_a]}$ implies that $a^{1/pqN_a}x \in IR^{1/pqN_a}$. By applying $f_a^{1/pqN_a}$ to both sides we get that $c^{1/q}x \in IR^{1/qN_a}$. Since $g_{N_a} : R^{1/N_a} \to R$ takes 1 to $c^2$, it follows that the $R^{1/q}$-module homomorphism $g_{N_a}^{1/q} : R^{1/qN_a} \to R^{1/q}$ takes 1 to $c^{2/q}$. Hence $c^{2/q}c^{1/q}x \in IR^{1/q}$. From this we have that $c^3x^q \in I^{[q]}$ for all large $q$. Thus $c^3$ is a test element for $R$.

(1) Let $d$ be the Krull dimension of $R$.

Suppose that there exist an integer $k$ and a non-zero $c \in R$ such that for all ideals primary to a maximal ideal and for all positive integers $n$, $c\overline{I^n} \subseteq I^{n-k}$. Then for an arbitrary ideal $I$,

$$c\overline{I^n} \subseteq c\bigcap_{\mathfrak{m},l} \overline{(I+\mathfrak{m}^l)^n} \subseteq \bigcap_{\mathfrak{m},l} (I+\mathfrak{m}^l)^{n-k} \subseteq \bigcap_{\mathfrak{m},l} (I^{n-k}+\mathfrak{m}^l) = I^{n-k}.$$

Thus it suffices to find a $c$ and a $k$ that work for zero-dimensional primary ideals only.

Let $\mathfrak{m}$ be a maximal ideal in $R$ and let $I$ be $\mathfrak{m}$-primary. By Exercise 4 in Ian Aberbach's notes on the Briançon–Skoda Theorem, $\overline{I^n R_\mathfrak{m}} \subseteq (I^{n-d+1}R_\mathfrak{m})^*$. We prove next that $\overline{I^n} \subseteq (I^{n-d+1})^*$. Let $x \in \overline{I^n}$. Let $c \in R$ such that the image of $c$ in $R_\mathfrak{m}$ is not contained in any minimal prime ideal and such that for all sufficiently large powers $q$ of $p$, $cx^q \in (I^{n-d+1})^{[q]}R_\mathfrak{m}$. If $c$ is contained in minimal prime ideals $P_1, \ldots, P_s$ but not in minimal prime ideals $Q_1, \ldots, Q_t$, then by assumption the $P_i$ are not contained in $\mathfrak{m}$ and at least one $Q_i$ is contained in $\mathfrak{m}$. Choose $e$ in $(Q_1 \cdots Q_t) \setminus (P_1 \cup \cdots \cup P_s)$. Since the minimal prime ideals of $R_\mathfrak{m}$ are the localizations of (some of) the $Q_j$, it follows that $e$ is nilpotent in $R_\mathfrak{m}$, and by replacing $e$ with its own high power we may assume that $e$ is 0 in $R_\mathfrak{m}$. Thus $c+e = c$ in $R_\mathfrak{m}$, and in $R$, $c+e$ is not in any minimal prime ideal of $R$. Thus without loss of generality we may assume that $c \in R^\circ$. Hence from $cx^q \in (I^{n-d+1})^{[q]}R_\mathfrak{m}$ we deduce that there exists $s_q \in R \setminus \mathfrak{m}$ such that $cs_qx^q \in (I^{n-d+1})^{[q]}$. But $(I^{n-d+1})^{[q]}$ is $\mathfrak{m}$-primary and $s_q \notin \mathfrak{m}$, so that $cx^q \in (I^{n-d+1})^{[q]}$. This proves that $x \in (I^{n-d+1})^*$ globally.

Now let $c$ be a test element. Then $c\overline{I^n} \subseteq c(I^{n-d+1})^* \subseteq I^{n-d+1}$. Thus $c \in T_{d-1}(R)$. □

The proof of the next theorem is also beyond the scope of these notes. The proof uses some deep results of Joe Lipman and Avinash Sathaye from [**15**].

THEOREM 4.2 (This is [**13**, Theorem 4.10]). *Let $R$ be a regular domain with infinite residue fields and whose field of fractions $L$ has characteristic 0. Then $T(S) \neq 0$ for any finitely generated $R$-algebra $S$ that contains $R$ and that is a domain of finite Krull dimension.*

If $T_k(R) = R$ for some $k$, then $T_k(U^{-1}R) = U^{-1}R$ for all multiplicatively closed subsets $U$ of $R$.

THEOREM 4.3 (This is approximately [**13**, Proposition 3.7]). *Let $R$ be a Noetherian reduced ring. Assume that the weak Uniform Artin–Rees Lemma holds for $R$ and that for all minimal prime ideals $P$ of $R$, $T(R/P)$ is not zero. Then there exists an integer $k$ such that for all ideals $I$ and for all integers $n$,*

$$\overline{I^n} \subseteq I^{n-k}.$$

PROOF. Let $P_1, \ldots, P_s$ be the minimal prime ideals of $R$. Let $d_i$ be non-zero in $T(R/P_i)$ for $i = 1, \ldots, s$. We lift $d_i$ to an element $c_i$ of $R$ that is not in any $P_1, \ldots, P_s$. (If a first lift $e_i$ of $d_i$ to $R$ is in $P_{j_1}, \ldots, P_{j_s}$ but not in $P_{i_1}, \ldots, P_{i_t}$, let $f_i$ be an element in $P_i \cap P_{i_1} \cap \cdots \cap P_{i_t} \setminus (P_{j_1} \cup \cdots \cup P_{j_s})$ and set $c_i = e_i + f_i$.) Let $l_i$ be a positive integer such that for all integers $n$, $c_i\overline{I^n(R/P_i)} \subseteq I^{n-l_i}(R/P_i)$. Set $l = \max\{l_1, \ldots, l_s\}$. Then for all integers $n$ and all $i = 1, \ldots, s$, $c_i\overline{I^n} \subseteq I^{n-l} + P_i$.

Let $g_i$ be in all $P_j$ except for $P_i$. Set $c = \sum_i g_ic_i$. Then

$$c\overline{I^n} \subseteq \sum_i g_ic_i\overline{I^n} \subseteq \sum_i g_i(I^{n-l}+P_i) \subseteq I^{n-l}$$

because $R$ is reduced. Hence

$$c\overline{I^n} \subseteq I^{n-l} \cap (c),$$

and because of the assumption that the Uniform Artin–Rees Lemma holds in $R$, there exists an integer $k$ such that for all integers $n$,

$$c\overline{I^n} \subseteq I^{n-l} \cap (c) \subseteq cI^{n-l-k}.$$

It follows that

$$\overline{I^n} \subseteq I^{n-l-k} + (0 : c),$$

but $c$ is not in any minimal prime ideal in a reduced ring, so that $0 : c = 0$. □

## 5. Cohen-Macaulayfiers

DEFINITION 5.1 (This is [**13**, (2.11)], it first appeared in [**11**, Discussion (9.7)]). Let

$$\mathbb{F} : 0 \to F_t \xrightarrow{f_t} F_{t-1} \to \cdots \to F_1 \xrightarrow{f_1} F_0$$

be a complex of finitely generated free $R$ modules. We say that $\mathbb{F}$ satisfies *the standard condition on rank* if $\operatorname{rank}(f_t) = \operatorname{rank} F_t$ and for $1 \le i < t$, $\operatorname{rank}(f_i) + \operatorname{rank}(f_{i+1}) = \operatorname{rank} F_i$.

Let $I(f_i)$ denote the ideal generated by the rank-size minors of $f_i$. We say that $\mathbb{F}$ satisfies *the standard condition on height* if $\operatorname{height}(I(f_i)) \ge i$ for all $i$.

DEFINITION 5.2 (This is [**13**, Definition 2.12]). We define $\operatorname{CM}(R)$ to be the set of all elements $x \in R$ such that for all complexes $\mathbb{F}$ satisfying the standard rank and height conditions, $xH_i(\mathbb{F}) = 0$ for all $i \ge 1$. We think of these elements as "Cohen-Macaulayfiers".

LEMMA 5.3 (This is [**13**, Lemma 3.3]). *Let $R$ be a Noetherian ring and $P$ a prime ideal in $R$. Let $c$ be an element of $R$ not contained in $P$ such that the image of $c$ in $R/P$ lies in* $\operatorname{CM}(R/P)$. *Let $J$ be any ideal in $R$ generated by elements $a_1, \dots, a_d$ such that* $\operatorname{height}(a_1, \dots, a_d)R/P \ge d$. *Suppose that there exists an integer $k$ such that $J^n \cap (P + cR) \subseteq J^{n-k}(P + cR)$ for all $n$. Then for all integers $n$,*

$$J^n \cap P \subseteq J^{n-k}P.$$

PROOF. We have that

$$J^n \cap P = J^n \cap (P + cR) \cap P \subseteq J^{n-k}(P + cR) \cap P = J^{n-k}P + J^{n-k}cR \cap P.$$

Thus it suffices to prove that $c(J^{n-k} \cap P) = J^{n-k}cR \cap P$ lies in $J^{n-k}P$, i.e., that $c$ multiplies $J^{n-k} \cap P$ into $J^{n-k}P$. By renaming, it suffices to prove that $c$ multiplies $J^n \cap P$ into $J^nP$.

Thus by properties of Cohen-Macaulayfiers it suffices to prove that $\frac{J^n \cap P}{J^nP}$ lies in some $H_i(\mathbb{F})$ for some $\mathbb{F}$ satisfying standard rank and height conditions and for some $i \ge 1$.

Let $(\mathbb{G}, \underline{g})$ be a minimal homogeneous free resolution of $(X_1, \dots, X_d)^n$ in the polynomial ring $T = \mathbb{Z}[X_1, \dots, X_d]$. More specifically, we take the Eagon–Northcott free resolution of $(X_1, \dots, X_d)^n$ (see [**3**, 2.14]), and then $\sqrt{I(g_i)} = (X_1, \dots, X_d)$ for all $i$. Let $\mathbb{F} = \mathbb{G} \otimes_T R$, where the map $T \to R$ takes $X_i$ to $a_i$.

We compare $\mathbb{F}$ with a free resolution $\mathbb{H}$ of $R/J^n$:

$$\begin{array}{ccccccccc}
\mathbb{F}: & \cdots & F_2 & \longrightarrow & R^{\binom{n+d-1}{d-1}} & \xrightarrow{f_1} & R & \longrightarrow & 0 \\
 & & \downarrow & & \| & & \| & & \\
\mathbb{H}: & \cdots & H_2 & \longrightarrow & R^{\binom{n+d-1}{d-1}} & \xrightarrow{f_1} & R & \longrightarrow & 0
\end{array}$$

where the leftmost vertical arrow exists because $F_2$ is free and $\mathbb{H}$ is exact. Now tensor both complexes with $\otimes_R R/P$: we get a surjective map from $H_1(\mathbb{F} \otimes_R R/P)$ to $H_1(\mathbb{H} \otimes_R R/P) = \operatorname{Tor}_1^R(R/J^n, R/P) = \frac{J^n \cap P}{J^n P}$. As by assumption $\mathbb{F} \otimes_R R/P$ satisfies the standard conditions on height and rank, then $cH_1(\mathbb{F} \otimes_R R/P) = 0$ by a simple diagram chase, so the lemma is proved. □

## 6. When is CM(R) non-zero

By definition (Definition 5.2), elements of $\mathrm{CM}(R)$ annihilate homologies of complexes of finitely generated free $R$-modules that satisfy standard conditions on rank and height. Hochster and Huneke proved in [**11**, (9.8) Theorem] the following *phantom acyclicity criterion*: For every Noetherian ring of characteristic $p$ that is a locally equidimensional homomorphic image of a Cohen–Macaulay ring, in every complex of finitely generated free modules that satisfy standard conditions on rank and height, the module of the cycles is contained in the tight closure of the module of the boundaries. This generalized in some contexts the Buchsbaum–Eisenbud exactness criterion from [**4**].

THEOREM 6.1 (This is [**13**, Proposition 4.5]). *Let $R$ be a locally equidimensional Noetherian ring. Then* $\mathrm{CM}(R)$ *is not contained in any of the minimal prime ideals of $R$ if any of the following conditions is satisfied:*

(1) *$R$ is Cohen–Macaulay.*
(2) *$R$ is a domain of positive characteristic with a test element and is a homomorphic image of a Cohen–Macaulay ring.*
(3) *$R$ is a domain such that $R_P$ is not Cohen-Macaulay for only finitely many prime ideals $P$, and* $\mathrm{CM}(R_P)$ *is non-zero for each such $P$.*
(4) *$R$ has characteristic $p$, is a domain, and $R^{1/p}$ is module-finite over $R$.*
(5) *$R$ is a homomorphic image of a Gorenstein ring of finite Krull dimension.*
(6) *$(R, \mathfrak{m})$ is local and the local cohomology modules $H^i_{\mathfrak{m}}(R)$ are finitely generated for all $i < \dim(R)$.*
(7) *$R$ is a two-dimensional domain whose integral closure is a finitely generated $R$-module.*

PROOF. (1) If $R$ is Cohen–Macaulay, then every complex of finitely generated $R$-modules that satisfy standard conditions on rank and height also satisfies the standard conditions on rank and depth, so it is exact by the Buchsbaum–Eisenbud criterion. Thus all homology modules are 0, so $\mathrm{CM}(R) = R$.

(2) By the phantom acyclicity criterion from [**11**, (9.8) Theorem], any test element annihilates the homology. Thus by assumption $\mathrm{CM}(R)$ is not empty.

(3) Let $c_P \in R$ be such that after localization at $P$ it is a non-zero element of $\mathrm{CM}(R_P)$. Let $c$ be the product of all these $c_P$ as $P$ varies of the finitely many primes in question. Let $\mathbb{F}$ be a complex of finitely generated free $R$-modules satisfying the standard height and rank conditions. We claim that $cH_i(\mathbb{F}) = 0$ for all $i$. This is true if and only if it is true after localization at any prime ideal $Q$. Since $R$ is a

domain, $\mathbb{F}_Q$ is a complex of finitely generated free $R_Q$-modules that satisfies the standard rank and height conditions. So $cH_i(\mathbb{F}) = 0$ for all $i$ after localizing at any of the $P$ for which $R_P$ is not Cohen–Macaulay, and it is $H_i(\mathbb{F}) = 0$ for all $i$ after localizing at any of the other $Q$ by (1).

(4) As in the proof of Theorem 4.1 (3), $R$ has a test element $c$. We want to prove that $cH_i(\mathbb{F}) = 0$ for all $i$ and all complexes $\mathbb{F}$ of finitely generated free modules satisfying standard conditions on height and rank. It suffices to prove this after localization at an arbitrary prime ideal $P$. After localization, $\mathbb{F}_P$ is a complex of finitely generated free $R_P$-modules that satisfies the standard rank and height conditions. What is not clear is that $c$ is still a test element in $R_P$ or that its image is a test element in the $P$-adic completion of $R_P$. A more careful proof as in Theorem 4.1 (3) (which was already in Kevin Tucker's list of questions, you may want to fill in as a non-trivial exercise) would show that there exists a test element $c$ in $R$ that remains a test element after localization at prime ideals and after completion. By [**14**], $R$ is also excellent, so that the completion of $R_P$ is a homomorphic image of a Cohen–Macaulay ring, plus it is equidimensional because $R$ is. But then as in (2), the test element $c$ annihilates the homology modules.

(5) is a consequence of [**11**, (11.8) Theorem]: any element $c$ for which $R_c$ is Cohen–Macaulay has a power that annihilates all homology modules in finite complexes of finitely generated free $R$-modules that satisfy the standard rank and height conditions. The full proof is too lengthy for these notes.

Proof of (6) uses Peter Schenzel's cohomological annihilators from [**21**]. No details are provided here.

(7) Since $\overline{R}$ is module-finite over $R$, the conductor ideal $R : \overline{R}$ contains a non-zero element $c$. Because $R$ is a domain, $R_P$ is Cohen–Macaulay for all prime ideals of height 0 or 1. Since $R_c$ is integrally closed, it satisfies Serre's condition $(S_2)$, so that $R_P$ is Cohen-Macaulay for all prime ideals $P$ that do not contain $c$. If $P$ contains $c$ and is not associated to $c$, then $P$ contains a regular sequence of length at least 2, so that $R_P$ is also Cohen–Macaulay. Thus $R_P$ is not Cohen–Macaulay for at most finitely many maximal ideals $P$, all of which are associated to $cR$. But $H^i_{PR_P}(R_P)$ are zero and hence finitely generated for all $i < 2$, so it follows from (6) that $\mathrm{CM}(R_P)$ is non-zero for all $P$ of height 2 that are associated to $cR$. But then by (2) we have the conclusion. □

## 7. Uniform Artin–Rees Lemmas

THEOREM 7.1 (The Weak Uniform Artin–Rees Lemma, [**13**, Theorem 3.4]). *Let $R$ be a Noetherian ring with infinite residue fields such that for all prime ideals $P$, $\mathrm{CM}(R/P)$ and $T(R/P)$ contain non-zero elements. Let $N \subseteq M$ be finitely generated $R$-modules. Then there exists a positive integer $k$ such that for all ideals $I$ and for all integers $n$,*

$$I^n M \cap N \subseteq I^{n-k} N.$$

PROOF. By reductions in Step 2 it suffices to prove that for any prime ideal $P$ there exists an integer $k$ such that for all integers $n$,

$$I^n \cap P \subseteq I^{n-k} P.$$

If $P$ is a maximal ideal, then $I^n \cap P \subseteq I^{n-1} P$ for all ideals $I$. So we may assume that $P$ is not a maximal ideal. By Noetherian induction we may assume that the conclusion holds for all prime ideals $Q$ properly containing $P$.

By assumption there exists $c \in R \setminus P$ such that the image of $c$ in $R/P$ is in $\mathrm{CM}(R/P) \cap T(R/P)$. All prime ideals needed for the prime filtration of $R/(P+(c))$ properly contain $P$ and thus satisfy the conclusion. Thus by reductions in Step 2 there exists a positive integer $k$ such that for all ideals $I$ and all integers $n$, $I^n \cap (P+(c)) \subseteq I^{n-k}(P+(c))$.

By reductions in Step 3 we may assume that $I$ is primary to a maximal ideal $\mathfrak{m}$. By the assumption on infinite residue fields, there exists an ideal $J \subseteq I$ generated by $\dim(R/P)_{\mathfrak{m}}$ elements such that $J(R/P)_{\mathfrak{m}}$ is a minimal reduction of $I(R/P)_{\mathfrak{m}}$. We may assume that the generators of $J$ are in $R$ (not just in $R_{\mathfrak{m}}$) and that modulo $P$ (in $R$, not just in $R_{\mathfrak{m}}$, they form an ideal of height equal to $\dim(R/P)_{\mathfrak{m}}$. Then by Lemma 5.3, $J^n \cap P \subseteq J^{n-k}P$ for all integers $n$.

Since $c \in T_t(R/P)$ for some $t$, we have locally at $\mathfrak{m}$ that

$$cI^n(R/P)_{\mathfrak{m}} \subseteq c\,\overline{(I(R/P))^n_{\mathfrak{m}}} = c\,\overline{(J(R/P))^n_{\mathfrak{m}}} \subseteq J^{n-t}(R/P)_{\mathfrak{m}}.$$

It follows that $cI^n(R/P)_{\mathfrak{m}} \subseteq J^{n-t}(R/P)_{\mathfrak{m}} \cap c(R/P)_{\mathfrak{m}}$ and so

$$\begin{aligned} cI^n_{\mathfrak{m}} &\subseteq (J^{n-t}+P)_{\mathfrak{m}} \cap (P+(c))_{\mathfrak{m}} + P_{\mathfrak{m}} \\ &= J^{n-t}_{\mathfrak{m}} \cap (P+(c))_{\mathfrak{m}} + P_{\mathfrak{m}} \\ &\subseteq J^{n-t-k}(P+(c))_{\mathfrak{m}} + P_{\mathfrak{m}} \\ &= J^{n-t-k}c_{\mathfrak{m}} + P_{\mathfrak{m}}. \end{aligned}$$

Since $c$ is a non-zerodivisor modulo $P$, we conclude that $I^n_{\mathfrak{m}} \subseteq J^{n-t-k}_{\mathfrak{m}} + P_{\mathfrak{m}}$.

Thus by Lemma 3.8, $I^n_{\mathfrak{m}} \cap P_{\mathfrak{m}} \subseteq I^{n-2k-t-1}P_{\mathfrak{m}}$ for all $n$, and by Lemma 3.7, $I^n \cap P \subseteq I^{n-2k-t-1}P$ for all $n$. □

THEOREM 7.2 (The Weak Uniform Artin–Rees Lemma, [**13**, Theorem 4.12]). *Let $R$ be a Noetherian ring satisfying at least one of the following conditions:*

(i) *$R$ is essentially of finite type over a Noetherian local ring.*
(ii) *$R$ is a ring of positive prime characteristic $p$ and is module-finite over $R^p$ ($R^p$ is the subring generated by $p$th powers of elements of $R$).*
(iii) *$R$ is essentially of finite type over $\mathbb{Z}$.*

*Let $N \subseteq M$ be finitely generated $R$-modules. Then there exists a positive integer $k$ such that for all ideals $I$ and for all integers $n$,*

$$I^nM \cap N \subseteq I^{n-k}N.$$

PROOF. (i) Let $R$ be essentially of finite type over a Noetherian local ring $(A, \mathfrak{n})$. Let $B$ be the completion of $A[t]_{\mathfrak{n}A[t]}$. So $B$ is a faithfully flat extension of $A$ and $R \otimes_A B$ is a faithfully flat extension of $R$ that is essentially of finite type over $B$. By Lemma 3.2 it suffices to prove that $R \otimes_A B$ has the weak Uniform Artin-Rees property. The advantage of this new ring is that all of its residue fields are infinite and that by the Cohen Structure Theorem it is a homomorphic image of a regular ring $S$ all of whose residue fields are infinite. For any prime ideal $P$ in $S$, by Theorem 4.1, $T(S/P)$ is non-zero in case $S/P$ has characteristic $p$, by Theorem 4.2, $T(S/P)$ is non-zero in case $S/P$ has characteristic 0, and by Theorem 6.1, $\mathrm{CM}(S/P)$ is non-zero. Thus by Theorem 7.1, $S$ has the uniform Artin-Rees property. By Exercise 9.2, then $R \otimes_A B$ has the uniform Artin-Rees property.

(ii) Proof of (ii) is similar: For all prime ideals $P$ in $R$, $T(R/P)$ is non-zero by Theorem 4.1 and $\mathrm{CM}(R/P)$ is non-zero by Theorem 6.1.

(iii) By Exercise 9.2, it suffices to prove it for $R$ finitely generated over $\mathbb{Z}$. Then for any prime ideal $P$ in $R$, $R/P$ is a homomorphic image of a finite-dimensional Gorenstein ring, so $\mathrm{CM}(R/P)$ is non-zero by Theorem 6.1. In case $R/P$ has prime characteristic $p$, then it is finitely generated over the excellent local ring $\mathbb{Z}/p\mathbb{Z}$, so $T(R/P)$ is non-zero by Theorem 4.1, and otherwise $R/P$ is finitely generated over the regular domain $\mathbb{Z}$, so $T(R/P)$ is non-zero by Theorem 4.2. □

## 8. More Uniform Artin–Rees Lemmas

THEOREM 8.1 (This is [**23**, Theorem 4.2]). *Let $R$ be a Noetherian ring satisfying at least one of the following conditions (same as in Theorem* 3.1*):*

(i) *$R$ is essentially of finite type over a Noetherian local ring.*
(ii) *$R$ is a ring of positive prime characteristic $p$ and is module-finite over $R^p$ ($R^p$ is the subring generated by $p$th powers of elements of $R$).*
(iii) *$R$ is essentially of finite type over $\mathbb{Z}$.*

*Then given any finitely generated $R$ modules $N \subseteq M$ there exists an integer $k$ such that for all integers $l$, for all ideals $J_1, \ldots, J_l$ in $R$ and all integers $n_i \geq k$, $i = 1, \ldots, l$,*

$$J_1^{n_1} \cdots J_l^{n_l} M \cap N \subseteq J_1^{n_1-k} \cdots J_l^{n_l-k} N.$$

THEOREM 8.2 (This is [**23**, Theorem 4.1]). *Let $I$ be an ideal in a Noetherian ring $R$. Then there exists an integer $k$ such that for all ideals $J$ in $R$, for all integers $n$ and for all $m \geq kn$,*

$$J^m \cap I^n \subseteq J^{m-kn} I^n.$$

THEOREM 8.3 (This is Planas-Vilanova [**19**, Theorem 1]). *Let $R$ be an excellent (or more generally, a J-2) ring. Then for any finitely generated $R$-modules $N \subseteq M$ such that $\dim(M/N) \leq 1$ there exists a positive integer $k$ such that for all positive integers $n \geq k$ and all ideals $I$ of $R$,*

$$I^n M \cap N = I^{n-k}(I^k M \cap N).$$

THEOREM 8.4 (This is Striuli [**22**, Theorem 1]; the Cohen–Macaulay case was proved by Dan Katz, as on page 579 of [**22**]). *Let $(R, \mathfrak{m})$ be a Noetherian local ring of dimension at most two, and let $M$ be a finitely generated $R$-module with a free resolution $(\mathbb{F}, \partial)$. Then there exists a positive integer $k$ such that for all integers $n$ and $i$ and all $\mathfrak{m}$-primary ideals $I$,*

$$I^n \mathbb{F}_i \cap \ker \partial_i \subseteq I^{n-k} \ker \partial_i.$$

THEOREM 8.5 (This is Giral, Planas-Vilanova [**9**, Theorem 1.2]). *Let $R$ be a Noetherian ring containing the field of rational numbers for which the set of reduction numbers of ideals with principal reductions is bounded above. Let $d - 1$ be the maximum. Then for any finitely generated $R$-modules $N \subseteq M$ and for any regular ideal $I$ in $R$ with a principal reduction and generated by a $d$-sequence with respect to $M/N$ and for any integer $n$,*

$$I^n M \cap N = I^{n-d}(I^d M \cap N).$$

THEOREM 8.6 (This is Aberbach, Hosry, Striuli [**1**, Main Theorem]). *Let $(R, \mathfrak{m})$ be a local Noetherian ring of dimension $d$. Then there exists a positive integer $k$ such that for all integers $n$ and $i \geq d$ all ideals $I$, and all free resolutions $(\mathbb{F}, \partial)$ of finitely generated $R$-modules,*

$$I^n \mathbb{F}_i \cap \operatorname{im} \partial_{i+1} \subseteq I^{n-k} \operatorname{im} \partial_{i+1}.$$

## 9. Exercises

About the exercises below: Some are more of a thinking or review type. I think it is important and helpful to think through some familiar results every once in a while because there is a strong likelihood that with experience we gain deeper insights and we can re-gain some of the old insights.

When there is an open question tacked onto an exercise, it is a good idea to re-think the proof of the known result yet again and to see where that proof fails to garner the needed traction for the open problem.

Exercise 9.0 is free for all: do you know of some other examples of usages of the Artin-Rees Lemma or of the uniform Artin-Rees Lemma that are not yet listed in these notes? Exercise 9.5 is also open-ended: do you have a crisp proof that you are willing to share?

A few of the problems are about the limits to the uniform behavior, so they require establishing some (counter)examples of more or less computational nature. I find it productive to think not only about the positive results but also where they fail.

Which exercises should you do? I do not think that you can or should do them all. Please do turn in Exercises 9.2, 9.6, 9.7, and work through some parts of Exercise 9.3.

EXERCISE 9.0. Expand on the list of uses on 141 of the classical Artin–Rees Lemma. Expand on the list of uses in Section 8 of the uniform Artin–Rees Lemma.

EXERCISE 9.1. Review a proof of the classical Artin–Rees Lemma and focus on how the $k$ is determined. How does it depend on $I$, $M$, $N$? Open question: Does there exist an Artin–Rees-lemma-type result for Frobenius powers?

EXERCISE 9.2. Suppose that $R$ has the Uniform Artin-Rees property.

(i) Prove that any homomorphic image of $R$ has the Uniform Artin-Rees property.
(ii) Prove that any localization of $R$ has the Uniform Artin-Rees property.

EXERCISE 9.3. (This example is from Eisenbud, Hochster [7].) Let $F$ be a field and $X_1, Y_1, X_2, Y_2, \ldots$ indeterminates over $F$. For each positive integer $n$ let

$$\begin{aligned} f_n &= X_n^n - Y_n^{n+1}, \\ I_n &= (f_2 - f_1, \ldots, f_n - f_1)F[X_1, Y_1, \ldots, X_n, Y_n], \\ S_n &= F[X_1, Y_1, \ldots, X_n, Y_n]/I_n, \\ U_n &= S_n \setminus \cup_{i=1}^n (X_i, Y_i)S_n. \end{aligned}$$

(i) Verity that $U_n$ is a multiplicatively closed subset of $S_n$. Set $R_n = U_n^{-1}S_n$.
(ii) Prove that there is an obvious injection $R_n \to R_{n+1}$.
(iii) Let $R = \lim_{\longrightarrow} R_n$ and let $f$ be the image of $f_i$ in $R$. Prove that $fR$ is a non-zero prime ideal. Prove that $f \in (X_i, Y_i)^i R$ for all $i$.
(iv) Prove that the maximal ideals of $R$ are precisely the ideals $(X_i, Y_i)R$.
(v) Prove that $R$ is a unique factorization domain.
(vi) Prove that other non-zero prime ideals are principal of height 1.
(vii) Prove that $R$ is a two-dimensional Noetherian ring.
(viii) Prove that $\cap_i (X_i, Y_i)R = (f)$.
(ix) Prove that there exists no positive integer $k$ such that for all positive integers $i$ and all $n \geq k$, $(X_i, Y_i)^n R \cap (f) \subseteq (X_i, Y_i)^{n-k}(f)$.

(x) Prove that the integral closure of $R/(f)$ is not a finitely generated $R/(f)$-module. (Thus $R$ is not pseudo-geometric.)

EXERCISE 9.4. Suppose that $R$ is a Cohen–Macaulay Noetherian ring. Prove that every complex of finitely generated $R$-modules that satisfies the standard rank and height conditions is exact. (Hint: Buchsbaum–Eisenbud exactness criterion, [**4**].)

EXERCISE 9.5. One of the results of Section 6 of the Hochster–Huneke paper *F-regularity, test elements, and smooth base change* ([**12**]) is that every Noetherian ring that is finitely generated over an excellent local ring has a test element. Section 6 has many pages and proves much more than that. Do you want to offer a more direct proof?

EXERCISE 9.6. [This is Wang [**24**, Example 6.1]] Let $(R, \mathfrak{m})$ be a three-dimensional regular local ring. Let $\mathfrak{m} = (x, y, z)$, $J = zR$ and $I_m = (x^m, y^m, x^{m-1}y + z^m)$. Prove that there exists no integer $k$ such that for all integers $n \geq k$ and all integers $m$,

$$I_m^n \cap J = I_m^{n-k}(I_m^k \cap J).$$

EXERCISE 9.7. Can the assumption $\dim(M/N) \leq 1$ in Theorem 8.3 be removed for the same conclusion?

EXERCISE 9.8. Let $R$ be a Noetherian ring of Krull dimension $d$.

(i) Prove that for any ideal $I$ there exists a $(d+1)$-generated ideal $I \subseteq J$ such that $\sqrt{I} = \sqrt{J}$.

(ii) Suppose that $\mathfrak{m}$ is a maximal ideal in $R$ of height $k$. Prove that every $\mathfrak{m}$-primary ideal has a $(d+k+1)$-generated reduction.

## 10. Solutions to Exercises by Kriti Goel

SOLUTION 9.2. (i) Let $J \subseteq R$ be an ideal and let $M, N$ be finitely generated $R/J$-modules. We need to show that there exists $k \in \mathbb{N}$ such that for all ideals $I/J \subseteq R/J$ and for all $n > k$, $(I^n + J)M \cap N(I^{nk} + J)N$. That is, $(I^n M \cap N) + J \subseteq I^{nk}N + J$. Note that $M, N$ are finitely generated $R$-modules via the natural map $\pi : R \to R/J$. Since $R$ has the uniform Artin-Rees property, there exists $k \in \mathbb{N}$ such that for all ideals $I \subseteq R$ and for all $n > k$, $I^n M \cap N \subseteq I^{nk}N$, where the multiplication happens via the map $\pi$. Hence, the result holds. (ii) Let $W \subset R$ be a multiplicatively closed set and let $M, N$ be finitely generated $W^1 R$-modules. We need to show that there exists $k \in N$ such that for all ideals $W^1 I \subseteq W^1 R$ and for all $n > k$, $(W^1 I)^n M \cap N(W^1 I)^{nk} N$. Note that there exist finitely generated $R$-modules $N \subseteq M$ such that $W^1 M = M$ and $W^1 N = N$. Since $R$ has the uniform Artin-Rees property, there exists $k \in \mathbb{N}$ such that for all ideals $I \subseteq R$ and for all $n > k$, $I^n M \cap N = I^{nk}N$. Since localization is exact, the result holds. □

SOLUTION 9.3. (i) Observe that for every positive integers $i \leq n$, $S_n/(X_i, Y_i)$ is isomorphic to

$$F[X_1, Y_1, \ldots, X_{i-1}, Y_{i-1}, X_{i+1}, Y_{i+1}, \ldots, X_n, Y_n]/((f_1, \ldots, f_{i-1}, f_{i+1}, \ldots, f_n).$$

This is a domain since each $f_i$ is an irreducible element and $f_1, \ldots, f_n$ use distinct variables. (The tensor product of finitely many domains over an algebraically closed field is a domain.) Thus $(X_i, Y_i)S_n$ is a prime ideal. Since $U_n$ is the complement of a union of finitely many prime ideals in $S_n$, it follows that $U_n$ is multiplicatively closed.

(ii) The kernel of the map $F[X_1, Y_1, \ldots, X_n, Y_n] \to F[X_1, Y_1, \ldots, X_{n+1}, Y_{n+1}] \to S_{n+1}$ equals $I_n$. so that $S_n \to S_{n+1}$ is injective. Since the pre-image of $U_{n+1}$ under this map is $U_n$, we get that $R_n \to R_{n+1}$ is injective.

Observe that $R$ is a domain. If $x, y \in R$ such that $xy = 0$, then since $x$ and $y$ both come from some $R_n$ and since the maps in the direct limit are injective, either $x = 0$ or $y = 0$.

(iii) Note that $f = f_i$ for all $i \in R_n$ and hence in $R$. Thus, $f$ is the image of $X_i^i - Y_i^{i+1}$ in $R$, and so $f \in (X_i, Y_i)^i R_n$ for all $i$. Also $R_n/(f)$ is a domain, implying that $fR \in \operatorname{Spec}(R)$.

(iv) Note that for all $n$, $(X_i, Y_i)R_n \in \operatorname{Spec}(R_n)$ for all $i \leq n$, and by the definition of $U_n$, $(X_i, Y_i)R_n$ is even a maximal ideal of $R_n$. These are all the maximal ideals of $R_n$. Furthermore, necessarily $(X_i, Y_i)R$ is a maximal prime ideal in $R$.

Now let $P$ be any maximal ideal of $R$. Note that all $(X_i, Y_i)R$ contain $f$, so $P \neq fR$ and so $P$ contains an element $g$ not in $fR$. Of necessity $g$ comes from some $R_n$. Then $P \cap R_n$ is a prime ideal in $R_n$, and there exists a subset $V_n$ of $\{1, \ldots, n\}$ such that $P \cap R_n \subseteq (X_i, Y_i)R_n$ exactly for $i \in V_n$. It follows that for all $m \geq n$, $P \cap R_m$ contains non-zero elements that live in those $(X_i, Y_i)R_m$. But those elements do not live in any other $(X_l, Y_l)R_m$ (because $g$ is not a multiple of $f$). Thus for all $m \geq n$, $V_{m+1} \subseteq V_m \subseteq V_n$. Necessarily $V = \cap_{m \geq n} V_m$ is not empty, and $P \subseteq (X_i, Y_i)R$ for all $i \in V$. But $P$ is maximal, so that $P = (X_i, Y_i)R$ for some $i$.

This proves that the only maximal ideals in $R$ are the $(X_i, Y_i)R$. Of these, $(X_1, Y_1)R = Y_1 R$ is principal, others are minimally generated by two elements.

(v and vi) By the Jacobian criterion, each $R_n$ is regular. We prove that any height-one prime ideal $Q$ in $R_n$ is principal. Since $R_n$ localized at $M_i = (X_i, Y_i)R_n$ is a two-dimensional regular local ring, it follows that there exists $g_i \in Q$ such that

$$g_i (R_n)_{M_i} = Q_{M_i}.$$

By letting $s_i$ be an element of all the maximal ideals $M_1, \ldots, M_n$ of $R_n$ except $M_i$, we see that $g = \sum_i s_i g_i$ generates $Q$ after localization at every maximal ideal of $R_n$, and hence $Q$ is principal in $R_n$. (For this we use that if $a$ generates an ideal $J$ in a local ring $(R, \mathfrak{m})$, so does $a + b$ for any $b \in \mathfrak{m}J$.) This proves that $R_n$ is a UFD. Hence $R$ is a UFD.

(vii) Thus all prime ideals in $R$ are finitely generated. Hence by a result of Cohen, all ideals are finitely generated (apply Zorn's lemma to the set of all non-finitely generated ideals). Thus $R$ is Noetherian. Furthermore, it is two-dimensional by the numbers of generators.

(viii) We know that $J = \cap_i (X_i, Y_i)$ is a radical ideal that contains the prime ideal $(f)$. We can write $J = (f, g_1, \ldots, g_l)$ for some elements $g_1, \ldots, g_l$. A primary decomposition of $J$ uses only finitely many prime ideals, and we know that an intersection of only finitely many of the maximal ideals is not equal to $J$. Thus at least one height-one prime ideal must be in the primary decomposition of $J$, so that the height of $J$ is 1. Since $(f)$ is a prime ideal of height one, necessarily all $g_i \in (f)$. Thus $J = (f)$.

(ix) Suppose that there does exist an integer $k$ as described. Then in particular for all $i = n \geq k$,

$$(f) = (X_n, Y_n)^n \cap (f) \subseteq (X_n, Y_n)^{n-k}(f),$$

which forces $n = k$.

(x) The integral closure $A$ of $R/(f)$ contains all $\frac{X_n}{Y_n}$ (it satisfies the equation of integral dependence $Z^n - Y_n = 0$). If $A$ were finitely generated as an $R/(f)$-module, then in particular the submodule $B$ of $A$ generated by all the $\frac{X_n}{Y_n}$ over $R/(f)$, as $n$ varies, would be finitely generated, say by the first $m$ of these fractions. Then

$$\frac{X_{m+1}}{Y_{m+1}} = \sum_{i=1}^{m} r_i \frac{X_i}{Y_i}$$

for some $r_i \in R/(f)$. By clearing denominators, since $X_1/Y_1 = X_1$, we get a relation in $R/(f)$ in which one monomial is $X_{m+1}Y_2 \cdots Y_m$, and the other terms are of the form $r_i X_i Y_{m+1} Y_2 \cdots Y_{i-1} Y_{i+1} \cdots Y_m$. When we lift to $R$, the difference of the two sides is not a multiple of $f$, since each variable appears to at most the first power.

DEFINITION. Let $R$ be a Noetherian ring, $I = (x_1, \ldots, x_m)$ be an ideal of $R$. The **relation type of I**, $\operatorname{rt} I$, is defined to be the least integer $r$ such that the kernel of the homomorphism $R[X_1, \ldots, X_m] \to R[It]$ defined by $X_i \mapsto x_i t$ is generated in degrees at most $r$.

LEMMA. *Let $(R, m)$ be a Noetherian local ring and $I = (x_1, \ldots, x_m)$ an ideal generated by an $R$-regular sequence. Let $J$ be an ideal of $R$ and $\overline{R} = R/J$. Assume that $I^n \cap J = I^{n-k}(I^k \cap J)$ for all $n \geq k$. The $\operatorname{rt}(I\overline{R}) \leq k$.*

PROOF. Let $F \in R[X_1, \ldots, X_m]$ be such that $\overline{F}$ is a relation on $\overline{x}_1, \ldots, \overline{x}_m$ of degree $n > k$. Then $F(x_1, \ldots, x_m) \in I^n \cap J = I^{n-k}(I^k \cap J)$. Therefore, there are polynomials $G_i \in R[X_1, \ldots, X_m]$ of degree $k$ and monomials $H_i$ of degree $n - k$ such that

$$F(x_1, \ldots, x_m) = \sum_i G_i(x_1, \ldots, x_m) H_i(x_1, \ldots, x_m),$$

and such that $G_i(x_1, \ldots, x_m) \in I^k \cap J$ for all $i$. It follows that $F - \sum_i G_i H_i$ is a relation on $x_1, \ldots, x_m$. Since $x_1, \ldots, x_m$ is a regular sequence, it follows that all the relations come from the Koszul relations, i.e., $F - \sum_i G_i H_i = \sum_j K_j L_j$ for some polynomials $K_j$ of degree $n - 1$ and some polynomials $L_j$ of degree 1 such that $L_j(x_1, \ldots, x_m) = 0$ for all $j$. It follows $\overline{F}$ can be written in terms of relations on $\overline{x}_1, \ldots, \overline{x}_m$ of degree at most $k$. □

SOLUTION 9.6. Let $\overline{R} = R/J$. Then $\overline{R}$ is a 2-dimensional regular local ring. We claim that $\operatorname{rt}(I^m \overline{R}) \geq m$, for all $m$. We need to prove that $\operatorname{rt}(x^m, y^m, x^{m1}y)$ is at least $m$. Indeed, we claim that the relation $F = X_1^{m1} X_2 X_3^m$ is not in the ideal generated by lower-degree relations. If it is, then there exists a relation 1 of strictly lower degree, say $G$, whose $X_2$-degree is either 0 or 1. If the $X_2$-degree is 0, then necessarily $G$ is a relation on $x^m$ and $x^{m1}y$ only, and so $G$ is a multiple of $xX_3yX_1$, but then no monomial in $F$ can be reduced modulo $G$. So necessarily the $X_2$-degree of $G$ is 1. But then the other monomial in $G$ must have $X_3$-degree equal to $m$, so that $G = F$. Thus, the claim holds. Since $x, y, z$ forms an $R$-regular sequence, then so does $x, y, z^m$. Thus, $x, y, x^{m1}y + z^m$ and hence $x^m, y^m, x^{m1}y + z^m$

is an $R$-regular sequence. If for some $k$, $I^n \cap J = I^{nk}(I^k \cap J)$ for all $n \geq k$,then it follows from the lemma above that $\mathrm{rt}(I^m\overline{R}) \leq k$. Hence no such $k$ exists.

SOLUTION 9.7. Let $R$ be as in Question 9.6. Take $M = R$ and $N = J$. Then $\dim(M/N) = \dim(R/J) = 2$ and as in Question 9.6, if $I_m = (x^m, y^m, x^{m1}y + z^m)$, then there exists no integer $k$ such that for all integers $n \geq k$ and all integers $m$, $I^nR \cap J = I^{nk}(I^kR \cap J)$.

## References


[1] Ian M. Aberbach, Aline Hosry, and Janet Striuli, *Uniform Artin-Rees bounds for syzygies*, Adv. Math. **285** (2015), 478–496, DOI 10.1016/j.aim.2015.07.031. MR3406507

[2] E. Bierstone and P. D. Milman, *Relations among analytic functions. I* (English, with French summary), Ann. Inst. Fourier (Grenoble) **37** (1987), no. 1, 187–239, DOI 10.5802/aif.1082. MR894566

[3] Winfried Bruns and Udo Vetter, *Determinantal rings*, Lecture Notes in Mathematics, vol. 1327, Springer-Verlag, Berlin, 1988, DOI 10.1007/BFb0080378. MR953963

[4] David A. Buchsbaum and David Eisenbud, *What makes a complex exact?*, J. Algebra **25** (1973), 259–268, DOI 10.1016/0021-8693(73)90044-6. MR314819

[5] A. J. Duncan and L. O'Carroll, *A full uniform Artin-Rees theorem*, J. Reine Angew. Math. **394** (1989), 203–207, DOI 10.1515/crll.1989.394.203. MR977443

[6] David Eisenbud, *Commutative algebra: With a view toward algebraic geometry*, Graduate Texts in Mathematics, vol. 150, Springer-Verlag, New York, 1995, DOI 10.1007/978-1-4612-5350-1. MR1322960

[7] David Eisenbud and Melvin Hochster, *A Nullstellensatz with nilpotents and Zariski's main lemma on holomorphic functions*, J. Algebra **58** (1979), no. 1, 157–161, DOI 10.1016/0021-8693(79)90196-0. MR535850

[8] Abdelhafed El Khadiri and Jean-Claude Tougeron, *Familles noethériennes de sous-modules de* $\mathbf{k}[[x]]^p$ *et applications. II* (French, with English summary), C. R. Acad. Sci. Paris Sér. I Math. **301** (1985), no. 10, 475–478. MR812561

[9] José M. Giral and Francesc Planas-Vilanova, *Integral degree of a ring, reduction numbers and uniform Artin-Rees numbers*, J. Algebra **319** (2008), no. 8, 3398–3418, DOI 10.1016/j.jalgebra.2007.10.007. MR2408325

[10] Melvin Hochster and Craig Huneke, *Tight closure and strong F-regularity*, Mém. Soc. Math. France (N.S.) **38** (1989), 119–133. Colloque en l'honneur de Pierre Samuel (Orsay, 1987). MR1044348

[11] Melvin Hochster and Craig Huneke, *Tight closure, invariant theory, and the Briançon-Skoda theorem*, J. Amer. Math. Soc. **3** (1990), no. 1, 31–116, DOI 10.2307/1990984. MR1017784

[12] Melvin Hochster and Craig Huneke, *F-regularity, test elements, and smooth base change*, Trans. Amer. Math. Soc. **346** (1994), no. 1, 1–62, DOI 10.2307/2154942. MR1273534

[13] Craig Huneke, *Uniform bounds in Noetherian rings*, Invent. Math. **107** (1992), no. 1, 203–223, DOI 10.1007/BF01231887. MR1135470

[14] Ernst Kunz, *On Noetherian rings of characteristic p*, Amer. J. Math. **98** (1976), no. 4, 999–1013, DOI 10.2307/2374038. MR432625

[15] Joseph Lipman and Avinash Sathaye, *Jacobian ideals and a theorem of Briançon-Skoda*, Michigan Math. J. **28** (1981), no. 2, 199–222. MR616270

[16] Hideyuki Matsumura, *Commutative ring theory*, Cambridge Studies in Advanced Mathematics, vol. 8, Cambridge University Press, Cambridge, 1986. Translated from the Japanese by M. Reid. MR879273

[17] Liam O'Carroll, *A uniform Artin-Rees theorem and Zariski's main lemma on holomorphic functions*, Invent. Math. **90** (1987), no. 3, 647–652, DOI 10.1007/BF01389183. MR914854

[18] L. O'Carroll, *A note on Artin-Rees numbers*, Bull. London Math. Soc. **23** (1991), no. 3, 209–212, DOI 10.1112/blms/23.3.209. MR1123327

[19] Francesc Planas-Vilanova, *The strong uniform Artin-Rees property in codimension one*, J. Reine Angew. Math. **527** (2000), 185–201, DOI 10.1515/crll.2000.081. MR1794022

[20] D. Rees, *Two classical theorems of ideal theory*, Proc. Cambridge Philos. Soc. **52** (1956), 155–157, DOI 10.1017/s0305004100031091. MR74386

[21] Peter Schenzel, *Cohomological annihilators*, Math. Proc. Cambridge Philos. Soc. **91** (1982), no. 3, 345–350, DOI 10.1017/S0305004100059417. MR654081

[22] Janet Striuli, *A uniform Artin-Rees property for syzygies in rings of dimension one and two*, J. Pure Appl. Algebra **210** (2007), no. 2, 577–588, DOI 10.1016/j.jpaa.2006.11.012. MR2320020

[23] Irena Swanson, *Powers of ideals. Primary decompositions, Artin-Rees lemma and regularity*, Math. Ann. **307** (1997), no. 2, 299–313, DOI 10.1007/s002080050035. MR1428875

[24] Hsin-Ju Wang, *Some uniform properties of 2-dimensional local rings*, J. Algebra **188** (1997), no. 1, 1–15, DOI 10.1006/jabr.1996.6825. MR1432344

[25] Oscar Zariski, *A fundamental lemma from the theory of holomorphic functions on an algebraic variety*, Ann. Mat. Pura Appl. (4) **29** (1949), 187–198, DOI 10.1007/BF02413926. MR41488

Department of Mathematics, Purdue University, West Lafayette, Indiana 47907
*Email address*: irena@purdue.edu

Indian Institute of Technology, Jodhpur, India
*Email address*: kritigoel.maths@gmail.com

Contemporary Mathematics
Volume **831**, 2025
https://doi.org/10.1090/conm/831/16635

# Lectures on the localization problem in tight closure

Wenliang Zhang, with exercise solutions by Sudeshna Roy

## 1. Lecture 1

Since tight closure was introduced in [**HH90**], one question had persisted:

QUESTION 1.1. Does tight closure commute with localization?

The answer, as shown in [**BM10**], is negative in general.

EXAMPLE 1.2 (Brenner-Monsky). Let $k$ be the algebraic closure of $\mathbb{Z}/2\mathbb{Z}$. Consider

$$R = \frac{k[x,y,z,t]}{(z^4 + xyz^2 + x^3z + y^3z + tx^2y^2)}, \quad I = (x^4, y^4, z^4), \quad W = k[t]\backslash\{0\}.$$

Then $y^3z^3 \in (W^{-1}I)^*$ and $y^3z^3 \notin I^*$.

**Takeaway:** Even though tight closure does not commute with localization in the full generality, there are still many interesting and important problems remaining open.

Our main references are [**Hun96**, §10, §12] (and [**HH90, Hoc07**]).

In these lectures we will focus on positive results and related open problems; the goal of these lectures is to justify our takeaway.

Our first lecture mainly is concerned with results on regular sequences and their generalizations to parameter ideals. We will start with the following result ([**HH94a**, Theorem 4.5]) due to Hochster and Huneke.

THEOREM 1.3 (Hochster-Huneke). *Let $R$ be a noetherian ring of prime characteristic $p$. Assume that an ideal $I$ is generated by a regular sequence. Then $(W^{-1}I)^* = W^{-1}(I^*)$ for every multiplicatively closed set $W$ in $R$.*

PROPOSITION 1.4 (Hochster-Huneke). *Let $R$ be a noetherian ring and $W$ a multiplicatively closed set in $R$.*

(a) *For each ideal $I$ there is a $u \in W$ such that $\cup_{w\in W}(I^n : w) = (I^n : u^n)$ for each $n \in \mathbb{N}$. ($u$ depends on $I$, but not on $n$.)*

(b) *Furthermore, assume that $R$ is a noetherian ring of prime characteristic $p$ and that $I$ is generated by a regular sequence with $\ell$ elements. Then, with $u$ as in part (a),*

$$\cup_{w\in W}(I^{[p^e]} : w) = (I^{[p^e]} : u^{(\ell+1)p^e})$$

*for every $e \in \mathbb{N}$.*

PROOF. We opt to sketch the main ideas of the proof.[1]

(a) Pick $u \in W$ such that $\mathrm{Ann}_G(u) = \mathrm{Ann}_G(uw)$ for every $w \in W$ where $G = gr_I R$ is the associated graded ring of $R$ with respect to the $I$-filtration.

(b) Fix $p^e$ and an element $r \in (I^{[p^e]} : w)$ for a $w \in W$. One can prove by induction on $\ell \geq 0$ that $u^{p^e+\ell} r \in I^{[p^e]} + I^{p^e+\ell}$. (Setting $\ell = p^e n$ produces $u^{(n+1)p^e} r \in I^{[p^e]} + I^{p^e(n+1)} \subseteq I^{[p^e]}$ when $I$ can be generated by $n$ elements.) □

PROOF OF THEOREM 1.3. Only one inclusion requires a proof: $(IW^{-1}R)^* \subseteq (I^*)W^{-1}R$. Pick any $z/v \in (IW^{-1}R)^*$ with $z \in R$ and $v \in W$. Then there is a $c_0 \in (W^{-1}R)^\circ$ such that $c_0(z/v)^{p^e} \in I^{[p^e]}W^{-1}R$. By Exercise 4.1, $c_0$ can be written as $c/w$ for some $c \in R^\circ$ and $w \in W$. Hence there are elements $w_e \in W$ such that $cw_e z^{p^e} \in I^{[p^e]}$ (which happens in $R$) for each $e \in \mathbb{N}$. (That is, $cz^{p^e} \in (\cup_{w \in W}(I^{[p^e]} : w))$.)

Then, with $u$ as in Proposition 1.4, one has $u^{(\ell+1)p^e} cz^{p^e} \in I^{[p^e]}$. Consequently, $u^{\ell+1} z \in I^*$ whence $z/1 \in (I^*)W^{-1}R$. □

There are at least two different approaches to generalizing Theorem 1.3 to parameter ideals:

(1) plus closure ([**Smi94**]);
(2) finite phantom projective dimension ([**AHH93**]).

DEFINITION 1.5. Let $R$ be a noetherian integral domain and let $K$ be its field of fractions. Let $R^+$ denote the integral closure of $R$ in the algebraic closure $\overline{K}$ of $K$. Let $N$ be a submodule of an $R$-module $M$. An element $z \in M$ is in the plus closure of $N$ in $M$, denoted by $N^+_M$, if the image of $z \otimes 1$ is in the image of $N \otimes_R R^+ \to M \otimes_R R^+$.

In the case $I \subseteq R$ is an ideal of $R$, we will write $I^+$ instead of $I^+_R$ whenever $R$ is clear form the context.

REMARK 1.6.

(1) Let $I$ be an ideal of $R$. One can check that $r \in R$ is in $I^+$ if and only if $r \in IR^+$.
(2) $r \in I^+$ if and only if there is a module-finite extension $R \hookrightarrow S$ such that $r \in IS$. Only one implication requires a proof. Assume that $r \in I^+$; there is
$$r = a_1 g_1 + \cdots + a_n g_n$$
where $a_i \in R^+$ and $\{g_1, \ldots, g_n\}$ is a set of generators of $I$. Since there only finitely many $a_i$, one can choose a module-finite extension $S$ that contains all $a_i$'s. Then $r \in IS$.
(3) Plus closure commutes with localization (Exercise 4.2).
(4) Assume, furthermore, that $R$ is a noetherian integral domain of characteristic $p$. Then $I^+ \subseteq I^*$. Here is one way to see it. Assume $r \in I^+$. Then there is a module-finite extension $R \hookrightarrow S$ such that $r \in IS$ and hence $r^{p^e} \in I^{[p^e]}S$ for all $e \geq 0$. Given any $R$-linear map $\phi : S \to R$, one has $\phi(1)r^{p^e} \in I^{[p^e]}$ for all $e \geq 0$. Hence it suffices to show that there is a $\phi \in \mathrm{Hom}_R(S, R)$ such that $\phi(1) \neq 0$. But this is clear passing to $K$ (the field of fractions of $R$).

[1] It was pointed out by Sudeshna Roy, whom we thank, that the proof had been assigned as an exercise during the lecture delivered by Florian Enescu.

The following theorem ([**Smi94**, Theorem 5.1]) extends Theorem 1.3 to parameter ideals via plus closure.

THEOREM 1.7 (Smith). *Let $R$ be a locally excellent domain of characteristic $p$. Then, for every parameter ideal $I$, one has*

$$I^* = I^+ := IR^+ \cap R.$$

Since plus closure commutes with localization (Exercise 4.2), an immediate consequence of Smith's theorem is that tight closure commutes with localization for parameter ideals in locally excellent domains. One of the key steps in the proof of 1.7 is to prove $0^*_{H^d_{\mathfrak{m}}(R)} = 0^+_{H^d_{\mathfrak{m}}(R)}$, which was alluded to by Thomas Polstra in his lectures.

In a different approach to generalizing results on regular sequences to those on parameter ideals, Aberbach, Hochster, and Huneke consider complexes that are exact up to tight closure. We will recall the definitions as follows.

DEFINITION 1.8. Let $R$ be a noetherian ring of prime characteristic $p$. Let $F^e_*R$ denote the $R$-module that has the same underlying abelian group as $R$, with the $R$-action defined via $r \cdot r' = r^{p^e}r'$ for each $r \in R$ and $r' \in F^e_*R$. The Peskine-Szpiro functor $\mathrm{F}^e(-)$ is defined by $\mathrm{F}^e(M) := F^e_*R \otimes_R M$ for each $R$-module $M$.

For each $z \in M$, write $z^{p^e}$ for the image of $z$ in $\mathrm{F}^e(M)$ under the natural map $M \xrightarrow{z \mapsto 1 \otimes z} \mathrm{F}^e(M)$. For $R$-modules $N \subseteq M$, define $N^{[p^e]}_M$ to be image of $\mathrm{F}^e(N) \to \mathrm{F}^e(M)$. We say that $z \in M$ is in the tight closure $N^*_M$ of $N$ in $M$ if there exists $c \in R^\circ$ such that $cz^{p^e} \in N^{[p^e]}_M$ for $e \gg 0$.

REMARK 1.9 (Remark 8.4 in [**HH90**]). Let $N$ be a submodule of $M$. Then an element $z$ of $M$ is in $N^*_M$ if and only if the image of $z \in M/N$ is in $0^*_{M/N}$. Hence the question about whether $z \in N^*_M$ can be reduced to the case when $N = 0$ by replacing $M$ by $M/N$, $z$ by its image in $M/N$, and $N$ by the 0 submodule of $M/N$.

DEFINITION 1.10. Let $R$ be a noetherian ring of prime characteristic $p$. A complex $C_\bullet$ of $R$-modules is called *phantom acyclic* if the kernel of $\delta_i : C_i \to C_{i-1}$ is contained in the tight closure[2] of the image of $\delta_{i+1} : C_{i+1} \to C_i$ for all $i > 0$. If $F^e(C_\bullet)$ is phantom acyclic for each $e \in \mathbb{N}$, then $C_\bullet$ is called *stably phantom acyclic.*

An $R$ module $M$ is said to have finite phantom projective dimension if there is a stably phantom acyclic complex of projectives $C_\bullet$ such that $H_0(C_\bullet) \cong M$ and $C_j = 0$ for all $j \gg 0$.

REMARK 1.11.

(1) Finite projective dimension clearly implies finite phantom projective dimension. Since every ideal generated by a regular sequence has finite projective dimension, it must have finite phantom projective dimension.
(2) A phantom acyclic complex may not be acyclic. Let $R = k[[x^4, x^3y, xy^3, y^4]]$ and let $I = (x^4, y^4)$. It is well-known (and left as an exercise to the interested reader) that $R/I$ does not have finite projective dimension. However, $R/I$ has finite phantom projective dimension. (*cf.* Theorem 1.12)
(3) Let $R = k[[x^2, x^3]]$. Then the Koszul complex $K(x^2, x^3)$ is phantom acyclic but not stably phantom acyclic. ([**Abe94**, 2.1.2])

[2]Both $\ker(\delta_i)$ and $\mathrm{Im}(\delta_{i+1})$ are viewed as submodules of $C_i$ and the tight closure is computed within $C_i$.

The following theorem ([**AHH93**, Theorem 8.1]) generalizes Theorem 1.3.

THEOREM 1.12 (Aberbach-Hochster-Huneke). *Let $R$ be a complete local ring*[3] *of characteristic $p$.*

(1) *If $R/I$ has finite phantom projective dimension, then*
$$(IW^{-1}R)^* = (I^*)W^{-1}R$$
*for every multiplicatively closed subset $W$ of $R$.*

(2) *If $I$ is generated by monomials in a (partial) system of parameters, then $R/I$ has finite phantom projective dimension.*

The key ingredient of the proof is the following Phantom Acyclicity Criterion (*cf.* [**Hun96**, Theorem 10.3]):

THEOREM 1.13. *Let $R$ be a complete local domain*[4] *of characteristic $p$. Let $G_\bullet$ be a complex of finitely generated free modules of the form:*
$$0 \to G_n \to \cdots \to G_0 \to 0.$$
*Denote the differential $G_i \to G_{i-1}$ by $\delta_i$ and set $r_i = \sum_{t=i}^{n}(-1)^{t-i}\operatorname{rank}(G_t)$. Let $\operatorname{rank}(\delta_i)$ denote the rank of the matrix associated with $\delta_i$. Assume that*

(1) $\operatorname{rank}(G_i) = \operatorname{rank}(\delta_{i+1}) + \operatorname{rank}(\delta_i)$ *for $1 \le i \le n$, and*

(2) $\operatorname{ht}(I_{r_i}(\delta_i)) \ge i$ *for $1 \le i \le n$.*

*Then $G_\bullet$ is phantom acyclic.*

The actual proof of Theorem 1.13 is fairly technical; the interested reader can find a proof in [**HH90**, §9].

## 2. Lecture 2

Even though tight closure does not commute with localization in general, there are still many related and interesting open problems. For instance,

QUESTION 2.1. Let $R$ be a noetherian ring of prime characteristic $p$ and let $f$ be an element in $R$. Is it true that $(IR_f)^* = I^*R_f$ for every ideal $I$?

One of the goals for this lecture is to provide some evidence that this is indeed an interesting problem. One notion that is closely related to the localization problem is $F$-regularity which we recall next:

DEFINITION 2.2. A Noetherian ring of positive characteristic in which each ideal is tightly closed is called *weakly $F$-regular.*

A Noetherian ring of positive characteristic such that $W^{-1}R$ is weakly $F$-regular for every multiplicatively closed subset $W \subseteq R$ is called *$F$-regular.*

REMARK 2.3. Let $R$ be a noetherian ring of prime characteristic $p$, $I$ be an ideal and $W$ be a multiplicatively closed set. If $W$ is disjoint from the union of associated primes of $I^{[p^e]}$ for all $e \ge 1$, then $(W^{-1}I)^* = W^{-1}(I^*)$. Sketch of proof: assume that $z/1 \in (W^{-1}I)^*$ and we wish to show $z \in I^*$. There exist $c \in R^\circ$ and $w_e \in W$ such that $cw_e z^{p^e} \in I^{[p^e]}$ for each $e$. Since $w_e$ is not in any associated prime of $I^{[p^e]}$, we must have $cz^{p^e} \in I^{[p^e]}$ whence $z \in I^*$.

[3]The actual hypothesis in [**AHH93**, Theorem 8.1] is weaker but more technical.

[4]The actual assumptions in [**Hun96**, Theorem 10.3] are weaker but more technical.

This observation can be used to recover results from [**Hun96**, Chapter 1] such as [**Hun96**, Theorem 1.5] which asserts that $R$ is weakly $F$-regular iff $R_{\mathfrak{m}}$ is weakly $F$-regular for every maximal ideal $\mathfrak{m}$ of $R$.

This brings us to the following question:

QUESTION 2.4. Let $R$ be a noetherian ring of prime characteristic $p$. Assume that $R$ is weakly $F$-regular. Is it true that $R_P$ is also weakly $F$-regular for every prime ideal $P$?

In the generality as stated in Question 2.4, the question is open; there is no known counterexample. For the rest of this lecture, we will focus on connections between Questions 2.1 and 2.4, some positive results and some sufficient conditions for positive answers to Questions 2.1 and 2.4.

We begin with the following result due to Murthy ([**Hun96**, Theorem 12.2])

THEOREM 2.5 (Murthy). *Let $k$ be an uncountable field of characteristic $p$ and let $R$ be a finitely generated $k$-algebra. Assume that $R$ is weakly $F$-regular. Then $W^{-1}R$ is weakly $F$-regular for every multiplicatively closed set $W$.*

PROOF. Assume otherwise; that is, there is an ideal $I$ such that $W^{-1}I$ is not tightly closed in $W^{-1}R$. Then there is an element $z \in (W^{-1}I)^* \backslash W^{-1}I$. We may assume that $z \in R$. There is an element $c \in R^\circ$ and elements $w_e \in W$ such that $cw_e z^{p^e} \in I^{[p^e]}$ for all $e$. Replacing $W$ by the multiplicatively closed set generated by $w_e$, we may assume $W$ is countable. Pick a maximal ideal $\mathfrak{m}$ in $W^{-1}R$ such that the image of $z$ in $(W^{-1}R)_{\mathfrak{m}}$ is not in $IR_{\mathfrak{m}}$. Hence $(W^{-1}R)_{\mathfrak{m}}$ is not weakly $F$-regular. Since $W$ is countable and $k$ is uncountable, it follows from Exercise 4.4 that the contraction of $\mathfrak{m}$ to $R$ must also be a maximal ideal. This implies that $R_{\mathfrak{m}} = (W^{-1}R)_{\mathfrak{m}}$ is not weakly $F$-regular, a contradiction, since weak $F$-regularity localizes at maximal ideals by Remark 2.3. □

The following observation indicates that Question 2.1 is an interesting and relevant question.

PROPOSITION 2.6. *A positive answer to Question* 2.1 *implies a positive answer to Question* 2.4.

PROOF. Assume that $R$ is weakly $F$-regular. One can localize $R$ at maximal ideals (still weakly $F$-regular) and then repeatedly localize at primes of dimension 1; this reduces the problem to proving that, *if $R$ is a weakly $F$-regular local ring and $P$ is a prime of dimension 1, then $R_P$ is weakly $F$-regular.*

Note that, a Noetherian local ring $(A, \mathfrak{m})$ is weakly $F$-regular if and only if every $\mathfrak{m}$-primary ideal is tightly closed. Hence it suffices to show that if $I$ is primary to $P$, then $I$ remains tightly closed in $R_P$. Let $f \in R$ be an element whose image in $R/P$ is a parameter; then $R_P$ is a localization of $R_f$ at the maximal ideal $PR_f$. Hence it suffices to show that $I$ remains tightly closed in $R_f$. By our hypothesis, $(IR_f)^* = I^*R_f$. Since $I = I^*$, this finishes the proof of our proposition. □

REMARK 2.7. Staring at Theorem 2.5 and Proposition 2.6, one may wonder whether tight closure commutes with localization at one element under the hypothesis in Theorem 2.5. Using a similar argument as in the proof of Theorem 2.5, one can prove the following statement:

*Let $R$ be a finitely generated $k$-algebra where $k$ is an uncountable field of characteristic $p$. Assuming that tight closure commutes with localization at arbitrary* *maximal ideals in $R$, then $(IR_f)^* = I^*R_f$ for every ideal $I$ and every element $f$ in $R$.*

REMARK 2.8. The proof of Proposition 2.6 also proves that, when tackling Question 2.4, it suffices to consider the following special case of Question 2.1: $I$ is primary to a dimension-1 prime $P$ and $f \in R$ becomes a parameter in $R/P$. The next open problem provides an approach to attacking this special case.

QUESTION 2.9 (Bounded annihilation of local cohomology). Let $(R, \mathfrak{m})$ be a noetherian local ring of characteristic $p$ and let $I$ be an ideal primary to a prime $P$ of dimension 1. Is there a constant $B$ (independent of $e$) such that, for every $e \geq 1$,

$$\mathfrak{m}^{Bp^e} \cdot H^0_{\mathfrak{m}}(R/I^{[p^e]}) = 0?$$

PROPOSITION 2.10. *A positive answer to Question 2.9 implies a positive answer to Question 2.4.*

PROOF. It follows from 2.8 that we are reduce to the case when $I$ is primary to a dimension-1 prime $P$ and $f \in R$ becomes a parameter in $R/P$.

Suppose that $z/1$ is in $(IR_f)^*$ and we wish to show that $z/1 \in I^*R_f$. We may pick $c \in R^\circ$ such that $cz^{p^e} \in I^{[p^e]}R_f$, meaning $cf^{n_e}z^{[p^e]} \in I^{[p^e]}$ for integers $n_e$. (A priori, $n_e$ depends on $e$; our goal here is prove that $n_e$ grows linearly with respect to $p^e$.)

Since $f^{n_e}(cz^{p^e}) \in I^{[p^e]}$, it follows that $cz^{p^e}$ belongs to $H^0_{\mathfrak{m}}(R/I^{[p^e]}) = H^0_{(f)}(R/I^{[p^e]})$. A positive answer to Question 2.9 implies that there is a constant $B$ such that $f^{Bp^e}H^0_{(f)}(R/I^{[p^e]}) = 0$; that is,

$$f^{Bp^e}(cz^{p^e}) \in I^{[p^e]} \quad \forall e \geq 1.$$

It follows that $c(f^Bz)^{[p^e]} \in I^{[p^e]}$ for every $e \geq 1$ and consequently $f^Bz \in I^*$. This proves that $z \in I^*R_f$ and completes the proof. □

It turns out that, in the graded case, Question 2.9 has a positive answer.

THEOREM 2.11. *Let $R = \oplus_{i\geq 0}R_i$ be an $\mathbb{N}$-graded noetherian ring where $R_0 = k$ is a field of positive characteristic $p$. Set $\mathfrak{m} = R_{>0}$. Assume that $R$ is equidimensional and $\dim(R/I) = 1$ for a homogeneous ideal $I$. Then there is a constant $C$ (independent of $e$) such that*

$$\mathfrak{m}^{Cp^e}H^0_{\mathfrak{m}}(R/I^{[p^e]}) = 0$$

*for all $e \geq 1$.*

Theorem 2.11 is proved independently by Huneke ([**Hun00**]) and Vraciu ([**Vra00**]); it also follows readily from the results in [**Zha15**].

There are different approaches to attacking Question 2.9 in literature. We will discuss here two of of them.

Let $(R, \mathfrak{m})$ be a noetherian ring of characteristic $p$ and $I$ be an ideal. In [**SS97**], Smith and Swanson consider the primary decomposition of the Frobenius powers $I^{[p^e]}$

$$I^{[p^e]} = Q_{e,1} \cap \cdots \cap Q_{e,n_e}$$

and ask the following question:

QUESTION 2.12. Does there exist a constant $\ell$ (independent of $e$) such that

$$(\sqrt{Q_{e,i}})^{\ell p^e} \subseteq Q_{e,i} \quad \forall e \geq 1?$$

One motivation behind studying Question 2.12 is the following: let $Q_e$ denote the $\mathfrak{m}$-primary component of $I^{[p^e]}$. Then $Q_e H^0_{\mathfrak{m}}(R/I^{[p^e]}) = 0$ (Exercise 4.5). Hence a positive answer to Question 2.12 produces a positive answer to Question 2.9. Only very partial results are known to Question 2.12, for instance when $R = k[x_1, \dots, x_n]/J$ where $J$ is a monomial ideal and $I$ is also generated by monomials.

One useful observation in the proof of Proposition 2.10 is that $H^0_{\mathfrak{m}}(R/I^{[p^e]}) = H^0_{(f)}(R/I^{[p^e]})$ for any parameter $f$ in $R/I$ when $\dim(R/I) = 1$. Hence one may consider the following variation of Question 2.9 due to Katzman ([**Kat98**, Conjecture 1]):

CONJECTURE 2.13 (Katzman). Let $R$ be a noetherian ring of characteristic $p$ and $I$ be an ideal of $R$. For every element $a \in R$, there exists a constant $\beta$ (depends on $a$, but not on $e$), such that

$$a^{\beta p^e} H^0_{(a)}(R/I^{[p^e]}) = 0, \quad \forall e \geq 1.$$

As stated in [**Kat98**], if Conjecture 2.13 holds, then tight closure commutes with localization at one element whence a positive answer to Question 2.1.

When one restricts attention to rings of finite type over a field (*i.e.* $R \cong k[x_1, \dots, x_n]/J$), one can restate Conjecture 2.13 in terms of the growth of degrees in $x_n$ of elements in a minimal Gröbner basis of $J + I^{[p^e]}$ for an ideal $I$ (*cf.* [**Kat98**, Conjecture 4]). Some partial results are known; for instance,

(1) when $J$ is a binomial ideal and $I$ is a monomial ideal ([**Kat98**, Corollary 11])
(2) when $J$ is a monomial ideal ([**Kat98**, Theorem 8]).

## 3. Lecture 3

In Lecture 2, we have considered the question whether weak $F$-regularity implies $F$-regularity. In this lecture, we will focus on the connection with strong $F$-regularity. Recall:

DEFINITION 3.1. Let $R$ be an $F$-finite[5] noetherian reduced ring of prime characteristic $p$. For each $r \in R$ and $e \geq 1$, let $\theta_{r,e} : R \to R^{1/p^e}$ denote the $R$-module morphism sending 1 to $r^{1/p^e}$.

$R$ is called *strongly $F$-regular* if for every $r \in R^\circ$ there exists $e \geq 1$ such that $\theta_{r,e}$ splits.

QUESTION 3.2 (Hochster-Huneke). Is a weakly $F$-regular ring also strongly $F$-regular?

Variations of this question can be found in the literature (*cf.* [**HH89**, **HH94b**, **LS99**]); it has been referred as the weak-imply-strong problem. A positive answer will certainly imply that weak $F$-regularity localizes since strong $F$-regularity does. In the generality as stated, this problem is wide open.

REMARK 3.3. It was proved by Hochster-Huneke ([**HH89**]) that

(1) Strong $F$-regularity localizes; that is, $R$ is strongly $F$-regular if and only if so is $W^{-1}R$ for every multiplicatively closed subset $W$ of $R$.

[5]Recall that a commutative ring $R$ of positive characteristic $p$ is called $F$-finite if $R$ is a finitely generated $R^p$-module.

(2) Strongly $F$-regular rings are $F$-regular (and are weakly $F$-regular).

We will focus on some known positive results and some approaches. To this end, we would like to characterize weak $F$-regularity and strong $F$-regularity.

THEOREM 3.4 (Theorem 8.23 in [**HH90**]). *Let $R$ be a noetherian ring of characteristic $p$. Then $R$ is weakly $F$-regular if and only if $N = N^*_M$ for every* **finitely generated** *$R$-module $M$ and its submodules $N$.*

PROOF. Only one implication requires a proof. Assume $R$ is weakly $F$-regular and there is a pair $N \subseteq M$ of finitely generated $R$-modules such that $N \neq N^*_M$; we will seek a contradiction. Pick $z \in N^*_M \backslash N$. Consider all the submodules of $M$ that do not contain $z$ but contain $N$ ($N$ certainly is one). There is a maximal element among them, call it $L$. Then $z \notin L$ and $z \in N^*_M \subseteq L^*_M$ (since $N \subseteq L$). We may replace $N$ by $L$ without affecting the problem. By Remark 1.9, we may replace the triple $(z, L, M)$ by $(z, 0, M/L)$ and we will denote this triple by $(z, 0, M)$; that is $z \neq 0$ and $z \in 0^*_M$. The construction of $L$ implies that $z$ is contained in every non-zero submodule of $M$; that is, $M$ is an essential extension of $R \cdot z$.

Let $P$ be an associated prime of $M$; there is an injection $R/P \hookrightarrow M$. View $R/P$ as a submodule of $M$. Since $z$ is in every non-zero submodule of $M$, it is in $R/P$, and it is even in every non-zero submodule of $R/P$. This forces $P$ to be a maximal ideal. A similar argument also shows that $P$ (a maximal ideal) is the only associated prime of $M$. Since $M$ is an essential extension of $R \cdot z$, it follows that $R \cdot z \cong R/P$ and $\mathrm{Socle}(M) \cong R/P$. Since $M$ is finitely generated and its only associated prime is a maximal ideal, $M$ must have finite length; consequently $P^n M = 0$ for some $n$.

Since $P$ is a maximal ideal, it follows from Remark 2.3 that we may replace $R$ by $R_P$ without affecting the problem at hand. Since $R$ is weakly $F$-regular, it must be normal ([**HH90**, 5.11]). By a theorem of Hochster ([**Hoc77**]), every power of $P$ contains an irreducible $P$-primary ideal. Let $Q$ be an irreducible $P$-primary ideal that is contained in $P^n$. Thus, $QM = 0$. View $M$ as an $R/Q$-module. Since $Q$ is irreducible and $P$-primary, $R/Q$ must be Gorenstein (self-injective). That is $R/Q$ is an injective module containing $R/P$. Since $M$ is an essential extension of $R/P \cong R \cdot z$, it follows $M \hookrightarrow R/Q$. Since $Q = Q^*$, it follows from Remark 1.9 that $0 = 0^*_{R/Q}$ which forces $0 = 0^*_M$, a contradiction. □

There is a characterization of strong $F$-regularity, parallel to Theorem 3.4, which can be found in [**Hoc07**, p. 164].

THEOREM 3.5. *Let $R$ be a reduced $F$-finite noetherian complete local ring of prime characteristic $p$. Then the following are equivalent:*

(1) *$R$ is strongly $F$-regular;*
(2) *$N = N^*_M$ for every (not necessarily finitely generated) $R$-module $M$ and every submodule $N$ of $M$.*
(3) *$0^*_E = 0$, where $E = E_R(R/\mathfrak{m})$.*

PROOF. We opt to only sketch a proof of $(c) \Rightarrow (a)$ here. Let $u$ denote a generator of the socle of $E$. Since $E$ is an essential extension of its socle, $0^*_E = 0$ if and only if $u \notin 0^*_E$. That is, for every $c \in R^\circ$, there exists $e$ such that[6]

$$(c^{1/p^e} \otimes u) \neq 0 \in R^{1/p^e} \otimes_R E$$

which is equivalent to, if we tensor $\theta_{c,e} : R \to R^{1/p^e}$ with $E$, the image of $u$ is not 0. This is equivalent to the injectivity of $\theta_{c,e} \otimes_R E : E \to R^{1/p^e} \otimes_R E$ since $E$ is an essential extension of its socle. Applying $\mathrm{Hom}_R(-, E)$, one can see that this condition is equivalent to the surjectivity of $\mathrm{Hom}_R(R^{1/p^e} \otimes_R E, E) \to \mathrm{Hom}_R(E, E)$, which becomes the condition that $\mathrm{Hom}_R(R^{1/p^e}, R) \to R$ is surjective. This amounts to the existence of an $R$-module morphism $\varphi : R^{1/p^e} \to R$ such that $\varphi \circ \theta_{c,e} = id_R$. This finishes the proof of $(c) \Rightarrow (a)$. □

These characterizations suggest an approach to attacking Question 3.2.

QUESTION 3.6. Let $(R, \mathfrak{m})$ be a local ring (or an $\mathbb{N}$-graded ring with $\mathfrak{m} = R_{\geq 1}$) of characteristic $p$. Let $E$ be the injective hull of $R/\mathfrak{m}$. Is it true that

$$0^*_E = \bigcup_{M \subseteq E \ M \text{ is f.g.}} 0^*_M =: 0^{*fg}_E?$$

Here $0^{*fg}_E$ denotes the finitistic tight closure of 0 in $E$ (the union of $0^*_M$ for all finitely generated submodules $M$ of $E$). Hochster and Huneke prove that a positive answer to Question 3.6 (the local case) is equivalent a positive one to Question 3.2.

We start with one of most substantial results to date in this direction, which is due to Lyubeznik and Smith in [**LS99**].

THEOREM 3.7 (Lyubeznik-Smith). *An $\mathbb{N}$-graded $F$-finite ring is weakly $F$-regular if and only if it is strongly $F$-regular.*

At the technical heart of the Lyubeznik-Smith's approach is the following theorem:

THEOREM 3.8 (Lyubeznik-Smith). *Let $R$ be an $F$-finite $\mathbb{N}$-graded ring. Let $M$ be a $\mathbb{Z}$-graded artinian $R$-module and $N$ be a $\mathbb{Z}$-graded submodule of $M$. Then*

$$N^*_M = \cup_{d \in \mathbb{Z}} (N_{\geq d})^*_{M_{\geq d}}.$$

One may mimic the Lyubeznik-Smith's approach in the local case as follows. Let $(R, \mathfrak{m})$ be a local ring of characteristic $p$. Set $E_n := \mathrm{Hom}_R(R/\mathfrak{m}^n, E)$ which has finite length. Consider the following question:

QUESTION 3.9. Is it true that $0^*_E = \cup_{n \in \mathbb{N}} 0^*_{E_n}$?

Question 3.9 remains wide open. The interested reader is referred to [**LS01**, §8] for related discussions and partial results. For instance, Lyubeznik and Smith ([**LS01**, Theorem 8.8]) proves that: *If $(R, \mathfrak{m})$ is Cohen-Macaulay with an isolated non-Gorenstein point, then $0^*_E = 0^{*fg}_E$.*

If one assumes that $R$ is $\mathbb{Q}$-Gorenstein, then Singh proves a much stronger result in [**Sin99**]: *Every $\mathbb{Q}$-Gorenstein splinter is strongly $F$-regular.*

It turns out that one can characterize strong $F$-regularity by considering a bigger (non-commutative) ring: ring of differential operators ([**Smi95**]) or ring of Frobenius operators ([**LS01**]).

---

[6] We may view $R \xrightarrow{r \mapsto r^{p^e}} R$ as $R \subseteq R^{1/p^e}$; then $\mathrm{F}^e(E)$ becomes $R^{1/p^e} \otimes_R E$ and $R$ acts via $r \mapsto r^{p^e}$.

THEOREM 3.10 (Smith). *Let $R$ be a reduced noetherian ring of characteristic $p$ and let $D$ denote the ring of $\mathbb{Z}$-linear differential operators on $R$. Then $R$ is strongly $F$-regular if and only if $R$ is $F$-split and is a finite product of $D$-simple rings.*

Let $R$ be a noetherian ring of characteristic $p$ and $M$ be an $R$-module. Set $\mathcal{F}^e(M)$ to be the set of all additive maps $\phi_e : M \to M$ such that $\phi_e(rm) = r^{p^e}\phi_e(m)$. Define $\mathcal{F}(M)$ as the graded associative (not necessarily commutative) ring

$$\mathcal{F}(M) = \mathcal{F}^0(M) \oplus \mathcal{F}^1(M) \oplus \cdots$$

which is called the ring of Frobenius operators on $M$ ([**LS01**, Definition 3.5]). Strong $F$-regularity can be characterized by the simplicity of $\mathcal{F}(E)$ as shown by Lyubeznik and Smith in [**LS01**] as follows:

THEOREM 3.11 (Lyubeznik-Smith). *Let $(R, \mathfrak{m})$ be a reduced $F$-finite local ring of characteristic $p$ and let $E$ be the injective hull of $R/\mathfrak{m}$. Then $R$ is strongly $F$-regular if and only if $E$ is a simple $\mathcal{F}(E)$-module.*

Hence, to tackle the weak-imply-strong problem, one may investigate the ring of differential operators or the ring of Frobenius operators.

On the other hand, the approach discussed in Lecture 2 indicates that one may consider some uniform behaviors. One such approach is the following result due to Williams in [**Wil95**].

THEOREM 3.12 (Williams). *Let $(R, \mathfrak{m})$ be a $d$-dimensional weakly $F$-regular $F$-finite domain that is an image of a Gorenstein ring. Let $\omega$ be a canonical ideal of $R$. Assume that*

(†) *there exist $c \in R^\circ$, a system of parameters $x_1, \dots, x_d$ and an integer $t_0 \geq 2$ such that for all $t \geq t_0$ and all $e \geq 1$ whenever*

$$z \in \ker(H^i(x_1^{p^e}, \dots, x_d^{p^e}; \omega^{[p^e]}) \to H^i(x_1^{tp^e}, \dots, x_d^{tp^e}; \omega^{[p^e]}))$$

*then*

$$cz \in \ker(H^i(x_1^{p^e}, \dots, x_d^{p^e}; \omega^{[p^e]}) \to H^i(x_1^{t_0p^e}, \dots, x_d^{t_0p^e}; \omega^{[p^e]}))$$

*for all $1 \leq i \leq d$.*

*Then $R$ is strongly $F$-regular.*

It is proved in [**Wil95**] that the condition (†) holds in dimension 3; consequently 3-dimensional weakly $F$-regular rings are strongly $F$-regular. Along a similar circle of ideas, Aberbach investigates the kernel of the natural map from Koszul cohomology to local cohomology in [**Abe02**] which is explored further in [**AP22**] and [**AHP23**]. Given the technical nature, we opt to include here one of the main results from [**AHP23**]; the interested reader is referred to [**AP22**] and [**AHP23**] for the details.

THEOREM 3.13 (Theorem B in [**AHP23**]). *Let $R$ be an excellent weakly $F$-regular local ring of characteristic $p$. Suppose that the anticanonical algebra[7] of $R$ is Noetherian on the punctured spectrum. Then $R$ is strongly $F$-regular.*

[7] An ideal $J$ of $R$ is an anticanonical ideal if it represents the inverse of the canonical divisor in the class group of $R$ and the anticanonical algebra of $R$ is the symbolic Rees algebra $R \oplus J \oplus J^{(2)} \oplus \cdots$

## EXERCISES WITH SOLUTIONS

by SUDESHNA ROY

EXERCISE 4.1. Let $R$ be a noetherian commutative ring and $W$ be a multiplicatively closed subset of $R$. Prove each element in $\left(W^{-1}R\right)^o$ can be written as $c/w$ where $c \in R^o$ and $w \in W$.

*Solution.* Let $c/w \in (W^{-1}R)^\circ$ with $c \in R$ and $w \in W$. We wish to find $c' \in R^\circ$ such that $c/w = c'/w$.

Set $\min R = \{\mathfrak{p}_1, \dots, \mathfrak{p}_s\}$. Since $c/w \in (W^{-1}R)^\circ$, it follows that $c$ is not contained in very $\mathfrak{p}_i$. After rearranging, we assume that $c \notin \mathfrak{p}_i$ for $i = 1, \dots, t$ for some $t \leq s$. If $t = s$, then clearly $c \in R^\circ$. Therefore, suppose that $t < s$ and $c \in \mathfrak{p}_i$ for $i = t+1, \dots, s$. Recall that

$$\min(W^{-1}R) = \{W^{-1}\mathfrak{p} \mid \mathfrak{p} \in \min(R) \text{ and } \mathfrak{p} \cap W = \phi\}.$$

Since $c/w \in \left(W^{-1}R\right)^o$,

$$\{\mathfrak{p} \in \min(R) \mid W^{-1}\mathfrak{p} \in \min(W^{-1}R)\} \subseteq \{\mathfrak{p}_1, \dots, \mathfrak{p}_t\}.$$

Consider the set

$$\left(\bigcap_{i=1}^{t} \mathfrak{p}_i\right) \setminus \left(\bigcup_{i=t+1}^{s} \mathfrak{p}_i\right).$$

We claim that the above set is nonempty and we reason as follows. Otherwise, $(\cap_{i=1}^t \mathfrak{p}_i) \subseteq \left(\cup_{i=t+1}^s \mathfrak{p}_i\right)$. Then $(\cap_{i=1}^t \mathfrak{p}_i) \subseteq \mathfrak{p}_j$ for some $j = t+1, \dots, s$ and hence $\mathfrak{p}_i \subseteq \mathfrak{p}_j$ for some $i = 1, \dots, t$ by [AM, Proposition 1.11], a contradiction.

Let $a$ be an element of $(\cap_{i=1}^t \mathfrak{p}_i) \setminus \left(\cup_{i=t+1}^s \mathfrak{p}_i\right)$. By our choice of $a$, the image $a/1$ of $a$ in $W^{-1}R$ is nilpotent. That is, $w'a^m = 0$ for some integer $m$ and $w' \in W$. Replace $a$ by $a^m$ (consequently $w'a = 0$). For $1 \leq i \leq t$, as $c \notin \mathfrak{p}_i$ and $a \in \mathfrak{p}_i$ so we get $c + a \notin \mathfrak{p}_i$. Again if $t+1 \leq i \leq s$, then as $c \in \mathfrak{p}_i$ and $a \notin \mathfrak{p}_i$ so we get $c + a \notin \mathfrak{p}_i$. Thus $c + a \in R^\circ$.

Set $c' = c + a$, then in $W^{-1}R$ we have

$$\frac{c'}{w} = \frac{w'c'}{w'w} = \frac{w'(c+a)}{w'w} = \frac{w'c}{w'w} = \frac{c}{w}.$$

This completes the proof. □

EXERCISE 4.2. Prove that plus closure commutes with localization for complete local domains.

*Solution.* Let $W$ be a multiplicatively closed subset of $R$. We first claim that $(R_W)^+ \cong (R^+)_W$. Observe that the fields of fractions of $R$ and $R_W$ are the same. We denote the common algebraic closure by $K$. Since $R^+$ is the integral closure of $R$ is $K$ so by [AM, Proposition 5.12] $(R^+)_W$ is the integral closure of $R_W$ in $K$. The claim follows.

Note that $R^+$ is an $R$-module. Thus by [AM, Corollary 3.4] we get that

$$\begin{aligned}(I^+)_W &= (IR^+ \cap R)_W \\ &= (IR^+)_W \cap R_W \\ &= I(R^+)_W \cap R_W \\ &\cong I(R_W)^+ \cap R_W \quad \text{since } (R_W)^+ \cong (R^+)_W \\ &= I_W(R_W)^+ \cap R_W \\ &= (I_W)^+ .\end{aligned}$$

□

EXERCISE 4.3. Let $R = \frac{\mathbb{F}_p[x,y,z]}{(x^3+y^3+z^3)}$ $(p > 5)$ and let $M = H^2_{(x,y,z)}(R)$. Compute $0^*_M$.

*Solution.* By [**Hun96**, Exercise 2.16], the test ideal of the ring $\mathbb{F}_p[x,y,z]/(x^3 + y^3 + z^3)$ is the maximal ideal. Notice that $\dim R = 2$. Suppose that $w_1, w_2$ is a system of parameters of $R$. By [Notes of F. Enescu, Proposition 3.10],

$$0^*_{H^2_\mathfrak{m}(R)} = \left\{\left[\frac{a}{\mathbf{w}^s}\right] : a \in (w_1^s, w_2^s)^* \text{ for } s \geq 1\right\}.$$

Moreover, if $w_1, w_2$ are test elements as well, then by [Notes of F. Enescu, Exercise 7],

$$(w_1^s, w_2^s)^* \subseteq (w_1^s, w_2^s) + \mathbf{w}^{s-1} \cdot (w_1, w_2)^*.$$

Hence

$$0^*_{H^2_\mathfrak{m}(R)} = \{[a/w_1w_2] \mid a \in (w_1, w_2)^*\}.$$

Now let $w_1 = y$ and $w_2 = z$. From [Notes of N. Epstein, Example 9.1] we have $(x,y)^* = (x,y,z^2)$. Hence $0^*_{H^2_\mathfrak{m}(R)} = [z^2/xy]$.

*Alternative solution.* We set $\deg x = \deg y = \deg z = 1$. Then $R$ is an $\mathbb{N}$-graded ring and $x, y$ is a homogeneous system of parameters. The $a$-invariant of $R$ is

$$\mathrm{a}(R) := \max\{n \mid H^2_\mathfrak{m}(R)_n \neq 0\} = 3 - (1+1+1) = 0,$$

see [Notes of F. Enescu, after Definition 5.2]. So $H^2_\mathfrak{m}(R)_{\geq 0} = H^2_\mathfrak{m}(R)_0$ is a proper $F$-stable submodule of $H^2_\mathfrak{m}(R)$. By [Notes of F. Enescu, Theorem 3.11], $0^*_{H^2_\mathfrak{m}(R)}$ is the unique maximal proper $F$-stable submodule of $H^2_\mathfrak{m}(R)$. Hence $H^2_\mathfrak{m}(R)_0 \subseteq 0^*_{H^2_\mathfrak{m}(R)}$. Note that $R$ is a Gorenstein ring. Thus $H^2_\mathfrak{m}(R) \cong E := E_R(k)$ and $\operatorname{soc}(E)$ is an 1-dimensional $k := \mathbb{F}_p$-vector space. Since $H^2_\mathfrak{m}(R)$ is an Artinian $R$-module, $\operatorname{soc}\left(H^2_\mathfrak{m}(R)\right) \subseteq H^2_\mathfrak{m}(R)$ is an essential extension. So $\operatorname{soc}\left(H^2_\mathfrak{m}(R)\right) \cap H^2_\mathfrak{m}(R)_0 \neq 0$ and therefore $\operatorname{soc}\left(H^2_\mathfrak{m}(R)\right)$ sits in degree zero. Since $R$ is a domain and $\mathfrak{m} := (x,y,z)$ is its test ideal, each non-zero $c \in \mathfrak{m}$ is a test element. Suppose that $\eta = [a/x^sy^s] \in 0^*_{H^2_\mathfrak{m}(R)}$. As $a \in (x^s, y^s)^*$ so for each non-zero $c \in \mathfrak{m}$ we have $ca^{p^e} \in (x^s, y^s)^{[p^e]}$ for all $e \geq 0$. In particular, $c \cdot \eta = c \cdot [a/x^sy^s] = [0]$. Hence $\eta \in \operatorname{soc}\left(H^2_\mathfrak{m}(R)\right) \subseteq H^2_\mathfrak{m}(R)_0$. It thus follows that $0^*_{H^2_\mathfrak{m}(R)} = H^2_\mathfrak{m}(R)_0$.

We now claim that $H^2_\mathfrak{m}(R)_0 = [z^2/xy]$. Take an element $\eta = [a/x^sy^s] \in H^2_\mathfrak{m}(R)_0$ for some $s \geq 1$, that is, $\deg a = 2s$. It is enough to assume that $a = x^{b_1}y^{b_2}z^{b_3}$, a monomial such that $b_1 + b_2 + b_3 = 2s$. We write $b_3 = 3b_0 + i$ for $0 \leq i \leq 2$. Then $z^{b_3} = (-x^3 - y^3)^{b_0} z^i$. So without loss of generality, we can assume that $0 \leq b_3 \leq 2$. If $b_3 = 0$ (resp. $b_3 = 1$), then $b_1 + b_2 = 2s$ (resp. $b_1 + b_2 = 2s - 1$). Therefore, $b_1 \geq s$ or $b_2 \geq s$. In either case, $a \in (x^s, y^s)$. So $\eta = [a/x^sy^s] = 0$. Next, let $b_3 = 2$.

So $b_1 + b_2 = 2s - 2$. If $b_1 \geq s$ or $b_2 \geq s$, then $\eta = 0$ by the above observation. Next, assume that $b_1, b_2 \leq s-1$. As $b_1 + b_2 = 2s-2$ so we get that $b_1 = s - 1 = b_2$. Therefore, $[x^{b_1}y^{b_2}z^2/x^s y^s] = [z^2/xy]$. The claim follows. □

NOTE. Observe that for the alternative solution, we don't have to explicitly compute $(x,y)^*$.

EXERCISE 4.4. Let $R$ be a finitely generated domain of positive Krull dimension over an uncountable field $k$. Prove that $W^{-1}R$ can not be a field for any countable multiplicatively closed subset $W$ of $R$.

*Solution* (Following the hints given in Hun96). Let $\dim R = d \geq 1$ and $A = k[x_1, \ldots, x_d]$ be a polynomial ring over $k$ such that $A \subseteq R$ is a finite extension. Then $R$ is integral over $A$ by [AM, Proposition 5.1]. Let $r \in R$. Suppose that $r$ satisfies an equation of the form $r^n + a_1 r^{n-1} + \cdots + a_n = 0$ where $a_i \in A$ for $i = 1, \ldots, n$. So

$$r\left(r^{n-1} + a_1 r^{n-2} + \cdots + a_{n-1}\right) = -a_n \in A.$$

If $n$ is the smallest degree of the monic polynomial, then $u(r) := r^{n-1} + a_1 r^{n-2} + \cdots + a_{n-1} \neq 0$. Thus we can say that every nonzero element in $R$ has a nonzero multiple in $A$. Thus the multiplicatively closed set $W \subseteq R$ induces a multiplicatively closed set $S \subseteq A$, for instance take $S := \langle \{u(r) \cdot r \mid r \in W\} \rangle$. Observe that $u(r_1 r_2)$ and $u(r_1)u(r_2)$ need to be the same (given our choice). The elements of $S$ are monomials in $u(r)$'s, that is, of the form $u(r_1) \cdots u(r_n)$ for some $n \geq 1$ and the number of such elements in $S$ is bounded above by $\mathbb{N}^n$. Thus the cardinality $|S|$ of $S$ is bounded above by $\bigcup_{n \geq 1} \mathbb{N}^n$ which is clearly countable. So we have the inclusions of rings $R \subseteq W^{-1}R \subseteq S^{-1}R$. Therefore, if $W^{-1}R$ is a field then so is $S^{-1}R$.

Next, notice that $A$ has uncountably many prime ideals (the assumption $d \geq 1$ is needed). For instance, the set $\mathcal{Z}$ of prime ideals in $A$ of the form $(a_1 x_1 + \cdots + a_d x_d + a_{d+1})$ with $a_i \in k$ is uncountable, as $k$ is uncountable. Since $S$ is countable, some prime ideals in $\mathcal{Z}$ must survive in $A_S$. So it can not be a field. On the other hand, $A_S \subseteq R_S$ is a finite extension. If $R_S$ is a field, then for any nonunit $f \in A$, $1/f^n \in R$ for each $n \geq 1$, which contradicts the finite generation of $R$ as an $A$-module. □

EXERCISE 4.5. Let $(R, \mathfrak{m})$ be a noetherian local ring and $I$ be an ideal. Let $Q_e$ denote the $\mathfrak{m}$-primary component of $I^{[p^e]}$. Prove that

$$Q_e H^0_{\mathfrak{m}}(R/I^{[p^e]}) = 0.$$

*Solution.* Let $\overline{r} \in H^0_{\mathfrak{m}}(R/I^{[p^e]})$ or some $r \in R$. So $\mathfrak{m}^t \cdot \overline{r} = \overline{0}$, i.e., $\mathfrak{m}^t r \in I^{[p^e]}$ for some $t \geq 1$. Suppose that $I^{[p^e]} = \cap_{i=1}^r \mathfrak{q}_i$ be an irredundant primary decomposition of $I^{[p^e]}$ where $\sqrt{\mathfrak{q}_i} = \mathfrak{p}_i \neq \mathfrak{m}$ for $i = 1, \ldots, r-1$ and $Q_e := q_r$ is the $\mathfrak{m}$-primary component. Then $\mathfrak{m}^t r \in \mathfrak{q}_i$ for $i = 1, \ldots, r$. Take $i < r$. So $\mathfrak{m}^t \subseteq (\mathfrak{q}_i : r)$ and hence $r \in \mathfrak{q}_i$, as otherwise $\sqrt{(\mathfrak{q}_i : r)} = \mathfrak{p}_i$ by [AM, Lemma 4.4]. Therefore, $r \in \cap_{i=1}^{r-1} \mathfrak{q}_i$. Hence $Q_e \cdot r \subseteq q_r \cdot \left(\cap_{i=1}^{r-1} \mathfrak{q}_i\right) \subseteq \cap_{i=1}^r \mathfrak{q}_i = I^{[p^e]}$. The result follows.

*Alternative solution* (Luís Duarte). We set $J := \cap_{i=1}^{r-1} \mathfrak{q}_i$. By [AM, Proposition 1.10], we have an injective map

$$0 \to \frac{R}{I^{[p^e]}} \to \frac{R}{J} \oplus \frac{R}{Q_e},$$

which induces an injective map

$$0 \to H^0_{\mathfrak{m}}\left(R/I^{p^e}\right) \to H^0_{\mathfrak{m}}(R/J) \oplus H^0_{\mathfrak{m}}(R/Q_e).$$

Observe that $\operatorname{Ass}_R R/J = \{\mathfrak{p}_1, \dots, \mathfrak{p}_{r-1}\}$. Thus depth $R/J > 0$ and hence $H^0_{\mathfrak{m}}(R/J) = 0$. Moreover, $R/Q_e$ is an Artinian ring. Therefore, $H^0_{\mathfrak{m}}\left(R/I^{p^e}\right) \hookrightarrow R/Q_e$. The statement follows. □

EXERCISE 4.6. Prove that a strongly $F$-regular ring is $F$-regular.

*Recall* (Notes of L. Ma, Definition 2). An $F$-finite ring $R$ of prime characteristic $p > 0$ is called *strongly $F$-regular* if for every $c \in R^o$, there exists $e > 0$ such that the map $R \to F^e_*(R)$ sending 1 to $F^e_*(c)$ splits as a map of $R$-modules.

*Solution.* Let $R$ be a strongly $F$-regular ring. To prove the statement, we need to show that $R_P$ is weakly $F$-regular for every $P \in \operatorname{Spec}(R)$. By [Notes of L. Ma, Lemma 5], $R_P$ is strongly $F$-regular for every $P \in \operatorname{Spec}(R)$. Thus it is enough to show that a strongly $F$-regular ring $R$ is weakly $F$-regular, that is, every ideal $I \subseteq R$ is tightly closed. Suppose that $I = (a_1, \dots, a_s)$ and $u \in I^*$. Then there is some $c \in R^0$ such that $cu^{p^e} \in I^{[p^e]}$ for all $e \gg 0$. Since $R$ is strongly $F$-regular, there exists $e(c) > 0$ such that the $R$-linear map $R \to F^{e(c)}_*(R)$ sending 1 to $F^{e(c)}_*(c)$ splits.

We now claim that if the $R$-linear map $R \to F^{e'}_*(R)$ sending 1 to $F^{e'}_*(c)$ splits then the $R$-linear map $R \to F^{e''}_*(R)$ sending 1 to $F^{e''}_*(c)$ splits for all $e'' \geq e'$. Let $\varphi : F^{e'}_*(R) \to R$ be the splitting map. Applying the Frobenius functor $F(-)$, we get the $R$-linear map $\varphi' : F^{e'+1}_*(R) \to F_*(R)$. Now consider commutative diagram

$$\begin{array}{ccc} F^{e'}_*(R) & \xrightarrow{\varphi} & R \\ \downarrow{\scriptstyle F} & & {\scriptstyle F}\downarrow \\ F^{e'+1}_*(R) & \xrightarrow{\varphi'} & F_*(R) \end{array}$$

Since by definition $R$ is a $F$-finite ring, the $R$-linear map $F : R \to F_*(R)$ splits by [Notes of L. Ma, Theorem 8]. From the above diagram we thus get an $R$-linear map $\theta : F^{e'+1}_*(R) \to R$ defined by $F^{e'+1}_*(c) \mapsto 1$. The claim follows.

Therefore, we can choose $e > 0$ such that $cu^{p^e} \in I^{[p^e]}$ and the $R$-linear map $R \to F^e_*(R)$ sending 1 to $F^e_*(c)$ splits. Suppose that $\theta : F^e_*(R) \to R$ is the splitting map and

$$cu^{p^e} = \sum_{i=1}^{s} r_i a_i^{p^e} \tag{1}$$

for some $r_i \in R$. We view (1) in $F^e_*(R)$ and apply $\theta$ on the both sides. Then

$$\theta\left(cu^{p^e}\right) = \sum_{i=1}^{s} \theta\left(r_i a_i^{p^e}\right).$$

So $u\theta(c) = \sum_{i=1}^{s} \theta(r_i) a_i$. Thus $u = \sum_{i=1}^{s} \theta(r_i) a_i \in I$ and hence $I^* \subseteq I$. □

EXERCISE 4.7. Prove that a (complete $F$-finite) weakly $F$-regular local ring is strongly $F$-regular.

*Recall* (Notes of F. Enescu, Definition 2.1). We say that a local ring $R$, which is a homomorphic image of a Cohen-Macaulay ring, is *$F$-rational* if and only if every ideal generated by system of parameters in $R$ is tightly closed.

*Solution.* Let $(R, \mathfrak{m}, k)$ be a (complete $F$-finite) Gorenstein weakly $F$-regular local ring with $\dim R = d$. Let $x_1, \ldots, x_d$ be a system of parameters of $R$. Let $\xi = [a/\mathbf{x}^s] \in 0^*_{H^d_{\mathfrak{m}}(R)}$. Then there exists $c \in R^o$ such that $c\eta^{p^e} = 0$ for all $e \gg 0$. Since $R$ is a Cohen-Macaulay ring, by [Notes of F. Enescu, Proposition 3.10], $c\eta^{p^e} = [ca^{p^e}/\mathbf{x}^{sp^e}] = 0$ if and only if $ca^{p^e} \in (x_1^{sp^e}, \ldots, x_d^{sp^e}) = (x_1^s, \ldots, x_d^s)^{[p^e]}$. It thus follows that $a \in (x_1^s, \ldots, x_d^s)^* = (x_1^s, \ldots, x_d^s)$, as $R$ is weakly $F$-regular. So $\xi = 0$ by [Notes of F. Enescu, Proposition 3.10]. Hence $0^*_{H^d_{\mathfrak{m}}(R)} = 0$. As $R$ is Gorenstein local so we have $H^d_{\mathfrak{m}}(R) \cong E_R(k) := E$, the injective hull of the residue field $k$ over $R$. Thus $0^*_E = 0$. Therefore, by [Notes of W. Zhang, Theorem 3.5 (2)], $R$ is strongly $F$-regular.

*Alternative solution* (Likun Xie). As $R$ is weakly $F$-regular so it is $F$-rational. Thus by [Notes of F. Enescu, Theorem 2.3], $R$ is a domain. From [Notes of T. Polstra, Theorem 5.2] it follows that

$$0^*_{H^d_{\mathfrak{m}}(R)} \cong \varinjlim \frac{(x_1, \ldots, x_d)^*}{(x_1, \ldots, x_d)} = 0.$$

*Alternative solution* (Following the hints given in [**Hun96**]). Take $c \in R^o$. Suppose that $\dim R = d$ and $x_1, \ldots, x_d$ is a system of parameters. Let $u$ be an element in $R$ such that its image in $R/(x_1, \ldots, x_d)$ is in the socle of this ring. (In fact, the image of $u$ is a generator of the socle, as $R/(x_1, \ldots, x_d)$ is an Gorenstein Artin ring.) Clearly, there are infinitely many $q$ such that $cu^q \notin (x_1^q, \ldots, x_d^q)$, as otherwise, $u \in (x_1, \ldots, x_d)^* = (x_1, \ldots, x_d)$. Take such $q$ and fix it. Then $c^{1/q}u \notin (x_1, \ldots, x_d)R^{1/q}$. Now consider the composition map $\theta : R \xrightarrow{F} R^{1/q} \xrightarrow{\cdot c^{1/q}} R^{1/q}$ sending 1 to $c^{1/q}$. Pick $t \geq 1$. Clearly, $\theta$ is an inective map, as $R$ is reduced and $c \notin R^o$ is a nonzerodivisor. Tensoring with $R/(x_1^t, \ldots, x_d^t)$, we get

$$\theta'_t : \frac{R}{(x_1^t, \ldots, x_d^t)} \xrightarrow{\cdot c^{1/q}} \frac{R^{1/q}}{(x_1^t, \ldots, x_d^t)R^{1/q}}.$$

We claim that $\theta'_t$ is injective. We prove our claim by contradiction. Suppose that $\ker \theta'_t \neq \overline{0}$. Then $\operatorname{soc}(R/(x_1^t, \ldots, x_d^t)) \cap \ker \theta'_t \neq \overline{0}$. Observe that image of $(x_1 \ldots x_d)^{t-1}u$ in $R/(x_1^t, \ldots, x_d^t)$ is in its socle. Since $R/(x_1^t, \ldots, x_d^t)$ is also a Gorenstein Artin ring, $\theta'_t\left(\overline{(x_1 \ldots x_d)^{t-1}u}\right) = \overline{0}$, i.e., $(x_1 \ldots x_d)^{t-1}uc^{1/q} \in (x_1^t, \ldots, x_d^t)R^{1/q}$. So $(x_1 \ldots x_d)^{q(t-1)}cu^q \in (x_1^{qt}, \ldots, x_d^{qt})R$. Since every ideal in $R$ which is generated by a system of parameters is tightly closed, by [Notes of N. Epstein, Theorem 6.1], it follows that $cu^q \in (x_1^q, \ldots, x_d^q)$. This contradicts our choice of $q$. As $R/(x_1^t, \ldots, x_d^t)$ is a Gorenstein Artin ring so it is self-injective. Thus there exists

$$\theta''_t : \frac{R^{1/q}}{(x_1^t, \ldots, x_d^t)R^{1/q}} \to \frac{R}{(x_1^t, \ldots, x_d^t)}$$

such that $\theta''_t \circ \theta'_t = \mathrm{Id}_R$. Due to the projection map $\pi_t : R^{1/q} \to R^{1/q}/(x_1^t, \ldots, x_d^t)R^{1/q}$, we have the composition

$$R \xrightarrow{\cdot c^{1/q}} R^{1/q} \xrightarrow{\theta''_t \circ \pi_t} \frac{R}{(x_1^t, \ldots, x_d^t)},$$

sending 1 to $1 + (x_1^t, \ldots, x_d^t)$ for each $t \geq 1$. This induces the map

$$R \xrightarrow{\cdot c^{1/q}} R^{1/q} \to \widehat{R}.$$

Since $R$ is a complete local ring, the statement follows. $\square$

## Acknowledgments

The author is grateful to the referee(s) for the helpful comments and suggestions. The author also thanks Dr. Sudeshna Roy for conducting the tutorial session and for providing the solutions to the exercises.

## References


[Abe94] I. M. Aberbach, *Finite phantom projective dimension*, Amer. J. Math. **116** (1994), no. 2, 447–477, DOI 10.2307/2374936. MR1269611

[AHP23] Ian Aberbach, Craig Huneke, and Thomas Polstra, *On the equality of test ideals*, Adv. Math. **441** (2024), Paper No. 109559, 31, DOI 10.1016/j.aim.2024.109559. MR4710865

[AP22] Ian Aberbach and Thomas Polstra, *Local cohomology bounds and the weak implies strong conjecture in dimension 4*, J. Algebra **605** (2022), 37–57, DOI 10.1016/j.jalgebra.2022.04.020. MR4418959

[Abe02] Ian M. Aberbach, *Some conditions for the equivalence of weak and strong $F$-regularity*, Comm. Algebra **30** (2002), no. 4, 1635–1651, DOI 10.1081/AGB-120013205. MR1894033

[AHH93] Ian M. Aberbach, Melvin Hochster, and Craig Huneke, *Localization of tight closure and modules of finite phantom projective dimension*, J. Reine Angew. Math. **434** (1993), 67–114, DOI 10.1515/crll.1993.434.67. MR1195691

[BM10] Holger Brenner and Paul Monsky, *Tight closure does not commute with localization*, Ann. of Math. (2) **171** (2010), no. 1, 571–588, DOI 10.4007/annals.2010.171.571. MR2630050

[Hoc07] M. Hochster, *Foundations of tight closure theory*, `https://dept.math.lsa.umich.edu/~hochster/711F07/fndtc.pdf`, 2007.

[Hoc77] Melvin Hochster, *Cyclic purity versus purity in excellent Noetherian rings*, Trans. Amer. Math. Soc. **231** (1977), no. 2, 463–488, DOI 10.2307/1997914. MR463152

[HH89] Melvin Hochster and Craig Huneke, *Tight closure and strong $F$-regularity*, Mém. Soc. Math. France (N.S.) **38** (1989), 119–133. Colloque en l'honneur de Pierre Samuel (Orsay, 1987). MR1044348

[HH90] Melvin Hochster and Craig Huneke, *Tight closure, invariant theory, and the Briançon-Skoda theorem*, J. Amer. Math. Soc. **3** (1990), no. 1, 31–116, DOI 10.2307/1990984. MR1017784

[HH94a] Melvin Hochster and Craig Huneke, *$F$-regularity, test elements, and smooth base change*, Trans. Amer. Math. Soc. **346** (1994), no. 1, 1–62, DOI 10.2307/2154942. MR1273534

[HH94b] Melvin Hochster and Craig Huneke, *Tight closure of parameter ideals and splitting in module-finite extensions*, J. Algebraic Geom. **3** (1994), no. 4, 599–670. MR1297848

[Hun96] Craig Huneke, *Tight closure and its applications*, CBMS Regional Conference Series in Mathematics, vol. 88, Published for the Conference Board of the Mathematical Sciences, Washington, DC; by the American Mathematical Society, Providence, RI, 1996. With an appendix by Melvin Hochster, DOI 10.1016/0167-4889(95)00136-0. MR1377268

[Hun00] Craig Huneke, *The saturation of Frobenius powers of ideals*, Comm. Algebra **28** (2000), no. 12, 5563–5572, DOI 10.1080/00927870008827175. Special issue in honor of Robin Hartshorne. MR1808589

[Kat98] Mordechai Katzman, *The complexity of Frobenius powers of ideals*, J. Algebra **203** (1998), no. 1, 211–225, DOI 10.1006/jabr.1997.7246. MR1620654

[LS99] Gennady Lyubeznik and Karen E. Smith, *Strong and weak $F$-regularity are equivalent for graded rings*, Amer. J. Math. **121** (1999), no. 6, 1279–1290. MR1719806

[LS01] Gennady Lyubeznik and Karen E. Smith, *On the commutation of the test ideal with localization and completion*, Trans. Amer. Math. Soc. **353** (2001), no. 8, 3149–3180, DOI 10.1090/S0002-9947-01-02643-5. MR1828602

[Sin99] Anurag K. Singh, *$\mathbf{Q}$-Gorenstein splinter rings of characteristic $p$ are $F$-regular*, Math. Proc. Cambridge Philos. Soc. **127** (1999), no. 2, 201–205, DOI 10.1017/S0305004199003710. MR1735920

[Smi94] K. E. Smith, *Tight closure of parameter ideals*, Invent. Math. **115** (1994), no. 1, 41–60, DOI 10.1007/BF01231753. MR1248078

[Smi95] Karen E. Smith, *The D-module structure of F-split rings*, Math. Res. Lett. **2** (1995), no. 4, 377–386, DOI 10.4310/MRL.1995.v2.n4.a1. MR1355702

[SS97] Karen E. Smith and Irena Swanson, *Linear bounds on growth of associated primes*, Comm. Algebra **25** (1997), no. 10, 3071–3079, DOI 10.1080/00927879708826041. MR1465103

[Vra00] Adela Vraciu, *Local cohomology of Frobenius images over graded affine algebras*, J. Algebra **228** (2000), no. 1, 347–356, DOI 10.1006/jabr.1999.8266. MR1760968

[Wil95] Lori J. Williams, *Uniform stability of kernels of Koszul cohomology indexed by the Frobenius endomorphism*, J. Algebra **172** (1995), no. 3, 721–743, DOI 10.1006/jabr.1995.1067. MR1324179

[Zha15] W. Zhang, *A note on the growth of regularity with respect to frobenius*, arXiv:1512.00049, 2015.

Department of Mathematics, Statistics, and Computer Science, University of Illinois at Chicago, 851 S. Morgan St., Chicago, Illinois 60607

*Email address*: wlzhang@uic.edu

# School on Commutative Algebra and Algebraic Geometry in Prime Characteristics, May 2023

Contemporary Mathematics
Volume **831**, 2025
https://doi.org/10.1090/conm/831/16637

# An invitation to equimultiplicity of F-invariants

Ilya Smirnov

*Dedicated to Craig and Mel for putting the wind in our sails.*

## 1. Introduction

This note grew from the lectures that I delivered at the ICTP school, but the content has changed a lot. Since there are several good introductions to F-invariants (such as [**9**, **23**, **32**]), I decided to omit the proofs of the basic results and to focus instead on the notion of *equimultiplicity* and its role in the theory of Hilbert–Kunz multiplicity and F-signature. My motivation was the lack of a good introduction to equimultiplicity in general and the feeling that there is still a lot to be done in the study of equimultiplicity for *F-invariants*, i.e., invariants derived from the Frobenius.

*Equimultiplicity* is defined for a numerical invariant of local ring, let $\epsilon$ be one of these. By localizing at primes, $\epsilon$ defines a function on the spectrum. A prime $\mathfrak{p} \in \operatorname{Spec} R$ is *$\epsilon$-equimultiple* if $\epsilon(\mathfrak{p}) = \epsilon(\mathfrak{q})$ for all $\mathfrak{p} \subseteq \mathfrak{q}$; this condition appears naturally from the stratification of the spectrum by the values of $\epsilon$. The study of $\epsilon$-equimultiplicity can be roughly understood as finding convenient necessary and sufficient conditions – perhaps a criterion – for $\mathfrak{p}$ to be $\epsilon$-equimultiple, conditions that make it a useful tool for other problems. In Section 2, I will discuss how equimultiplicity gives an approach to the following three problems:

(1) showing that an extremal value (which is 1 for the invariants considered here) of $\epsilon$ determines whether a ring is regular,
(2) proving that $\epsilon$ is strongly upper semicontinuous,
(3) resolving singularities.

This note pursues the first two applications for Hilbert–Kunz multiplicity and F-signature. It should be stressed that equimultiplicity theories for multiplicity and the Hilbert–Samuel function were crucial in the resolution of singularities in characteristic 0 and it could happen that one day F-invariants will serve a similar role in the resolution of singularities in positive characteristic. Although F-signature and Hilbert–Kunz multiplicity do not behave well after blowing up the maximal ideal of an isolated singularity (see Section 5 of [**33**]), they still can be a part of a

This research was supported by Spanish Ministry of Science, Innovation and Universities under Grants RYC2020-028976-I funded by MICIU/AEI/10.13039/501100011033 and EI ESF "ESF Investing in your future", PID2021-125052GA-I00 funded by MICIU/AEI/10.13039/501100011033 and FEDER, UE, and EUR2023-143443 funded by MICIU/AEI/10.13039/501100011033 and by the European Union NextGenerationEU/PRTR..

more complex resolving invariant and this is a good enough reason to study their equimultiplicity. With this point of view, the applications in this note are a useful byproduct and a test of tools.

**The structure of this note.** In Section 2, I will start with a brief overview of equimultiplicity and its main uses. Section 3 introduces F-signature and Hilbert–Kunz multiplicity and can be safely skipped by experienced readers. Section 4 considers equimultiplicity for F-signature, it is largely taken from my work with Polstra [**39**].

The following two sections are more novel. In Section 5 I will prove a uniform convergence estimate for Hilbert–Kunz multiplicity and present its applications; the treatment here is a bit different and inspired by Huneke's survey [**23**]. Building on the tools obtained through uniform convergence, Section 6 develops a theory of Hilbert–Kunz equimultiplicity for primes of dimension 1. An important difference with [**43**] is that it is not assumed that $R/\mathfrak{p}$ is regular. This leads to two novel applications: a rigidity theorem for Hilbert–Kunz multiplicity in weakly F-regular rings ([**39**] dealt with strongly F-regular rings) and a different proof of the Watanabe–Yoshida theorem. This proof is a modification of the proof given by Huneke and Yao in [**26**] – I replaced their use of the symbolic power $\mathfrak{p}^{(n)}$ by $(\mathfrak{p}^{[p^e]})^*$, this ideal is $\mathfrak{p}$-primary due to $\mathrm{e}_{HK}$-equimultiplicity. This change makes my proof more technical, but it also makes the proof more *conceptual* as it fits in the framework that can be applied to other invariants. The note is finished with suggestions for further research.

## 2. Equimultiplicity

The purpose of this section is to briefly discuss the notion of equimultiplicity and its potential applications. In what follows, a local ring $(R, \mathfrak{m})$ is defined as a Noetherian unital ring such that the set of all non-invertible elements forms an ideal $\mathfrak{m}$. Let $\epsilon$ be an invariant of local rings, it defines a function on the spectrum of a Noetherian ring $R$ by setting $\epsilon(\mathfrak{p}) = \epsilon(R_\mathfrak{p})$.

If $\epsilon$ takes values in a totally ordered set $\Lambda$, it is desirable that $\epsilon\colon \operatorname{Spec} R \to \Lambda$ is a map of posets. This property comes from the intuitive expectation that if $\mathfrak{p} \subseteq \mathfrak{q} \in \operatorname{Spec} R$, then the singularity of $R_\mathfrak{p}$ cannot be worse than the singularity of $R_\mathfrak{q}$. For example, the Hilbert–Samuel multiplicity, $\mathrm{e}(R)$, is a positive integer, so the inequality becomes $\mathrm{e}(\mathfrak{p}) \leq \mathrm{e}(\mathfrak{q})$. For some other invariants, such as F-signature, the inequality should be opposite.

A prime $\mathfrak{p}$ is $\epsilon$-equimultiple if $\epsilon(\mathfrak{p}) = \epsilon(\mathfrak{q})$ for all $\mathfrak{p} \subseteq \mathfrak{q}$; the intuitive meaning is that the singularity, as measured by $\epsilon$, does not change along $\mathfrak{p}$. In order to work with this condition, it is necessary to characterize it in a more intrinsic way.

Classical cases of equimultiplicity concern the Hilbert–Samuel multiplicity, the Hilbert function, which takes values in sequences of positive integers with an appropriate order [**4**], and the Hilbert polynomial. The following theorem presents the classical case of Hilbert–Samuel multiplicity, see also [**16**].

THEOREM 2.1 ([**10**, Theorem 2], see also [**30**, Corollary, page 121]). *Let $(R, \mathfrak{m})$ be a formally equidimensional local ring and $\mathfrak{p}$ be a prime ideal such that $R/\mathfrak{p}$ is regular. Then $\mathrm{e}(\mathfrak{m}) = \mathrm{e}(\mathfrak{p})$ if and only if the analytic spread of $\mathfrak{p}$, $\ell(\mathfrak{p})$, is minimal, i.e., the dimension of the fiber cone $\oplus \mathfrak{p}^n/\mathfrak{m}\mathfrak{p}^n$ is equal to* $\operatorname{ht} \mathfrak{p}$.

REMARK 2.2. In literature, any ideal $I$ such that $\ell(I) = \mathrm{ht}(I)$ is called equimultiple. While this condition has a connection with multiplicities (see [**30**, Theorem 4]), it is not equivalent to e-equimultiplicity if $R/\mathfrak{p}$ is not regular. When $R/\mathfrak{p}$ is not regular, I do not know of any necessary condition for Hilbert–Samuel equimultiplicity.

For example, any prime ideal in a regular ring is e-equimultiple by the definition, but the height and the analytic spread need not be equal. A family of examples is given by defining ideals of a monomial curve with 3 generators, such as the defining prime $\mathfrak{p} = (x^3 - yz, y^2 - xz, z^5 - xy^3) \subset k[x, y, z]$ of the monomial curve $k[t^3, t^4, t^5]$. As discussed on page 258 of [**22**] the defining primes are generated by a $d$-sequence, so the fiber cone is isomorphic to a polynomial ring in 3 variables by [**22**, Theorem 2.2].

### 2.1. Connection with semicontinuity.

DEFINITION 2.3. Let $\epsilon\colon X \to \mathbb{R}$ be a function on a topological space $X$. We say that $\epsilon$ is *upper semicontinuous* if the subset $X_{<a} := \{x \in X \mid \epsilon(x) < a\}$ is open for all $a \in \mathbb{R}$. We say that $\epsilon$ is *strongly upper semicontinuous* if any set of the form $X_{\leq a} := \{x \in X \mid \epsilon(x) \leq a\}$ is open. Since $X_{<a} = \cup_{b<a} X_{\leq b}$, strong upper semicontinuity is indeed a stronger notion.

Lower semicontinuity and strong lower semicontinuity are defined similarly, by using $X_{>a}$ and $X_{\geq a}$.

Semicontinuity is a notion of continuity appropriate for algebraic geometry and it has a number of important consequences in the theory of F-invariants, see for example [**13, 15, 45**] and the discussion in 2.2. In addition, strong semicontinuity provides a constructible stratification, *i.e.*, the sets $X_{=a} := X_{\leq a} \cap X_{\geq a}$ are locally closed.

In practice, it might be easier to establish strong semicontinuity, for example, due to the following connection with equimultiplicity that comes from Nagata's criterion for openness. The criterion asserts that a set $U \subseteq \operatorname{Spec} R$ is open if and only if for all pairs $\mathfrak{p} \subseteq \mathfrak{q} \in \operatorname{Spec} R$ the following conditions hold

(1) if $\mathfrak{q} \in U$ then $\mathfrak{p} \in U$, and
(2) if $\mathfrak{p} \in U$ there exists an element $s \notin \mathfrak{p}$ such that whenever $s \notin \mathfrak{q}$ then $\mathfrak{q} \in U$.

By applying this to $X_{\leq a}$, the following characterization follows immediately.

PROPOSITION 2.4. *The function $\epsilon\colon \operatorname{Spec} R \to \mathbb{R}$ is strongly upper semicontinuous if and only if $\epsilon$ satisfies the localization inequality,* i.e., *$\epsilon(R_\mathfrak{p}) \leq \epsilon(R_\mathfrak{q})$ when $\mathfrak{p} \subseteq \mathfrak{q}$, and for any prime $\mathfrak{p}$ there is $s \notin \mathfrak{p}$ such that $\mathfrak{p}R_s$ is $\epsilon$-equimultiple.*

Given a convenient sufficient condition for $\epsilon$-equimultiplicity, such as the one in Theorem 2.1, Proposition 2.4 can be easily used to show semicontinuity.

COROLLARY 2.5. *Multiplicity is upper semicontinuous on the spectrum of an excellent ring.*

PROOF. It was shown in [**10**] that multiplicity localizes if $R$ is excellent[1]. Furthermore, for any prime $\mathfrak{p}$ the quotient $R/\mathfrak{p}$ is still excellent, so its regular locus is open by the definition. Thus, there is an element $u \notin \mathfrak{p}$ such that $(R/\mathfrak{p})_u$ is regular.

[1]By a surprising result of [**29**], the localization property for all rings is equivalent to the notorious conjecture of Lech on multiplicity in faithfully flat extensions.

Second, the analytic spread of $\mathfrak{p}R_\mathfrak{p}$ is $\dim R_\mathfrak{p} = \operatorname{ht}\mathfrak{p}$, so it is possible to find $v \notin \mathfrak{p}$ so that the same equality holds in $R_v$. It remains to apply Theorem 2.1 in $R_{uv}$. □

On the other hand, Corollary 6.17 will use Proposition 2.4 to show that Hilbert–Kunz multiplicity is *not* strongly upper semicontinuous.

**2.2. Connection to resolution of singularities.** The notion of equimultiplicity is also closely connected to the main approach to the resolution of singularities, and essentially originates, in the case of Hilbert function, from Hironaka's work [**17**]. I will briefly outline this connection, and, for simplicity, I will assume that $\epsilon(R)$ is upper semicontinuous (a lower semicontinuous function is completely analogous).

Since $X = \operatorname{Spec} R$ is quasi-compact, the open cover $X = \cup X_{<a}$ is finite, hence $\epsilon$ must attain its supremum on $X$. Let $M$ to denote the maximal value of $\epsilon$ on $X$. By semicontinuity, the maximal locus of $\epsilon$, i.e., the set $X_{\max} := \{\mathfrak{p} \mid \epsilon(\mathfrak{p}) = M\} = X \setminus X_{<M}$, is closed. As measured by $\epsilon$, $X_{\max}$ is the locus of the worst singularity, thus a natural attempt to improve the singularity is to blow up $X_{\max}$ or its component. The key difficulty is to show that singularity improves after blowing up.

Because any prime ideal in $X_{\max}$ is $\epsilon$-equimultiple, $\epsilon$-equimultiplicity should help to understand the effect of the blowing up. The theories of equimultiplicity for Hilbert–Samuel multiplicity and Hilbert functions were developed for this reason. As a guideline consider the result of Dade that controls the multiplicity after blowing up an e-equimultiple ideal.

THEOREM 2.6 ([**10**, Theorem 3], [**16**, Corollary 31.2]). *Let $(R, \mathfrak{m})$ be a formally equidimensional local ring. If $\mathfrak{p}$ is a prime ideal such that $\ell(\mathfrak{p}) = \operatorname{ht}\mathfrak{p}$ then any local ring $S$ of the blowing up[2] of $\mathfrak{p}$ that dominates $R$ satisfies the inequality*

$$\mathrm{e}(S) \leq \mathrm{e}(R/\mathfrak{p})\,\mathrm{e}(R_\mathfrak{p}).$$

This theorem only says that the singularity does not get worse, but, in fact, in characteristic 0 one can build an invariant $\epsilon$ so that a resolution of singularities is obtained by consecutively blowing up the maximal locus of $\epsilon$. Loosely speaking, the constructions originating from Hironaka's work complete the Hilbert–Samuel function with additional invariants until the resulting sequence of invariants $\epsilon$ decreases with each blow-up. I refer to [**16**] for more details about equimultiplicity in the Hilbert–Samuel theory.

**2.3. Detection of singularities.** Equimultiplicity also provides an approach to showing that a given invariant $\epsilon$ detects singularities. For simplicity, assume $\epsilon(R) \in [1, \infty)$, so that the goal is to show that if $(R, \mathfrak{m})$ a local ring (perhaps with extra assumptions) and $\epsilon(\mathfrak{m}) = 1$, then $R$ is regular. If $\epsilon$ localizes well, then the condition $\epsilon(R) = 1$ implies that every prime ideal in $R$ is $\epsilon$-equimultiple. This opens an approach by induction on the dimension of $R$: we may take a suitable $\mathfrak{p} \neq \mathfrak{m}$, then $R_\mathfrak{p}$ is regular by induction. Thus the goal becomes to show that if $\mathfrak{p}$ is $\epsilon$-equimultiple and $R_\mathfrak{p}$ is regular, then $R$ must be regular.

In this note, this strategy is applied to Hilbert–Kunz multiplicity and F-signature. Interestingly, as Theorem 2.1 is limited to the situation where $R/\mathfrak{p}$ is regular, one cannot use the same strategy to recover Nagata's original criterion of $\mathrm{e}(R) = 1$.

[2]In other words, $S$ is a homogeneous localization of the Rees algebra at a relevant prime that contracts to $\mathfrak{m}$.

## 3. An introduction to F-invariants

The purpose of this section is to discuss the basics of the positive characteristic commutative algebra and introduce Hilbert–Kunz multiplicity and F-signature. The proofs and more comprehensive treatments can be found in [**9**, **23**, **32**].

A key result for the study of singularities in positive characteristic is the characterization of regular rings due to Kunz. Essentially all proofs showing that a certain invariant detects singularities utilize this famous theorem. In the theorem, $F_*^e R$ denotes an $R$-module obtained as the target of the $e$th iterate of the Frobenius morphism $F^e \colon R \to R$.

THEOREM 3.1 ([**27**]). *Let $(R, \mathfrak{m})$ be a local ring of prime characteristic $p > 0$. Then $R$ is regular if and only if $F_*^e R$ is a flat $R$-module for some (equivalently, all) positive integer $e$.*

If, in addition, $R$ is *F-finite* (that is, $F_*^e R$ is finitely generated), then flatness is upgraded to freeness and the "failure of freeness" of $F_*^e R$ can be measured numerically to provide criteria for regularity. As an example, assume that $R$ is a domain with the field of fractions $L$ and define the generic rank of $M$ as its rank at the generic point, $\dim_L M \otimes_R L$. Consider two functions

$$f(M) = \frac{\min\{N \mid \text{there is } R^{\oplus N} \to M \to 0\}}{\text{generic rank of } M} = \frac{\text{minimal number of generators of } M}{\text{generic rank of } M}$$

and

$$g(M) = \frac{\max\{N \mid \text{there is } M \to R^{\oplus N} \to 0\}}{\text{generic rank of } M} = \frac{\max\{N \mid M = \oplus^N R \oplus M'\}}{\text{generic rank of } M}.$$

It is easy to see that $f(M) \in [1, \infty)$, $g(M) \in [0, 1]$, and $f(M) = 1$ if and only if $M$ is free if and only if $g(M) = 1$.

Kunz's theorem gives numerical tests for singularity through the application of either function to $F_*^e R$ for any $e$. But it happens that the values change with $e$ in a controllable way leading to the following definitions.

DEFINITION 3.2 (Kunz [**28**], Monsky [**35**]). Let $(R, \mathfrak{m})$ be an F-finite local domain of characteristic $p > 0$. Then the Hilbert–Kunz multiplicity of $R$ is

$$\mathrm{e}_{HK}(R) = \lim_{e \to \infty} \frac{\text{minimal number of generators of } F_*^e R}{\text{generic rank of } F_*^e R}$$

DEFINITION 3.3 (Smith–Van der Bergh [**46**], Huneke–Leuschke [**24**]). Let $(R, \mathfrak{m})$ be an F-finite local domain of characteristic $p > 0$. Then the F-signature of $R$ is

$$\mathrm{s}(R) = \lim_{e \to \infty} \frac{\max\{N \mid F_*^e R = \oplus^N R \oplus M\}}{\text{generic rank of } F_*^e R}.$$

It requires work to show that these limits exist and also detect singularity. This will be discussed in Section 5.

**3.1. The generic rank theorem.** In order to work with these invariants it is convenient to restate them in a different way. The starting place is another fundamental result of Kunz [**28**] that computes the generic rank.

THEOREM 3.4 (Kunz). *Let $(R, \mathfrak{m}, k)$ be an F-finite local domain of dimension $d$. Then for any $e > 0$ the generic rank of $F_*^e R$ is $p^{ed}[k^{1/p^e} : k]$.*

Proof. The idea of the proof is that it reduces, via completion and Cohen's structure theorem, to the easiest case where $R = k[[x_1, \ldots, x_d]]$. In this case the computation is very explicit, because $F_*^e R \cong R^{1/p^e} = k^{1/p^e}[[x_1^{1/p^e}, \ldots, x_d^{1/p^e}]]$ as $R$-algebras by mapping $F_*^e r \mapsto r^{1/p^e}$. Let $B = k^{1/p^e}[[x_1, \ldots, x_d]]$. Since $k^{1/p^e}$ is free over $k$, $B$ is a free $R$-module of rank $[k^{1/p^e} : k]$. Furthermore, $k^{1/p^e}[[x_1^{1/p^e}, \ldots, x_d^{1/p^e}]]$ is a free $B$-module of rank $p^{ed}$ by giving explicitly a basis $x_1^{a_1/p^e} \cdots x_d^{a_d/p^e}$ where $0 \leq a_i \leq p^e - 1$. □

Corollary 3.5. *Let $R$ be an F-finite ring and $\mathfrak{p} \subsetneq \mathfrak{q}$ be prime ideals. Then for any $e > 1$*

$$[k(\mathfrak{p})^{1/p^e} : k(\mathfrak{p})] = p^{e \dim R_\mathfrak{q}/\mathfrak{p}R_\mathfrak{q}}[k(\mathfrak{q})^{1/p^e} : k(\mathfrak{q})].$$

Proof. Follows from the theorem by passing to $R_\mathfrak{q}/\mathfrak{p}R_\mathfrak{q}$. □

**3.2. Hilbert–Kunz multiplicity via colengths.** In order to employ ideal-theoretic techniques, it is convenient to restate the first definition of Hilbert–Kunz multiplicity.

In the rest of the paper $\lg_R(M)$ (or $\lg(M)$ if there is no risk of ambiguity) will be used the length of an $R$-module $M$.

Lemma 3.6. *Let $(R, \mathfrak{m})$ be an F-finite local ring with the residue field $k$. Then for any integer $e \geq 0$ the minimal number of generators of $F_*^e R$ is $[k^{1/p^e} : k] \lg(R/\mathfrak{m}^{[p^e]})$, where $\mathfrak{m}^{[p^e]}$ is the ideal generated by $\{x^{p^e} \mid x \in \mathfrak{m}\}$.*

Proof. Since $R$ is local the minimal number of generators of any module is $\lg(M/\mathfrak{m}M)$. Then $\mathfrak{m} \cdot F_*^e R$ consists of elements of the form $F_* \left(\sum_i r_i m_i^{p^e}\right)$, hence

$$\lg(F_*^e R/\mathfrak{m} F_*^e R) = \lg(F_*^e R/\mathfrak{m}^{[p^e]}).$$

By the definition of length, for $L = \lg(R/\mathfrak{m}^{[p^e]})$ there is a filtration

$$\mathfrak{m}^{[p^e]} = M_0 \subsetneq \cdots \subsetneq M_L = R$$

with $M_{i+1}/M_i \cong k$. Since $F_*^e\square$ is exact, it follows that $\lg(F_*^e R/\mathfrak{m} F_*^e R) = [F_*^e k : k]L$. Since $k$ is a field, $F_*^e k \cong k^{1/p^e}$ as $k$-algebras. □

It follows from Theorem 3.4 and the definition that

$$\frac{\text{number of generators of } F_*^e R}{\text{generic rank of } F_*^e R} = \frac{[k^{1/p^e} : k] \lg(R/\mathfrak{m}^{[p^e]})}{p^{e \dim R}[k^{1/p^e} : k]} = \frac{\lg(R/\mathfrak{m}^{[p^e]})}{p^{e \dim R}}.$$

This immediately extends the definition of Hilbert–Kunz multiplicity to all local rings of positive characteristic and all $\mathfrak{m}$-primary ideals.

Definition 3.7. If $(R, \mathfrak{m})$ is a local ring of characteristic $p > 0$, $I$ an $\mathfrak{m}$-primary ideal, and $M$ a finitely generated $R$-module, then the Hilbert–Kunz multiplicity of $I$ on $M$ is

$$\mathrm{e}_{HK}(I; M) = \lim_{e \to \infty} \frac{\lg(M/I^{[p^e]}M)}{p^{e \dim R}}.$$

It is a custom to denote $\mathrm{e}_{HK}(I) := \mathrm{e}_{HK}(I; R)$, moreover, in line with the previous definition of Hilbert–Kunz multiplicity and to highlight its meaning as a measure of singularity of $R$, it is common to denote $\mathrm{e}_{HK}(R) := \mathrm{e}_{HK}(\mathfrak{m})$. Unfortunately, this may cause a certain ambiguity, but I will not use $I = R$ in Definition 3.7 as it is not $\mathfrak{m}$-primary.

It is now time to record several needed properties of Hilbert–Kunz multiplicity.

PROPOSITION 3.8 (Some properties of Hilbert–Kunz multiplicity). *Let $I$ be an $\mathfrak{m}$-primary ideal. Then the following properties hold.*

(1) $\mathrm{e}_{HK}(I; M) = 0$ *if* $\dim M < \dim R$.
(2) $\mathrm{e}_{HK}(I^{[p]}; M) = p^d\, \mathrm{e}_{HK}(I; M)$.
(3) *(Additive) If* $0 \to L \to M \to N \to 0$ *is an exact sequence of finitely generated $R$-modules, then* $\mathrm{e}_{HK}(I; M) = \mathrm{e}_{HK}(I; L) + \mathrm{e}_{HK}(I; N)$.
(4) *(Associativity)* $\mathrm{e}_{HK}(I; M) = \sum_{\mathfrak{p}} \mathrm{e}_{HK}(I; R/\mathfrak{p}) \lg(M_{\mathfrak{p}})$, *where the sum ranges over primes such that* $\dim R/\mathfrak{p} = \dim R$.
(5) *(Lech-type inequality)* $\mathrm{e}_{HK}(I) \leq \lg(R/I)\, \mathrm{e}_{HK}(\mathfrak{m})$.
(6) *Similarly,* $\mathrm{e}_{HK}(I) \leq \lg(J/I)\, \mathrm{e}_{HK}(\mathfrak{m}) + \mathrm{e}_{HK}(J)$ *for any ideal* $I \subseteq J \subset R$.

PROOF. Most of there properties were observed in [**48**]. I want to comment on the last three.

First, recall that $M$ has a prime filtration: $0 = M_0 \subsetneq M_1 \subsetneq \cdots \subsetneq M_L = M$ such that $M_{i+1}/M_i \cong R/\mathfrak{p}_i$. Thus $\mathrm{e}_{HK}(I; M) = \sum_i \mathrm{e}_{HK}(I; R/\mathfrak{p}_i)$ by (3). However, by the first property, $\mathrm{e}_{HK}(I; R/\mathfrak{p}_i) = 0$ unless $\dim R/\mathfrak{p}_i = \dim R$. Since any prime $\mathfrak{p}$ such that $\dim R/\mathfrak{p} = d$ is minimal, in the filtration $(M_i)_{\mathfrak{p}}$ only remain the quotients such that $R/\mathfrak{p}_i = R/\mathfrak{p}$. Thus the localized filtration is a composition series of $M_{\mathfrak{p}}$, its length must be $\lg(M_{\mathfrak{p}})$.

The proof of the last two properties is similar. Take a composition series of $J/I$, where $J = R$ in (5): a sequence $I = I_0 \subsetneq I_1 \subsetneq \cdots \subsetneq I_\ell = J$ where $I_{k+1}/I_k \cong R/\mathfrak{m}$, i.e., $I_{k+1} = (I_k, u_k)$ where $\mathfrak{m} u_k \subseteq I_k$. It follows that $\mathfrak{m}^{[p^e]} u_k^{p^e} \subseteq {I_k}^{[p^e]}$, so

$$\lg(R/I^{[p^e]}) = \sum_{k=0}^{\ell-1} \lg({I_{k+1}}^{[p^e]}/{I_k}^{[p^e]}) + \lg(R/J^{[p^e]}) \leq \sum_{k=0}^{\ell-1} \lg(R/\mathfrak{m}^{[p^e]}) + \lg(R/J^{[p^e]}).$$

As $\lg(R/R^{[p^e]}) = 0$, both (5) and (6) follow after dividing by $p^{e \dim R}$ and taking limits. □

## 4. Equimultiplicity for F-signature

I will now present the results on F-signature that were obtained in [**39**]. As a first step, we should verify that the invariant satisfies the localization property. In order to shorten Definition 3.3, I will denote $a_e(R) = \max\{N \mid F_*^e R = \oplus^N R \oplus M\}$.

LEMMA 4.1. *If $(R, \mathfrak{m})$ is an F-finite local domain then F-signature satisfies the localization property: if $\mathfrak{p}$ is a prime ideal then* $\mathrm{s}(R_{\mathfrak{p}}) \geq \mathrm{s}(R)$.

PROOF. If $M$ is an arbitrary finitely generated $R$-module then the free rank of $M_{\mathfrak{p}}$ may only increase: if $M = \oplus^n R \oplus M'$ then $M'_{\mathfrak{p}}$ may gain free summands. Hence, after confirming that $(F_*^e R)_{\mathfrak{p}} \cong F_*^e R_{\mathfrak{p}}$ one immediately sees that

$$\mathrm{s}(R) = \lim_{e\to\infty} \frac{a_e(R)}{\text{generic rank of } F_*^e R} \leq \lim_{e\to\infty} \frac{a_e(R_{\mathfrak{p}})}{\text{generic rank of } F_*^e R} = \mathrm{s}(R_{\mathfrak{p}}).$$

□

The key insight of the section is the following lemma that produces an adequate number of direct summands.

LEMMA 4.2. *Suppose $(R, \mathfrak{m})$ is an $F$-finite domain and $\mathrm{s}(R) > 0$. Suppose $M$ is a finitely generated $R$-module and $e_0$ is a positive integer such that there is a decomposition*

$$F_*^{e_0} R \cong M \oplus N_0.$$

*If $M$ has no free direct summand, then for any $e \geq e_0$ there is a decomposition $F_*^e R \cong R^{\oplus a_e(R)} \oplus M^{\oplus b_e} \oplus N_e$ such that*

$$\lim_{e \to \infty} \frac{b_e}{\operatorname{rank}(F_*^e R)} = \frac{\mathrm{s}(R)}{\operatorname{rank}(F_*^{e_0} R)} > 0.$$

PROOF. For simplicity, I will write $a_e$ instead of $a_e(R)$. For any $e \in \mathbb{N}$ write $F_*^e R \cong R^{\oplus a_e} \oplus M_e$. Then

$$F_*^{e+e_0} R \cong F_*^{e_0} R^{\oplus a_e} \oplus F_*^{e_0} M_e \cong \oplus^{a_e} (M \oplus N_0) \oplus F_*^{e_0} M_e \cong \oplus^{a_e} M \oplus (\oplus^{a_e} N_0 \oplus F_*^{e_0} M_e).$$

Therefore via this decomposition

$$\lim_{e \to \infty} \frac{b_e}{\operatorname{rank}(F_*^e R)} = \lim_{e \to \infty} \frac{a_{e-e_0}}{\operatorname{rank}(F_*^e R)} = \frac{\mathrm{s}(R)}{\operatorname{rank}(F_*^{e_0} R)}.$$

It remains to observe that the maximal rank of a free summand of any finitely generated $R$-module is an invariant independent of the choice of a direct sum decomposition. This is because $R$ is a direct summand of $N$ if and only if $\hat{R}$ is a direct summand of $\hat{N}$ and a complete local ring satisfies the Krull–Schmidt condition. Therefore, as $M$ has no free summand, it is possible to decompose $\oplus^{a_e} N_0 \oplus F_*^{e_0} M_e \cong R^{\oplus a_{e+e_0}} \oplus N_{e+e_0}$. □

The proof of the localization property leads quite naturally to the following observation on s-equimultiplicity. While a truly intrinsic characterization of s-equimultiplicity is not known, the result still suffices for showing that F-signature detects singularity.

THEOREM 4.3 (Rigidity property). *Let $(R, \mathfrak{m})$ be an $F$-finite local domain and $\mathfrak{p} \in \operatorname{Spec} R$ be a prime ideal. Suppose that $\mathrm{s}(R) > 0$. Then $\mathrm{s}(R) = \mathrm{s}(R_\mathfrak{p})$ if and only if $a_e(R) = a_e(R_\mathfrak{p})$ for every $e \in \mathbb{N}$.*

PROOF. If $a_e(R) = a_e(R_\mathfrak{p})$ for every $e \in \mathbb{N}$ then clearly $\mathrm{s}(R) = \mathrm{s}(R_\mathfrak{p})$. Otherwise, suppose that $a_{e_0}(R_\mathfrak{p}) > a_{e_0}(R)$ for some $e_0$ and write $F_*^{e_0} R \cong R^{\oplus a_{e_0}} \oplus M_{e_0}$. Then $(M_{e_0})_\mathfrak{p}$ has a free $R_\mathfrak{p}$-summand. By Lemma 4.2 for each $e \in \mathbb{N}$ we may decompose

$$F_*^e R \cong R^{\oplus a_e(R)} \oplus M_{e_0}^{\oplus b_e} \oplus N_e$$

where $b_e$ grows as fast as $\operatorname{rank}(F_*^e R)$. Localizing at the prime $P$ we see that $a_e(R_\mathfrak{p}) \geq a_e(R) + b_e$ and

$$\mathrm{s}(R_\mathfrak{p}) = \lim_{e \to \infty} \frac{a_e(R_\mathfrak{p})}{\operatorname{rank}(F_*^e R)} \geq \lim_{e \to \infty} \frac{a_e(R) + b_e}{\operatorname{rank}(F_*^e R)} = \mathrm{s}(R) \frac{1 + \operatorname{rank}(F_*^{e_0} R)}{\operatorname{rank}(F_*^{e_0} R)} > \mathrm{s}(R).$$

□

As an application we get a test useful for computer experiments: if there is a single $e$ such that $a_e(R) \neq a_e(R_\mathfrak{p})$ then $\mathrm{s}(R) \neq \mathrm{s}(R_\mathfrak{p})$.

REMARK 4.4. In the case where $\mathrm{s}(R) = 0$ the F-signature functions can be different. As an example, take $R = \mathbb{F}_p[[x, y, z]]/(x^2 - y^2 z)$ with $p > 2$. It was computed in [**5**, 4.3.2] that $a_e(R) \approx p^e/2$. On the other hand, the prime ideal $\mathfrak{p} = (x, y)$ is the F-splitting prime of $R$ in the sense of [**1**]. Then by [**1**, Proposition 3.6, Corollary 3.4] $a_e(R_\mathfrak{p}) = p^e$.

**4.1. F-signature and non-singularity.** In contrast with the multiplicity-like invariants, F-signature has two distinguished values, 0 and 1. It was shown by Aberbach–Leuschke in [**2**] (see also [**40**] for alternative proofs) that F-signature is positive if and only if $R$ is *strongly F-regular*. It will be used in the proof that a strongly F-regular local ring is a domain, [**32**, Lemma 3.2].

THEOREM 4.5 ([**24**, Corollary 16]). *Let $(R, \mathfrak{m})$ be an F-finite local ring of prime characteristic $p$. Then $\mathrm{s}(R) = 1$ if and only if $R$ is a regular local ring.*

PROOF. Since $\mathrm{s}(R) > 0$, $R$ is a domain, so $R_{(0)}$ is a field and $a_e(R_{(0)}) = \operatorname{rank}(F_*^e R)$. By the localization inequality, $\mathrm{s}(R_{(0)}) = \mathrm{s}(R)$, so $a_e(R) = a_e(R_{(0)}) = \operatorname{rank}(F_*^e R)$ by Theorem 4.3. Therefore $F_*^e R$ is a free $R$-module and $R$ is a regular local ring by Theorem 3.1. □

## 5. Uniform convergence of F-invariants

The theory of $\mathrm{e}_{HK}$-equimultiplicity requires a result interchanging the order of limits. Following [**43**] the formula will be derived from a uniform convergence estimate. This type of convergence estimates first appeared in [**47**]. The proof presented here borrows from Huneke's survey [**23**], these arguments are slightly different from [**43, 47**].

**5.1. Convergence estimates.** One way to think through the proof of existence of Hilbert–Kunz multiplicity is through an equivalence relation on modules, where $M \sim N$ if there exists a module $T$ such that $\dim T < \dim R$ and $|\lg(M/IN) - \lg(N/IN)| < \lg(T/IT)$ for all $\mathfrak{m}$-primary ideals $I$. The uniform convergence argument imposes a further control on the difference module $T$.

LEMMA 5.1. *Let $(R, \mathfrak{m})$ be a local ring and $M, N$ be finitely generated $R$-modules. Suppose that $M_\mathfrak{p} \cong N_\mathfrak{p}$ for every prime $\mathfrak{p}$ with $\dim R/\mathfrak{p} = \dim R$. Then there exist exact sequences*

$$M \to N \to T_1 \to 0 \text{ and } N \to M \to T_2 \to 0$$

*such that $\dim T_1, \dim T_2 < \dim R$. In particular, if $T = T_1 \oplus T_2$, then for any $\mathfrak{m}$-primary ideal $I$*

$$|\lg(M/IN) - \lg(N/IN)| < \lg(T/IT).$$

PROOF. There are only finitely many primes $\mathfrak{p}_i$ with $\dim R/\mathfrak{p}_i = \dim R$. Set $S = R \setminus \cup \mathfrak{p}_i$. By the assumption $S^{-1}M \cong S^{-1}N$. Since Hom commutes with localization for finitely presented modules, there are maps $\psi \colon M \to N$ and $\phi \colon N \to M$ which become isomorphisms after localizing in $S$. Hence $\dim \operatorname{coker} \phi, \operatorname{coker} \psi < \dim R$ and the assertion follows. □

COROLLARY 5.2. *Let $(R, \mathfrak{m}, k)$ be an F-finite reduced local ring and $M$ be a finitely generated $R$-module. Then for $p^\gamma = p^{\dim R}[k^{1/p} : k]$ we have exact sequences*

$$F_* M \to \oplus^{p^\gamma} M \to T_1 \to 0 \text{ and}$$

$$\oplus^{p^\gamma} M \to F_* M \to T_2 \to 0,$$

*where* $\dim T_1, \dim T_2 < \dim R$.

PROOF. Let $\mathfrak{p} \in \operatorname{Minh}(R)$, i.e., $\dim R/\mathfrak{p} = \dim R$. Then $R_{\mathfrak{p}}$ is a field, $k(\mathfrak{p})$, so $M_{\mathfrak{p}}$ is a vector space over it. If $r$ is the rank of this vector space then we compute

$$\dim_{k(\mathfrak{p})}(F_* M)_{\mathfrak{p}} = \dim_{k(\mathfrak{p})} F_* M_{\mathfrak{p}} = r \dim_{k(\mathfrak{p})} F_* k(\mathfrak{p}) = rp^{\gamma},$$

by Theorem 3.4. Therefore $\oplus^{p^{\gamma}} M_{\mathfrak{p}}$ and $F_* M_{\mathfrak{p}}$ are isomorphic for any $\mathfrak{p} \in \operatorname{Minh}(R)$. $\square$

The following result is very much inspired by the treatment of Dutta's lemma ([**14**]) in [**23**, Lemma 3.4].

LEMMA 5.3. *Let $(R, \mathfrak{m}, k)$ be an F-finite local ring with a residue field $k$ and let $M$ be a finitely generated $R$-module. Denote $p^{\dim M}[k : k^p] = p^{\gamma}$. Then there exists a finitely generated $R$-module $T$ such that* $\operatorname{Supp} M = \operatorname{Supp} T$ *and for all $\mathfrak{m}$-primary ideals $I$ and $e \geq 0$*

$$\lg(R/I \otimes_R F_*^e M) \leq p^{e\gamma} \lg(T/IT).$$

PROOF. Note that for all $e \gg 0$ the annihilator of $F_*^e M$ is a radical ideal. We may replace $M$ with $F_*^{e_0} M$: if $T'$ is such that $\lg(R/I \otimes_R F_*^e M) \leq p^{e\gamma} \lg(T'/IT')$ for $e \geq e_0$, then the assertion holds for $T = \oplus_{e=0}^{e_0} F_*^e M \oplus T'$. Hence by replacing $R$ by $R/\operatorname{Ann} M$, it suffices to assume that $R$ is reduced and $\operatorname{Supp} M = \operatorname{Spec} R$.

The proof proceeds by induction on $\dim M$. In the base case, $\dim M = 0$, $M$ is a $k$-vector space. Let $m$ be its dimension. Then

$$\lg(R/I \otimes_R F_*^e M) = \lg(F_*^e M) = m[k^{1/p^e} : k] = mp^{e\gamma}$$

and the base case follows by taking $T = M$.

In the general case, Corollary 5.2 provides a module $N$ such that $\dim N < \dim R$ and there is an exact sequence $\oplus^{p^{\gamma}} M \to F_* M \to N \to 0$. It follows that the inequality

$$\lg(R/I \otimes_R F_*^{e+1} M) \leq p^{\gamma} \lg(R/I \otimes_R F_*^e M) + \lg(R/I \otimes_R F_*^e N)$$

holds for all $e \geq 0$. Adding up these inequalities gives the bound

$$\lg(R/I \otimes F_*^e M) \leq p^{e\gamma} \lg(M/IM) + \sum_{n=0}^{e-1} p^{(e-1-n)\gamma} \lg(R/I \otimes F_*^n N). \tag{1}$$

Since $\dim N \leq \dim R - 1$, by induction there is a module $T'$ such that $\lg(R/I \otimes F_*^n N) \leq p^{n(\gamma-1)} \lg(T'/IT')$, plugging this estimate in (1) yields that

$$\begin{aligned}
\lg(R/I \otimes F_*^e M) &\leq p^{e\gamma} \lg(M/IM) + \sum_{n=0}^{e-1} p^{(e-1-n)\gamma} \lg(R/I \otimes F_*^n N) \\
&\leq p^{e\gamma} \lg(M/IM) + \sum_{n=0}^{e-1} p^{e\gamma - n} \lg(T'/IT') \\
&\leq p^{e\gamma} \left( \lg(M/IM) + \lg(T'/IT') \sum_{n=0}^{e-1} p^{-n} \right) \\
&\leq p^{e\gamma} \left( \lg(M/IM) + 2\lg(T'/IT') \right).
\end{aligned}$$

Since $\operatorname{Spec} R = \operatorname{Supp} M \subseteq \operatorname{Supp}(M \oplus T' \oplus T')$ the assertion follows for $T = M \oplus T' \oplus T'$. $\square$

COROLLARY 5.4. *Let* $(R, \mathfrak{m}, k)$ *be an F-finite local ring and* $M$ *be a finitely generated* $R$*-module. Denote* $p^{\dim M}[k : k^p] = p^\gamma$*. There exists a positive integer* $e_0$ *(which can be taken to be* $0$ *if* $R$ *is reduced) and a finitely generated* $R$*-module* $D$ *such that* $\dim D < \dim R$ *and for any* $\mathfrak{m}$*-primary ideal* $I$ *and all* $e \geq 1$

$$\left|\lg(R/I \otimes_R F_*^{e+e_0} M) - p^{e\gamma} \lg(R/I \otimes_R F_*^{e_0} M)\right| \leq p^{e\gamma} \lg(D/ID).$$

PROOF. There is no harm in assuming that $\dim M = \dim R$. Fix any $e_0$ such that $\sqrt{0}^{[p^{e_0}]} = 0$. Then $F_*^{e+e_0} M$ is an $R_{\mathrm{red}}$-module for any $e \geq 0$. Hence Corollary 5.2 applies and yields that

$$\left|\lg(R/I \otimes_R F_*^{e+1+e_0} M) - p^\gamma \lg(R/I \otimes_R F_*^{e+e_0} M)\right| \leq \lg(R/I \otimes_R F_*^e T),$$

where $\dim T < \dim R$. Adding these equations up bounds the difference

$$\left|\lg(R/I \otimes_R F_*^{e+e_0} M) - p^{e\gamma} \lg(R/I \otimes_R F_*^{e_0} M)\right| \leq \sum_{i=0}^{e-1} p^{(e-i)\gamma} \lg(R/I \otimes_R F_*^i T).$$

Lemma 5.3 provides $T'$ such that $\lg(R/I \otimes_R F_*^i T) \leq p^{i(\gamma-1)} \lg(T'/IT')$ and $\dim T' = \dim T < \dim R$. Set $D = T' \oplus T'$. It follows that

$$\begin{aligned}\left|\lg(R/I \otimes_R F_*^{e+e_0} M) - p^{e\gamma} \lg(R/I \otimes_R F_*^{e_0} M)\right| &\leq \sum_{i=0}^{e-1} \frac{p^{e\gamma - i}}{2} \lg(D/ID) \\ &\leq p^{e\gamma} \lg(D/ID).\end{aligned}$$

□

This corollary, combined with the computation in Lemma 3.6, immediately shows that the sequence $\lg(R/I^{[p^e]})/p^{e \dim R}$, appearing in the definition of Hilbert–Kunz multiplicity, converges by the Cauchy criterion. Even more, by passing to the limit and using the standard field extension result, it leads to the following convergence estimate.

COROLLARY 5.5. *Let* $(R, \mathfrak{m}, k)$ *be a local ring of characteristic* $p > 0$*. There exists a faithfully flat local* $R$*-algebra* $S$*, such that* $S/\mathfrak{m}S$ *is an algebraically closed field, a finitely generated* $S$*-module* $T$ *with* $\dim T < \dim R$*, and a non-negative integer* $e_0$ *(which can be taken to be* $0$ *if* $R$ *is F-finite and reduced) with the following property: for any* $\mathfrak{m}$*-primary ideal* $I$

$$\left|\mathrm{e}_{HK}(I) - p^{-e_0 \dim R} \lg_R(R/I^{[p^{e_0}]})\right| \leq \lg_S(T/IT).$$

PROOF. By Cohen's structure theorem, it is possible to write $\widehat{R}$ as a quotient of a power series ring $R = k[[x_1, \ldots, x_n]]/J$. Denote $S = \overline{k}[[x_1, \ldots, x_n]]/J\overline{k}[[x_1, \ldots, x_n]]$. By tensoring a composition series of a finite length $R$-module $M$ with $S$, it is easy to see that $\lg_R(M) = \lg_S(M \otimes_R S)$. In particular, $\mathrm{e}_{HK}(IS) = \mathrm{e}_{HK}(I)$ for any $\mathfrak{m}$-primary ideal $I$. Hence, it suffices to replace $R$ with the F-finite ring $S$, then the assertion follows from Corollary 5.4 after dividing by $p^{(e+e_0)\gamma}$ and passing to the limit as $e \to \infty$. □

**5.2. Applications.**

5.2.1. *Existence of F-signature.* The estimate in Corollary 5.5 provides uniform convergence of various families of ideals. This applies to F-signature as it can be expressed as the limit of a sequence of ideals that was introduced in [**1**, **49**].

LEMMA 5.6. *Let $(R, \mathfrak{m}, k)$ be an F-finite local domain of characteristic $p > 0$ and denote*

$$I_e = \{x \in R \mid \phi(F_*^e x) \in \mathfrak{m} \text{ for all } \phi \in \operatorname{Hom}(F_*^e R, R)\}.$$

*Then*

$$\frac{\lg(R/I_e)}{p^{ed}} = \frac{\max\{N \mid F_*^e R = \oplus^N R \oplus M\}}{\text{generic rank of } F_*^e R}.$$

*In particular,*

$$\mathrm{s}(R) = \lim_{e \to \infty} \frac{\lg(R/I_e)}{p^{ed}}.$$

PROOF. Let $a_e = a_e(R)$ and take a decomposition $F_*^e R = R^{\oplus a_e} \oplus M_e$. The key observation is that for any $\phi \in \operatorname{Hom}(F_*^e R, R)$ there is a containment $\phi(M_e) \subseteq \mathfrak{m}$, otherwise there will be at least one more splitting. Therefore, $\mathfrak{m}F_*^e R + M_e = \mathfrak{m}^{\oplus a_e} \oplus M_e \subseteq F_*^e I_e$, so $R^{\oplus a_e} \otimes_R R/\mathfrak{m}$ surjects onto $F_*^e(R/I_e)$ bounding its length from above. On the other hand, the minimal generators of $R^{\oplus a_e}$ are clearly not in $F_*^e I_e$, so

$$a_e = \lg_R(F_*^e(R/I_e)) = [k^{1/p^e} : k] \lg(R/I_e)$$

and the assertion follows from Theorem 3.4. □

THEOREM 5.7 (Tucker, [**47**]). *The limit in the definition of* $\mathrm{s}(R)$ *exists.*

PROOF. It is easy to see from the definition that $I_e^{[p]} \subseteq I_{e+1}$, hence $\mathfrak{m}^{[p^e]} \subseteq I_e$. Furthermore, if $\mathfrak{m}$ can be generated by $\mu$ elements, then $\mathfrak{m}^{\mu p^e} \subseteq \mathfrak{m}^{[p^e]}$ and it follows from Corollary 5.5 that $|\mathrm{e}_{HK}(I_e) - \lg(R/I_e)| \leq \lg(T/\mathfrak{m}^{\mu p^e} T)$. The existence of the Hilbert–Samuel polynomial, $n \mapsto \lg(T/\mathfrak{m}^n T)$ for $n \gg 0$, allows to find a constant $C$ such that the bound $\lg(T/\mathfrak{m}^{\mu p^e} T) < C p^{e \dim T}$ holds for all $e \gg 0$. The proof is finished after noting that $\lim_{e \to \infty} \mathrm{e}_{HK}(I_e)/p^{e \dim R}$ exists as this sequence is monotone due to the containment $I_e^{[p]} \subseteq I_{e+1}$. □

5.2.2. *The localization inequality for Hilbert–Kunz multiplicity as a swap of limits.* The main application needed in the next section is a formula presenting the Hilbert–Kunz multiplicity of $R_\mathfrak{p}$ as a limit of $\mathfrak{m}$-primary ideals. For this, consider a two-parameter family of ideals $(I^{[p^e]}, x^{np^e})$ where $I$ is an ideal such that $\dim R/I = 1$ and $x \in \mathfrak{m}$ is such that $(I, x)$ is $\mathfrak{m}$-primary. The first step is to get control on the right-hand side of Corollary 5.5.

LEMMA 5.8. *Let $(R, \mathfrak{m})$ be a local ring of characteristic $p > 0$, $I$ be an ideal such that $\dim R/I = 1$, and $M$ be a finitely generated $R$-module of dimension $d$. If $x \in \mathfrak{m}$ is such that $(I, x)$ is $\mathfrak{m}$-primary then there exists a constant $C$ such that $\lg(M/(I^{[p^e]}, x^{np^e})M) < C p^{ed} n$.*

PROOF. Since $M$ can be written as a quotient of a free $(R/\operatorname{Ann} M)$-module, it suffices to show the statement for $M = R$. In $A = R/I^{[p^e]}$ we have exact sequences

$$0 \to A/(xA + 0 :_A x^n) \to A/x^{n+1}A \to A/x^n A \to 0$$

so $\lg(A/x^{n+1}A) \leq \lg(A/x^nA) + \lg(A/xA)$. Hence, by induction

$$\lg(R/(I^{[p^e]}, x^{np^e})R) \leq n \lg(R/(I^{[p^e]}, x^{p^e})R)$$

and it remains to note that the Hilbert–Kunz function of $(I, x)$ is bounded by some $Cp^{ed}$ as discussed before. □

If $x$ is an element and $I$ is an ideal of a ring, then, with a slight abuse of notation, I will use $xR/I$ to denote the ideal generated by the image of $x$ in $R/I$.

THEOREM 5.9. *Let* $(R, \mathfrak{m})$ *be a local ring of characteristic* $p > 0$, $I$ *be an ideal such that* $\dim R/I = 1$ *and* $\operatorname{ht} I = \dim R - 1$. *Then we have*

$$\lim_{n\to\infty} \frac{\mathrm{e}_{HK}((I, x^n))}{n} = \sum_{\mathfrak{p}\in\Lambda} \mathrm{e}(xR/\mathfrak{p})\, \mathrm{e}_{HK}(IR_\mathfrak{p}),$$

*where the sum is taken over minimal primes* $\mathfrak{p}$ *of* $I$ *such that* $\dim R/\mathfrak{p} = 1$ *and* $\dim R_\mathfrak{p} = \dim R - 1$.

PROOF. Plugging Lemma 5.8 into Corollary 5.5 gives a convergence estimate

$$\left| \mathrm{e}_{HK}((I, x^n)) - \frac{1}{p^{ed}} \lg(R/(I^{[p^e]}, x^{np^e})) \right| < n\frac{C}{p^{e-e_0}} \leq n\frac{C}{p^e}.$$

Therefore, the bisequence $a_{n,e} = \frac{1}{np^{ed}} \lg(R/(I^{[p^e]}, x^{np^e}))$ converges uniformly in $e$. Since the limits $\lim_{e\to\infty} a_{n,e}$ and $\lim_{n\to\infty} a_{n,e} = \mathrm{e}(x^{p^e}R/I^{[p^e]})/p^{ed}$ exist, the standard result says that the double limits exist and are interchangeable. Hence, by the associativity formula for multiplicity applied in $R/I^{[p^e]}$

$$\begin{aligned}
&\lim_{e\to\infty}\lim_{n\to\infty} a_{e,n} \\
&= \lim_{e\to\infty} \frac{1}{p^{e(d-1)}} \mathrm{e}((x)R/I^{[p^e]}) = \lim_{e\to\infty} \frac{1}{p^{e(d-1)}} \sum_{\mathfrak{p}\in \operatorname{Minh}(R/I^{[p^e]})} \mathrm{e}(xR/\mathfrak{p}) \lg(R_\mathfrak{p}/I^{[p^e]}R_\mathfrak{p}) \\
&= \sum_{\mathfrak{p}\in \operatorname{Minh}(R/I)} \mathrm{e}(xR/\mathfrak{p}) \lim_{e\to\infty} \frac{1}{p^{e(d-1)}} \lg(R_\mathfrak{p}/I^{[p^e]}R_\mathfrak{p}) = \sum_{\mathfrak{p}\in\Lambda} \mathrm{e}(xR/\mathfrak{p})\, \mathrm{e}_{HK}(IR_\mathfrak{p}),
\end{aligned}$$

where it was first used that $\operatorname{Minh}(R/I^{[p^e]}) = \operatorname{Minh}(R/I)$ and, second, that this set can be cut down to $\Lambda$ because the limit was taken over $p^{e(d-1)}$. □

An easy consequence is a proof of the localization property for Hilbert–Kunz multiplicity, originally this property was observed in [**28**, Corollary 3.8].

COROLLARY 5.10. *Let* $(R, \mathfrak{m})$ *be a local ring of characteristic* $p > 0$, *and* $\mathfrak{p}$ *be a prime ideal such that* $\dim R_\mathfrak{p} + \dim R/\mathfrak{p} = \dim R$. *Then* $\mathrm{e}_{HK}(R_\mathfrak{p}) \leq \mathrm{e}_{HK}(R)$.

PROOF. By localizing at an appropriate prime $\mathfrak{p} \subsetneq \mathfrak{q} \subseteq \mathfrak{m}$, it is enough to consider the case $\dim R/\mathfrak{p} = 1$. Let $x$ be any element such that $(\mathfrak{p}, x)$ is $\mathfrak{m}$-primary. By Theorem 5.9

$$\lim_{n\to\infty} \frac{\mathrm{e}_{HK}((\mathfrak{p}, x^n))}{n} = \mathrm{e}(xR/\mathfrak{p})\, \mathrm{e}_{HK}(R_\mathfrak{p}).$$

On the other hand, by (5) of Proposition 3.8

$$\mathrm{e}_{HK}((\mathfrak{p}, x^n)) \leq \lg(R/(\mathfrak{p}, x^n))\, \mathrm{e}_{HK}(R).$$

Thus, after taking the limit as $n\to\infty$ we obtain $\mathrm{e}(xR/\mathfrak{p})\, \mathrm{e}_{HK}(R_\mathfrak{p}) \leq \mathrm{e}(xR/\mathfrak{p})\, \mathrm{e}_{HK}(R)$. □

## 6. Equimultiplicity for Hilbert–Kunz multiplicity

I will describe a necessary condition for $e_{HK}$-equimultiplicity in the special case where $\dim R/\mathfrak{p} = 1$. This case is easier to treat but still suffices for the applications. The ideas originate from [**43**], however it was assumed there that $R/\mathfrak{p}$ is regular as in Theorem 2.1.

### 6.1. A review of tight closure.

The characterization of equimultiplicity will be given in terms of tight closure, so let me list the needed facts from [**19**]. A good introduction to the tight closure theory is [**18**].

THEOREM 6.1. *Let $R$ be a Noetherian ring of characteristic $p > 0$.*

(1) *$x \in I^*$ if and only if $cx^{p^e} \in I^{[p^e]}$ for all $e \gg 0$ and $c \in R$ which is not contained in any minimal prime.*
(2) *If $R$ is complete and reduced, then by* [**20**] *it has a* test element*: an element $c$ such that $c$ is not contained in any minimal prime and the containment $cx^{p^e} \in I^{[p^e]}$ holds for all $I$, all $x \in I^*$, and all $e \geq 1$. Test elements exist if $R$ is complete and reduced by* [**20**].
(3) *If $R$ is formally unmixed (i.e., $\operatorname{Ass}(\widehat{R}) = \operatorname{Minh}(\widehat{R})$) then for a pair of $\mathfrak{m}$-primary ideals $I \subseteq J$ we have $e_{HK}(I) = e_{HK}(J)$ if and only if $J \subseteq I^*$.*

LEMMA 6.2. *Suppose that $(R, \mathfrak{m})$ has a test element $c$. Let $I$ be an ideal and $x \in \mathfrak{m}$ be an element. Then*

$$\cap_{n \geq 1} (I + x^n)^* = I^*.$$

PROOF. Take an element $z$ in the intersection on the left. Since $c$ is a test element, $cz^{p^e} \in (I^{[p^e]}, x^{np^e})$ for all $e, n \geq 1$. But $I^{[p^e]} = \cap_n (I^{[p^e]}, x^{np^e})$ by Krull's intersection theorem in $R/I^{[p^e]}$, so it follows that $z \in I^*$. The opposite containment is clear. □

LEMMA 6.3. *Let $(R, \mathfrak{m})$ be a formally unmixed local ring of dimension $d$ and characteristic $p > 0$ and $I$ be an $\mathfrak{m}$-primary ideal. Suppose that $L_e$ is a sequence of ideals such that*

(1) $I^{[p^e]} \subseteq L_e$,
(2) $L_e^{[p]} \subseteq L_{e+1}$,
(3) $\lim_{e \to \infty} \lg(R/L_e)/p^{ed} = e_{HK}(I)$.

*Then $L_e \subseteq (I^{[p^e]})^*$.*

PROOF. The assumptions give containments $I^{[p^{e+e'}]} \subseteq L_e^{[p^{e'}]} \subseteq L_{e+e'}$. Then it is easy to see that $e_{HK}(L_e) = e_{HK}(I^{[p^e]})$ and the assertion follows from Theorem 6.1. □

There is also a converse.

LEMMA 6.4. *Let $(R, \mathfrak{m})$ be a local ring of characteristic $p > 0$ and $I$ be an $\mathfrak{m}$-primary ideal. Let $I_e$ be a sequence of ideals such that $I^{[p^e]} \subseteq I_e \subseteq (I^{[p^e]})^*$. If $R$ has a test element $c$, then*

$$\lim_{e \to \infty} \frac{1}{p^{e \dim R}} \lg(R/I_e) = e_{HK}(I).$$

PROOF. We tensor the exact sequence $R \xrightarrow{c} R \to R/(c) \to 0$ with $R/I_e$. Since $cI_e \subseteq c(I^{[p^e]})^* \subseteq I^{[p^e]}$, it is immediate that the sequence

$$R/I_e \xrightarrow{c} R/I^{[p^e]} \to R/(c, I^{[p^e]}) \to 0$$

is still exact. From the sequence once gets the estimate

$$\lg(R/I_e) \leq \lg(R/I^{[p^e]}) \leq \lg(R/I_e) + \lg(R/(c, I^{[p^e]})R).$$

Note that $\dim R/cR < \dim R$ since $c$ is not contained in any minimal prime, so the estimate gives that $0 = \lim_{e\to\infty} \frac{1}{p^{e \dim R}} \left(\lg(R/I_e) - \lg(R/I^{[p^e]})\right)$ and the assertion follows. □

**6.2. Equimultiplicity.** As a starting point, recall that Theorem 5.9 gives a descent formula for Hilbert–Kunz multiplicity: if $\dim R/\mathfrak{p} = 1$ and $\dim R_\mathfrak{p} = \dim R - 1$ then

$$\lim_{n\to\infty} \frac{\mathrm{e}_{HK}((\mathfrak{p}, x^n))}{n} = \mathrm{e}(xR/\mathfrak{p})\, \mathrm{e}_{HK}(R_\mathfrak{p}).$$

A key observation is that the sequence on the left-hand side is non-increasing.

LEMMA 6.5. *Suppose that* $\dim R/\mathfrak{p} = 1$ *and* $x \notin \mathfrak{p}$. *Then*

$$\frac{\mathrm{e}_{HK}((\mathfrak{p}, x^n))}{n} \geq \frac{\mathrm{e}_{HK}((\mathfrak{p}, x^{n+1}))}{n+1}$$

*Moreover, the equality holds if and only if*

$$\mathrm{e}_{HK}((\mathfrak{p}, x^n)) = n \lim_{e\to\infty} \frac{\lg(R/(x^{p^e}R + \mathfrak{p}^{[p^e]} :_R x^{np^e}))}{p^{e \dim R}}.$$

PROOF. First, take $I$ to be any ideal such that $\dim R/I = 1$ and the image of $x$ is a parameter in $R/I$. We have the exact sequence

$$0 \to R/(xR + I :_R x^k) \to R/(I, x^{k+1}) \to R/(I, x^k) \to 0. \tag{2}$$

These exact sequences can be iterated to deduce that

$$\lg(R/(I, x^n)) = \lg(R/(I, x^{n-1})) + \lg(R/(xR+I :_R x^{n-1})) = \sum_{i=0}^{n-1} \lg(R/(xR+I :_R x^i)). \tag{3}$$

It immediately follows that $\lg(R/(I, x^n)) \geq n \lg(R/(xR + I :_R x^n))$ which allows to deduce from (2) the estimate

$$\begin{aligned}\lg(R/(I, x^{n+1})) &= \lg(R/(I, x^n)) + \lg(R/(xR + I :_R x^n)) \\ &\leq \lg(R/(I, x^n)) + \frac{1}{n} \lg(R/(I, x^n)).\end{aligned}$$

Plugging $I = \mathfrak{p}^{[p^e]}$ and replacing $x$ by $x^{p^e}$ yields the estimate

$$\begin{aligned}\lg(R/(\mathfrak{p}^{[p^e]}, x^{p^e(n+1)})) &= \lg(R/(\mathfrak{p}^{[p^e]}, x^{np^e})) + \lg(R/(x^{p^e}R + \mathfrak{p}^{[p^e]} :_R x^{np^e})) \\ &\leq \frac{n+1}{n} \lg(R/(\mathfrak{p}^{[p^e]}, x^{np^e}))\end{aligned}$$

which implies both assertions after dividing by $p^{e \dim R}$ and taking the limits. □

REMARK 6.6. Before the next proof let me remind some properties of multiplicity. In what follows $S$ is a one-dimensional local ring $S$ and $x$ is a parameter. First, set $I = 0$ in (3) and note that $\lg(S/(xS + 0 :_S x^i)) \leq \lg(S/xS)$. It is then immediate that $\mathrm{e}(xS) \leq \lg(S/xS)$ and that the equality holds if and only if $0 :_S x^i = 0$ for all $i$, *i.e.*, $S$ is Cohen-Macaulay.

In the following proof we will be given $S = R/I$, in which case $\mathrm{e}(xS) = \lg(S/xS)$ if and only if $\mathfrak{m}$ is not an associated prime of $I$, *i.e.*, $I :_R \mathfrak{m}^\infty := \cup_n I :_R \mathfrak{m}^n = I$.

THEOREM 6.7. *Let $(R, \mathfrak{m})$ be a formally unmixed local ring of characteristic $p > 0$ and dimension $d$, $c \in$ be a test element, and $\mathfrak{p}$ be a prime ideal such that* $\dim R/\mathfrak{p} = 1$ *and* $\dim R_\mathfrak{p} = d - 1$. *If* $\mathrm{e}_{HK}(R) = \mathrm{e}_{HK}(R_\mathfrak{p})$ *then* $(\mathfrak{p}^{[p^e]})^* :_R \mathfrak{m}^\infty = (\mathfrak{p}^{[p^e]})^*$ *for all integers $e \geq 0$.*

PROOF. Let $x \in \mathfrak{m} \setminus \mathfrak{p}$ be an arbitrary element. Note that by (5) of Proposition 3.8

$$\mathrm{e}_{HK}((\mathfrak{p}, x)) \leq \mathrm{e}_{HK}(R) \lg(R/\mathfrak{p} + xR) = \mathrm{e}_{HK}(R)\, \mathrm{e}(xR/\mathfrak{p})$$

where the last equality holds by Remark 6.6. Thus by Lemma 6.5 for any positive integer $n$ there is a bound

$$\begin{aligned} \mathrm{e}_{HK}(R)\, \mathrm{e}(xR/\mathfrak{p}) \geq \mathrm{e}_{HK}((\mathfrak{p}, x)) &\geq \frac{\mathrm{e}_{HK}((\mathfrak{p}, x^n))}{n} \\ &\geq \frac{\mathrm{e}_{HK}((\mathfrak{p}, x^{n+1}))}{n+1} \geq \mathrm{e}(xR/\mathfrak{p})\, \mathrm{e}_{HK}(R_\mathfrak{p}). \end{aligned} \tag{4}$$

Since $\mathrm{e}_{HK}(R) = \mathrm{e}_{HK}(R_\mathfrak{p})$ there should be equality throughout, so Lemma 6.5 gives that for any $n$

$$\mathrm{e}_{HK}((\mathfrak{p}, x)) = \frac{1}{n} \mathrm{e}_{HK}((\mathfrak{p}, x^n)) = \lim_{e \to \infty} \frac{\lg(R/(x^{p^e} R + \mathfrak{p}^{[p^e]} :_R x^{np^e}))}{p^{e \dim R}}. \tag{5}$$

It is then easy to check that for any $n$ the sequence of ideals $L_e = x^{p^e} R + \mathfrak{p}^{[p^e]} :_R x^{np^e}$ satisfies the assumptions of Lemma 6.3 for $I = (\mathfrak{p}, x)$. Therefore, for all $e$ and $n$

$$(x^{p^e} R + \mathfrak{p}^{[p^e]} :_R x^{np^e}) \subseteq (x^{p^e} R + \mathfrak{p}^{[p^e]})^*.$$

After letting $n$ grow, it follows that $\mathfrak{p}^{[p^e]} :_R \mathfrak{m}^\infty = \mathfrak{p}^{[p^e]} :_R x^\infty \subseteq (x^{p^e} R + \mathfrak{p}^{[p^e]})^*$. Because $x$ was chosen arbitrarily, we might have taken instead any power $x^k$ in the argument above, so by Lemma 6.2 for any positive integer $e$ there is a containment

$$\mathfrak{p}^{[p^e]} :_R \mathfrak{m}^\infty \subseteq \cap_n (x^{np^e} R + \mathfrak{p}^{[p^e]})^* = (\mathfrak{p}^{[p^e]})^*.$$

Now I claim that $(\mathfrak{p}^{[p^e]})^* : x \subseteq (\mathfrak{p}^{[p^e]})^*$ for any element $x \in \mathfrak{m} \setminus \mathfrak{p}$. If not, there is $a \in \mathfrak{m}$ such that $ax \in (\mathfrak{p}^{[p^e]})^*$, i.e., $ca^{p^{e'}} x^{p^{e'}} \in \mathfrak{p}^{[p^{e+e'}]}$ for all $e' \gg 0$. Hence, $ca^{p^{e'}} \in \mathfrak{p}^{[p^{e+e'}]} : \mathfrak{m}^\infty \subseteq (\mathfrak{p}^{[p^{e+e'}]})^*$, hence $c^2 a^{p^{e'}} \subseteq \mathfrak{p}^{[p^{e+e'}]}$ proving that $a \in (\mathfrak{p}^{[p^e]})^*$. As we may again replace $x$ with any power $x^k$, $(\mathfrak{p}^{[p^e]})^* : x^k \subseteq (\mathfrak{p}^{[p^e]})^*$ for all $k$ and the assertion follows. □

COROLLARY 6.8. *Let $(R, \mathfrak{m})$ be a formally unmixed local ring of characteristic $p > 0$ and dimension $d$, $c$ be a test element, and $\mathfrak{p}$ be a prime ideal such that* $\dim R/\mathfrak{p} = 1$ *and* $\dim R_\mathfrak{p} = d - 1$. *If* $\mathrm{e}_{HK}(R) = \mathrm{e}_{HK}(R_\mathfrak{p})$ *then for any $x \in \mathfrak{m} \setminus \mathfrak{p}$ we have equality*

$$\lg(R/((\mathfrak{p}^{[p^e]})^* + xR)) = \mathrm{e}(xR/(\mathfrak{p}^{[p^e]})^*) = \mathrm{e}(xR/\mathfrak{p}) \lg(R_\mathfrak{p}/(\mathfrak{p}^{[p^e]})^* R_\mathfrak{p}).$$

PROOF. Because $\mathfrak{m}$ is not an associated prime of $(\mathfrak{p}^{[p^e]})^*$ by Theorem 6.7, the quotient ring $S = R/(\mathfrak{p}^{[p^e]})^*$ is a one-dimensional Cohen-Macaulay local ring with the only minimal prime $\mathfrak{p}S$. Thus Remark 6.6 and the associativity formula for the multiplicity immediately demonstrate that $\lg(S/xS) = \mathrm{e}(xS) = \mathrm{e}(xS/\mathfrak{p}) \lg(S_\mathfrak{p})$. □

In the case where $R/\mathfrak{p}$ is regular, the tight closure condition is not just necessary but is also sufficient.

COROLLARY 6.9. *Let $(R, \mathfrak{m})$ be a formally unmixed local ring of characteristic $p > 0$ and dimension $d$, $c$ be a test element, and $\mathfrak{p}$ be a prime ideal such that $R/\mathfrak{p}$ is a one-dimensional regular ring and $\dim R_\mathfrak{p} = d - 1$. Then $\mathrm{e}_{HK}(R) = \mathrm{e}_{HK}(R_\mathfrak{p})$ if and only if $(\mathfrak{p}^{[p^e]})^* :_R \mathfrak{m}^\infty = (\mathfrak{p}^{[p^e]})^*$ for all non-negative integers $e$.*

PROOF. The goal is to show the converse to Theorem 6.7. By the assumption, there is an element $x \in R$ such that $\mathfrak{m} = \mathfrak{p} + xR$ and the image of $x$ is a regular element in all $R/(\mathfrak{p}^{[p^e]})^*$, a one-dimensional Cohen–Macaulay ring. Hence by Lemma 6.4

$$\mathrm{e}_{HK}(\mathfrak{m}) = \lim_{e\to\infty} \frac{\lg(R/((\mathfrak{p}^{[p^e]})^* + x^{p^e}R))}{p^{ed}} = \lim_{e\to\infty} \frac{\mathrm{e}(x^{p^e}R/(\mathfrak{p}^{[p^e]})^*)}{p^{ed}}$$
$$= \lim_{e\to\infty} \frac{\mathrm{e}(x^{p^e}R/\mathfrak{p}) \lg(R_\mathfrak{p}/(\mathfrak{p}^{[p^e]})^* R_\mathfrak{p})}{p^{ed}} = \lim_{e\to\infty} \frac{\lg(R_\mathfrak{p}/(\mathfrak{p}^{[p^e]})^* R_\mathfrak{p})}{p^{e(d-1)}} = \mathrm{e}_{HK}(R_\mathfrak{p}).$$

□

Although it looks different, this formula is a positive characteristic analogue of Theorem 2.1. Namely, by the characterization of the analytic spread in terms of integral closure given by Burch in [**8**], Theorem 2.1 asserts that if $R/\mathfrak{p}$ is a DVR then $\mathrm{e}(R_\mathfrak{p}) = \mathrm{e}(R)$ if and only if $\overline{\mathfrak{p}^n}$ is $\mathfrak{p}$-primary for all $n$.

REMARK 6.10. In the case where $\dim R/\mathfrak{p} > 1$, the characterization is slightly more complicated. Loosely speaking, if $\mathfrak{p}$ is an $\mathrm{e}_{HK}$-equimultiple prime then $\mathfrak{p}^{[p^e]}$ should be "Cohen-Macaulay up to tight closure": if $m = \dim R/\mathfrak{p}$, then there exist elements $x_1, \ldots, x_m$ such that $x_{i+1}$ is not a zerodivisor modulo $(\mathfrak{p}^{[p^e]}, x_1^{p^e}, \ldots, x_i^{p^e})^*$ for all $e$ and all $i < m$. See [**43**].

### 6.3. Applications.

6.3.1. *Rigidity in weakly F-regular rings.* Theorem 4.3 shows that F-signature is rigid. A similar result holds for Hilbert–Kunz multiplicity provided that tight closure is a trivial operation.

DEFINITION 6.11. We say that a ring $R$ is weakly F-regular if $I^* = I$ for all ideals $I$.

PROPOSITION 6.12 (Rigidity). *Let $(R, \mathfrak{m})$ be weakly F-regular local ring and $\mathfrak{p}$ be a prime ideal such that $\dim R/\mathfrak{p} + \dim R_\mathfrak{p} = \dim R$. Then $\mathrm{e}_{HK}(R) = \mathrm{e}_{HK}(R_\mathfrak{p})$ if and only if $\lg(R/\mathfrak{m}^{[p^e]}) = p^{e \dim R/\mathfrak{p}} \lg(R_\mathfrak{p}/\mathfrak{p}^{[p^e]}R_\mathfrak{p})$ for all $e \geq 1$.*

*In particular, if $R$ is a weakly F-regular F-finite domain then $\mathrm{e}_{HK}(R) = \mathrm{e}_{HK}(R_\mathfrak{p})$ if and only if the numbers of minimal generators of $F^e_* R$ and $F^e_* R_\mathfrak{p}$ are equal for all $e \geq 1$.*

PROOF. Suppose that $\mathrm{e}_{HK}(R) = \mathrm{e}_{HK}(R_\mathfrak{p})$ and use induction on $\dim R/\mathfrak{p}$ to reduce to the case where $\dim R/\mathfrak{p} = 1$. Let $x \in \mathfrak{m}$ be such that $\mathfrak{p} + xR$ is $\mathfrak{m}$-primary.

By Corollary 6.8 for all $e \geq 1$

$$\lg(R/(\mathfrak{p}^{[p^e]} + x^{p^e} R)) = p^e \operatorname{e}(xR/\mathfrak{p}) \lg(R_\mathfrak{p}/\mathfrak{p}^{[p^e]} R_\mathfrak{p}).$$

In particular, $\operatorname{e}_{HK}(\mathfrak{p} + xR) = \operatorname{e}(xR/\mathfrak{p}) \operatorname{e}_{HK}(R_\mathfrak{p}) = \operatorname{e}(xR/\mathfrak{p}) \operatorname{e}_{HK}(\mathfrak{m})$.

Now, proceed as in the proof of Proposition 3.8 and take a composition series of $R/(\mathfrak{p} + xR)$:

$$I_0 = \mathfrak{p} + xR \subsetneq I_1 \subsetneq \cdots \subsetneq I_\ell = R,$$

where $I_{k+1} = (I_k, u_k)$ with $\mathfrak{m} u_k \subseteq I_k$. Raising to Frobenius powers gives a filtration $\mathfrak{p}^{[p^e]} + x^{p^e} R \subsetneq I_1^{[p^e]} \subsetneq \cdots$ which computes that

$$\lg(R/(\mathfrak{p}^{[p^e]} + x^{p^e} R)) = \sum_{k=0}^{\ell-1} \lg(R/I_k^{[p^e]} : u_k^{p^e}) \leq \lg(R/(\mathfrak{p} + xR)) \lg(R/\mathfrak{m}^{[p^e]}).$$

Since the limits of the left-hand side and the right-hand side agree, we must have that

$$\lim_{e \to \infty} \frac{\lg(R/I_k^{[p^e]} : u_k^{p^e})}{p^{e \dim R}} = \operatorname{e}_{HK}(\mathfrak{m}) \text{ for all } 0 \leq k \leq \ell - 1.$$

By Lemma 6.3 $\mathfrak{m}^{[p^e]} \subseteq I_k^{[p^e]} : u_k^{p^e} \subseteq (\mathfrak{m}^{[p^e]})^* = \mathfrak{m}^{[p^e]}$, hence the equality holds and it follows that

$$p^e \operatorname{e}(xR/\mathfrak{p}) \lg(R_\mathfrak{p}/\mathfrak{p}^{[p^e]} R_\mathfrak{p}) = \lg(R/(\mathfrak{p}^{[p^e]} + x^{p^e} R)) = \lg(R/(\mathfrak{p} + xR)) \lg(R/\mathfrak{m}^{[p^e]})$$

and one direction follows. The converse is clear. □

6.3.2. *Hilbert–Kunz multiplicity and singularity.* I will now present a new proof of the celebrated theorem of Watanabe and Yoshida which opened the use of Hilbert–Kunz theory in the study of singularities.

THEOREM 6.13 (Watanabe–Yoshida, [**48**, Theorem 1.5]). *Let $(R, \mathfrak{m})$ be a local ring of characteristic $p > 0$. Then $R$ is regular if and only if $\operatorname{e}_{HK}(R) = 1$ and $R$ is formally unmixed (i.e., $\operatorname{Ass} \widehat{R} = \operatorname{Minh}(\widehat{R})$).*

It seems that all proofs of this statement pass through the following crucial lemma which reduces the assertion to the Kunz theorem.

LEMMA 6.14 (Watanabe–Yoshida). *Let $(R, \mathfrak{m})$ be a local ring such that $\operatorname{e}_{HK}(R) = 1$. Suppose that there exists an ideal $I \subseteq \mathfrak{m}^{[p]}$ such that $\operatorname{e}_{HK}(I) \geq \lg(R/I)$. Then $R$ is regular.*

PROOF. By (6) of Proposition 3.8, $\operatorname{e}_{HK}(I) \leq \lg(\mathfrak{m}^{[p]}/I) \operatorname{e}_{HK}(R) + \operatorname{e}_{HK}(\mathfrak{m}^{[p]}) = \lg(\mathfrak{m}^{[p]}/I) + p^d$. Therefore,

$$\lg(\mathfrak{m}^{[p]}/I) + p^d \geq \operatorname{e}_{HK}(I) \geq \lg(R/I) = \lg(\mathfrak{m}^{[p]}/I) + \lg(R/\mathfrak{m}^{[p]})$$

and we obtain that $p^d \geq \lg(R/\mathfrak{m}^{[p]})$. This implies that $R$ is regular by Theorem 3.1. □

Thus the Watanabe–Yoshida theorem amounts to finding an ideal $I$ fitting the assumptions of the lemma. There are several possible options. The original proof of Watanabe and Yoshida uses a *parameter* ideal, but the inequality $\operatorname{e}_{HK}(I) \geq \lg(R/I)$ requires $R$ to be Cohen-Macaulay and this is hard to show. Second, in [**34**] we showed that $\operatorname{e}_{HK}(I) \geq \lg(R/I)$ holds for any *integrally closed* ideal, so one can take for example $I = \overline{\mathfrak{m}^n}$ for any $n \gg 0$. The third construction, due to Huneke and Yao in [**26**], is the simplest of all proofs that I know. In this case, $I = \mathfrak{p}^{(n)} + xR$ where

$\mathfrak{p}$ is a prime such that $\dim R/\mathfrak{p} = 1$ and $R_\mathfrak{p}$ is regular, $x \in \mathfrak{m}^{[p]} \setminus \mathfrak{p}$, and $n$ is taken sufficiently large for $I$ to be contained in $\mathfrak{m}^{[p]}$, such $n$ exists by Chevalley's lemma. Such primes exist: if $\mathfrak{p}$ is such that $\dim R/\mathfrak{p} = 1$ and $\dim R_\mathfrak{p} = \dim R - 1$, then a) $e_{HK}(R_\mathfrak{p}) = e_{HK}(R)$ by the localization property, b) $R_\mathfrak{p}$ is regular by induction on $\dim R$.

The proof below also starts with a prime $\mathfrak{p}$ with the above properties, but it uses $e_{HK}$-equimultiplicity to construct the ideal $I$. This approach conceptualizes the construction of Huneke and Yao, although it requires a lot more work to build the equimultiplicity machinery.

Last, I want to note that if $R$ is strongly F-regular, then one can avoid the Watanabe–Yoshida lemma and modify the proof of Theorem 4.5 to get an easy proof, see [**39**]. However, there does not seem to be an easy way for showing that a complete domain with $e_{HK}(R) = 1$ is strongly F-regular.

THE PROOF OF THE WATANABE–YOSHIDA THEOREM. If $R$ is regular, then $e_{HK}(R) = 1$ by Theorem 3.1, the main direction is the converse. The proof is by induction on $\dim R$. The base case $\dim R = 0$ is trivial since $1 = e_{HK}(R) = \lg(R)$, so $R$ is a field.

It is enough to assume that $R$ is complete, as the assumptions pass to the completion. It also suffices to assume that $R$ is reduced. Namely, $e_{HK}(R) \geq e_{HK}(R_{\mathrm{red}}) \geq 1$, so $e_{HK}(R_{\mathrm{red}}) = 1$. Now, if $R_{\mathrm{red}}$ is regular, then $\mathfrak{p} = \sqrt{0}$ is prime and, by the associativity formula,

$$1 = e_{HK}(R) = e_{HK}(R/\mathfrak{p}) \lg(R_\mathfrak{p}) = \lg(R_\mathfrak{p}).$$

Hence, $R_\mathfrak{p}$ is a field. Because $\operatorname{Ass}(R) = \{\mathfrak{p}\}$, $R \setminus \mathfrak{p}$ consists of non zerodivisors, so $\mathfrak{p} = 0$ and $R = R_{\mathrm{red}}$ is regular.

It is known by [**20**] that a complete reduced ring has a test element. As explained before the proof, there is a prime ideal $\mathfrak{p}$ such that $\dim R/\mathfrak{p} = 1$ and $e_{HK}(R_\mathfrak{p}) = 1$, so $R_\mathfrak{p}$ is regular by induction.

CLAIM 1. For $L_e = (\mathfrak{p}^{[p^e]})^* + x^{p^e}R$ we have $\lg(R/L_e) = e_{HK}(L_e)$.

PROOF. First, by Corollary 6.8

$$\lg(R/((\mathfrak{p}^{[p^e]})^* + x^{p^e}R)) = e(x^{p^e}R/(\mathfrak{p}^{[p^e]})^*) = p^e\, e(xR/\mathfrak{p}) \lg(R_\mathfrak{p}/(\mathfrak{p}^{[p^e]})^* R_\mathfrak{p}).$$

Since $R_\mathfrak{p}$ is regular, we compute $\lg(R_\mathfrak{p}/(\mathfrak{p}^{[p^e]})^* R_\mathfrak{p}) = \lg(R_\mathfrak{p}/\mathfrak{p}^{[p^e]} R_\mathfrak{p}) = p^{e \dim R_\mathfrak{p}}$. Therefore

$$\lg(R/((\mathfrak{p}^{[p^e]})^* + x^{p^e}R)) = p^{e \dim R}\, e(xR/\mathfrak{p}).$$

Second, $e_{HK}((\mathfrak{p}^{[p^e]})^* + x^{p^e}R) = e_{HK}((\mathfrak{p} + xR)^{[p^e]}) = p^{e \dim R} e_{HK}(\mathfrak{p} + xR)$, because $(\mathfrak{p}^{[p^e]})^* + x^{p^e}R$ is contained in the tight closure of $(\mathfrak{p} + xR)^{[p^e]}$. But then the inequalities (4) in the proof of Theorem 6.7 show that $e_{HK}(\mathfrak{p} + xR)) = e_{HK}(xR/\mathfrak{p})\, e_{HK}(R_\mathfrak{p}) = e_{HK}(xR/\mathfrak{p})$ and the claim follows. $\square$

It remains to show that $L_e \subseteq \mathfrak{m}^{[p]}$ for a large $e$. This follows from Chevalley's lemma or from the properties of integral closure: there a constant $c$ such that $L_e \subseteq \overline{(\mathfrak{p}, x)^e} \subseteq (\mathfrak{p}, x)^{e-c}$ because $R$ is a complete domain, see [**25**, Proposition 5.3.4]. The assertion now follows from Lemma 6.14. $\square$

6.3.3. *An application to semicontinuity.* The study of semicontinuity in the theory of F-invariants originates from [**28**] and [**15**]. It is known that Hilbert–Kunz multiplicity is an upper semicontinuous function ([**31, 42**]) and F-signature is a lower semicontinuous function ([**31, 38, 40**]) on the spectrum of an excellent ring, see also [**13**] for results on Frobenius Betti numbers. However, it was originally believed that Hilbert–Kunz multiplicity would be strong upper semicontinuous just as multiplicity. I will now present an argument from [**43**] that shows that this is not the case. The counter-example is based on Monsky's computation of the Hilbert–Kunz multiplicity of a family of quartics in 3 variables in [**37**].

THEOREM 6.15 (Monsky, [**37**]). *Let $K$ be an algebraically closed field of characteristic 2. For $\alpha \in K$ let $Q_\alpha = K[[x,y,z]]/(z^4 + xyz^2 + (x^3+y^3)z + \alpha x^2y^2)$. Then*

(1) $\mathrm{e}_{HK}(Q_\alpha) = 3 + \frac{1}{2}$, *if* $\alpha = 0$,
(2) $\mathrm{e}_{HK}(Q_\alpha) = 3 + 4^{-m}$, *if $\alpha \neq 0$ is algebraic over $\mathbb{Z}/2\mathbb{Z}$, where* $m = [\mathbb{Z}/2\mathbb{Z}(\lambda) : \mathbb{Z}/2\mathbb{Z}]$ *for $\lambda$ such that* $\alpha = \lambda^2 + \lambda$
(3) $\mathrm{e}_{HK}(Q_\alpha) = 3$ *if $\alpha$ is transcendental over* $\mathbb{Z}/2\mathbb{Z}$.

PROOF. The last two cases are computed by Monsky in [**37**]. In the first case there is a factorization

$$\begin{aligned} z^4 + xyz^2 + (x^3+y^3)z &= z(x+y+z)(x^2+y^2+z^2+zx+zy+xy) \\ &= z(x+y+z)((x+y+z)^2+zy). \end{aligned}$$

Hence, by the associativity formula

$$\begin{aligned} \mathrm{e}_{HK}(Q_0) &= \mathrm{e}_{HK}(K[x,y]) + \mathrm{e}_{HK}(K[x,y,z]/(x+y+z)) \\ &\quad + \mathrm{e}_{HK}(K[x,y,z]/((x+y+z)^2+zy)) \\ &= 2 + \mathrm{e}_{HK}(K[x,y,z]/(x^2+zy)) = 3\frac{1}{2}, \end{aligned}$$

as the last equation, $(A_1)$-singularity, was computed in [**48**, Theorem 5.4]. □

Building on this computation, in [**7**] Brenner and Monsky showed that tight closure does not commute with localization on the hypersurface[3]

$$R = F[x,y,z,t]/(z^4 + xyz^2 + (x^3+y^3)z + tx^2y^2),$$

where $F$ is an algebraic closure of $\mathbb{Z}/2\mathbb{Z}$. By the Jacobian criterion, the singular locus of this hypersurface is $V((x,y,z))$ and $\mathfrak{p} = (x,y,z)$ is prime, in fact $R/\mathfrak{p} \cong F[t]$.

PROPOSITION 6.16. *Let $F$ be the algebraic closure of $\mathbb{Z}/2\mathbb{Z}$,*

$$R = F[x,y,z,t]/(z^4 + xyz^2 + (x^3+y^3)z + tx^2y^2),$$

*and $\mathfrak{p} = (x,y,z)$, a prime ideal of $R$. Then $\mathrm{e}_{HK}(\mathfrak{p}) = 3$, but $\mathrm{e}_{HK}(\mathfrak{m}) > 3$ for any maximal ideal $\mathfrak{m}$ containing $\mathfrak{p}$.*

---

[3]Monsky showed that quartics split into two families and studied their Hilbert–Kunz multiplicities in [**36, 37**]. The hypersurface of Brenner and Monsky parametrizes the family appearing in [**37**]. Yet, a recent work [**6**] establishes the failure of localization in the second family [**36**].

PROOF. The residue field of $R_{\mathfrak{p}}$ is $K = F(t)$, so if we define $Q_t$ as in Theorem 6.15, then $\widehat{R_{\mathfrak{p}}} \cong Q_t$ by Cohen's structure theorem. Thus $e_{HK}(R_{\mathfrak{p}}) = 3$. Since $F$ is algebraically closed, all maximal ideals containing $\mathfrak{p}$ are of the form $\mathfrak{m}_\alpha = (\mathfrak{p}, t-\alpha)$ for $\alpha \in F$. By the way of contradiction, suppose that $e_{HK}(\mathfrak{m}_\alpha) = e_{HK}(\mathfrak{p})$. Then by Corollary 6.8

$$\lg(R_{\mathfrak{m}_\alpha}/(t-\alpha, (\mathfrak{p}^{[p^e]})^*)) = \lg(R_{\mathfrak{p}}/(\mathfrak{p}^{[p^e]})^* R_{\mathfrak{p}})$$

for all $e \geq 1$. Now, Lemma 6.4 computes that

$$\begin{aligned} e_{HK}(R_{\mathfrak{m}_\alpha}/(t-\alpha)R_{\mathfrak{m}_\alpha}) &= \lim_{e\to\infty} \frac{\lg(R_{\mathfrak{m}_\alpha}/(t-\alpha, (\mathfrak{p}^{[p^e]})^*))}{p^{2e}} \\ &= \lim_{e\to\infty} \frac{\lg(R_{\mathfrak{p}}/(\mathfrak{p}^{[p^e]})^* R_{\mathfrak{p}})}{p^{2e}} = 3 \end{aligned}$$

contradicting Theorem 6.15 where it was computed that $e_{HK}(R_{\mathfrak{m}_\alpha}/(t-\alpha)R_{\mathfrak{m}_\alpha}) > 3$. □

COROLLARY 6.17. *Let $R = F[x,y,z,t]/(z^4 + xyz^2 + (x^3+y^3)z + tx^2y^2)$ where $F$ is a field of characteristic 2. Then the set $\{\mathfrak{q} \in \operatorname{Spec} R \mid e_{HK}(\mathfrak{q}) \leq 3\}$ is not open. In particular, $e_{HK}$ is not strongly upper semicontinuous.*

PROOF. If the set was open, its intersection with $V((x,y,z))$ should be open and non-empty. In particular, all but finitely many maximal ideals $\mathfrak{m}$ containing $(x,y,z)$ should belong to the open set. □

## 7. Further directions

There are many invariants derived from Frobenius, but equimultiplicity was only studied for Hilbert–Kunz multiplicity and F-signature. In fact, we are still lacking an intrinsic characterization of equimultiplicity for F-signature similar to Theorem 6.7. Among the remaining invariants, the most promising are dual F-signature [**21**, **41**, **45**], the Frobenius–Betti numbers [**3**, **11**], the Frobenius–Euler characteristics [**13**], and the defect of the F-pure threshold [**12**] because they are already known to be semicontinuous. In fact, the defect of the F-pure threshold is known to be strongly upper semicontinuous [**12**, Theorem 4.7] in the $\mathbb{Q}$-Gorenstein case; a theory of equimultiplicity could extend this result to the full generality.

Second, there is a related notion of equimultiplicity for singularities in families, see [**30**]. As of now, this side has not been touched for F-invariants, it is only known that Hilbert–Kunz multiplicity is upper semicontinuous [**44**].

## Acknowledgments

I thank Linquan Ma for many fruitful conversations, Thomas Polstra for collaborating on [**39**], and Craig Huneke, Austyn Simpson, and the anonymous referee for helping me improve this manuscript. I am indebted to Craig Huneke and Mel Hochster for their mathematics and mentorship through the years. Like most papers in "characteristic $p$ methods", this note has something from Mel and Craig on almost every page. I want to highlight that the presented necessary condition for Hilbert–Kunz equimultiplicity is given in terms of *tight closure*.

This note originated from my thesis, specifically the portion developed while Craig was still in Kansas, and it reflects the influence of every weekly meeting we had. In particular, I clearly remember one meeting in which I showed him that equimultiplicity implies the saturation of Frobenius powers is contained in their tight closures. Craig suggested that this should imply the tight closure itself is saturated, see Theorem 6.7. During that time, Craig also introduced me to Lipman's article on equimultiplicity [**30**], which has had a significant influence on me over the years.

## References


[1] Ian M. Aberbach and Florian Enescu, *The structure of F-pure rings*, Math. Z. **250** (2005), no. 4, 791–806, DOI 10.1007/s00209-005-0776-y. MR2180375

[2] Ian M. Aberbach and Graham J. Leuschke, *The $F$-signature and strong $F$-regularity*, Math. Res. Lett. **10** (2003), no. 1, 51–56, DOI 10.4310/MRL.2003.v10.n1.a6. MR1960123

[3] Ian M. Aberbach and Jinjia Li, *Asymptotic vanishing conditions which force regularity in local rings of prime characteristic*, Math. Res. Lett. **15** (2008), no. 4, 815–820, DOI 10.4310/MRL.2008.v15.n4.a17. MR2424915

[4] Bruce Michael Bennett, *On the characteristic functions of a local ring*, Ann. of Math. (2) **91** (1970), 25–87, DOI 10.2307/1970601. MR252388

[5] Manuel Blickle, Karl Schwede, and Kevin Tucker, *$F$-signature of pairs and the asymptotic behavior of Frobenius splittings*, Adv. Math. **231** (2012), no. 6, 3232–3258, DOI 10.1016/j.aim.2012.09.007. MR2980498

[6] Levi Borevitz, Naima Nader, Theodore J. Sandstrom, Amelia Shapiro, Austyn Simpson, and Jenna Zomback, *On localization of tight closure in line-$S_4$ quartics*, J. Pure Appl. Algebra **228** (2024), no. 9, Paper No. 107682, 17, DOI 10.1016/j.jpaa.2024.107682. MR4731180

[7] Holger Brenner and Paul Monsky, *Tight closure does not commute with localization*, Ann. of Math. (2) **171** (2010), no. 1, 571–588, DOI 10.4007/annals.2010.171.571. MR2630050

[8] Lindsay Burch, *Codimension and analytic spread*, Proc. Cambridge Philos. Soc. **72** (1972), 369–373, DOI 10.1017/s0305004100047198. MR304377

[9] Alessio Caminata and Alessandro De Stefani, *Frobenius, tight closure, and F-singularities*, Lecture notes available at `https://www.dima.unige.it/~destefani/files/Notes.pdf`.

[10] Everett C. Dade, *Multiplicity and monoidal transformations*, ProQuest LLC, Ann Arbor, MI, 1960. Thesis (Ph.D.)–Princeton University. MR2613132

[11] Alessandro De Stefani, Craig Huneke, and Luis Núñez-Betancourt, *Frobenius Betti numbers and modules of finite projective dimension*, J. Commut. Algebra **9** (2017), no. 4, 455–490, DOI 10.1216/JCA-2017-9-4-455. MR3713524

[12] Alessandro De Stefani, Luis Núñez-Betancourt, and Ilya Smirnov, *The defect of the F-pure threshold*, Available at `arXiv:2501.13613`.

[13] Alessandro De Stefani, Thomas Polstra, and Yongwei Yao, *Global Frobenius Betti numbers and Frobenius Euler characteristics*, Michigan Math. J. **71** (2022), no. 3, 533–552, DOI 10.1307/mmj/20195824. MR4574363

[14] Sankar P. Dutta, *Frobenius and multiplicities*, J. Algebra **85** (1983), no. 2, 424–448, DOI 10.1016/0021-8693(83)90106-0. MR725094

[15] Florian Enescu and Kazuma Shimomoto, *On the upper semi-continuity of the Hilbert-Kunz multiplicity*, J. Algebra **285** (2005), no. 1, 222–237, DOI 10.1016/j.jalgebra.2004.11.014. MR2119113

[16] M. Herrmann, S. Ikeda, and U. Orbanz, *Equimultiplicity and blowing up*, Springer-Verlag, Berlin, 1988. An algebraic study; With an appendix by B. Moonen, DOI 10.1007/978-3-642-61349-4. MR954831

[17] Heisuke Hironaka, *Resolution of singularities of an algebraic variety over a field of characteristic zero. I, II*, Ann. of Math. (2) **79** (1964), 109–203; 79 (1964), 205–326, DOI 10.2307/1970547. MR199184

[18] Melvin Hochster, *Foundations of tight closure*, Lecture notes for Math 711, Fall 2007 available at `http://www.math.lsa.umich.edu/~hochster/711F07/711.html`.

[19] Melvin Hochster and Craig Huneke, *Tight closure, invariant theory, and the Briançon-Skoda theorem*, J. Amer. Math. Soc. **3** (1990), no. 1, 31–116, DOI 10.2307/1990984. MR1017784

[20] Melvin Hochster and Craig Huneke, *$F$-regularity, test elements, and smooth base change*, Trans. Amer. Math. Soc. **346** (1994), no. 1, 1–62, DOI 10.2307/2154942. MR1273534

[21] Melvin Hochster and Yongwei Yao, *The F-rational signature and drops in the Hilbert-Kunz multiplicity*, Algebra Number Theory **16** (2022), no. 8, 1777–1809, DOI 10.2140/ant.2022.16.1777. MR4516193

[22] Craig Huneke, *The theory of d-sequences and powers of ideals*, Adv. in Math. **46** (1982), no. 3, 249–279, DOI 10.1016/0001-8708(82)90045-7. MR683201

[23] Craig Huneke, *Hilbert-Kunz multiplicity and the F-signature*, Commutative algebra, Springer, New York, 2013, pp. 485–525, DOI 10.1007/978-1-4614-5292-8_15. MR3051383

[24] Craig Huneke and Graham J. Leuschke, *Two theorems about maximal Cohen-Macaulay modules*, Math. Ann. **324** (2002), no. 2, 391–404, DOI 10.1007/s00208-002-0343-3. MR1933863

[25] Craig Huneke and Irena Swanson, *Integral closure of ideals, rings, and modules*, London Mathematical Society Lecture Note Series, vol. 336, Cambridge University Press, Cambridge, 2006. MR2266432

[26] Craig Huneke and Yongwei Yao, *Unmixed local rings with minimal Hilbert-Kunz multiplicity are regular*, Proc. Amer. Math. Soc. **130** (2002), no. 3, 661–665, DOI 10.1090/S0002-9939-01-06113-5. MR1866016

[27] Ernst Kunz, *Characterizations of regular local rings of characteristic p*, Amer. J. Math. **91** (1969), 772–784, DOI 10.2307/2373351. MR252389

[28] Ernst Kunz, *On Noetherian rings of characteristic p*, Amer. J. Math. **98** (1976), no. 4, 999–1013, DOI 10.2307/2374038. MR432625

[29] Tomas Larfeldt and Christer Lech, *Analytic ramifications and flat couples of local rings*, Acta Math. **146** (1981), no. 3-4, 201–208, DOI 10.1007/BF02392463. MR611383

[30] Joseph Lipman, *Equimultiplicity, reduction, and blowing up*, Commutative algebra (Fairfax, Va., 1979), Lect. Notes Pure Appl. Math., vol. 68, Dekker, New York, 1982, pp. 111–147. MR655801

[31] Shiji Lyu, *Uniform bounds in excellent $\boldsymbol{F}_p$-algebras and applications to semicontinuity*, Preprint, `arXiv:2503.13846`.

[32] Linquan Ma and Thomas Polstra, *F-singularities: a commutative algebra approach*, Lecture notes available at `https://www.math.purdue.edu/~ma326/F-singularitiesBook.pdf`.

[33] Linquan Ma, Thomas Polstra, Karl Schwede, and Kevin Tucker, *F-signature under birational morphisms*, Forum Math. Sigma **7** (2019), Paper No. e11, 20, DOI 10.1017/fms.2019.6. MR3940231

[34] Linquan Ma, Pham Hung Quy, and Ilya Smirnov, *Colength, multiplicity, and ideal closure operations*, Comm. Algebra **48** (2020), no. 4, 1601–1607, DOI 10.1080/00927872.2019.1691577. MR4079331

[35] P. Monsky, *The Hilbert-Kunz function*, Math. Ann. **263** (1983), no. 1, 43–49, DOI 10.1007/BF01457082. MR697329

[36] Paul Monsky, *Hilbert-Kunz functions in a family: line-$S_4$ quartics*, J. Algebra **208** (1998), no. 1, 359–371, DOI 10.1006/jabr.1998.7517. MR1644023

[37] Paul Monsky, *Hilbert-Kunz functions in a family: point-$S_4$ quartics*, J. Algebra **208** (1998), no. 1, 343–358, DOI 10.1006/jabr.1998.7500. MR1644019

[38] Thomas Polstra, *Uniform bounds in F-finite rings and lower semi-continuity of the F-signature*, Trans. Amer. Math. Soc. **370** (2018), no. 5, 3147–3169, DOI 10.1090/tran/7030. MR3766845

[39] Thomas Polstra and Ilya Smirnov, *Equimultiplicity theory of strongly F-regular rings*, Michigan Math. J. **70** (2021), no. 4, 837–856, DOI 10.1307/mmj/1600913073. MR4332680

[40] Thomas Polstra and Kevin Tucker, *F-signature and Hilbert-Kunz multiplicity: a combined approach and comparison*, Algebra Number Theory **12** (2018), no. 1, 61–97, DOI 10.2140/ant.2018.12.61. MR3781433

[41] Akiyoshi Sannai, *On dual F-signature*, Int. Math. Res. Not. IMRN **1** (2015), 197–211, DOI 10.1093/imrn/rnt162. MR3340299

[42] Ilya Smirnov, *Upper semi-continuity of the Hilbert-Kunz multiplicity*, Compos. Math. **152** (2016), no. 3, 477–488, DOI 10.1112/S0010437X15007800. MR3477638

[43] Ilya Smirnov, *Equimultiplicity in Hilbert-Kunz theory*, Math. Z. **291** (2019), no. 1-2, 245–278, DOI 10.1007/s00209-018-2082-5. MR3936069

[44] Ilya Smirnov, *On semicontinuity of multiplicities in families*, Doc. Math. **25** (2020), 381–400. MR4106897

[45] Ilya Smirnov and Kevin Tucker, *The theory of F-rational signature*, J. Reine Angew. Math. **812** (2024), 1–58, DOI 10.1515/crelle-2024-0010. MR4767385

[46] Karen E. Smith and Michel Van den Bergh, *Simplicity of rings of differential operators in prime characteristic*, Proc. London Math. Soc. (3) **75** (1997), no. 1, 32–62, DOI 10.1112/S0024611597000257. MR1444312

[47] Kevin Tucker, *F-signature exists*, Invent. Math. **190** (2012), no. 3, 743–765, DOI 10.1007/s00222-012-0389-0. MR2995185

[48] Kei-ichi Watanabe and Ken-ichi Yoshida, *Hilbert-Kunz multiplicity and an inequality between multiplicity and colength*, J. Algebra **230** (2000), no. 1, 295–317, DOI 10.1006/jabr.1999.7956. MR1774769

[49] Yongwei Yao, *Observations on the F-signature of local rings of characteristic p*, J. Algebra **299** (2006), no. 1, 198–218, DOI 10.1016/j.jalgebra.2005.08.013. MR2225772

BCAM – Basque Center for Applied Mathematics, Bilbao, Spain; and IKERBASQUE, Basque Foundation for Science, Bilbao, Spain

*Email address*: ismirnov@bcamath.org

Contemporary Mathematics
Volume **831**, 2025
https://doi.org/10.1090/conm/831/16638

# Vector bundles and tight closure

Holger Brenner

This are the notes of a course held at the ICTP in Trieste in May 2023 within the school on Commutative Algebra and Algebraic Geometry in Prime Characteristics. There is a substantial overlap of the material presented here with courses given by myself on other occasions. The material covered here stems from [**2**], [**5**] [**4**], [**3**], [**6**], [**7**]. The notes were written on Wikiversity and are under the CC-by-sa 3.0 license.

## 1. Lecture - Hilbert-Kunz theory and vector bundles

**1.1. Hilbert-Kunz theory.** In 1969, Kunz was the first to consider the following function and the corresponding limit [**17**].

DEFINITION 1.1. Let $K$ denote a field of positive characteristic $p$, let $K \subseteq R$ be a noetherian ring and let $I \subseteq R$ be an ideal which is primary to some maximal ideal. Then the *Hilbert-Kunz function* is the function

$$\varphi_I \colon \mathbb{N} \longrightarrow \mathbb{N}, e \longmapsto \varphi_I(e) = \operatorname{length}(R/I^{[p^e]}),$$

where $I^{[p^e]}$ is the extended ideal under the $e$-th iteration of the Frobenius homomorphism

$$R \longrightarrow R, f \longmapsto f^{p^e}.$$

DEFINITION 1.2. Let $K$ denote a field of positive characteristic $p$, let $K \subseteq R$ be a noetherian ring and let $I \subseteq R$ be an ideal which is primary to some maximal ideal of height $d$. Then the *Hilbert-Kunz multiplicity* of $I$ is the limit (if it exists)

$$\lim_{e \to \infty} \frac{\operatorname{length}(R/I^{[p^e]})}{p^{ed}} = \lim_{e \to \infty} \frac{\varphi_I(e)}{p^{ed}}.$$

The Hilbert-Kunz multiplicity of the maximal ideal of a local noetherian ring $R$ is called the *Hilbert-Kunz multiplicity* of $R$. The existence of Hilbert-Kunz multiplicity was proven by Monsky [**20**].

THEOREM 1.3. *Let $K$ denote a field of positive characteristic $p$, let $K \subseteq R$ be a noetherian ring and let $I \subseteq R$ be an ideal which is primary to some maximal ideal. Then the Hilbert-Kunz multiplicity $e_{HK}(I)$ exists and is a positive real number.*

With the help of the Hilbert-Kunz multiplicity of a local noetherian ring one may characterize when $R$ is regular, as the following theorem shows (which was initiated by Kunz in 1969 but finally proven by Watanabe and Yoshida in 2000 [**24**]).

THEOREM 1.4. *Let $R$ be a local noetherian ring of positive characteristic. Then the following hold.*

(1) *The Hilbert-Kunz multiplicity of $R$ is $e_{HK}(R) \geq 1$.*
(2) *If $R$ is unmixed, then $e_{HK}(R) = 1$ if and only if $R$ is regular.*

**1.2. Vector bundles.** We will have a look at Hilbert-Kunz theory and tight closure (to be introduced in the next lecture) from the viewpoint of vector bundles. To motivate this concept, which exists in algebraic geometry, differential geometry, topology, mathematical physics, we go back to linear algebra. Let $K$ be a field. We consider a system of linear homogeneous equations over $K$,

$$\begin{aligned} f_{11}t_1 + \cdots + f_{1n}t_n &= 0, \\ f_{21}t_1 + \cdots + f_{2n}t_n &= 0, \\ &\vdots \\ f_{m1}t_1 + \cdots + f_{mn}t_n &= 0, \end{aligned}$$

where the $f_{ij}$ are elements in $K$. The solution set to this system of homogeneous equations is a vector space $V$ over $K$ (a linear subspace of $K^n$), its dimension is $n - \operatorname{rk}(A)$, where

$$A = (f_{ij})_{ij}$$

is the matrix given by these elements. Suppose now that $X$ is a geometric object (a topological space, a manifold, a variety, a scheme, the spectrum of a ring) and that instead of elements in the field $K$ we have functions

$$f_{ij} \colon X \longrightarrow K$$

on $X$ (which are continuous, or differentiable, or algebraic). We form the matrix of functions $A = (f_{ij})_{ij}$, which yields for every point $P \in X$ a matrix $A(P)$ over $K$. Then we get from these data the space

$$V = \left\{ (P; t_1, \ldots, t_n) \mid A(P) \begin{pmatrix} t_1 \\ \vdots \\ t_n \end{pmatrix} = 0 \right\} \subseteq X \times K^n$$

together with the projection to $X$. For a fixed point $P \in X$, the fiber $V_P$ of $V$ over $P$ is the solution space in $K^n$ to the corresponding system of homogeneous linear equations given by inserting $P$ into $f_{ij}$. In particular, all fibers of the map

$$V \longrightarrow X\,,$$

are vector spaces (maybe of non-constant dimension). These vector space structures yield an addition

$$V \times_X V \longrightarrow V, (P; s_1, \ldots, s_n; t_1, \ldots, t_n) \longmapsto (P; s_1 + t_1, \ldots, s_n + t_n)\,,$$

($V \times_x V$ is the fiber product of $V \to X$ with itself, only points in the same fiber can be added). The mapping

$$X \longrightarrow V, P \longmapsto (P; 0, \ldots, 0)\,,$$

is called the *zero-section* of the bundle. If we have just one equation with functions $f_1, \ldots, f_n$ and if $U \subseteq X$ denotes the open subset where not all $f_i$ vanish, then we get a short exact sequence

$$0 \longrightarrow V|_U \longrightarrow U \times K^n \longrightarrow U \times K \longrightarrow 0 .$$

If $D(f_i) \subseteq U$ denotes the locus where $f_i$ does not vanish, then we get a linear isomorphism

$$V|_{D(f_i)} \longrightarrow D(f_i) \times K^{n-1}, (t_1, \ldots, t_n) \longmapsto (t_1, \ldots, t_{i-1}, t_{i+1}, , \ldots, , t_n) ,$$

as we can reconstruct

$$t_i = -\frac{1}{f_i}(t_1 f_1 + \cdots + t_{i-1} f_{i-1} + t_{i+1} f_{i+1} + \cdots + t_n f_n)$$

from the other variables. This local trivialization also shows that, on the intersection $D(f_i) \cap D(f_j)$, the transformation between two trivializations is linear. So locally, the object $V|_U$ is trivial, but globally it might be complicated.

We now consider the scheme version of a vector bundle and in particular of a syzygy bundle (kernel bundle), trying to keep the idea that we are dealing with objects from linear algebra, but over a varying base. Let $R$ denote a commutative ring, let $I$ denote an ideal and fix generators $I = (f_1, \ldots, f_n)$. This defines a short exact sequence of $R$-modules

$$0 \longrightarrow \operatorname{Syz}(f_1, \ldots, f_n) \longrightarrow R^n \stackrel{f_1, \ldots, f_n}{\longrightarrow} I \longrightarrow 0 .$$

The syzygy module $\operatorname{Syz}(f_1, \ldots, f_n)$ is in general not locally free on $X = \operatorname{Spec}(R)$. However, on the open subset

$$U = D(I) = \bigcup_{i=1}^{n} D(f_i) \subseteq X$$

defined by the ideal, this module will be locally free, since it is free on each $D(f_i)$, using the same trivialization as above. Moreover, on $U$ we get the short exact sequence

$$0 \longrightarrow \operatorname{Syz}(f_1, \ldots, f_n)|_U \longrightarrow \mathcal{O}_U^n \stackrel{f_1, \ldots, f_n}{\longrightarrow} \mathcal{O}_U \longrightarrow 0$$

of coherent sheaves. Later on, $I$ will be an $\mathfrak{m}$-primary ideal in a local noetherian ring $R$ and then

$$U = X \setminus \{\mathfrak{m}\}$$

will be the punctured spectrum of $R$. The sheaves occurring in the last sequence are locally free in the following sense.

DEFINITION 1.5. A coherent $\mathcal{O}_X$-module $\mathcal{F}$ on a scheme $X$ is called *locally free* of rank $r$, if there exists an open covering $X = \bigcup_{i \in I} U_i$ and $\mathcal{O}_{U_i}$-module-isomorphisms $\mathcal{F}|_{U_i} \cong (\mathcal{O}_{U_i})^r$ for every $i \in I$.

An equivalent concept of a locally free sheaf is the concept of a geometric vector bundle, see [**11**, Exercise II.5.18]. The free sheaf $\mathcal{O}_X^r$ corresponds to affine space $\mathbb{A}_X^r = X \times \mathbb{A}^r$ over $X$.

DEFINITION 1.6. Let $X$ denote a scheme. A scheme $V$ equipped with a morphism

$$p \colon V \longrightarrow X$$

is called a *geometric vector bundle* of rank $r$ over $X$ if there exists an open covering $X = \bigcup_{i \in I} U_i$ and $U_i$-isomorphisms

$$\psi_i \colon U_i \times \mathbb{A}^r = \mathbb{A}^r_{U_i} \longrightarrow V|_{U_i} = p^{-1}(U_i)$$

such that for every open affine subset $U \subseteq U_i \cap U_j$, the transition mappings

$$\psi_j^{-1} \circ \psi_i \colon \mathbb{A}^r_{U_i}|_U \longrightarrow \mathbb{A}^r_{U_j}|_U$$

are linear automorphisms, i.e. they are induced by an automorphism of the polynomial ring $\Gamma(U, \mathcal{O}_X)[T_1, \ldots, T_r]$ given by $T_i \mapsto \sum_{j=1}^r a_{ij} T_j$.

We will work with both concepts and switch between them as needed.

**1.3. The graded case.** We will restrict now to the standard-graded case in order to work on the corresponding projective variety. Let $R$ be a standard-graded normal domain over an algebraically closed field $K$. Let

$$Y = \operatorname{Proj}(R)$$

be the corresponding projective variety and let

$$I = (f_1, \ldots, f_n)$$

be an $R_+$-primary homogeneous ideal with generators of degrees $d_1, \ldots, d_n$. Then we get on the projective variety $Y$ the short exact sequence

$$0 \longrightarrow \operatorname{Syz}(f_1, \ldots, f_n)(m) \longrightarrow \bigoplus_{i=1}^n \mathcal{O}_Y(m - d_i) \stackrel{f_1, \ldots, f_n}{\longrightarrow} \mathcal{O}_Y(m) \longrightarrow 0\,.$$

Here $\operatorname{Syz}(f_1, \ldots, f_n)(m)$ is a vector bundle, called the *syzygy bundle*, its rank is $n - 1$.

Our approach to the computation of the Hilbert-Kunz multiplicity is by using the presenting sequence of the ideal

$$0 \longrightarrow \operatorname{Syz}(f_1, \ldots, f_n) \longrightarrow \bigoplus_{i=1}^n \mathcal{O}_Y(-d_i) \stackrel{f_1, \ldots, f_n}{\longrightarrow} \mathcal{O}_Y \longrightarrow 0$$

and twists of its $e$-th Frobenius pull-backs(under the $e$th absolute Frobenius morphism on $Y$), that is

$$0 \longrightarrow \operatorname{Syz}(f_1^q, \ldots, f_n^q)(m) \longrightarrow \bigoplus_{i=1}^n \mathcal{O}_Y(m - qd_i) \stackrel{f_1^q, \ldots, f_n^q}{\longrightarrow} \mathcal{O}_Y(m) \longrightarrow 0$$

(where $q = p^e$; this is also the presenting sequence of the ideal $I^{[q]}$, considered on $Y$), and to relate the asymptotic behavior of

$$\operatorname{length}(R/I^{[q]}) = \dim_K(R/I^{[q]}) = \sum_{m=0}^{\infty} \dim_K(R/I^{[q]})_m$$

to the asymptotic behavior of the global sections of the Frobenius pull-backs

$$(F^{e*}(\operatorname{Syz}(f_1, \ldots, f_n))(m) = \operatorname{Syz}(f_1^q, \ldots, f_n^q)(m).$$

What we want to compute is just the cokernel of the complex of global sections of the above sequence, namely $\dim_K (R/I^{[q]})_m$, which is

$$h^0(Y, \mathcal{O}_Y(m)) - \sum_{i=1}^n h^0(Y, \mathcal{O}_Y(m - qd_i)) + h^0(Y, \operatorname{Syz}(f_1^q, \ldots, f_n^q)(m)).$$

Here, $H^0(Y, \mathcal{S})$ is the vector space of global sections of a sheaf $\mathcal{S}$ on $Y$, and $h^0(Y, \mathcal{S})$ denotes its vector space dimension. The infinite summation over $m$ above is in fact finite, though the range depends on $q$, and the terms

$$h^0(Y, \mathcal{O}_Y(m)) = \dim_K \Gamma(Y, \mathcal{O}_Y(m)) = \dim_K R_m$$

are easy to control, so we need to understand the behavior of the global syzygies

$$H^0(Y, \operatorname{Syz}(f_1^q, \dots, f_n^q)(m))$$

for all $q$ and $m$, at least asymptotically. This is a Frobenius-Riemann-Roch problem.

By this translation of Hilbert-Kunz theory into a projective setting, we gain the following.

(1) We work with projective varieties; if we look at local rings with an isolated singularity, we even work on smooth projective varieties.
(2) We work with locally free sheaves. Taking Frobenius pull-backs is then exact. The mentioned Frobenius-Riemann-Roch problem is not specific for syzygy bundles, but should be addressed in general.
(3) We can use the advanced methods of algebraic geometry, like intersection theory, Riemann-Roch theorem, vanishing theorems, ampleness, cohomology, moduli spaces.

This is still a difficult problem in general. However, if the local normal ring has dimension two and the corresponding variety is a smooth projective curve, then our understanding is good enough to solve the main problems from Hilbert-Kunz theory. The main advantages in the curve case compared to the case of higher-dimensional varieties are the following.

(1) The degree of a vector bundle is independent of a polarization.
(2) There are only the 0th and the first cohomology, which are directly related by Serre-duality.
(3) The Riemann-Roch theorem relates these notions.
(4) Semistability gives good criteria for having no global sections.

We introduce these concepts on a smooth projective curve $C$ over an algebraically closed field $K$.

Definition 1.7. Let $C$ denote a smooth projective curve over an algebraically closed field $K$. For a locally free sheaf $\mathcal{G}$ on $C$ of rank $r$ we define its *degree* by the degree of the determinant sheaf $\bigwedge^r \mathcal{G}$.

The determinant sheaf is an invertible sheaf and corresponds therefore to a Weil divisor $D$, say

$$D = \sum_{P \in C} n_P P,$$

its degree is defined by $\sum_{P \in C} n_P$. The degree of the curve itself is defined as the degree of $\mathcal{O}_C(1)$. The degree of bundles is additive on short exact sequences of locally free sheaves. Applying additivity to

$$0 \longrightarrow \operatorname{Syz}(f_1, \dots, f_n)(m) \longrightarrow \bigoplus_{i=1}^n \mathcal{O}_C(m - d_i) \stackrel{f_1, \dots, f_n}{\longrightarrow} \mathcal{O}_C(m) \longrightarrow 0$$

we get

$$\deg(\operatorname{Syz}(f_1, \dots, f_n)(m)) = ((n-1)m - \sum_{i=1}^n d_i) \deg(C).$$

In fact,

$$\bigwedge^{n-1} (\operatorname{Syz}(f_1, \ldots, f_n)(m)) \cong \mathcal{O}_C((n-1)m - \sum_{i=1}^{n} d_i)\,.$$

The rational number $\mu(\mathcal{S}) = \frac{\deg(\mathcal{S})}{\operatorname{rk}(\mathcal{S})}$ is called the *slope* of a vector bundle.

DEFINITION 1.8. Let $\mathcal{S}$ be a vector bundle on a smooth projective curve $C$. It is called *semistable*, if $\mu(\mathcal{T}) \leq \mu(\mathcal{S})$ for all subbundles $\mathcal{T} \subseteq \mathcal{S}$.

Suppose that the base field has positive characteristic $p > 0$. Then $\mathcal{S}$ is called *strongly semistable*, if all (absolute) Frobenius pull-backs $F^{e*}(\mathcal{S})$ are semistable.

An important property of a semistable bundle of negative degree is that it can not have any global section $\neq 0$. The semistable bundles are those for which there exists a moduli space.

EXAMPLE 1.9. Let $R = K[x, y, z]/(x^3 + y^3 + z^3)$, where $K$ is a field of positive characteristic $p \neq 3$, $I = (x^2, y^2, z^2)$, and

$$C = \operatorname{Proj}(R).$$

The equation $x^3 + y^3 + z^3 = 0$ yields a global section of $\operatorname{Syz}(x^2, y^2, z^2)(3)$ and thus the short exact sequence

$$0 \longrightarrow \mathcal{O}_C \longrightarrow \operatorname{Syz}(x^2, y^2, z^2)(3) \longrightarrow \mathcal{O}_C \longrightarrow 0\,.$$

The quotient bundle is trivial since the determinant bundle of the syzygy bundle is trivial by its defining sequence. This shows that $\operatorname{Syz}(x^2, y^2, z^2)$ is strongly semistable, as this is true for all bundles $\mathcal{S}$ appearing as an extension

$$0 \longrightarrow \mathcal{O}_C \longrightarrow \mathcal{S} \longrightarrow \mathcal{O}_C \longrightarrow 0.$$

This extension property is preserved by Frobenius pull-backs, so we only need to show that $\mathcal{S}$ is semistable. Because $\mathcal{S}$ has degree 0 and rank 2, we only have to show that $\mathcal{S}$ does not contain an invertible sheaf of positive degree. Suppose that $\mathcal{L}$ is an invertible sheaf of positive degree and $\mathcal{L} \to \mathcal{S}$ is a morphism. Then the composed morphism to the structure sheaf on the right is 0. But then the morphism factors through the structure sheaf on the left, so again the morphism is 0.

EXAMPLE 1.10. Let $C$ be the smooth Fermat quartic given by $x^4 + y^4 + z^4$, and consider on it the syzygy bundle $\operatorname{Syz}(x, y, z)$ (which is also the restricted cotangent bundle from the projective plane). Suppose that the characteristic is 3. Then its Frobenius pull-back is $\operatorname{Syz}(x^3, y^3, z^3)$. The curve equation gives a global non-trivial section of this bundle of total degree 4. But the degree of $\operatorname{Syz}(x^3, y^3, z^3)(4)$ is negative, hence it can not be semistable. We get the short exact sequence

$$0 \longrightarrow \mathcal{O}_C(-4) \longrightarrow \operatorname{Syz}(x^3, y^3, z^3) \longrightarrow \mathcal{O}_C(-5) \longrightarrow 0.$$

This also shows that the bundle $\operatorname{Syz}(x, y, z)$ itself is semistable, since a destabilizing invertible subsheaf $\mathcal{L} \subset \operatorname{Syz}(x, y, z)$ would pull-back to $\mathcal{O}_C(-4) \subset \operatorname{Syz}(x^3, y^3, z^3)$, which is not possible, since the degree of $\mathcal{O}_C(-4)$ is not divisible by 3.

DEFINITION 1.11. Let $\mathcal{S}$ be a vector bundle on a smooth projective curve $C$ over an algebraically closed field $K$. Then the (uniquely determined) filtration

$$0 = \mathcal{S}_0 \subset \mathcal{S}_1 \subset \ldots \subset \mathcal{S}_{t-1} \subset \mathcal{S}_t = \mathcal{S}$$

of subbundles such that all quotient bundles $\mathcal{S}_k/\mathcal{S}_{k-1}$ are semistable with decreasing slopes $\mu_k = \mu(\mathcal{S}_k/\mathcal{S}_{k-1})$, is called the *Harder-Narasimhan filtration* of $\mathcal{S}$.

For a semistable bundle, the Harder-Narasimhan filtration consists just of the bundle itself. In Example 1.10, $0 \subset \mathcal{O}_C(-4) \subset \operatorname{Syz}\left(x^3, y^3, z^3\right)$ is the Harder-Narasimhan filtration of $\operatorname{Syz}\left(x^3, y^3, z^3\right)$. In positive characteristic, a Harder-Narasimhan filtration is called *strong*, if all its succesive quotients are strongly semistable. The following theorem shows that some Frobenius pull-back of a bundle has a strong Harder-Narasimhan filtration.

THEOREM 1.12. *Let $C$ denote a smooth projective curve over an algebraically closed field of positive characteristic $p$, and let $\mathcal{S}$ be a vector bundle on $C$. Then there exists a natural number $e \in \mathbb{N}$ such that the Harder-Narasimhan filtration of the eth Frobenius pull-back $F^{e*}(\mathcal{S})$, say*

$$0 = \mathcal{S}_0 \subset \mathcal{S}_1 \subset \ldots \subset \mathcal{S}_{t-1} \subset \mathcal{S}_t = F^{e*}(\mathcal{S})$$

*has the property that the quotients $\mathcal{S}_k/\mathcal{S}_{k-1}$ are strongly semistable.*

This theorem is due to A. Langer [**19**] and holds also in higher dimension. An immediate consequence of this is that the Harder-Narasimhan filtration of all higher Frobenius pull-backs are just the pull-backs of this filtration. With these filtrations we can at least Frobenius-asymptotically control the global sections of the pull-backs and hence also the Hilbert-Kunz multiplicity. This implies the following theorem [**3**], [**23**].

THEOREM 1.13. *Let $R$ be a two-dimensional standard-graded normal domain over an algebraically closed field of positive characteristic. Let $I = (f_1, \ldots, f_n)$ be a homogeneous $R_+$-primary ideal with homogeneous generators of degree $d_i$. Let $\mathcal{S} = \operatorname{Syz}(f_1, \ldots, f_n)$ be the syzygy bundle on $C = \operatorname{Proj}(R)$ and suppose that the Harder-Narasimhan filtration of $F^{e*}(\mathcal{S})$ is strong, and let $\mu_k$, $k = 1, \ldots, t$, be the corresponding slopes. We set $\nu_k = \frac{-\mu_k}{\deg(C)p^e}$ and $r_k = \operatorname{rk}(\mathcal{S}_k/\mathcal{S}_{k-1})$. Then the Hilbert-Kunz multiplicity of $I$ is*

$$e_{HK}(I) = \frac{\deg(C)}{2}\left(\sum_{k=1}^{t} r_k \nu_k^2 - \sum_{i=1}^{n} d_i^2\right).$$

*In particular, it is a rational number.*

In the situation of the maximal ideal on a hypersurface, this formula becomes

$$e_{HK}(\mathfrak{m}) = \frac{\deg(C)}{2}\left(\nu_1^2 + \nu_2^2 - 3\right),$$

where $\nu_1 \le \nu_2$ and $\nu_1 + \nu_2 = 3$ and where $\nu_1 = \nu_2$ corresponds to the case where $\operatorname{Syz}(x, y, z)$ is strongly semistable. This gives the following.

COROLLARY 1.14. *Let $R = K[x, y, z]/(H)$ be a normal homogeneous hypersurface domain of dimension two and degree $\delta$ over an algebraically closed field of positive characteristic. Then there exists a rational number $\nu_2$, $\frac{3}{2} \le \nu_2 \le 2$, such that the Hilbert-Kunz multiplicity of $R$ is*

$$e_{HK}(R) = \delta\left(\nu_2^2 - 3\nu_2 + 3\right).$$

## 2. Lecture - Tight closure and torsors

### 2.1. Tight closure.

DEFINITION 2.1. Let $R$ be a noetherian domain of positive characteristic and let $I \subseteq R$ be an ideal. The *tight closure* of $I$ is the ideal

$$I^* = \left\{ f \in R \mid \text{ there exists } z \neq 0 \text{ such that } zf^q \in I^{[q]} \text{ for all } q = p^e \right\}.$$

This theory was introduced by M. Hochster and C. Huneke (see [**13**], [**14**], [**15**], [**16**]). There is a direct relation between Hilbert-Kunz multiplicity and tight closure.

THEOREM 2.2. *Let $(R, \mathfrak{m})$ be an analytically unramified and formally equidimensional local noetherian ring of positive characteristic, let $I \subseteq R$ be an $\mathfrak{m}$-primary ideal. Let $f \in R$. Then*

$$f \in I^* \text{ if and only if } e_{HK}((I, f)) = e_{HK}(I)\,.$$

We try to understand tight closure from the perspective of bundles and will have again a look at the syzygy bundle. Let $R$ denote a noetherian normal domain and let $I = (f_1, \ldots, f_n)$ denote an ideal of height $I$ at least 2 (think of a local normal domain of dimension at least 2 and an $\mathfrak{m}$-primary ideal $I$, or the graded version of this). Let $U = D(I) \subseteq \operatorname{Spec}(R)$ and consider again the short exact sequence

$$0 \longrightarrow \operatorname{Syz}(f_1, \ldots, f_n)|_U \longrightarrow \mathcal{O}_U^n \stackrel{f_1, \ldots, f_n}{\longrightarrow} \mathcal{O}_U \longrightarrow 0$$

of locally free sheaves on $U$. Another element $f \in R = \Gamma(U, \mathcal{O}_U)$ (because of the height condition) defines via the connecting homomorphism $\delta$ inside the long exact sequence of cohomology the cohomology class $c = \delta(f) \in H^1(U, \operatorname{Syz}(f_1, \ldots, f_n))$. When $R$ contains a field of positive characteristic, we try to understand tight closure in terms of this cohomology class. Quite directly, we have the $e$th absolute Frobenius on $U$. As the sheaves are locally free, we have

$$F^{e*}(\operatorname{Syz}(f_1, \ldots, f_n)) = \operatorname{Syz}(f_1^q, \ldots, f_n^q),$$

and the $e$th Frobenius pull-back of the cohomology class is

$$F^{e*}(c) \in H^1(U, F^{e*}(\operatorname{Syz}(f_1, \ldots, f_n)) \cong H^1(U, \operatorname{Syz}(f_1^q, \ldots, f_n^q))$$

($q = p^e$), and this is the cohomology class corresponding to $f^q$. By the height assumption, we have $zF^{e*}(c) = 0$ if and only if $zf^q \in (f_1^q, \ldots, f_n^q)$, and this holds for all $e$ if and only if $f \in I^*$ by definition. This shows already that under the given conditions, tight closure does only depend on the cohomology class. In the graded case, we can also translate the tight closure question $f \in I^*$ for homogeneous data into the question whether the corresponding cohomology class $c \in H^1(Y, \operatorname{Syz}(f_1, \ldots, f_n)(m))$ on $\operatorname{Proj}(R)$ is tightly zero in the sense that $zF^{e*}(c) = 0$ holds for some homogenous $z \neq 0$ ($z$ considered inside some ample invertible sheaf $\mathcal{O}_Y(\ell)$). This property of being tightly zero is relevant for every cohomology class in any locally free sheaf. Here, this translation is in particular helpful for inclusion results. For exclusion results we need to go another way and consider torsors.

**2.2. Torsors and forcing algebras.** We come back to the situation of a system of linear homogeneous equations over a field $K$ with which we tried to motivate the concept of a vector bundle. However, we now consider a system of linear inhomogeneous equations,

$$\begin{aligned} f_{11}t_1 + \cdots + f_{1n}t_n &= f_1, \\ f_{21}t_1 + \cdots + f_{2n}t_n &= f_2, \\ &\vdots \\ f_{m1}t_1 + \cdots + f_{mn}t_n &= f_m. \end{aligned}$$

The solution set $T$ of this inhomogeneous system may be empty, but nevertheless it is tightly related to the solution space $V$ of the homogeneous system. First of all, there exists an action

$$V \times T \longrightarrow T, (v, t) \longmapsto v + t,$$

because the sum of a solution of the homogeneous system and of a solution of the inhomogeneous system is again a solution of the inhomogeneous system. This action is a group action of the group $(V, +, 0)$ on the set $T$. Moreover, if we fix one solution $t_0 \in T$ (supposing that at least one solution exists), then there exists a bijection

$$V \longrightarrow T, v \longmapsto v + t_0 .$$

This means that the group $V$ acts simply transitive on $T$, and so $T$ can be identified with the vector space $V$, however not in a canonical way.

Suppose now that $X$ is a geometric object and we have functions

$$f_{ij}, f_i \colon X \longrightarrow K$$

on $X$ (which are continuous, or differentiable, or algebraic). As before in Section 1.2, we get for the $f_{ij}$ a bundle with an addition and such that the fibers are vector spaces.

Then we can form the set

$$T = \left\{ (P; t_1, \ldots, t_n) \mid A(P) \begin{pmatrix} t_1 \\ \vdots \\ t_n \end{pmatrix} - \begin{pmatrix} f_1(P) \\ \vdots \\ f_n(P) \end{pmatrix} \right\} \subseteq X \times K^n$$

with the projection to $X$. Again, every fiber $T_P$ of $T$ over a point $P \in X$ is the solution set to the system of inhomogeneous linear equations which arises by inserting $P$ into $f_{ij}$ and $f_i$. The actions of the fibers $V_P$ on $T_P$ (coming from linear algebra) extend to an action

$$V \times_X T \longrightarrow T, (P; t_1, \ldots, t_n; s_1, \ldots, s_n) \longmapsto (P; t_1 + s_1, \ldots, t_n + s_n) .$$

Also, if a (continuous, differentiable, algebraic) map

$$s \colon X \longrightarrow T$$

with $s(P) \in T_P$ exists, then we can construct a (continuous, differentiable, algebraic) isomorphism between $V$ and $T$. However, different from the situation in linear algebra (which corresponds to the situation where $X$ is just one point), such a section does rarely exist.

These objects $T$ have new and sometimes difficult global properties which we try to understand. We will work mainly in an algebraic setting and restrict to the situation where just one equation

$$f_1 t_1 + \cdots + f_n t_n = f$$

is given. Then in the homogeneous case ( $f = 0$) the fibers are vector spaces of dimension $n-1$ or $n$, and the latter holds exactly for the points $P \in X$ where $f_1(P) = \ldots = f_n(P) = 0$. In the inhomogeneous case the fibers are either empty or of dimension $n-1$ or $n$. We give a typical example.

EXAMPLE 2.3. Let $X$ denote a plane (like $K^2, \mathbb{R}^2, \mathbb{A}^2_K$) with coordinate functions $x$ and $y$. We consider an inhomogeneous linear equation of type

$$x^a t_1 + y^b t_2 = x^c y^d .$$

The fiber of the solution set $T$ over a point $\neq (0,0)$ is one-dimensional, whereas the fiber over $(0,0)$ has dimension two (for $a, b, c, d \geq 1$). Many properties of $T$ depend on these four exponents.

In (most of) these example, we can observe the following behavior. On an open subset, the dimension of the fibers is constant and equals $n-1$, whereas the fiber over some special points degenerates to an $n$-dimensional solution set (or becomes empty).

DEFINITION 2.4. Let $V$ denote a geometric vector bundle over a scheme $X$. A scheme $T \to X$ together with an action

$$\beta \colon V \times_X T \longrightarrow T$$

is called a geometric (Zariski)-*torsor* for $V$ (or a *principal fiber bundle* or a *principal homogeneous space*) if there exists an open covering $X = \bigcup_{i \in I} U_i$ and isomorphisms

$$\varphi_i \colon T|_{U_i} \longrightarrow V|_{U_i}$$

such that the diagrams (we set $U = U_i$ and $\varphi = \varphi_i$)

$$\begin{array}{ccc} V|_U \times_U T|_U & \xrightarrow{\beta} & T|_U \\ {\scriptstyle \mathrm{Id} \times \varphi} \downarrow & & \downarrow {\scriptstyle \varphi} \\ V|_U \times_U V|_U & \xrightarrow{\alpha} & V|_U \end{array}$$

commute, where $\alpha$ is the addition on the vector bundle.

The torsors of vector bundles can be classified in the following way, see [**2**, Proposition 2.3.3].

PROPOSITION 2.5. *Let $X$ denote a noetherian separated scheme and let*

$$p \colon V \longrightarrow X$$

*denote a geometric vector bundle on $X$ with sheaf of sections $\mathcal{S}$. Then there exists a correspondence between first cohomology classes $c \in H^1(X, \mathcal{S})$ and geometric $V$-torsors.*

REMARK 2.6. Let $\mathcal{S}$ denote a locally free sheaf on a scheme $X$. For a cohomology class $c \in H^1(X, \mathcal{S})$ one can construct a geometric object: Because of $H^1(X, \mathcal{S}) \cong \mathrm{Ext}^1(\mathcal{O}_X, \mathcal{S})$, the class defines an extension

$$0 \longrightarrow \mathcal{S} \longrightarrow \mathcal{S}' \longrightarrow \mathcal{O}_X \longrightarrow 0 .$$

This extension is such that under the connecting homomorphism of cohomology, $1 \in \Gamma(X, \mathcal{O}_X)$ is sent to $c \in H^1(X, \mathcal{S})$. The extension yields a projective subbundle[1]

$$\mathbb{P}(\mathcal{S}^\vee) \subset \mathbb{P}(\mathcal{S}'^\vee).$$

If $V$ is the corresponding geometric vector bundle of $\mathcal{S}$, one may think of $\mathbb{P}(\mathcal{S}^\vee)$ as $\mathbb{P}(V)$ which consists for every base point $x \in X$ of all the lines in the fiber $V_x$ passing through the origin. The projective subbundle $\mathbb{P}(V)$ has codimension one inside $\mathbb{P}(V')$, for every point it is a projective space lying (linearly) inside a projective space of one dimension higher. The complement is then over every point an affine space. One can show that the global complement

$$T = \mathbb{P}(\mathcal{S}'^\vee) \setminus \mathbb{P}(\mathcal{S}^\vee)$$

is another model (concrete realization) for the torsor given by the cohomology class. The advantage of this viewpoint is that we may work, in particular, when $X$ is projective, in an entirely projective setting.

Within the algebraic setting, torsors can also be realized as open subsets of spectra of forcing algebras.

DEFINITION 2.7. Let $R$ be a commutative ring and let $f_1, \ldots, f_n$ and $f$ be elements in $R$. Then the $R$-algebra

$$R[T_1, \ldots, T_n]/(f_1T_1 + \cdots + f_nT_n - f)$$

is called the *forcing algebra* of these elements (or these data).

THEOREM 2.8. *Let $R$ denote a noetherian ring, let $I = (f_1, \ldots, f_n)$ denote an ideal and let $f \in R$ be another element. Let*

$$c = \delta(f) \in H^1(D(I), \operatorname{Syz}(f_1, \ldots, f_n))$$

*be the corresponding cohomology class and let*

$$B = R[T_1, \ldots, T_n]/(f_1T_1 + \cdots + f_nT_n - f)$$

*denote the forcing algebra for these data. Then the scheme* $\operatorname{Spec}(B)|_{D(I)}$ *together with the natural action of the syzygy bundle on it is isomorphic to the torsor given by $c$.*

Forcing algebras provide a natural framework for closure operations in general, it is however a special feature of tight closure that the induced torsor contains the relevant information.

**2.3. Tight closure and solid closure.** Forcing algebras occurred in the work of Hochster on solid closure. The following theorem of Hochster [**12**, Theorem 8.6] gives a characterization of tight closure in terms of forcing algebra and local cohomology.

THEOREM 2.9. *Let $R$ be a normal excellent local domain with maximal ideal $\mathfrak{m}$ over a field of positive characteristic. Let $f_1, \ldots, f_n$ generate an $\mathfrak{m}$-primary ideal $I$ and let $f$ be another element in $R$. Then $f \in I^*$ if and only if $H^{\dim(R)}_{\mathfrak{m}}(B) \neq 0$, where $B = R[T_1, \ldots, T_n]/(f_1T_1 + \cdots + f_nT_n - f)$ denotes the forcing algebra of these elements.*

[1] $\mathcal{S}^\vee$ denotes the dual bundle. According to our convention, the geometric vector bundle corresponding to a locally free sheaf $\mathcal{T}$ is given by $\operatorname{Spec}(\oplus_{k \geq 0} S^k(\mathcal{T}))$ and the projective bundle is $\operatorname{Proj}(\oplus_{k \geq 0} S^k(\mathcal{T}))$, where $S^k$ denotes the $k$th symmetric power.

If the dimension $d$ is at least two, then

$$H^d_{\mathfrak{m}}(R) \longrightarrow H^d_{\mathfrak{m}}(B) \cong H^d_{\mathfrak{m}B}(B) \cong H^{d-1}(D(\mathfrak{m}B), \mathcal{O}_B).$$

This means that we have to look at the cohomological properties of the complement of the exceptional fiber over the closed point, i.e. the torsor given by these data. If $H^{d-1}(D(\mathfrak{m}B), \mathcal{O}_B) = 0$ then this is true for all quasicoherent sheaves instead of just the structure sheaf. This property can be expressed by saying that the *cohomological dimension* of $D(\mathfrak{m}B)$ is $\leq d-2$ and thus smaller than the cohomological dimension of the punctured spectrum $D(\mathfrak{m})$, which is exactly $d-1$. So belonging to tight closure can be rephrased by saying that the formation of the corresponding torsor does not change the cohomological dimension.

If the dimension is two, then we need to look whether the first cohomology of the structure sheaf vanishes. This is true (by Serre's cohomological criterion for affineness) if and only if the open subset $D(\mathfrak{m}B)$ is an *affine scheme* (the spectrum of a ring).

The right hand side of the equivalence in Theorem 2.9 (the non-vanishing of the top-dimensional local cohomology) is independent of any characteristic assumption, and can be taken as the basis for the definition of another closure operation, called *solid closure*. So the theorem above says that in positive characteristic, tight closure and solid closure coincide. There is also a definition of tight closure for algebras over a field of characteristic 0 by reduction to positive characteristic, see [**15**, Appendix 1].

**2.4. The graded two-dimensional case.** In the situation of a forcing algebra of homogeneous elements, this torsor $T$ can also be obtained as $\operatorname{Proj}(B)$, where $B$ is the (not necessarily positively) graded forcing algebra. In particular, it follows that the containment $f \in I^*$ is equivalent to the property that $T$ is not an affine variety. For this property, positivity (ampleness) properties of the syzygy bundle are crucial. We need again the concept of semistability introduced in the first lecture.

For a strongly semistable vector bundle $\mathcal{S}$ on $C$ and a cohomology class $c \in H^1(C, \mathcal{S})$ with corresponding torsor we obtain the following affineness criterion. In characteristic 0, we mean by strongly semistable just semistable.

THEOREM 2.10. *Let $C$ denote a smooth projective curve over an algebraically closed field $K$ and let $\mathcal{S}$ be a strongly semistable vector bundle over $C$ together with a cohomology class $c \in H^1(C, \mathcal{S})$. Then the torsor $T(c)$ is an affine scheme if and only if* $\deg(\mathcal{S}) < 0$ *and $c \neq 0$ ($F^e(c) \neq 0$ for all $e$ in positive characteristic[2]).*

This result rests on the ampleness of $\mathcal{S}'^\vee$ occuring in the dual(to the sequel in Remark 2.6) exact sequence $0 \to \mathcal{O}_C \to \mathcal{S}'^\vee \to \mathcal{S}^\vee \to 0$ given by $c$ (this rests on work of Gieseker and Hartshorne(see [**9**], [**10**]). It implies for a strongly semistable syzygy bundle the following *degree formula* for tight closure.

THEOREM 2.11. *Let $I = (f_1, \ldots, f_n)$ be an $R_+$-primary homogeneous ideal in a two-dimensional standard-graded normal domain over an algebraically closed field*

[2]Here one has to check only finitely many $e$s and there exist good estimates how far one has to go based on [**9**, Lemma 2.2]. Also, in a relative situation over $\operatorname{Spec}(\mathbb{Z})$, under the condition that the bundle is semistable in characteristic 0, this is only an extra condition for finitely many prime numbers, see [**4**, Remark 2.2]

*of positive characteristic. Suppose that* $\mathrm{Syz}\,(f_1,\ldots,f_n)$ *is strongly semistable on* $C=\mathrm{Proj}(R)$. *Then*

$$R_m \subseteq I^* \text{ for } m \geq \frac{\sum d_i}{n-1} R_m \cap I^* \subseteq I^F \text{ for } m < \frac{\sum d_i}{n-1}.$$

Here, in the exclusion result on the right hand side, $I^F$ denotes the *Frobenius closure* of the ideal. For fixed positive characteristic, one can not replace $I^F$ by $I$. In a relative setting over $\mathrm{Spec}\,(\mathbb{Z})$ however, for $p$ large enough, we can replace $I^F$ by $I$ and get thus a precise degree description for tight closure.

We indicate the proof of the inclusion result. The degree condition implies that $c=\delta(f)\in H^1(C,\mathcal{S})$ is such that $\mathcal{S}=\mathrm{Syz}\,(f_1,\ldots,f_n)(m)$ has non-negative degree, and all Frobenius pull-backs $F^*(\mathcal{S})$ have non-negative degree. Let $\mathcal{L}=\mathcal{O}(k)$ be a twist of the tautological line bundle on $C$ such that its degree is larger than the degree of $\omega_C^{-1}$, the dual of the canonical sheaf. Let $z\in H^0(Y,\mathcal{L})$ be a non-zero element. Then $zF^{e*}(c)\in H^1(C,F^{e*}(\mathcal{S})\otimes\mathcal{L})$, and by Serre duality we have

$$H^1(C,F^{e*}(\mathcal{S})\otimes\mathcal{L}) \cong H^0(F^{e*}(\mathcal{S}^\vee)\otimes\mathcal{L}^{-1}\otimes\omega_C)^\vee.$$

On the right hand side we have a semistable sheaf of negative degree, which can not have a non-trivial section. Hence

$$zF^{e*}(c)=0,$$

and therefore $f$ belongs to the tight closure.

In general, there exists the following exact criterion for the affineness of the torsor $T(c)$ depending on $c$ and the *strong Harder-Narasimhan filtration* of $\mathcal{S}$ (see [**4**, Theorem 2.3] for the proof).

THEOREM 2.12. *Let $C$ denote a smooth projective curve over an algebraically closed field $K$ of positive characteristic and let $\mathcal{S}$ be a vector bundle over $C$ together with a cohomology class $c\in H^1(C,\mathcal{S})$. Let*

$$\mathcal{S}_1 \subset \mathcal{S}_2 \subset \ldots \subset \mathcal{S}_{t-1} \subset \mathcal{S}_t = F^{e*}(\mathcal{S})$$

*be a strong Harder-Narasimhan filtration. We choose $i$ such that $\mathcal{S}_i/\mathcal{S}_{i-1}$ has degree $\geq 0$ and that $\mathcal{S}_{i+1}/\mathcal{S}_i$ has degree $<0$. We set $\mathcal{Q}=F^{e*}(\mathcal{S})/\mathcal{S}_i$. Then the following are equivalent.*

(1) *The torsor $T(c)$ is not an affine scheme.*
(2) *Some Frobenius power of the image of $F^{e*}(c)$ inside $H^1(X,\mathcal{Q})$ is* 0.

## 3. Lecture - Plus closure and trivializable bundles

**3.1. Plus closure.** For an ideal $I\subseteq R$ in a domain $R$ define its *plus closure* by

$$I^+ = \{f\in R \mid \text{there exists a finite domain extension } R\subseteq T \text{ such that } f\in IT\}.$$

Equivalent: Let $R^+$ be the *absolute integral closure* of $R$. This is the integral closure of $R$ in an algebraic closure of the quotient field $Q(R)$ (first considered by Artin [**1**]). Then

$$f\in I^+ \text{ if and only if } f\in IR^+.$$

The plus closure commutes with localization.

We also have the inclusion $I^+\subseteq I^*$. Here the question arises:

Question: Is $I^+=I^*$?

This question is known as the *tantalizing question* in tight closure theory.

In terms of forcing algebras and their torsors, the containment inside the plus closure means that there exists a $d$-dimensional closed subscheme inside the torsor which meets the exceptional fiber (the fiber over the maximal ideal) in isolated points, and this means that the so-called superheight of the extended ideal is $d$. In this case the local cohomological dimension of the torsor must be $d$ as well, since it contains a closed subscheme with this cohomological dimension. So also the plus closure depends only on the torsor.

In characteristic zero, the plus closure behaves very differently compared to the case of positive characteristic. If $R$ is a normal domain of characteristic 0, then the trace map shows that the plus closure is trivial, $I^+ = I$ for every ideal $I$.

**3.2. Plus closure in dimension two.** Let $K$ be a field and let $R$ be a normal two-dimensional standard-graded domain over $K$ with corresponding smooth projective curve $C$. A homogeneous $\mathfrak{m}$-primary ideal with homogeneous ideal generators $f_1, \ldots, f_n$ and another homogeneous element $f$ of degree $m$ yield a cohomology class

$$c = \delta(f) = H^1(C, \operatorname{Syz}(f_1, \ldots, f_n)(m)).$$

Let $T(c)$ be the corresponding torsor. We have seen that the affineness of this torsor over $C$ is equivalent to the affineness of the corresponding torsor over $D(\mathfrak{m}) \subseteq \operatorname{Spec}(R)$ (and to the property of not belonging to the tight closure). Now we want to understand what the property $f \in I^+$ means for $c$ and for $T(c)$. Instead of the plus closure we will work with the *graded plus closure* $I^{+\mathrm{gr}}$, where $f \in I^{+\mathrm{gr}}$ holds if and only if there exists a finite graded extension $R \subseteq S$ such that $f \in IS$. The existence of such an $S$ translates into the existence of a finite morphism

$$\varphi\colon C' = \operatorname{Proj}(S) \longrightarrow \operatorname{Proj}(R) = C$$

such that $\varphi^*(c) = 0$. Here we may assume that $C'$ is also smooth. Therefore, we discuss the more general question when a cohomology class $c \in H^1(C, \mathcal{S})$, where $\mathcal{S}$ is a locally free sheaf on $C$, can be annihilated by a finite morphism

$$C' \longrightarrow C$$

of smooth projective curves. The advantage of this more general approach is that we may work with short exact sequences (in particular, the sequences coming from the Harder-Narasimhan filtration) in order to reduce the problem to semistable bundles which do not necessarily come from an ideal situation.

LEMMA 3.1. *Let $C$ denote a smooth projective curve over an algebraically closed field $K$, let $\mathcal{S}$ be a locally free sheaf on $C$ and let $c \in H^1(C, \mathcal{S})$ be a cohomology class with corresponding torsor $T \to C$. Then the following conditions are equivalent.*

(1) *There exists a finite morphism*

$$\varphi\colon C' \longrightarrow C$$

*from a smooth projective curve $C'$ such that $\varphi^*(c) = 0$.*

(2) *There exists a projective curve $Z \subseteq T$.*

PROOF. If (1) holds, then the pull-back $\varphi^*(T) = T \times_C C'$ is trivial (as a torsor), as it equals the torsor given by $\varphi^*(c) = 0$. Hence $\varphi^*(T)$ is isomorphic to a vector bundle and contains in particular a copy of $C'$. The image $Z$ of this copy is a projective curve inside $T$.

If (2) holds, then let $C'$ be the normalization of $Z$. Since $Z$ dominates $C$, the resulting morphism

$$\varphi\colon C' \longrightarrow C$$

is finite. Since this morphism factors through $T$ and since $T$ annihilates the cohomology class by which it is defined, it follows that $\varphi^*(c) = 0$. □

We want to show that the cohomological criterion for (non)-affineness of a torsor along the Harder-Narasimhan filtration of the vector bundle also holds for the existence of projective curves inside the torsor, under the condition that the projective curve is defined over a finite field. This implies that tight closure is (graded) plus closure for graded $\mathfrak{m}$-primary ideals in a two-dimensional graded domain over a finite field.

**3.3. Annihilation of cohomology for strongly semistable sheaves.** We deal first with the situation of a strongly semistable sheaf $\mathcal{S}$ of degree 0. The following two results are due to Lange and Stuhler [**18**]. We say that a locally free sheaf is *étale trivializable* if there exists a finite étale morphism $\varphi\colon C' \to C$ such that $\varphi^*(\mathcal{S}) \cong \mathcal{O}_{C'}^r$. Such bundles are directly related to linear representations of the étale fundamental group.

LEMMA 3.2. *Let $K$ denote a finite field (or the algebraic closure of a finite field) and let $X$ be a smooth projective curve over $K$. Let $\mathcal{S}$ be a locally free sheaf over $X$. Then $\mathcal{S}$ is étale trivializable if and only if there exists some $n$ such that $F^{n*}\mathcal{S} \cong \mathcal{S}$.*

THEOREM 3.3. *Let $K$ denote a finite field (or the algebraic closure of a finite field) and let $X$ be a smooth projective curve over $K$. Let $\mathcal{S}$ be a strongly semistable locally free sheaf over $X$ of degree 0. Then there exists a finite morphism*

$$\varphi\colon Y \longrightarrow X$$

*such that $\varphi^*(\mathcal{S})$ is trivial.*

PROOF. We consider the family of locally free sheaves $F^{e*}(\mathcal{S})$, $e \in \mathbb{N}$. Because these are all semistable of degree 0, and defined over the same finite field, we must have (by the existence of the moduli space for vector bundles) a repetition, i.e.

$$F^{e*}(\mathcal{S}) \cong F^{e'*}(\mathcal{S})$$

for some $e' > e$. By Lemma 3.2, the bundle $F^{e*}(\mathcal{S})$ admits an étale trivialization $\varphi\colon Y \to X$. Hence the finite map $F^e \circ \varphi$ trivializes the bundle. □

THEOREM 3.4. *Let $K$ denote a finite field (or the algebraic closure of a finite field) and let $X$ be a smooth projective curve over $K$. Let $\mathcal{S}$ be a strongly semistable locally free sheaf over $X$ of nonnegative degree and let $c \in H^1(X, \mathcal{S})$ denote a cohomology class. Then there exists a finite morphism*

$$\varphi\colon Y \longrightarrow X$$

*such that $\varphi^*(c)$ is trivial.*

PROOF. If the degree of $\mathcal{S}$ is positive, then for $e$ sufficiently large, the Frobenius pull-back $F^{e*}(\mathcal{S})$ has arbitrary large degree and is still semistable. By Serre duality we get that $H^1(X, F^{e*}(\mathcal{S})) = H^0(X, F^{e*}(\mathcal{S}^\vee) \otimes \omega_X) = 0$, as $F^{e*}(\mathcal{S}^\vee) \otimes \omega_X$ has then negative degree and is also semistable. So in this case we can annihilate the class by an iteration of the Frobenius alone.

So suppose that the degree is 0. Then there exists by Theorem 3.3 a finite morphism which trivializes the bundle. So we may assume that $\mathcal{S} \cong \mathcal{O}_X^r$. Then the cohomology class has several components $c_i \in H^1(X, \mathcal{O}_X)$ and it is enough to annihilate them separately by finite morphisms. But this is possible by the parameter theorem of K. Smith [**22**] (or directly using Frobenius and Artin-Schreier extensions). □

**3.4. The general case.** We look now at an arbitrary locally free sheaf $\mathcal{S}$ on $C$, a smooth projective curve over a finite field. We want to show that the same numerical criterion (formulated in terms of the Harder-Narasimhan filtration) for non-affineness of a torsor holds also for the finite annihilation of the corresponding cohomomology class (or the existence of a projective curve inside the torsor).

THEOREM 3.5. *Let $K$ denote a finite field (or the algebraic closure of a finite field) and let $X$ be a smooth projective curve over $K$. Let $\mathcal{S}$ be a locally free sheaf over $X$ and let $c \in H^1(X, \mathcal{S})$ denote a cohomology class. Let $\mathcal{S}_1 \subset \ldots \subset \mathcal{S}_t$ be a strong Harder-Narasimhan filtration of $F^{e*}(\mathcal{S})$. We choose $i$ such that $\mathcal{S}_i/\mathcal{S}_{i-1}$ has degree $\geq 0$ and that $\mathcal{S}_{i+1}/\mathcal{S}_i$ has degree $< 0$. We set $\mathcal{Q} = F^{e*}(\mathcal{S})/\mathcal{S}_i$. Then the following are equivalent.*

(1) *The class $c$ can be annihilated by a finite morphism $Y \to X$.*
(2) *Some Frobenius power of the image of $F^{e*}(c)$ inside $H^1(X, \mathcal{Q})$ is 0.*

PROOF. Suppose that (1) holds. Then the torsor is not affine and hence by Theorem 2.12 the assertion (2) holds.

Suppose (2) holds. By applying a certain power of the Frobenius, we may assume that the image of the cohomology class in $\mathcal{Q}$ is 0. Hence the class stems from a cohomology class $c_i \in H^1(X, \mathcal{S}_i)$. We look at the short exact sequence

$$0 \longrightarrow \mathcal{S}_{i-1} \longrightarrow \mathcal{S}_i \longrightarrow \mathcal{S}_i/\mathcal{S}_{i-1} \longrightarrow 0,$$

where the sheaf on the right hand side has a nonnegative degree. Therefore the image of $c_i$ in $H^1(X, \mathcal{S}_i/\mathcal{S}_{i-1})$ can be annihilated by a finite morphism due to Theorem 3.4. Hence, after applying a finite morphism, we may assume that $c_i$ stems from a cohomology class $c_{i-1} \in H^1(X, \mathcal{S}_{i-1})$. Going on inductively we see that $c$ can be annihilated by a finite morphism. □

THEOREM 3.6. *Let $C$ denote a smooth projective curve over the algebraic closure of a finite field $K$, let $\mathcal{S}$ be a locally free sheaf on $C$ and let $c \in H^1(C, \mathcal{S})$ be a cohomology class with corresponding torsor $T \to C$. Then $T$ is affine if and only if it does not contain any projective curve.*

PROOF. Due to Theorem 2.12 and Theorem 3.5, for both properties the same numerical criterion does hold. □

These results imply the following theorem in the setting of a two-dimensional graded ring.

THEOREM 3.7. *Let $R$ be a standard-graded, two-dimensional normal domain over (the algebraic closure of) a finite field. Let $I$ be an $R_+$-primary graded ideal. Then*

$$I^* = I^+.$$

This is also true for non-primary graded ideals, see [**4**, Theorem 4.2] and also for submodules in finitely generated graded submodules, since this situation may also be reduced to Theorem 3.6. Moreover, G. Dietz [**8**] has shown that one can get rid also of the graded assumption (of the ideal or module, but not of the ring).

## 4. Lecture - Deformations and localization problem

After having understood tight closure and plus closure in the two-dimensional situation we proceed to a special three-dimensional situation, namely families of two-dimensional rings parametrized by a one-dimensional base scheme.

**4.1. Affineness under deformations.** We consider a base scheme $B$ and a morphism

$$Z \longrightarrow B$$

together with an open subscheme $W \subseteq Z$. For every base point $b \in B$ we get the open subset

$$W_b \subseteq Z_b$$

inside the fiber $Z_b$. It is a natural question to ask how properties of $W_b$ vary with $b$. In particular, we may ask how the cohomological dimension of $W_b$ varies with $b$. Recall that the cohomological dimension of a scheme $X$ is the maximal number $i$ such that $H^i(X, \mathcal{F}) \neq 0$ for some quasicoherent sheaf $\mathcal{F}$. A noetherian scheme is affine if and only if its cohomological dimension is 0. Tight closure can be characterized by the cohomological dimension of torsors.

In the algebraic setting, we have a commutative $K$-algebra $D$, a commutative $D$-algebra $S$ and an ideal $\mathfrak{a} \subseteq S$ (so $B = \operatorname{Spec}(D)$ $Z = \operatorname{Spec}(S)$ and $W = D(\mathfrak{a})$) which defines for every prime ideal $\mathfrak{p} \in \operatorname{Spec}(D)$ the extended ideal $\mathfrak{a}_{\mathfrak{p}}$ in $S \otimes_D \kappa(\mathfrak{p})$. Then in this situation, $D(\mathfrak{a}_{\mathfrak{p}}) \subseteq \operatorname{Spec}(S \otimes_D \kappa(\mathfrak{p}))$ is the fiber over $\mathfrak{p}$.

This question is already interesting when $B = \operatorname{Spec}(D)$ is an affine one-dimensional integral scheme, in particular in the following two situations.

(1) $B = \operatorname{Spec}(\mathbb{Z})$. Then we speak of an *arithmetic deformation* and want to know how affineness varies with the characteristic and what the relation is to characteristic zero.

(2) $B = \mathbb{A}^1_K = \operatorname{Spec}(K[t])$, where $K$ is a field. Then we speak of a *geometric deformation* and want to know how affineness varies with the parameter $t$, in particular how the behavior over the special points where the residue class field is algebraic over $K$ is related to the behavior over the generic point.

It is fairly easy to show that if the open subset in the generic fiber is affine, then also the open subsets are affine for almost all special points.

We deal with this question where $W$ is a torsor over a family of smooth projective curves (or a torsor over a punctured two-dimensional spectrum). The arithmetic as well as the geometric variant of this question are directly related to questions in tight closure theory. Because of the above mentioned degree criteria in the strongly semistable case (see Theorem 2.11), a weird behavior of the affineness property of torsors is only possible if we have a weird behavior of strong semistability.

**4.2. Arithmetic deformations.** We start with the arithmetic situation, the following example is due to Brenner and Katzman [**6**].

EXAMPLE 4.1. Consider $\mathbb{Z}[x,y,z]/(x^7+y^7+z^7)$ and take the ideal $I=(x^4,y^4,z^4)$ and the element $f = x^3y^3$. Consider reductions $\mathbb{Z} \to \mathbb{Z}/(p)$. Then

$$f \in I^* \text{ holds in } \mathbb{Z}/(p)[x,y,z]/(x^7+y^7+z^7) \text{ for } p \equiv 3 \bmod 7$$

and

$$f \notin I^* \text{ holds in } \mathbb{Z}/(p)[x,y,z]/(x^7+y^7+z^7) \text{ for } p \equiv 2 \bmod 7\,.$$

In particular, the bundle $\operatorname{Syz}(x^4,y^4,z^4)$ is semistable (see [**6**, Corollary 4.3]) in the generic fiber, but not strongly semistable for any reduction $p \equiv 2 \bmod 7$ by Theorem 2.11. The corresponding torsor is an affine scheme for infinitely many prime reductions and not an affine scheme for infinitely many prime reductions.

In terms of affineness (or local cohomology) of quasiaffine schemes, this example has the following properties: the open subset given by the ideal

$$(x,y,z) \subseteq \mathbb{Z}/(p)[x,y,z,s_1,s_2,s_3]/(x^7+y^7+z^7, s_1x^4+s_2y^4+s_3z^4+x^3y^3)$$

has cohomological dimension 1 if $p = 3 \mod 7$ and has cohomological dimension 0 (equivalently, $D(x,y,z)$ is an affine scheme) if $p = 2 \mod 7$.

**4.3. Geometric deformations - A counterexample to the localization problem.** Let $S \subseteq R$ be a multiplicative system and $I$ an ideal in $R$. Then the *localization problem* of tight closure is the question whether the identity

$$(I^*)_S = (IR_S)^*$$

holds.

Here the inclusion $\subseteq$ is always true and $\supseteq$ is the problem. The problem means explicitly:

If $f \in (IR_S)^*$, can we find an $h \in S$ such that $hf \in I^*$ holds in $R$?

PROPOSITION 4.2. *Let $\mathbb{Z}/(p) \subset D$ be a one-dimensional domain, $D \subseteq R$ of finite type and $I$ an ideal in $R$. Suppose that localization holds and that*

$$f \in I^* \text{ holds in } R \otimes_D Q(D) = R_{D^*} = R_{Q(D)}$$

*($S = D^* = D\setminus\{0\}$ is the multiplicative system). Then $f \in I^*$ holds in $R\otimes_D \kappa(\mathfrak{p})$ for almost all $\mathfrak{p}$ in Spec $D$.*

PROOF. By localization, there exists $h \in D$, $h \neq 0$, such that $hf \in I^*$ in $R$. By persistence of tight closure (under a ring homomorphism), we get

$$hf \in I^* \text{ in } R_{\kappa(\mathfrak{p})}\,.$$

The element $h$ does not belong to $\mathfrak{p}$ for almost all $\mathfrak{p}$, so $h$ is a unit in $R_{\kappa(\mathfrak{p})}$ and hence

$$f \in I^* \text{ in } R_{\kappa(\mathfrak{p})}$$

for almost all $\mathfrak{p}$. □

In order to get a counterexample to the localization property we will look now at geometric deformations:

$$D = \mathbb{F}_p[t] \subset \mathbb{F}_p[t][x,y,z]/(g) = S,$$

where $t$ has degree 0 and $x, y, z$ have degree 1 and $g$ is homogeneous. Then (for every homomorphism $\mathbb{F}_p[t] \to K$, where $K$ is a field)

$$S \otimes_{\mathbb{F}_p[t]} K$$

is a two-dimensional standard-graded ring over $K$. For the residue class fields of points of $\mathbb{A}^1_{\mathbb{F}_p} = \operatorname{Spec}(\mathbb{F}_p[t])$ we have basically two possibilities.

- $K = \mathbb{F}_p(t)$, the function field. This is the *generic* or *transcendental* case.
- $K = \mathbb{F}_q$, the *special* or *algebraic* or *finite* case.

How does $f \in I^*$ vary with $K$? To analyze the behavior of tight closure in such a family we can use what we know in the two-dimensional standard-graded situation.

In order to establish an example where tight closure does not behave uniformly under a geometric deformation, we first need a situation where strong semistability does not behave uniformly. Such an example was given, in terms of Hilbert-Kunz theory, by Paul Monsky in 1998 [**21**].

EXAMPLE 4.3. Let

$$g = z^4 + z^2xy + z(x^3 + y^3) + (t + t^2)x^2y^2.$$

Consider

$$S = \mathbb{F}_2[t, x, y, z]/(g).$$

Then Monsky proved the following results on the *Hilbert-Kunz multiplicity* of the maximal ideal $(x, y, z)$ in $S \otimes_{\mathbb{F}_2[t]} L$, $L$ a field:

$$e_{HK}(S \otimes_{\mathbb{F}_2[t]} L) = \begin{cases} 3 \text{ for } L = \mathbb{F}_2(t) \\ 3 + \frac{1}{4^d} \text{ for } L = \mathbb{F}_q = \mathbb{F}_2(\alpha),\ (t \mapsto \alpha,\ d = \deg(\alpha)) . \end{cases}$$

We consider $S$ as an $\mathbb{F}_2[t]$-algebra, the corresponding morphism $\operatorname{Spec}(S) \to \mathbb{A}^1_{\mathbb{F}_2}$ and the corresponding smooth projective relative curve $C = \operatorname{Proj}(S) \to \mathbb{A}^1_{\mathbb{F}_2}$. The fibers are $\operatorname{Spec}\left(S_{\kappa(\mathfrak{p})}\right)$ and $C_{\kappa(\mathfrak{p})}$ respectively.

By the geometric interpretation of Hilbert-Kunz theory, in particular Corollary 1.14 the computations mentioned in Example 4.3 mean that the restricted cotangent bundle

$$\operatorname{Syz}(x, y, z) = (\Omega_{\mathbb{P}^2})|_C$$

is strongly semistable in the transcendental case, but not strongly semistable in the algebraic case. In fact, for $d = \deg(\alpha)$, $t \mapsto \alpha$, where $L = \mathbb{F}_2(\alpha)$, the $d$-th Frobenius pull-back destabilizes (meaning that it is not semistable anymore).

The maximal ideal $(x, y, z)$ can not be used directly, as it is tightly closed. However, we look at the second Frobenius pull-back which is (characteristic two) just

$$I = \left(x^4, y^4, z^4\right).$$

By the degree formula, we need to look for an element of degree 6. Let's take $f = y^3z^3$. This is our example ($x^3y^3$ does not work). First, by strong semistability in the transcendental case, we have

$$f \in I^* \text{ in } S \otimes \mathbb{F}_2(t)$$

by the degree formula. If localization holds, then by Proposition 4.2, $f$ belongs to the tight closure of $I$ for almost all algebraic instances $\mathbb{F}_q = \mathbb{F}_2(\alpha)$, $t \mapsto \alpha$.

Contrary to that we show that for all algebraic instances, the element $f$ never belongs to the tight closure of $I$.

LEMMA 4.4. *Let* $\mathbb{F}_q = \mathbb{F}_p(\alpha)$, $t \mapsto \alpha$, $\deg(\alpha) = d$. *Set* $Q = 2^{d-1}$. *Then*

$$xyf^Q \notin I^{[Q]}.$$

PROOF. This is an elementary but tedious computation. □

THEOREM 4.5. *Tight closure does not commute with localization.*

PROOF. One knows in our situation that $xy$ is a so-called test element. Hence Lemma 4.4 shows that $f \notin I^*$, see [**7**] for details. □

In terms of affineness of quasiaffine schemes (or local cohomology), this example has the following properties: the open subset given by the ideal

$$(x, y, z) \subseteq \mathbb{F}_2(t)[x, y, z, s_1, s_2, s_3]/\big(g, s_1x^4 + s_2y^4 + s_3z^4 + y^3z^3\big)$$

has cohomological dimension 1 if $t$ is transcendental and has cohomological dimension 0 (equivalently, $D(x, y, z)$ is an affine scheme) if $t$ is algebraic.

COROLLARY 4.6. *Tight closure is not necessarily plus closure in graded dimension two for fields with transcendental elements.*

PROOF. Consider

$$R = \mathbb{F}_2(t)[x, y, z]/(g).$$

In this ring $y^3z^3 \in I^*$, but it can not belong to the plus closure. Else there would be a curve morphism

$$Y \longrightarrow C_{\mathbb{F}_2(t)}$$

which annihilates the cohomology class $c$ and this would extend to a morphism of relative curves over $\mathbb{A}^1_{\mathbb{F}_2}$ almost everywhere. □

COROLLARY 4.7. *There is an example of a smooth variety $Z$ and an effective divisor $D \subset Z$ and a projective morphism*

$$Z \longrightarrow \mathbb{A}^1_{\mathbb{F}_2}$$

*such that $(Z \setminus D)_\eta$ is not an affine variety over the generic point $\eta$, but for every algebraic point $x$ the fiber $(Z \setminus D)_x$ is an affine variety.*

PROOF. Take $C \to \mathbb{A}^1_{\mathbb{F}_2}$ to be the Monsky quartic of Example 4.3 and consider the syzygy bundle

$$\mathcal{S} = \operatorname{Syz}\big(x^4, y^4, z^4\big)(6)$$

together with the cohomology class $c$ determined by $f = y^3z^3$. This class defines an extension

$$0 \longrightarrow \mathcal{S} \longrightarrow \mathcal{S}' \longrightarrow \mathcal{O}_C \longrightarrow 0$$

and hence $\mathbb{P}(\mathcal{S}^\vee) \subset \mathbb{P}(\mathcal{S}'^\vee)$. Then $\mathbb{P}(\mathcal{S}'^\vee) \setminus \mathbb{P}(\mathcal{S}^\vee)$ is an example with the stated properties by the previous results. □

It is an open question whether such an example exists in characteristic zero.

## Acknowledgments

I thank the organizers for the opportunity to give this lecture series in Trieste and the ICTP for its hospitality. I thank the two referees for many valuable suggestions.

## References


[1] M. Artin, *On the joins of Hensel rings*, Advances in Math. **7** (1971), 282–296 (1971), DOI 10.1016/S0001-8708(71)80007-5. MR289501

[2] Holger Brenner, *The theory of tight closure from the viewpoint of vector bundles*, Habilitationsschrift, Ruhr-Universität Bochum, 2003.

[3] Holger Brenner, *The rationality of the Hilbert-Kunz multiplicity in graded dimension two*, Math. Ann. **334** (2006), no. 1, 91–110, DOI 10.1007/s00208-005-0703-x. MR2208950

[4] Holger Brenner, *Tight closure and plus closure in dimension two*, Amer. J. Math. **128** (2006), no. 2, 531–539. MR2214902

[5] Holger Brenner, *Tight closure and vector bundles*, Three lectures on commutative algebra, Univ. Lecture Ser., vol. 42, Amer. Math. Soc., Providence, RI, 2008, pp. 1–71, DOI 10.1090/ulect/042/01. MR2410690

[6] Holger Brenner and Mordechai Katzman, *On the arithmetic of tight closure*, J. Amer. Math. Soc. **19** (2006), no. 3, 659–672, DOI 10.1090/S0894-0347-05-00514-X. MR2220102

[7] Holger Brenner and Paul Monsky, *Tight closure does not commute with localization*, Ann. of Math. (2) **171** (2010), no. 1, 571–588, DOI 10.4007/annals.2010.171.571. MR2630050

[8] Geoffrey D. Dietz, *Tight closure of finite length modules in graded rings*, J. Algebra **306** (2006), no. 2, 520–534, DOI 10.1016/j.jalgebra.2006.05.009. MR2271350

[9] David Gieseker, *p-ample bundles and their Chern classes*, Nagoya Math. J. **43** (1971), 91–116. MR296078

[10] Robin Hartshorne, *Ample vector bundles on curves*, Nagoya Math. J. **43** (1971), 73–89. MR292847

[11] Robin Hartshorne, *Algebraic geometry*, Graduate Texts in Mathematics, vol. No. 52, Springer-Verlag, New York-Heidelberg, 1977. MR0463157

[12] Melvin Hochster, *Solid closure*, Commutative algebra: syzygies, multiplicities, and birational algebra (South Hadley, MA, 1992), Contemp. Math., vol. 159, Amer. Math. Soc., Providence, RI, 1994, pp. 103–172, DOI 10.1090/conm/159/01508. MR1266182

[13] Melvin Hochster and Craig Huneke, *Tight closure*, Commutative algebra (Berkeley, CA, 1987), Math. Sci. Res. Inst. Publ., vol. 15, Springer, New York, 1989, pp. 305–324, DOI 10.1007/978-1-4612-3660-3_15. MR1015524

[14] Melvin Hochster and Craig Huneke, *Tight closure, invariant theory, and the Briançon-Skoda theorem*, J. Amer. Math. Soc. **3** (1990), no. 1, 31–116, DOI 10.2307/1990984. MR1017784

[15] Craig Huneke, *Tight closure and its applications*, CBMS Regional Conference Series in Mathematics, vol. 88, Published for the Conference Board of the Mathematical Sciences, Washington, DC; by the American Mathematical Society, Providence, RI, 1996. With an appendix by Melvin Hochster, DOI 10.1016/0167-4889(95)00136-0. MR1377268

[16] Craig Huneke, *Tight closure, parameter ideals, and geometry*, Six lectures on commutative algebra (Bellaterra, 1996), Progr. Math., vol. 166, Birkhäuser, Basel, 1998, pp. 187–239. MR1648666

[17] Ernst Kunz, *Characterizations of regular local rings of characteristic p*, Amer. J. Math. **91** (1969), 772–784, DOI 10.2307/2373351. MR252389

[18] Herbert Lange and Ulrich Stuhler, *Vektorbündel auf Kurven und Darstellungen der algebraischen Fundamentalgruppe* (German), Math. Z. **156** (1977), no. 1, 73–83, DOI 10.1007/BF01215129. MR472827

[19] Adrian Langer, *Semistable sheaves in positive characteristic*, Ann. of Math. (2) **159** (2004), no. 1, 251–276, DOI 10.4007/annals.2004.159.251. MR2051393

[20] P. Monsky, *The Hilbert-Kunz function*, Math. Ann. **263** (1983), no. 1, 43–49, DOI 10.1007/BF01457082. MR697329

[21] Paul Monsky, *Hilbert-Kunz functions in a family: point-$S_4$ quartics*, J. Algebra **208** (1998), no. 1, 343–358, DOI 10.1006/jabr.1998.7500. MR1644019

[22] K. E. Smith, *Tight closure of parameter ideals*, Invent. Math. **115** (1994), no. 1, 41–60, DOI 10.1007/BF01231753. MR1248078

[23] V. Trivedi, *Semistability and Hilbert-Kunz multiplicities for curves*, J. Algebra **284** (2005), no. 2, 627–644, DOI 10.1016/j.jalgebra.2004.10.016. MR2114572

[24] Kei-ichi Watanabe and Ken-ichi Yoshida, *Hilbert-Kunz multiplicity and an inequality between multiplicity and colength*, J. Algebra **230** (2000), no. 1, 295–317, DOI 10.1006/jabr.1999.7956. MR1774769

University of Osnabrück, 49069 Osnabrück, Germany

Contemporary Mathematics
Volume **831**, 2025
https://doi.org/10.1090/conm/831/16639

# F-singularities; characteristic $p$ methods in commutative ring theory and algebraic geometry

Kei-ichi Watanabe

*Dedicated to Mel Hochster and Craig Huneke who made wonderful world of commutative ring theory and gave me wonderful tools for my researches*

## Introduction

In 1985, M. Hochster and C. Huneke invented the notion of **tight closures** of the ideals in rings of characteristic $p$. This method enables us to make many things clearer and sometimes to generalize in broader settings.

This manuscript is written in the occasion of "ICTP School on Commutative Algebra and Algebraic Geometry in Prime Characteristics, 2 - 5 May 2023".

Curiously, it happened that the Frobenius splitting proved to be a very useful tool for classifying singularities appearing in MMP (Minimal Model Program) in algebraic geometry over a field of characteristic 0.

The main idea of my lectures starts from "tight closure theory" and then introduces "characteristic $p$ method" used in algebraic geometry to the audience, who are supposed to be commutative algebraists and not very familiar to language of algebraic geometry.

This manuscript consists of 4 sections corresponding to 4 lectures given at ICTP. Since this volume contains also lectures on related topics, I omit proofs of such results.

I also provide an appendix containing some needed results.

## 1. Tight closure of ideals

In this section, we fix a prime number $p$ and assume all rings are Noetherian and of characteristic $p$.

If the characteristic of a ring $A$ is $p$ and $a, b \in A$, then we have $(a+b)^p = a^p + b^b$. Hence the **Frobenius map**

$$F : A \to A, \quad F(a) = a^p$$

is a ring homomorphism. In the following, the letter $q$ will denote always a power of $p$ ($q = p^e$ for some $e$).

Let us review the definition of tight closure.

DEFINITION 1.1. Let $I$ be an ideal of $A$.

1. We denote $A^\circ$ the set of elements of $A$ which are not in any minimal prime ideals of $A$. For example, if $A$ is an integral domain, then $A^\circ = A \setminus \{0\}$.

2. We put $I^{[q]} = (a^q \mid a \in I)$.
3. We put $I^* = \{x \in A \mid \exists c \in A^\circ, cx^q \in I^{[q]} (\forall q \gg 1)\}$. We call $I^*$ the **tight closure of** $I$.

**1.1. F-regular rings and Theorem of Briançon-Skoda.** Let us list some fundamental properties of the tight closure. For the proofs, we refer [**Hu2**].

PROPOSITION 1.2. *Let $I \subset J$ be ideals of $A$ and $x \in A$.*

1. *$I^*$ is an ideal of $A$ containing $I$ and we have $(I^*)^* = I^*$.*
2. *If $I \subset J$, then $I^* \subset J^*$.*
3. *$x \in I^*$ if and only if for every minimal prime $\mathfrak{p}$ of $A$ we have $x + \mathfrak{p} \in ((I + \mathfrak{p})/\mathfrak{p})^*$ in $A/\mathfrak{p}$.*

REMARK 1.3. In the definition of the tight closure; "$\exists c \in A^0, cx^q \in I^{[q]}$ $(\forall q = p^n \gg 1)$", it is natural to ask "Can we change "$\forall q = p^n \gg 1$" to "$\forall q = p^n \geq 1$" ? Actually, if $A$ is reduced or $\mathrm{ht}(I) > 0$, we can change the definition to "$\exists c \in A^0, cx^q \in I^{[q]}$ $(\forall q = p^n)$". In fact, if for finite $q$, $cx^q \notin I^{[q]}$, let $q_0$ be the maximal of such $q$ and if $\mathrm{ht}(I) > 0$, take $c' \in I \cap A^0$. then we have $c(c')^{q_0} x^q \in I^{[q]}$ for all $q$.

On the other hand, if $A$ has non zero nilpotent elements and if $\mathrm{Ass}(A) = \mathrm{Assh}(A)$, then since the elements of $A^0$ are non zero divisors, a nilpotent element $x \neq 0$ does not satisfy the condition "$\exists c \in A^0$ such that $cx^q = 0$ for all $q = p^e$". But certainly a nilpotent element $x$ is in $(0)^*$ by Definition 1.1.

Let us present some examples.

EXAMPLE 1.4. Let $k$ be a field of characteristic $p$.

1. Let $A = k[X, Y]$ and $I = (X^2, Y^2)$. Then $XY \notin I^*$ and since $(X^2, XY, Y^2)$ is the unique minimal ideal strictly containing $I$, we have $I^* = I$.
2. Let $A = k[X, Y, Z]/(X^n + Y^n + Z^n)$ with $(n, p) = 1$. If $I = (y, z)$, then $x^2 \in I^*$, where $x, y, z$ are the images of $X, Y, Z$,respectively.

If $(A, \mathfrak{m})$ is a regular local ring, then for every ideal $I$ of $A$ we have $I^* = I$. To prove that, the following Theorem by E. Kunz is essential.

THEOREM 1.5. *(E. Kunz) For a Noetherian local ring $(A, \mathfrak{m})$ of dimension $d$, the following conditions are equivalent.*

1. *$A$ is regular.*
2. *$F : A \to A$ is a flat map.*
3. *$\forall q$, $l_A(A/\mathfrak{m}^{[q]}) = q^d$*
4. *$\exists q$, $l_A(A/\mathfrak{m}^{[q]}) = q^d$*

EXERCISE 1.6. Let $(A, \mathfrak{m})$ be a Noetherian local ring. We say a set $\{f_1, \ldots, f_s\}$ $\subset A$ is **independent** if it satisfies the condition (#).

(#) If $\{a_1, \ldots, a_s\} \subset A$ and $\sum_{i=1}^s a_i f_i = 0$, then $a_1, \ldots a_s \in (f_1, \ldots, f_s)$.

(1) Let $\{x_1, \ldots, x_n\}$ be a minimal generator system of $\mathfrak{m}$. If $A \to A^{1/q}$ is flat, then show that $(x_1, \ldots, x_n)$ is independent.

(2) If $\{f_1, \ldots, f_s\}$ is independent and if $f_1 \in (g_1)$, show that $\{g_1, f_2, \ldots, f_s\}$ is independent and show also that $(f_2, \ldots, f_s) : g_1 \subset (f_1, f_2, \ldots, f_s)$.

(3) Show that if $\{x_1, \ldots, x_n\}$ is a set of minimal generators of $\mathfrak{m}$ and if $\{x_1, \ldots, x_n\}$ is independent, show that $\ell_A(A/(x_1^q, \ldots, x_n^q)) = q^d$.

The following fundamental fact of tight closure theory follows from Theorem 1.5.

THEOREM 1.7. *If $(A, \mathfrak{m})$ is a regular local ring, then $I^* = I$ for every ideal $I$ of $A$.*

From this Theorem we define $F$-regular rings, which behave like regular rings with respect to tight closure.

DEFINITION 1.8. A Noetherian ring $A$ of characteristic $p > 0$ is called a **weakly $F$-regular** ring if $I^* = I$ holds for every ideal $I$ of $A$. We say $A$ is **$F$-regular** if every localization $S^{-1}A$ of $A$ is weakly $F$-regular.

Compatibility of tight closure with localization is a subtle subject and has been a big issue of the theory.

REMARK 1.9. Let $A$ be a Noetherian ring and $S \subset A$ be a multiplicatively closed set.

1. For every ideal $I$ of $A$ it is easy to show that $S^{-1}I^* \subset (S^{-1}I)^*$.
2. The converse implication of (1) was a celebrated open problem of tight closure theory for over 20 years. Then H. Brenner and P. Monsky gave a counterexample in the case $p = 2$ in 2007 ([**BM**]). The counterexample is a hypersurface of dimension 3 given by $k[X, Y, Z, T]/(Z^4 + XYZ^2 + X^3Z + Y^3Z + TX^2Y^2)$, where $k$ is the algebraic closure of $\mathbb{F}_2$. But it seems that no counterexamples appeared for $p \geq 3$.
3. If $\sqrt{I} = \mathfrak{m}$ is a maximal ideal, we can show $I^*A_{\mathfrak{m}} = (IA_{\mathfrak{m}})^*$.
4. We can show that $A$ is weakly $F$-regular if and only if for every maximal ideal $\mathfrak{m}$ and for every $\mathfrak{m}$-primary ideal $I$ we have $I^* = I$. Hence $A$ is weakly $F$-regular if and only if so is $A_{\mathfrak{m}}$ for every maximal ideal $\mathfrak{m}$ of $A$.
5. If $A$ is regular, then $A$ is $F$-regular.

Next, we will see the relationship between tight closure and integral closure.

PROPOSITION 1.10. *We denote by $\overline{I}$ the integral closure of $I$. If $I$ is an ideal of $A$ with* $\operatorname{ht}(I) > 0$, *then the following statements hold.*

1. *$x \in \overline{I}$ if and only if there exist $c \in A^\circ$ and an infinite subset $\Lambda$ of $\mathbb{N}$ such that $\forall N \in \Lambda, cx^N \in I^N$.*
2. *$I^* \subset \overline{I}$.*

In many cases, $I^*$ is much smaller than $\overline{I}$ (although, in some cases, $I^*$ is much bigger than $I$). This fact is the reason of the name "tight" closure.

The next Theorem is called "Briançon-Skoda Theorem" and is very important explaining the behavior of integral closure. The proof of this Theorem, which used to be very difficult, became surprisingly easier by introducing the notion of tight closure.

THEOREM 1.11 (Theorem of Briançon-Skoda). *Let $A$ be a Noetherian ring and $I$ be an ideal of $A$, generated by $n$ elements. Then for every integer $w \geq 1$ we have*

$$\overline{I^{n+w-1}} \subset (I^w)^*.$$

In particular, if $n = w = 1$, since $\overline{I} \subset I^*$, we have the following.

PROPOSITION 1.12. *If $x \in A^\circ$ and if $I = (x)$, we have $(x)^* = \overline{(x)}$.*

As a corollary of this Proposition we get the following important result.

COROLLARY 1.13. *$F$-regular rings are normal.*

PROOF. Assume that $A$ is $F$-regular. If $x \in A$ is a non 0 divisor of $A$ and if for $y \in A$, $y/x$ is integral over $A$, then since $y \in \overline{(x)} = (x)$, we have $y/x \in A$. Hence $A$ is normal. □

Next Theorem is called "Boutot's Theorem", which is very important for theory of rational singularities in algebraic geometry over a field of characteristic 0. Again, in the following form, the proof is almost trivial.

THEOREM 1.14. *Let $A$ be a pure subring of an $F$-regular ring $B$ and assume that $A^\circ \subset B^\circ$. Then $A$ is also $F$-regular.*

PROOF. Let $I$ be any ideal of $A$. By our assumption, we have $I^* \subset (IB)^* = IB$ since $B$ is $F$-regular. Then $I^* \subset (IB)^* \cap A = IB \cap A = I$, since $A \subset B$ is pure. □

In the next section, we will show that F-regular rings are Cohen-Macaulay. Then we get a simple and beautiful proof of the fact that "a pure subring of a regular ring is Cohen-Macaulay".

**1.2. Colon capturing, F-rational rings.** Let $(A, \mathfrak{m})$ be a Noetherian local ring and $(x_1, \ldots, x_d)$ be a system of parameters (which we abbreviate as **s.o.p.**) of $A$. Then it is very important to know how much bigger is $(x_1, \ldots, x_i) : x_{i+1}$ compared to $(x_1, \ldots, x_i)$. For example, $A$ is Cohen-Macaulay if and only if $(x_1, \ldots, x_i) : x_{i+1} = (x_1, \ldots, x_i)$ for every (some) s.o.p. $(x_1, \ldots, x_d)$.

We will show that $(x_1, \ldots, x_i) : x_{i+1} \subset (x_1, \ldots, x_i)^*$. We call this property of s.o.p. "colon capturing" property. Then we define an **F-rational** ring to be a ring whose parameter ideals are tightly closed. Then we show that F-rational rings are normal and Cohen-Macaulay. Also, we will show later that F-rational rings are closely related to the notion of **rational singularities** in characteristic 0.

Let us begin with "colon capturing" of s.o.p..

THEOREM 1.15. *Let $R \subset A$ be an extension of rings and assume that $R$ is a regular domain and $A$ is finitely generated as $R$ module. Then for every ideals $I, J$ of $R$, we have*

$$IA : JA \subset ((I : J)A)^*.$$

*In particular, if $(x_1, \ldots, x_d)$ is an s.o.p. of $R$, then $(x_1, \ldots, x_i)A : x_{i+1} \subset ((x_1, \ldots, x_i)A)^*$ for every $i$.*

PROOF. Let $K$ be the quotient field of $R$ and assume $A \otimes_R K \cong K^r$. Take a basis $\{a_1, \ldots, a_r\} \subset A$ of $A \otimes_R K$ over $K$.

Let $F$ be the free $R$ submodule of $A$ generated by $\{a_1, \ldots, a_r\}$. Then, since $A$ is finitely generated as $R$ module, we can take $c \in R, c \neq 0$ such that $cA \subset F$. Also, since the Frobenius map on $R$ is flat, we have $I^{[q]} : J^{[q]} = [I : J]^{[q]}$.

Now, let $x \in A, x \in IA : JA$. Since $xJ \subset IA$, we have $x^q J^{[q]} \subset I^{[q]}A$. If we multiply $c$ to both sides, we get $cx^q J^{[q]} \subset I^{[q]}F$. Since $F$ is a free $R$ module, we have $cx^q \in (I^{[q]}F : J^{[q]}) = (I : J)^{[q]}F \subset ((I : J)A)^{[q]}$ and we get the claim of the Theorem. □

"Colon capturing" holds if $A$ is a homomorphic image of a Cohen-Macaulay local ring.

THEOREM 1.16. *Assume that a Noetherian local ring $A$ is a homomorphic image of a Cohen-Macaulay local ring $(B, \mathfrak{n})$ and* $\dim A = \dim A/\mathfrak{p}$ *holds for every* *minimal prime ideal* $\mathfrak{p}$ *of* $A$. *Then for every s.o.p.* $(x_1, \ldots, x_d)$ *and for every* $i$, $0 \le i \le d-1$, *we have* $(x_1, \ldots, x_i) : x_{i+1} \subset (x_1, \ldots, x_i)^*$.

REMARK 1.17. The condition "$\dim A = \dim A/\mathfrak{p}$ for every minimal prime ideal $\mathfrak{p}$ of $A$" in Theorem 1.16 is necessary.

For example, let $k$ be a field and let $A = k[[X, Y, Z]]/(XY, XZ)$. We let $x, y, z$ be the images of $X, Y, Z$ in $A$, respectively. Then $(y, x-z)$ is a s.o.p. of $A$. In this case we have $x \in (0) : y$. But since $A$ is reduced, we have $(0)^* = (0)$ and $(0) : y \subset (0)^*$ does not hold. In this case, the minimal primes of $A$ are $(x)$ and $(y, z)$ with $\dim A/(x) = 2 \neq \dim A/(y, z) = 1$.

In Theorem 1.16, we can assert a slightly stronger result $(x_1, \ldots, x_i)^* : x_{i+1} \subset (x_1, \ldots, x_i)^*$, which will be necessary later.

THEOREM 1.18. *Let* $(B, \mathfrak{n})$ *be a Cohen-Macaulay local ring and* $A = B/\mathfrak{a}$ *as in* 1.16. *We also assume that* $\dim A = \dim A/\mathfrak{p}$ *holds for every minimal prime ideal* $\mathfrak{p}$ *of* $A$. *Then for every s.o.p.* $(x_1, \ldots, x_d)$ *and for every* $i$ $(0 \le i \le d-1)$, *we have* $(x_1, \ldots, x_i)^* : x_{i+1} \subset (x_1, \ldots, x_i)^*$.

Let us give the definition of **F-rational** rings. The name "F-rational" means "characteristic $p$ version of rational singularities in characteristic 0" and actually, this "equivalence" was proved by K. Smith, N. Hara and V. Mehta - V. Srinivas.

DEFINITION 1.19. Let $(A, \mathfrak{m})$ be a Noetherian ring containing a field of characteristic $p > 0$ and we also assume that $A$ is a homomorphic image of a Cohen-Macaulay ring.

1. If $A$ is a local ring, we say $A$ is $F$-rational if $A$ is equidimensional and for every s.o.p. $(x_1, \ldots, x_d)$ of $A$, $(x_1, \ldots, x_d)^* = (x_1, \ldots, x_d)$ holds.
2. We call $A$ $F$-rational if for every prime ideal $\mathfrak{p}$ of $A$, $A_{\mathfrak{p}}$ is F-rational in the sense of (1).

$F$-rational rings have many nice properties.

THEOREM 1.20. *Let* $(A, \mathfrak{m})$ *be a local ring such that for some s.o.p.* $(x_1, \ldots, x_d)$ *of* $A$, $(x_1, \ldots, x_d)^* = (x_1, \ldots, x_d)$ *holds. Then we have the following properties;*

1. *For every* $i, 0 \le i \le d-1$, *we have* $(x_1, \ldots, x_i)^* = (x_1, \ldots, x_i)$.
2. *$A$ is Cohen - Macaulay.*
3. $(y_1, \ldots, y_d)^* = (y_1, \ldots, y_d)$ *holds for every s.o.p. of* $A$. $(y_1, \ldots, y_d)$ *of* $A$.
4. *$A$ is an integrally closed integral domain (the arguments of Proposition* 1.12 *and Theorem* 1.13 *work in our case, too.)*
5. *(Y. Nakamura,Hochster-Huneke) For every prime ideal* $\mathfrak{p}$ *of* $A$, $A_{\mathfrak{p}}$ *is also an $F$-rational ring.*
6. *If $A$ is Gorenstein and if $A$ is $F$-rational, then $A$ is $F$-regular. Namely, for Gorenstein rings, "$F$-rational" and "$F$-regular" are equivalent conditions.*

LEMMA 1.21. *Let $I$ be an ideal of $A$. If* $I^* = I$, *then for every ideal* $\mathfrak{a}$ *of* $A$, *we have* $[I : \mathfrak{a}]^* = [I : \mathfrak{a}]$.

PROOF. If $\mathfrak{a} = (a_1, \ldots, a_n)$, then $[I : \mathfrak{a}] = \cap_{i=1}^n [I : (a_i)]$. Hence we can assume that $\mathfrak{a} = (a)$. If $u \in [I : (a)]^*$ and if $cu^q \in (I : (a))^{[q]}$, then $cu^q = v^q$ for some $v \in [I : (a)]$ and we have $c(ua)^q \in I^{[q]}$. Hence we have $ua \in I^* = I$, $u \in I : (a)$. □

The notion of $F$-injective rings is a weaker version of $F$-rational rings.

DEFINITION 1.22. Let $(A, \mathfrak{m})$ be a local ring and let $J = (x_1, \ldots, x_d)$ be an s.o.p. of $A$. If $A$ is Cohen-Macaulay, we say that $A$ is $F$**-injective** if the multiplication of $(x_1 \cdots x_d)^{q-1} : A/J \to A/J^{[q]}$ is injective. (This condition is equivalent to saying that the Frobenius morphism on $\mathrm{H}^d_{\mathfrak{m}}(A)$ is injective (see §3).)

PROBLEM 1.23 (Fedder). Let $(A, \mathfrak{m})$ be a local ring and let $x \in \mathfrak{m}$ be a non zero divisor of $A$. Assume that $A/xA$ is Cohen-Macaulay, normal, $F$- injective and that $A_x$ is $F$-rational. Then show that $A$ is $F$-rational.

PROBLEM 1.24 (Huneke). Let $x, T$ be variables over $k$ and put

$$R = k[T, xT^4, x^{-1}T^4, (x-1)^{-1}T^4].$$

Then show the following statements.

1. $R$ is $F$-rational.
2. $R$ is not $F$ -regular.
3. Put $I = (xT^4, x^{-1}T^4)$. Then show that $T^7 \notin I$ and $T^7 \in (xT^4, x^{-1}T^4)^*$.

PROBLEM 1.25. (cf. [**Wa97**]) Let $A = k[[X, Y, Z]]/(X^2 + Y^3 + Z^4)$ and $\mathfrak{m}$ be the maximal ideal of $A$. Let $R = A[\mathfrak{m}T] \subset A[T]$ be the Rees algebra of $\mathfrak{m}$ over $A$. Then show the following statements.

1. Show that for every prime $p$, $R$ is $F$-rational.
2. Show that $A$ is $F$-rational for $p \geq 5$ and not $F$-rational for $p = 2, 3$.
3. Show that $A$ is a pure subring of $R$. Consequently, the statement "If $A$ is a pure subring of $B$ and if $B$ is $F$-rational, then so is $A$" ("$F$-rational version of Boutot's Theorem for rational singularities) does not hold true.

## 2. Tight closure of modules, $F$ pure rings

Let $A$ be a ring of characteristic $p > 0$. We define Frobenius functors $F^e$ on the category of $A$-modules and in a fixed $A$-module $M$, we define tight closure $N^*_M$ of a submodule $N$ of $M$. In particular, the tight closure of submodules of local cohomology module $\mathrm{H}^i_{\mathfrak{m}}(A)$ and injective envelope $E_A(A/\mathfrak{m})$ are very important.

This concept is applied to theory of complexes over the ring to get "phantom" results as in Theorem 2.15.

In this section, the rings are Noetherian rings of characteristic $p > 0$ and we put $q = p^e$.

DEFINITION 2.1.

1. We write ${}^eA$ the set $A$ made as $A$-module via the homomorphism $F^e : A \to A, F^e(a) = a^q$. Namely, we define
$$a.x = a^q x$$
for $a \in A$ and $x \in {}^eA$.
2. For an $A$-module $M$, we define $F^e(M) = M \otimes_A {}^eA$. For an $A$-module homomorphism $u : M \to M'$, we define $F^e(u) : F^e(M) \to F^e(M')$ by $F^e(u)(x \otimes 1) = u(x) \otimes 1$.
3. For an $A$ submodule $N$ of $M$, we define
$$N^{[q]} = \mathrm{Im}(F^e(i) : F^e(N) \to F^e(M)).$$
Here $i : N \to M$ is the inclusion map of $N$ in $M$. (Since $F^e$ is not an exact functor, note that $F^e(i)$ is not injective in general). In particular, if $N = I$

is an ideal of $A = M, N^{[q]} = I^{[q]}$ coincides with our definition in Definition 1.1.

4. Let $x \in M$. We denote by $x^q$ the image of $x$ in $F^e(M)$ by the Frobenius morphism $F^e(x) = x \otimes 1$. Then we have equalities

$$(x+y)^q = x^q + y^q, \text{ and } (ax)^q = a^q x^q$$

for $x, y \in M$ and $a \in A$.

EXERCISE 2.2. Let $M$ be a finitely generated $A$-module.

1. If $M$ is represented by the following exact sequence ($(a_{ij})$ is an $m \times n$ matrix)

$$A^m \xrightarrow{(a_{ij})} A^n \longrightarrow M \longrightarrow 0$$

then tensoring ${}^eA$ to this exact sequence (put $q = p^e$) we have the exact sequence

$$A^m \otimes_A {}^eA \xrightarrow{(a_{ij}^q)} A^n \otimes_A {}^eA \longrightarrow F^e(M) \longrightarrow 0.$$

2. In particular, for an ideal $I$ of $A$, $F^e(A/I) \cong A/I^{[q]}$.

Let us note the following Theorem proved using the Theorem 5.2.

THEOREM 2.3. *Let*

$$G_\bullet : 0 \to G_n \overset{\varphi_n}{\to} G_{n-1} \overset{\varphi_{n-1}}{\to} G_{n-2} \to \cdots \to G_1 \overset{\varphi_1}{\to} G_0$$

*be a complex of finitely generated free $A$-modules. Then $G$ is acyclic if and only if so is $F^e(G_.)$. In particular, if* $\operatorname{hd}_A I < \infty$ *if and only if* $\operatorname{hd}_A I^{[q]} < \infty$ *for every* $q = p^e$.

PROOF. The condition (2) of 5.2 for $G_\bullet$ and $F^e(G_\bullet)$ are clearly equivalent. □

Let us define the tight closure of a submodule.

DEFINITION 2.4. Let $M$ be an $A$-module and $N$ be a submodule of $M$. We define the **tight closure $N_M^*$ of $N$ in $M$** by

$$x \in N_M^* \iff \exists c \in A^\circ, \forall q \gg 1, cx^q \in N^{[q]}.$$

$N_M^*$ is an $A$ submodule of $M$ containing $N$.

REMARK 2.5. For $x \in M$, $x \in N_M^*$ if and only if the image of $x$ in $M/N$ is an element of $(0)_{M/N}^*$. Namely, we have $N_M^*/N \cong (0)_{M/N}^*$.

In the applications of tight closure of modules, the tight closures in local cohomology modules and injective envelopes are most typical.

EXERCISE 2.6. Let $(A, \mathfrak{m})$ be a local ring with $\dim A = d$ and let $\underline{x} = (x_1, \ldots, x_d)$ be an s.o.p. of $A$. Also we denote $C^\bullet = C^\bullet(\underline{x})$ be the Čech complex as from Proposition 5.1.

1. By Proposition 5.1, $\mathrm{H}_\mathfrak{m}^d(A) = \mathrm{H}_{\underline{x}}^d(A) \cong A_{x_1 x_2 \cdots x_d} / \operatorname{Im}[C^{d-1} \to C^d]$. Hence we have $F^e(\mathrm{H}_\mathfrak{m}^d(A)) = \mathrm{H}_{\underline{x}^q}^d(A) \cong \mathrm{H}_\mathfrak{m}^d(A)$.

   If we write by $[\frac{s}{(x_1 x_2 \cdots x_d)}]$ the image of $\frac{s}{(x_1 x_2 \cdots x_d)} \in C^d$ in $\mathrm{H}_{\underline{x}}^d(A)$, then for $\xi = [\frac{s}{(x_1 x_2 \cdots x_d)}]$, we have $\xi^q = [\frac{s^q}{(x_1 x_2 \cdots x_d)^q}]$.

2. If $A$ is Cohen-Macaulay, $A$ is $F$ rational if and only if $(0)^*_{\mathrm{H}^d_{\mathfrak{m}}(A)} = 0$. In fact, for $\xi = [\frac{s}{(x_1x_2\cdots x_d)}]$ and $c \in A^\circ$, $c\,\xi^q = 0$ in $\mathrm{H}^d_{\mathfrak{m}}(A)$ is equivalent to saying that $cs^q \in (\underline{x})^q$. Hence $\xi \in (0)^*_{\mathrm{H}^d_{\mathfrak{m}}(A)} \iff s \in (\underline{x})^*$.
3. We will show in Theorem 3.6 that $A$ is $F$-rational if and only if $\mathrm{H}^d_{\mathfrak{m}}(A)$ has no proper submodule which is stable under Frobenius action (we say that $N \subset M$ is stable under Frobenius action if $F^1(N) \subset N$).

The action of Frobenius endomorphism characterizes $F$-regular rings.

THEOREM 2.7. *Assume that a Noetherian local ring $(A, \mathfrak{m})$, satisfies the condition*

*(∗) For every $n > 0$, there exists an* **irreducible** *$\mathfrak{m}$-primary ideal $J \subset \mathfrak{m}^n$.*
*Then the following conditions on $A$ are equivalent.*

1. *$A$ is weakly $F$-regular.*
2. *For every finitely generated $A$-module $M$ and for any submodule $N$ of $M$, we have $N^*_M = N$.*
3. *Let $E = E_A(A/\mathfrak{m})$ be the injective envelope of $A/\mathfrak{m}$ as $A$ module. Then we have $(0)^{*fg}_E = 0$, where $(0)^{*fg}_E$ is the sum of $(0)^*_M$ for all finitely generated $A$-submodules of $E$.*

PROOF. We have shown in Remark 1.9 that $A$ is weakly $F$-regular if for every $\mathfrak{m}$-primary ideal $I$, we have $I^* = I$. Since any $\mathfrak{m}$-primary ideal is an intersection of irreducible $\mathfrak{m}$-primary ideals, it suffices to assume that $I$ is irreducible. Then there is an injection $A/I \to E_A(A/\mathfrak{m})$. Then assumption (3) implies $(0)^*_{A/I} = (0)$, or $I^* = I$.

Next, we show (1)$\Longrightarrow$ (2). Assume that $x \in N^*_M, x \notin N$. Then assuming that $N$ is a maximal submodule with $x \notin N$, we may assume that $\mathfrak{m}x \subset N$ and $M/N \subset E$. Put $I = \mathrm{Ann}_A(M/N)$ and take an irreducible $\mathfrak{m}$ primary ideal $J \subset I$. since $A/J \to A/I$ is surjective, taking Matlis dual, $M/N \cong (A/I)^\vee \subset (A/J)^\vee \cong A/J$. Then we have $x \in J^*/J$, which implies $A$ is not weakly $F$-regular. Thus we show (1)$\Longrightarrow$ (2). (2)$\Longrightarrow$ (3) is obvious. □

Next we will introduce the notion of $F$-pure rings. This concept is historically the oldest one among the notions in this section.

DEFINITION 2.8. $A$ is called **$F$-pure**, if $A$ is a pure subring of ${}^1A$ by $F^1 : A \to {}^1A$.

PROBLEM 2.9. For a local ring $(A, \mathfrak{m})$, let $E = E_A(A/\mathfrak{m})$ be the injective envelope of $A/\mathfrak{m}$. Show that $A$ is $F$-pure if and only if $F^1 : E \to E \otimes_A {}^1A$ is injective.

REMARK 2.10. By Problem 2.9, if $A$ is $F$-regular, then $A$ is $F$-pure. There is no implications between "$F$-pure" and "$F$-rational".

PROBLEM 2.11. Let $R$ be a graded ring over a field $k = R_0$ of characteristic $p > 0$.

1. Show that a face ring (or a Stanley-Reisner ring; see [**BH**], Chapter 5, §1) $k[\Delta]$ is $F$-pure.
2. Show that if $R$ is $F$-pure (resp. $F$-rational), then $a(R) \le 0$ (resp. $a(R) < 0$).

3. Moreover, assume that $k$ is algebraically closed. If $\dim R = 1$ and $R$ is $F$-pure, then $R$ is isomorphic to a face ring (see also [**GW2**]).

PROBLEM 2.12 (Fedder's criterion). Let $(B, \mathfrak{n})$ be a regular local ring and $A = B/I$.

1. If we put $E = E_B(B/\mathfrak{n})$, then show that $E_A(A/\mathfrak{m}) \cong [0 :_E I]$ and $E_A(A/\mathfrak{m}) \otimes {}^1A \cong [0 :_E I^{[p]}]/I[0 :_E I^{[p]}]$.
2. (Fedder's criterion) Show that $A$ is $F$-pure if and only if $[I^{[p]} : I] \not\subset \mathfrak{n}^{[p]}$.
3. In particular, if $I$ is generated by a regular sequence $(f_1, \dots, f_r)$, then $A$ is $F$-pure if and only if $(f_1 f_2 \cdots f_r)^{p-1} \notin \mathfrak{n}^{[p]}$.
4. Let $A = k[X, Y, Z]/(X^3 + Y^3 + Z^3)$ $(p \neq 3)$. Then $A$ is $F$-pure if and only if $p \equiv 1 \pmod 3$. Also, show that $A = k[X, Y, Z]/(X^2 + Y^4 + Z^4)$ $(p \neq 2)$ is $F$-pure if and only if $p \equiv 1 \pmod 4$.

EXAMPLE 2.13. ([**Wa87**], [**Wa91**]) Let $x, T$ be variables over a field $k$ of $\mathrm{char}(k) = p > 0$ and put $R = k[T, xT^a, x^{-1}T^b, (x-1)^{-1}T^c]$, where $a, b, c \geq 2$ are positive integers. Then the following statements are true. Give examples of $F$-rational rings (or rational singularities) which are not $F$-pure.

1. For any $a, b, c$, $R$ is $F$-rational.
2. $R/TR \cong k[X, Y, Z]/(XY, YZ, ZX)$ and hence is $F$-pure.
3. If $R$ is $F$-pure (resp. $F$-regular), then $1/a + 1/b + 1/c \geq 1$ (resp. $> 1$).
4. Conversely, if $1/a + 1/b + 1/c > 1$ and $p > 5$, then $R$ is $F$-regular and if $1/a + 1/b + 1/c = 1$, then $R$ is $F$-pure for infinitely many $p$.

Let us introduce the notion of Phantom homology.

DEFINITION 2.14. Let $\mathbf{G}$ be a complex of finitely generated $A$ modules. Then we say $\mathbf{G}$ has **Phantom Homology** if $\mathrm{Ker}(G_i \to G_{i-1}) \subset [\mathrm{Im}(G_{i+1} \to G_i)]^*$. for every $i$.

The following Theorem is a generalization of a Theorem of Buchsbaum-Eisenbud 5.2 in terms of tight closure.

THEOREM 2.15 (Phantom homology). *Let* $G_\bullet : 0 \to G_n \overset{\varphi_n}{\to} G_{n-1} \overset{\varphi_{n-1}}{\to} G_{n-2} \to \cdots \to G_1 \overset{\varphi_1}{\to} G_0$ *be a complex of finitely generated free $A$-modules with* $G_i \neq (0)$ $(1 < i < n)$. *Then the following conditions for* $G_\bullet$ *are equivalent.*

1. $G_\bullet$ *has phantom homology.*
2. *For every $i$, $1 \leq i \leq n$, the following 2 conditions are satisfied.*
   (a) $\mathrm{ht}(I(\varphi_i)) \geq i$.
   (b) $\mathrm{rank}(\varphi_i, M) + \mathrm{rank}(\varphi_{i+1}, M) = \mathrm{rank}_A F_i$ *(here we put* $\mathrm{rank}(\varphi_{n+1}, M) = 0$*).*

From this Theorem, we have the following Corollary, since it is obvious that the conditions (a), (b) is preserved by applying $F^e$.

COROLLARY 2.16. *Let $M$ be a finitely generated $A$-module. If $M$ has finite projective dimension, so does $F^e(M)$.*

This was a very quick overview of tight closure in positive characteristics. If you are interested to know more of the theory, refer to the lecture notes of C. Huneke [**Hu2**].

This theory is known to be closely related to the geometric properties of singularities of algebraic varieties defined over a field of characteristic 0. For example, a rational singularity $A$ over a field of characteristic 0 is characterized by the property "for infinitely many $p$, $A_p$, the reduction modulo $p$ of $A$ are $F$-rational (K. Smith, N. Hara and V.Mehta-V.Srinivas). By using Perfectoid, it turned out that if $A_p$ is $F$-rational for a single $p$, then $A$ is $F$-rational ([**MaSch**]). Also, "log-terminal singularities" are characterized by "$\mathbb{Q}$-Gorenstein and $A_p$ is $F$-regular for infinitely many $p$". Furthemore, the "multiplier ideals in characteristic 0 are also known to be computable by characteristic $p$ method by N. Hara - K. Yoshida and S. Takagi. (The precise definition of the notation $A_p$ will appear in Definition 3.7.)

In the next section, we will explain how to use these characteristic $p$ methods to analyze singularities defined over a field of characteristic 0.

## 3. Reduction modulo $p$, $F$-rational rings and rational singularities in characteristic 0

First, let us recall the definition of **rational singularities**.

DEFINITION 3.1. Let $k$ be a field of characteristic 0 and let $(A, \mathfrak{m})$ be a normal local ring which is a localization of a finitely generated algebra over $k$. We put $\Sigma = \{\mathfrak{p} \in \mathrm{Spec}(A) \mid A_\mathfrak{p} \text{ is not regular } \}$ be the singular locus of $\mathrm{Spec}(A)$. It is known that $\Sigma$ is closed in $\mathrm{Spec}(A)$.

1. A $k$-morphism $f : X \to \mathrm{Spec}(A)$ is called a **resolution of singularities of** $\mathrm{Spec}(A)$ if $f$ is a projective birational morphism and $f$ is isomorphism over $\mathrm{Spec}(A) \setminus \Sigma$.
2. $A$ is called a **rational singularity** if $\mathrm{H}^i(X, \mathcal{O}_X) = 0$ for every $i > 0$.

The following Theorem is very important for the theory of rational singularities.

THEOREM 3.2 (Grauert-Riemanschneider vanishing theorem). *Let $(A, \mathfrak{m})$ be a local ring essentially of finite type over a field of characteristic* 0. *If $f : X \to \mathrm{Spec}(A)$ is a resolution of singularities and if $\omega_X$ is the dualizing sheaf on $X$, then $H^i(X, \omega_X) = 0$ for $i > 0$.*

REMARK 3.3.

1. Using Theorem 3.2 we can prove that $A$ is a rational singularity if and only if the following conditions (a), (b) hold.
   (a) $A$ is Cohen-Macaulay
   (b) $\mathrm{H}^0(X, \omega_X) = K_A$.
2. If the field $k$ has positive characteristic, Theorem 3.2 fails in general. But we cam say if $f : X \to \mathrm{Spec}(A)$ is a projective birational morphism, where $X$ is a normal Cohen-Macaulay variety and if $\mathrm{H}^i(X, \omega_X) = 0$ for every $i > 0$, the equivalence in (1) above holds.
3. In positive characteristics in dimension $\geq 3$, there are counterexamples of Theorem 3.2 and the Cohen-Macaulay property for rational singularities also fails. See Remark 4.10 for references for counterexamples.
4. An important property for rational singularities is Boutot's Theorem ([**Bou**]) saying that a pure subring of a rational singularity is again a rational singularity. It turns out that this is not true for $F$-rational rings.
5. ([**Fl**], [**Wa83**]) If $R = \bigoplus_{n \geq 0} R_n$ is a normal graded ring with $R_0 = k$ is a field of characteristic 0, then $R$ is a rational singularity if $R$ satisfies the following three properties; (1) $\mathrm{Spec}(R) \setminus \{\mathfrak{m}\}$ has rational singularities, (2) $R$ is Cohen-Macaulay, (3) $a(R) < 0$.

To show that $F$-rational sings are rational singularities, it is convenient to introduce the notion of pseudo-rational rings, which was introduced in [**LT**].

DEFINITION 3.4. Let $(A, \mathfrak{m})$ be a Cohen-Macaulay local ring of $\dim A = d$. We say that $A$ is a **pseudo-rational** ring if it satisfies the following condition (†).

(†) For every projective birational morphism $f : W \to \operatorname{Spec}(A)$ where $W$ is normal, the natural surjective map $\delta_f : \mathrm{H}^d_{\mathfrak{m}}(A) \to \mathrm{H}^d_{\mathbb{E}}(\mathcal{O}_W)$ is injective, where we put $\mathbb{E} = f^{-1}(\mathfrak{m})$.

It is stated in [**LT**], p. 102, that a normal Cohen-Macaulay ring $A$ is pseudo-rational if and only if for every $f$ as above, there exists projective birational morphism $g : W' \to W$ such that $\delta_{fg}$ is injective.

Note that if $f$ is isomorphism over $\operatorname{Spec}(A) \setminus \{\mathfrak{m}\}$, from local cohomology long exact sequence, we have an exact sequence

$$\mathrm{H}^{d-1}_{\mathbb{E}}(\mathcal{O}_X) \to \mathrm{H}^{d-1}(X, \mathcal{O}_X) \to \mathrm{H}^{d-1}(X - \mathbb{E}, \mathcal{O}_X) \cong \mathrm{H}^d_{\mathfrak{m}}(A) \to \mathrm{H}^d_{\mathbb{E}}(\mathcal{O}_X) \to 0.$$

REMARK 3.5.

(1) If $A$ is essentially of finite type over a field $k$ of characteristic 0, "pseudo rational" and "rational singularity" are equivalent conditions.

(2) We saw Briançon-Skoda Theorm in 1.11 using tight closure. We have also Briançon-Skoda Theorem for pseudo-rational rings ([**LT**] Theorem 2.1). Namely if $(A, \mathfrak{m})$ is a pseudo-rational local ring of dimension $d$, then for every ideal $I$ in $A$, we have $\overline{I^d} \subset I$, where $\overline{I^d}$ is the integral closure of $I^d$.

We can show that $F$-rational rings are pseudo-rational.

THEOREM 3.6 (K. Smith). *Let $(A, \mathfrak{m})$ be a complete normal Cohen-Macaulay local ring of characteristic $p > 0$, $\dim A = d$. Then;*

1. *$A$ is $F$-rational if and only if $\mathrm{H}^d_{\mathfrak{m}}(A)$ has no proper submodule closed under the Frobenius action. (We say that an $A$ submodule $N$ of $\mathrm{H}^d_{\mathfrak{m}}(A)$ is closed under the action of Frobenius if for every $x \in N$ we have $x^q \in N$.)*
2. *If $A$ is $F$-rational, then $A$ is pseudo-rational.*

PROOF. (1) If $x \in (0)^*_{\mathrm{H}^d_{\mathfrak{m}}(A)}$, then $cx^q = 0$ for some $c \in A^\circ$ and for all $q = p^e \gg 1$. Hence if $x \in (0)^*_{\mathrm{H}^d_{\mathfrak{m}}(A)}$ and satisfies the condition above, certainly $c(x^p)^q = 0$ for all $q = p^e \gg 1$ and $x^p \in (0)^*_{\mathrm{H}^d_{\mathfrak{m}}(A)}$. This shows that $(0)^*_{\mathrm{H}^d_{\mathfrak{m}}(A)}$ is closed under Frobenius action. Hence if $\mathrm{H}^d_{\mathfrak{m}}(A)$ has no proper submodule closed under the action of Frobenius, then $(0)^*_{\mathrm{H}^d_{\mathfrak{m}}(A)}$ should be $(0)$ and $A$ is $F$-rational by Exercise 2.6.

Conversely, if $A$ is $F$-rational and if $N$ is a proper $A$ submodule of $\mathrm{H}^d_{\mathfrak{m}}(A)$, the Matlis dual $N^\vee$ of $N$ is a proper quotient of $K_{\hat{A}}$, the canonical module of $\hat{A}$. Since $K_{\hat{A}}$ is a torsion free $\hat{A}$-module of rank 1, any proper quotient of it is a torsion module and we can take $c \in A^\circ$ such that $cN^\vee = 0$, which imply that $\forall x \in N, cx^q = 0$ and then $N \subset (0)^*_{\mathrm{H}^d_{\mathfrak{m}}(A)}$.

(2) Let $f : X \to \operatorname{Spec}(A)$ be a birational morphism with $X$ normal. Then $X$ has Frobenius action $\mathcal{O}_X \to \mathcal{O}_X^{1/p}$. Put $E = f^{-1}(\mathfrak{m})$. Since the map $\delta_f : \mathrm{H}^d_{\mathfrak{m}}(A) \to \mathrm{H}^d_E(\mathcal{O}_X)$ is compatible with Frobenius action, $\operatorname{Ker}(\delta_f)$ is an $F$-stable submodule of $\mathrm{H}^d_{\mathfrak{m}}(A)$ and hence must be $(0)$. □

Now we introduce the notion of "reduction modulo $p$" for varieties of finite type over a field $k$ of characteristic 0. For more precise informations, see [**TW**] and literatures referred there.

DEFINITION 3.7. Let $k$ be a field of characteristic 0 and consider a ring $A$ which is essentially of finite type over $k$. There is an affine $k$-algebra $R = k[x_1, \dots, x_n]/(f_1 \dots, f_m)$ such that $(A, \mathfrak{m})$ is a localization of $R$. Since the coefficients of $f_i, \dots, f_m$ are finite, we can take a finitely generated $\mathbb{Z}$-subalgebra $B$ of $k$ and finitely generated $B$-algebra $R_B$ such that $R \cong R_B \otimes_B k$. Moreover, by "generic flat theorem", we can take $a \in A$ so that $R_A[a^{-1}]$ is flat over $A_a$. Then considering $A_a$ instead of $A$, we can assume that $R_A$ is flat over $A$. We say $R_A$ a model of $R$ over $A$. Since $A$ is finitely generated over $Z$, for any maximal ideal $\mu$ of $A$, $A/\mu$ is a field finitely generated over $\mathbb{Z}$ and hence is a finite field and in particular, has positive characteristic. If the characteristic of $A/\mu$ is $p$, we say $R_A \otimes A/\mu$ a **reduction of $R$ modulo $p$**.

EXAMPLE 3.8. If $R = k[x, y, z]/(x^a + y^b + z^c)$, we can take $A = \mathbb{Z}, R_A = \mathbb{Z}[x, y, z]/(x^a + y^b + z^c)$ and $\mathbb{Z}/(p)[x, y, z]/(x^a + y^b + z^c)$ is a reduction modulo $p$ of $R$.

Using the notion of reduction modulo $p$, we can define the notion of **$F$-P-type** for a property **P** of a ring. For most properties **P** and extension $k \subset k'$ of fields, $R$ satisfies **P** if and only if $R \otimes_k k'$ satisfies **P** and hence this property depends only on the characteristic $p$. In this sense, we refer to a reduction $R_A \otimes_A A/\mu$ simply $R_p$ if $\operatorname{char}(A/\mu) = p$.

DEFINITION 3.9. Let $R$ be a ring essentially of finite type over a field $k$ of characteristic zero. Suppose we are given a model $R_A$ of $R$ over a finitely generated $\mathbb{Z}$-subalgebra $A$ of $k$. We say that $R$ is of **open** $F$-**P**-type (resp. **dense**) $F$-**P**-type if $R_p$ satisfies **P** for all but finitely many $p$ (resp. for infinitely many) $p$.

These definitions are independent of the choice of the model $R_A$.

EXAMPLE 3.10.

(1) Let $R = \mathbb{C}[X, Y, Z]/(X^3 - YZ(Y+Z))$. Then the ring $R_{\mathbb{Z}} = \mathbb{Z}[X, Y, Z]/(X^3 - YZ(Y+Z))$ is a model of $R$ over $\mathbb{Z}$. We take the dense subset $S$ of $\operatorname{Spec}\mathbb{Z}$ to be $\{p \in \operatorname{Spec}\mathbb{Z} \mid p \equiv 1 \bmod 3\}$. (Show that by Fedder's criterion 2.12, $k[X, Y, Z]/(X^3 - YZ(Y+Z))$ is $F$-pure if and only if $p \equiv 1 \pmod 3$.) Hence the ring $R$ is of dense $F$-pure type and not of open $F$-pure type.

(2) Let $R = \mathbb{C}[X, Y, Z]/(X^2 + Y^3 + Z^5)$. The ring $R_{\mathbb{Z}} = \mathbb{Z}[X, Y, Z]/(X^2 + Y^3 + Z^5)$ is a model of $R$ over $\mathbb{Z}$. It is easy to see that $x \in (y, z)^*$ if and only if $p = 2, 3, 5$. Since $S = \{p \in \operatorname{Spec}\mathbb{Z} \mid p \geq 7\} \cup \{0\}$ is a dense open subset of $\operatorname{Spec}\mathbb{Z}$, the ring $R$ is of open strongly $F$-regular type (and of open $F$-pure type).

We will show that if $R$ has dense $F$-rational type, then $R$ is a rational singularity.

THEOREM 3.11. *Let $(A, \mathfrak{m})$ be a local ring essentially of finite type over a field $k$ of characteristic 0. Then $A$ is a rational singularity if and only if $A$ is of open $F$-rational type.*

PROOF. First, since rational singularities and $F$-rational rings are both Cohen-Macaulay, we may assume $A$ is a Cohen-Macaulay ring. Assume that $A$ is not a rational singularity. Take a resolution of singularity $f : X \to \operatorname{Spec}(A)$. Take the support $Y = \bigcup_{i \geq 1} \operatorname{Supp}_A(\mathrm{H}^i(X, \mathcal{O}_X))$. If $\mathfrak{p}$ is a generic point of $Y$, since localizing at $\mathfrak{p}$ preserves resolution, we can assume that $Y = \{\mathfrak{m}\}$. Then by Grothendieck duality, we have $H^{d-1}(X, \mathcal{O}_X) \cong [\omega_A / f_*(\omega_X)]' \neq 0$. Since $A_p$ is $F$-rational for all but finitely many $p$, we can assume that $A_p$ is $F$-rational. But then $\mathrm{H}^{d-1}(X, \mathcal{O}_X)$

is a $F$- stable subset of $H^d_{\mathfrak{m}}(A)$, which contradicts Smith's Theorem 3.6. Hence $A$ should be a rational singularity.

The converse needs deeper Theorems of algebraic geometry and proved by N. Hara ([**Hara**]) and V. B. Mehta - V. Srinivas ([**MeSr**]). □

## 4. Log terminal, log canonical singularities and $F$-regular, $F$-pure rings

Log terminal (LT) and log canonical (LC) singularities play very important role in Minimal Model Program (MMP) in algebraic geometry over a field of characteristic 0. Those singularities are defined by the **discrepancy** of canonical divisors.

First we define $\mathbb{Q}$-Gorenstein local rings.

DEFINITION 4.1. (See [**KolMo**]) Let $(A, \mathfrak{m})$ be a normal local ring which has a canonical module $K_A$. We call $A$ is **$\mathbb{Q}$-Gorenstein** if $\mathrm{cl}(K_A) \in \mathrm{Cl}(A)$ is (0 or) a torsion element [1]. Also, if the order of $\mathrm{cl}(K_A)$ is $r$, we say that $A$ is **$r$-Gorenstein**. For a normal local ring $A$, $A$ is Gorenstein if $A$ is 1-Gorenstein and Cohen-Macaulay.

For a normal algebraic variety $X$, we say "$X$ is **$r$-Gorenstein** if $r$ is the least positive integer such that $\mathcal{O}_X(rK_X)$ is an invertible $\mathcal{O}_X$-Module.

Let us also define the concept of **SNC (Simple Normal Crossings)**.

DEFINITION 4.2. Let $X$ be a regular scheme of dimension $n$. We say that a closed subscheme $Y$ of $X$ satisfies the condition **(SNC)** if it satisfies the following condition.

**(SNC)** At any closed point $x$ of $X$, we can take a regular s.o.p. $(t_1, \ldots, t_d)$ of $\mathcal{O}_{X,x}$ so that the defining equation of $Y$ at $x$ is $t_1 t_2 \cdots t_r$ for some $r$.

By Hironaka's resolution of singularities over a field of characteristic 0, if $X$ is any algebraic variety over a field of characteristic 0, we can take a resolution $f : \widetilde{X} \to X$ of singularities of $X$ so that the exceptional set $E$ of $f$ satisfies the condition (SNC) in $\widetilde{X}$.

DEFINITION 4.3. We write $\mathrm{Irr}^1(X)$ to be the set of irreducible (reduced) closed subschemes of $X$.

Now we define the discrepancy of a birational morphism with respect to an element $E \in \mathrm{Irr}^1(X)$.

DEFINITION 4.4. Let $k$ be a field, $X, Y$ be normal algebraic varieties of dimension $n$ and assume $Y$ is $r$-Gorenstein. Let $f : X \to Y$ be a birational morphism and $E \in \mathrm{Irr}^1(X)$ with $\dim f(E) < n - 1$. Then we call $E$ an **exceptional divisor** of $f$.

We define the **discrepancy** $a(E, X)$ of $E$ as follows;

Let $z \in E$ be a general point of $E$ and $(x_1, \ldots, x_n)$ be a regular s.o.p. of $X$ at $z$. Also let $y$ be the defining equation of $E$ at $z$.

On the other hand, at $w = f(z) \in Y$, let $\mathcal{O}_Y(rK_Y)_w = t\mathcal{O}_{Y,w}$. If

$$f^*(t) = uy^s(dx_1 \wedge dx_2 \wedge \ldots \wedge dx_n)^r \tag{4.5}$$

at $z$ (where $u$ is a unit of $\mathcal{O}_{X,z}$), we write

$$a(E, X) = \frac{1}{r}s$$

[1] For a Noetherian normal ring or a Krull ring $A$, $\mathrm{Cl}(A)$ denotes the divisor class group of $A$. See [Eis], §10 for example. For a fractional reflexive ideal $\mathfrak{a}$ of $A$, $\mathrm{cl}(\mathfrak{a}) \in \mathrm{Cl}(A)$ denotes the equivalence (isomorphism) class of $\mathfrak{a}$.

and call it the **discrepancy** of $E$. Since $f^*(\mathcal{O}(rK_Y)) \cong \mathcal{O}_X(rK_X - sE)$ near $z$, we can write

$$K_X = f^*(K_Y) + \sum_E a(X,E)E,$$

where $E$ moves over all exceptional irreducible divisors of $f$. (We show some computations in Example 4.8.)

REMARK 4.5. A divisor $E$ defines a valuation $v_E$ of $X$. Then the discrepancy $E$ depends only on $v_E$. Namely, for birational normal morphisms $f : X \to Y$ and $f' : X' \to Y$ and $E$ on $X$ and $E'$ on $X'$, if the valuations $v_E$ and $v_{E'}$ are the same, then we have $a(X,E) = a(X',E')$.

DEFINITION 4.6. Assume that $(Y,y)$ be normal and $r$-Gorenstein. We call $A = \mathcal{O}_{Y,y}$ is a **terminal singularity** (resp. **canonical singularity**, **log terminal singularity**, **log canonical singularity**), if for any normal variety $X$, any birational morphism $f : X \to Y$ and any exceptional irreducible divisor $E$ with $y \in f(E)$, we have $a(E,X) > 0$ (resp. $a(E,X) \geq 0$, $a(E,X) > -1$, $a(E,X) \geq -1$).

REMARK 4.7. To show that $A = \mathcal{O}_{Y,y}$ is one of the singularities defined in 4.6, it suffices to show the condition for a resolution of singularities $f : X \to \mathrm{Spec}(A)$.

EXAMPLE 4.8. We give two simple examples of discrepancy.

1. Let $\mathcal{L}$ be an ample invertible sheaf on a normal projective variety $V$ and assume that $\omega_V = \mathcal{O}_V(K_V) \cong \mathcal{L}^{\otimes a}$. Let $R = \oplus_{n\geq 0} \mathrm{H}^0(V, \mathcal{L}^{\otimes n})$ be a cone over $V$. Then a resolution $f : X \to Y = \mathrm{Spec}(R)$ is given by the blowing-up of the graded maximal ideal of $R$ and the exceptional set is $E \cong V$. In this case we have $a(E,X) = a+1$.
2. Let $A = k[x^3, x^2y, xy^2, y^3] \cong k[y, xy, x^2y, x^3y] = k[\sigma^\vee \cap \mathbb{Z}^2]$, where $\sigma^\vee$ is the cone in $\mathbb{R}^2$ generated by $(0,1)$ and $(3,1)$. Also, $K_A$ is generated by $\mathrm{Int}(\sigma^\vee) \cap \mathbb{Z}^2$ (where $\mathrm{Int}(\sigma^\vee)$ is the relative interior of $\sigma^\vee$ and generated by $(xy, x^2y)$. Since $(xy, x^2y)^3 = (x^3y^2)(y, xy, x^2y, x^3y)$, $(K_A)^{(3)}$ is a principal module generated by $x^3y^2$.

   We use terminology of "toric geometry". Let $\sigma$ be a convex cone generated by $(-1,3)$ and $(1,0)$ in $\mathbb{R}^2$ so that the dual cone

   $$\sigma = (\sigma^\vee)^\vee = \{x \in \mathbb{R}^2(x', y') \mid 3x' + y' \geq 0 \text{ and } y' \geq 0\},$$

   which shows that $\sigma$ is generated by $(-1,3)$ and $(1,0)$.

   Now, the resolution of $\mathrm{Spec}(A)$ is defined by the unimodular decomposition

   $$\sigma = \sigma_1 \cup \sigma_2,$$

   where $\sigma_1 =< (-1,3), (0,1) >, \sigma_2 =< (0,1), (1,0) >$ and then the resolution $X$ of $\mathrm{Spec}(\sigma^\vee)$ is given by union of 2 affine open sets $X = U_1 \cup U_2$, $U_1 = \mathrm{Spec}(\sigma_1^\vee) = \mathrm{Spec}(k[x, x^3y])$ and $U_2 = \mathrm{Spec}(k[x,y])$.

   Now, $t$ in equation (4.1) corresponds to $x^3y^2$ and on $U_1$, $x^3y^2 = [x \cdot (x^3y)]^3 \cdot x^{-9}y^{-1}$ and on $U_2$, $x^3y^2 = (xy)^3y^{-1}$. Since $E$ is the exceptional curve corresponding to $(0,1)$, this shows that $a(E,X) = -1/3$. Then we conclude that $\mathrm{Spec}(A)$ is a log terminal singularity but not a canonical singularity.

Properties of $\mathbb{Q}$-Gorenstein rings are frequently reduced to those of 1-Gorenstein rings using **index 1 cover**.

DEFINITION 4.9. Let $A$ be a normal ring and assume that $K_A \cong I$, which has pure height 1. If $I^{(r)} = wA$, we put

$$B = \bigoplus_{i \in \mathbb{Z}/r\mathbb{Z}} I^{(i)} u^i,$$

where $u^r = w^{-1}$. Then $B$ is an $A$-algebra, which we call an **index 1 cover of** $A$.

Notice that the construction of $B$ depends on the choice of $w$ and there may exist non-isomorphic index 1 covers. Since $K_B \cong \mathrm{Hom}_A(B, K_A) \cong \bigoplus_{i \in \mathbb{Z}/r\mathbb{Z}} I^{(1-i)} u^{-i} \cong u^{-1}B$, $B$ is Gorenstein if $B$ is Cohen-Macaulay.

REMARK 4.10. We could define the notions defined in Definition 4.6, also in positive characteristic. Then it is natural to ask if they are Cohen-Macaulay, since a rational singularity is Cohen-Macaulay only if in dimension 2 or when GR vanishing holds for a resolution. But GR vanishing fails in positive characteristic in dimension $\geq 3$ and examples of "terminal singularity" in positive characteristic which is **not** Cohen-Macaulay appear in [**Kov**], [**La-R**], [**Tot**].

EXAMPLE 4.11. Let $k$ be a field and put $A = k[x^3, x^2y, xy^2, y^3]$ be the Veronese subring of $k[x, y]$. In this case, since $K_A = (xy \cdot k[x, y])^{(3)} = (x^2y, xy^2)$, we can put $I = (xy^2, y^3)$. Then $I^{(3)} = (y^6)$ and index 1 cover is

$$B = A \oplus (x, y)A \oplus (x, y)^2 A \cong k[x, y].$$

The property of $\mathbb{Q}$-Gorenstein rings can be discussed using **index 1 cover**.

THEOREM 4.12. *(cf.* [**Re**], *Proposition 1.7) Let $A$ be a $\mathbb{Q}$-Gorenstein normal ring essentially of finite type over a field of characteristic 0 and $B$ be an index 1 cover of $A$. Then we have the following:*

1. *If $A$ is log-terminal (resp. terminal, canonical, log-canonical), so is $B$.*
2. *If $A$ is log-terminal, then $A$ is a rational singularity and hence is Cohen-Macaulay.*

The discrepancy of $B$ is an integer and hence if $A$ is log-terminal, then $B$ is canonical.

THEOREM 4.13 ([**El**], [**Fl**]). *If $A$ is a canonical singularity, essentially of finite type over a field of characteristic 0 then $A$ is a rational singularity and hence is Cohen-Macaulay.*

DEFINITION 4.14. We say that $A$ is **strongly $F$-regular**, if for every $c \in A^\circ$, there exists $q = p^e$ such that the map $A \to A^{1/q}$ sending 1 to $c^{1/q}$ splits as $A$-module homomorphism. Namely, there is a $A$-module homomorphism $A^{1/q} \to A$ such that $g(c^{1/q}) = 1$.

Next, we review $F$-regular and $F$-pure rings from the viewpoint of Frobenius splittings.

DEFINITION 4.15. Let $A$ be a Noetherian ring containing a field $k$ of characteristic $p > 0$. We say that $A$ is **$F$-finite** if $A^{1/p}$ is a finitely generated $A$-module. It is easy to show that if $A$ is essentially of finite type over $k$ or $(A, \mathfrak{m})$ is complete local and $A/\mathfrak{m}$ is perfect, then $A$ is $F$-finite. It is shown by E. Kunz that $F$-finite rings are excellent rings. Note that $A$ is $F$-pure if and only if $A \to A^{1/q}$ splits.

REMARK 4.16. If $A$ is strongly $F$-regular, then $A$ is weakly $F$-regular. In fact, if $x \in I^*$ and $cx^q \in I^{[q]}$ for any $q$, take $q$ so that there exists $g : A^{1/q} \to A$ such that $g(c^{1/q}) = 1$. Then $x = xg(c^{1/q}) = g((cx^q)^{1/q}) \in g((I^{[q]})^{1/q}) = I$. Hence we have $I^* = I$. Actually, for quite many F-finite cases, weakly and strongly $F$-regular rings are known to be equivalent and as far as I know, there is no example of $F$-finite weakly $F$-regular ring which is not strongly $F$-regular.

The next is the most important Lemma in what follows.

LEMMA 4.17. *Let $A$ be an $F$-finite normal ring of characteristic $p > 0$ with the canonical module $K_A$ (resp. $X$ be a normal algebraic variety over a field $k$ with* $\mathrm{char}(k) = p$ *with canonical divisor $K_X$). Then for $q = p^e$, we have*

$$\mathrm{Hom}_A(A^{1/q}, A) \cong [K_A^{(1-q)}]^{1/q},$$

$$(\textit{ resp. } \ \mathcal{H}om_{\mathcal{O}_X}(\mathcal{O}_X^{1/q}, \mathcal{O}_X) \cong [\mathcal{O}_X((1-q)K_X)]^{1/q}).$$

PROOF. In general, if $A \subset B$ is a ring extension and if $B$ is finitely generated as an $A$-module, then we have

$$K_B \cong \mathrm{Hom}_A(B, K_A).$$

Putting $B = A^{1/q}$, we have $K_{A^{1/q}} \cong \mathrm{Hom}_A(A^{1/q}, K_A)$. Hence $\mathrm{Hom}_A(A^{1/q}, A)$ $\cong [\mathrm{Hom}_A(A^{1/q}, K_A) \otimes_A K_A^{(-1)}]^{**} \cong [(K_A \otimes_A K_A^{(-q)})^{**}]^{1/q} \cong (K_A^{(1-q)})^{1/q}$. □

THEOREM 4.18 ([**HaWa**]). *Let $(A, \mathfrak{m})$ be an $F$-finite normal local ring of dimension $d$, essentially of finite type over a field $k$ of characteristic $p > 0$ and assume that $K_A^{(r)}$ is principal for a positive integer $r$. Let $f : X \to Y =$* $\mathrm{Spec}(A)$ *be a birational morphism and assume that $X$ is normal. If $E$ is an irreducible reduced closed subvariety of $X$ with* $\dim f(E) < d - 1$. *then we have*

1. *If $A$ is $F$-pure, then we have $a(E, X) \geq -1$.*
2. *If $A$ is strongly $F$-regular, then $a(E, X) > -1$.*

PROOF. Assume that $A$ is $F$-pure and let $\phi : A^{1/q} \to A$ satisfy $\phi \circ i = 1_A$, where $i$ is the injection $A \to A^{1/q}$. Let $L$ be the quotient field of $A$. Then we can extend $\phi$ to $\phi : L^{1/q} \to L$. Hence $\phi$ defines $\tilde{\phi} : \mathcal{O}_X^{1/q} \to \mathcal{O}_X$, for which $\tilde{\phi} \circ \tilde{i} = 1_{\mathcal{O}_X}$, where $\tilde{i} : \mathcal{O}_X \to \mathcal{O}_X^{1/q}$ is the injection.

Now, assume $a(E, X) < -1$ and deduce a contradiction. At the generic point $\xi$ of $E$, $\phi \circ i$ is the identity. On the other hand, by Lemma 4.17, $\tilde{\phi} \in \mathcal{H}om_{\mathcal{O}_X}(\mathcal{O}_X^{1/q}, \mathcal{O}_X) \cong [\mathcal{O}_X((1-q)K_X)]^{1/q}$ and since $\tilde{\phi}$ is a pull back of $\phi$,

$$\tilde{\phi} \in f^*(\mathcal{O}_Y((1-q)K_Y)) = \mathcal{O}_X((1-q)K_X)(-(1-q)a(E, X)E).$$

Since for large enough $q$, $(1-q)a(E, X) \geq q$, $\tilde{\phi}(\mathcal{O}_{X,\xi}^{1/q})$ is included in the maximal ideal of $\mathcal{O}_{X,\xi}$ and contratticts the fact that $\tilde{\phi} \cdot \tilde{i}$ is the identity. □

From this we show;

THEOREM 4.19. *Let $A$ be a local ring essentially of finite type over a field of characteristic* 0. *Assume also that $K_A^{(r)}$ is principal for some $r$.*

1. *If $A$ is of dense $F$-pure type, then $A$ is log-canonical.*
2. *$A$ is log-terminal if and only if $A$ is of dense $F$-regular type* [2].

[2] By the result of [**MaSch**], if $A_p$ is $F$-regular for single $p$, then $A$ is log-terminal.

REMARK 4.20. The converse of (1) above is a long standing conjecture and holds in dimension $\leq 3$ (cf. §4 of [**TW**]).

## 5. Appendix. Referred theorems

PROPOSITION 5.1. *Let $R$ be a Noetherian ring, $\mathfrak{a}$ be an ideal of $R$. If $\underline{f} = (f_1, f_2, \cdots, f_n) \subset R$ satisfies $\sqrt{\mathfrak{a}} = \sqrt{(f_1, f_2, \ldots, f_n)}$, then for every $p \in \mathbb{Z}$, we have an isomorphism*

$$\mathrm{H}^p_{\mathfrak{a}}(*) = \mathrm{H}^p_{\underline{f}}(*) := \mathrm{H}^p(C^\bullet),$$

*where the Čech complex $C^\bullet = C^\bullet(\underline{f})$ is defined by*

$$C^\bullet = [0 \to C^0 \to C^1 \to \ldots \to C^n \to 0],$$

*with $C^r = \oplus_{\sharp I = r} \bigwedge R_{f_I} e_I, f_I = \prod_{i \in I} f_i$.*

THEOREM 5.2. *(Buchsbaum-Eisenbud, [**BE**]) Let $A$ be a Noetherian ring, $M$ be a finitely generated $A$-module and*

$$F_\bullet : 0 \to F_n \overset{\varphi_n}{\to} F_{n-1} \overset{\varphi_{n-1}}{\to} F_{n-2} \to \cdots \to F_1 \overset{\varphi_1}{\to} F_0$$

*be a complex of finitely generated $A$ free modules and assume $\varphi_i \otimes_A M \neq 0$ $(1 \leq i \leq n)$. Then the following conditions are equivalent.*

1. *The complex*

$$0 \to F_n \otimes M \overset{\varphi_n \otimes M}{\longrightarrow} F_{n-1} \otimes M \overset{\varphi_{n-1} \otimes M}{\longrightarrow} F_{n-2} \otimes M \longrightarrow \cdots \longrightarrow F_1 \otimes M \overset{\varphi_1 \otimes M}{\longrightarrow} F_0 \otimes M$$

   *is exact.*
2. *For every $i, 1 \leq i \leq n$, the following 2 conditions are satisfied.*
   (a) $\mathrm{depth}(\mathrm{I}(\varphi_i, M), M) \geq i$.
   (b) $\mathrm{rank}(\varphi_i, M) + \mathrm{rank}(\varphi_{i+1}, M) = \mathrm{rank}_A F_i$, *where we put* $\mathrm{rank}(\varphi_{n+1}, M) = 0$.

## References


[Dem] Michel Demazure, *Anneaux gradués normaux* (French), Introduction à la théorie des singularités, II, Travaux en Cours, vol. 37, Hermann, Paris, 1988, pp. 35–68. MR1074589

[Bou] Jean-François Boutot, *Singularités rationnelles et quotients par les groupes réductifs* (French), Invent. Math. **88** (1987), no. 1, 65–68, DOI 10.1007/BF01405091. MR877006

[BE] David A. Buchsbaum and David Eisenbud, *What makes a complex exact?*, J. Algebra **25** (1973), 259–268, DOI 10.1016/0021-8693(73)90044-6. MR314819

[BH] Winfried Bruns and Jürgen Herzog, *Cohen-Macaulay rings*, Cambridge Studies in Advanced Mathematics, vol. 39, Cambridge University Press, Cambridge, Revised Edition, 1997. MR1251956

[BM] Holger Brenner and Paul Monsky, *Tight closure does not commute with localization*, Ann. of Math. (2) **171** (2010), no. 1, 571–588, DOI 10.4007/annals.2010.171.571. MR2630050

[Eis] David Eisenbud, *Commutative algebra: With a view toward algebraic geometry*, Graduate Texts in Mathematics, vol. 150, Springer-Verlag, New York, 1995, DOI 10.1007/978-1-4612-5350-1. MR1322960

[El] Renée Elkik, *Rationalité des singularités canoniques* (French), Invent. Math. **64** (1981), no. 1, 1–6, DOI 10.1007/BF01393930. MR621766

[Fl] Hubert Flenner, *Rationale quasihomogene Singularitäten* (German), Arch. Math. (Basel) **36** (1981), no. 1, 35–44, DOI 10.1007/BF01223666. MR612234

[GW1] Shiro Goto and Kei-ichi Watanabe, *On graded rings. I*, J. Math. Soc. Japan **30** (1978), no. 2, 179–213, DOI 10.2969/jmsj/03020179. MR494707

[GW2] Shiro Goto and Kei-ichi Watanabe, *The structure of one-dimensional F-pure rings*, J. Algebra **49** (1977), no. 2, 415–421, DOI 10.1016/0021-8693(77)90250-2. MR453729

[Hara] Nobuo Hara, *A characterization of rational singularities in terms of injectivity of Frobenius maps*, Amer. J. Math. **120** (1998), no. 5, 981–996. MR1646049

[HaWa] Nobuo Hara and Kei-ichi Watanabe, *F-regular and F-pure rings vs. log terminal and log canonical singularities*, J. Algebraic Geom. **11** (2002), no. 2, 363–392, DOI 10.1090/S1056-3911-01-00306-X. MR1874118

[HH1] Melvin Hochster and Craig Huneke, *Tight closure, invariant theory, and the Briançon-Skoda theorem*, J. Amer. Math. Soc. **3** (1990), no. 1, 31–116, DOI 10.2307/1990984. MR1017784

[Hu2] Craig Huneke, *Tight closure and its applications*, CBMS Regional Conference Series in Mathematics, vol. 88, Published for the Conference Board of the Mathematical Sciences, Washington, DC; by the American Mathematical Society, Providence, RI, 1996. With an appendix by Melvin Hochster, DOI 10.1016/0167-4889(95)00136-0. MR1377268

[KKMSD73] G. Kempf, Finn Faye Knudsen, D. Mumford, and B. Saint-Donat, *Toroidal embeddings. I*, Lecture Notes in Mathematics, Vol. 339, Springer-Verlag, Berlin-New York, 1973. MR335518

[KolMo] János Kollár and Shigefumi Mori, *Birational geometry of algebraic varieties*, Cambridge Tracts in Mathematics, vol. 134, Cambridge University Press, Cambridge, 1998. With the collaboration of C. H. Clemens and A. Corti; Translated from the 1998 Japanese original, DOI 10.1017/CBO9780511662560. MR1658959

[Kov] Sándor J. Kovács, *Non-Cohen-Macaulay canonical singularities*, Local and global methods in algebraic geometry, Contemp. Math., vol. 712, Amer. Math. Soc., [Providence], RI, [2018] ©2018, pp. 251–259, DOI 10.1090/conm/712/14349. MR3832406

[Ku] Ernst Kunz, *On Noetherian rings of characteristic p*, Amer. J. Math. **98** (1976), no. 4, 999–1013, DOI 10.2307/2374038. MR432625

[La-R] N. Lauritzen and A. P. Rao, *Elementary counterexamples to Kodaira vanishing in prime characteristic*, Proc. Indian Acad. Sci. Math. Sci. **107** (1997), no. 1, 21–25, DOI 10.1007/BF02840470. MR1453823

[Li] Joseph Lipman, *Rational singularities, with applications to algebraic surfaces and unique factorization*, Inst. Hautes Études Sci. Publ. Math. **36** (1969), 195–279. MR276239

[Li74] Joseph Lipman, *Unique factorization in complete local rings*, Algebraic geometry (Proc. Sympos. Pure Math., Vol. 29, Humboldt State Univ., Arcata, Calif., 1974), Proc. Sympos. Pure Math., Vol. 29, Amer. Math. Soc., Providence, RI, 1975, pp. 531–546. MR374125

[LT] Joseph Lipman and Bernard Teissier, *Pseudorational local rings and a theorem of Briançon-Skoda about integral closures of ideals*, Michigan Math. J. **28** (1981), no. 1, 97–116. MR600418

[MaSch] Linquan Ma and Karl Schwede, *Singularities in mixed characteristic via perfectoid big Cohen-Macaulay algebras*, Duke Math. J. **170** (2021), no. 13, 2815–2890, DOI 10.1215/00127094-2020-0082. MR4312190

[MeSr] V. B. Mehta and V. Srinivas, *A characterization of rational singularities*, Asian J. Math. **1** (1997), no. 2, 249–271, DOI 10.4310/AJM.1997.v1.n2.a4. MR1491985

[MuSr] Mircea Mustaţă and Vasudevan Srinivas, *Ordinary varieties and the comparison between multiplier ideals and test ideals*, Nagoya Math. J. **204** (2011), 125–157, DOI 10.1215/00277630-1431849. MR2863367

[Nak] Yukio Nakamura, *Local rings of Cohen-Macaulay F-rational rings are F-rational*, Tokyo J. Math. **14** (1991), no. 1, 41–44, DOI 10.3836/tjm/1270130485. MR1108152

[Re] Miles Reid, *Canonical 3-folds*, Journées de Géometrie Algébrique d'Angers, Juillet 1979/Algebraic Geometry, Angers, 1979, Sijthoff & Noordhoff, Alphen aan den Rijn—Germantown, Md., 1980, pp. 273–310. MR605348

[Tot] Burt Totaro, *The failure of Kodaira vanishing for Fano varieties, and terminal singularities that are not Cohen-Macaulay*, J. Algebraic Geom. **28** (2019), no. 4, 751–771, DOI 10.1090/jag/724. MR3994312

[TW] Shunsuke Takagi and Kei-ichi Watanabe, *$F$-singularities: applications of characteristic p methods to singularity theory [translation of MR3135334]*, Sugaku Expositions **31** (2018), no. 1, 1–42, DOI 10.1090/suga/427. MR3784697

[Wa83] Kei-ichi Watanabe, *Rational singularities with $k^*$-action*, Commutative algebra (Trento, 1981), Heidelberger Taschenbücher, Dekker, New York, 1983, pp. 339–351. MR686954

[Wa87] Kei-ichi Watanabe, *Study of $F$-purity in dimension two*, Algebraic geometry and commutative algebra, Vol. II, Kinokuniya, Tokyo, 1988, pp. 791–800. MR977783

[Wa91] Kei-ichi Watanabe, *$F$-regular and $F$-pure normal graded rings*, J. Pure Appl. Algebra **71** (1991), no. 2-3, 341–350, DOI 10.1016/0022-4049(91)90157-W. MR1117644

[Wa97] Kei-ichi Watanabe, *$F$-rationality of certain Rees algebras and counterexamples to "Boutot's theorem" for $F$-rational rings*, J. Pure Appl. Algebra **122** (1997), no. 3, 323–328, DOI 10.1016/S0022-4049(97)00064-9. MR1481095

Department of Mathematics, College of Humanities and Sciences, Nihon University, Setagaya-ku Tokyo, 156–8550; and Organization for the Strategic Coordination of Research and Intellectual Properties, Meiji University Japan

*Email address*: watnbkei@gmail.com

Contemporary Mathematics
Volume **831**, 2025
https://doi.org/10.1090/conm/831/16641

# An introduction to singularities in commutative algebra via perfectoid big Cohen-Macaulay algebras

Linquan Ma and Kevin Tucker

## 1. Perfect rings

Throughout these notes, we will assume all rings are commutative with unity, and $p > 0$ is a fixed prime number. We write $\mathbb{F}_p$ for the finite field with $p$ elements, and $\mathbb{Z}_p$ for the $p$-adic integers. We will denote a Noetherian local ring as a triple $(R, \mathfrak{m}, k)$ where $R$ is a Noetherian local ring with unique maximal ideal $\mathfrak{m}$ and $k = R/\mathfrak{m}$ is the corresponding residue field. It will often be assumed that $p \in \mathfrak{m}$, giving rise to two possible situations to be contrasted throughout:

- the (equal) characteristic $p$ setting, where $\operatorname{char} R = \operatorname{char} k = p$, *e.g.* (quotients of) $R = \mathbb{F}_p[[x_1, \ldots, x_n]]$, or;
- the mixed characteristic setting, where $\operatorname{char} R \neq p = \operatorname{char} k$, *e.g.* $R = \mathbb{Z}_p[[x_1, \ldots, x_n]]$. In this case, $\operatorname{char} R$ is necessarily 0 if $R$ is a domain, and is otherwise either 0 or $p^e$ for some integer $e > 1$.

DEFINITION 1.1. We say that a ring $R$ is *perfect* if $\operatorname{char} R = p$ and the Frobenius endomorphism $F\colon R \xrightarrow{r \mapsto r^p} R$ is an isomorphism. When the Frobenius endomorphism is merely surjective, we say $R$ is *semi-perfect*.

In general, when $\operatorname{char} R = p$, the Frobenius endomorphism is injective if and only if $R$ is reduced. In this case, after identifying $R$ with its image, Frobenius is equivalent to the inclusion $R^p \subseteq R$ or alternatively $R \subseteq R^{1/p}$. Iterating gives an infinite string

$$\cdots \subseteq R^{p^e} \subseteq \cdots \subseteq R^{p^2} \subseteq R^p \subseteq R \subseteq R^{1/p} \subseteq R^{1/p^2} \subseteq \cdots \subseteq R^{1/p^e} \subseteq \cdots$$

where any $e$ consecutive inclusions are identified with $F^e\colon R \longrightarrow R$.

Interesting perfect rings are necessarily non-Noetherian, as the only perfect Noetherian rings are finite products of perfect fields. Indeed, for any $s \in S$ where $S$ is a perfect ring, it is easily verified that

$$\sqrt{(s)} = (s^{1/p^\infty}) := (s, s^{1/p}, s^{1/p^2}, \ldots)$$

---

The first author was supported in part by NSF Grant DMS #2302430 and a fellowship from the Sloan Foundation. The second author was supported in part by NSF Grant DMS #2200716.

from which it follows $\sqrt{(s)} = (s^{1/p^n}) = (s^{1/p^{n+1}})$ for $n \gg 0$ if also $S$ is Noetherian. In particular, we have that $s^{1/p^{n+1}} = s^{1/p^n} t$ for some $t \in S$, hence

$$s^{1/p^{n+1}}(1 - s^{1/p^n - 1/p^{n+1}} t) = 0$$

and so we see every non-zero divisor is a unit. Note that, more generally, for any ideal $I$ of a perfect ring $R$ we have

$$\sqrt{I} = (f^{1/p^e} \mid f \in I, e \geq 0)$$

and in particular $\sqrt{(f_1, \ldots, f_m)} = (f_1^{1/p^\infty}, \ldots, f_m^{1/p^\infty}) := \sum_{i=1}^m (f_i^{1/p^\infty})$ for any $f_1, \ldots, f_m \in R$.

Given an arbitrary ring $R$ of characteristic $p$, there are two important ways to limit over the iterates of Frobenius to get a perfect ring. The first and easier of the two is to take the *perfection*

$$R_{\mathrm{perf}} := \varinjlim_{r \mapsto r^p} R = \varinjlim \left( R \xrightarrow{F} R \xrightarrow{F} R \xrightarrow{F} \cdots \right).$$

which comes bundled together with the mapping $R \longrightarrow R_{\mathrm{perf}}$ onto the first copy of $R$ in the sequence. One checks that $R_{\mathrm{perf}}$ is indeed perfect, and is in a sense the smallest perfect $R$-algebra in that it is initial in the category of perfect $R$-algebras. Explicitly, this means that any ring homomorphism $R \longrightarrow A$ where $A$ is perfect factors uniquely as $R \longrightarrow R_{\mathrm{perf}} \longrightarrow A$. When $R$ is reduced, we have

$$R_{\mathrm{perf}} = \varinjlim \left( R \subseteq R^{1/p} \subseteq R^{1/p^2} \subseteq \cdots \right) = \bigcup_{e \geq 0} R^{1/p^e}$$

which in particular for $R$ a domain can be viewed as a subring of an algebraic closure of the fraction field $K = \operatorname{Frac} R$. More generally, for arbitrary $R$, one has $R_{\mathrm{perf}} = (R_{\mathrm{red}})_{\mathrm{perf}}$. Important examples to keep in mind include:

- $\mathbb{F}_p[x_1, \ldots, x_d]_{\mathrm{perf}} = \mathbb{F}_p[x_1^{1/p^\infty}, \ldots, x_d^{1/p^\infty}]$;
- $\mathbb{F}_p[[x_1, \ldots, x_d]]_{\mathrm{perf}} = \mathbb{F}_p[[x_1, \ldots, x_d]][x_1^{1/p^\infty}, \ldots, x_d^{1/p^\infty}]$, which it is important to note is not $(x_1, \ldots, x_d)$-adically complete;
- $(\mathbb{F}_p[[x_1, \ldots, x_d]]/(f_1, \ldots, f_m))_{\mathrm{perf}}$
  $= \mathbb{F}_p[[x_1, \ldots, x_d]]_{\mathrm{perf}} / \sqrt{(f_1, \ldots, f_m)}$
  $= \mathbb{F}_p[[x_1, \ldots, x_d]][x_1^{1/p^\infty}, \ldots, x_d^{1/p^\infty}]/(f_1^{1/p^\infty}, \ldots, f_m^{1/p^\infty})$.

Note that an element of $R$ is a unit if and only if so too is its image in $R_{\mathrm{perf}}$, and that $R \twoheadrightarrow R_{\mathrm{perf}} = R_{\mathrm{red}}$ when $R$ is semi-perfect. Perfection gives a functor from the category of rings of characteristic $p$ to its full subcategory of perfect rings.

In contrast to the perfection, limiting over the iterates of Frobenius in the other direction when $\operatorname{char} R = p$ gives the *tilt*

$$R^\flat := \varprojlim_{r \mapsto r^p} R = \varprojlim \left( R \xleftarrow{F} R \xleftarrow{F} R \xleftarrow{F} \cdots \right) \tag{1}$$

so that an element of $R^\flat$ is a sequence $(r_0, r_1, r_2, \ldots)$ of elements of $R$ satisfying $r_{n+1}^p = r_n$ for all $n \geq 0$. Once more, one checks that $R^\flat$ is perfect and is the largest perfect ring mapping to $R$ in the sense that any ring homomorphism $A \longrightarrow R$ where $A$ is perfect factors through the ring homomorphism

$$\mathrm{pr}_0 \colon R^\flat \xrightarrow[r = (r_0, r_1, r_2, \ldots) \longmapsto r_0 (= r_n^{p^n})]{} R \quad .$$

When $R$ is reduced, we have that

$$R^\flat = \varprojlim \left( \cdots \subseteq R^{p^2} \subseteq R^p \subseteq R \right) = \bigcap_{e \geq 0} R^{p^e}$$

is the subring of elements admitting arbitrary $p$-power roots. In comparison, $\mathrm{pr}_0 \colon R^\flat \longrightarrow R$ is surjective when $R$ is semi-perfect. When $R$ is a perfect ring and $I \subseteq R$ an ideal, it is an instructive exercise to verify that the map of inverse systems

$$\begin{array}{ccccccccc}
\Big( \cdots & \xrightarrow{F} & R/I & \xrightarrow{F} & R/I & \xrightarrow{F} & R/I & \xrightarrow{F} & R/I \Big) \\
 & & \big\downarrow{\scriptstyle F^3} & & \big\downarrow{\scriptstyle F^2} & & \big\downarrow{\scriptstyle F} & & \big\downarrow{\scriptstyle =} \\
\Big( \cdots & \twoheadrightarrow & R/I^{[p^3]} & \twoheadrightarrow & R/I^{[p^2]} & \twoheadrightarrow & R/I^{[p]} & \twoheadrightarrow & R/I \Big)
\end{array}$$

identifies $(R/I)^\flat$ with the completion of $R$ along the sequence of ideals $I^{[p^e]} := F^e(I)R = (f^{p^e} \mid f \in I)$. In particular, if $(f_1, \ldots, f_m)$ is a finitely generated ideal of a perfect ring $R$, $(R/(f_1, \ldots, f_m))^\flat$ agrees with the $(f_1, \ldots, f_m)$-adic completion of $R$. As a consequence, if $d \in R^\flat$ is in the kernel of $\mathrm{pr}_0$, it also follows that $R^\flat = (R^\flat/(d))^\flat = (R^\flat)^{\wedge d}$ is necessarily $d$-complete and separated. In general, note that $u \in R^\flat$ is a unit if and only if $\mathrm{pr}_0(r) \in R$ is a unit. Important examples of tilting include:

- $\left(\mathbb{F}_p[x_1, \ldots, x_d]\right)^\flat = \left(\mathbb{F}_p[[x_1, \ldots, x_d]]\right)^\flat = \mathbb{F}_p$;
- $\left(\mathbb{F}_p[t^{1/p^\infty}]/(t)\right)^\flat = \left(\mathbb{F}_p[[t]][t^{1/p^\infty}]/(t)\right)^\flat = \mathbb{F}_p[t^{1/p^\infty}]^{\wedge t}$;
- $\left(\mathbb{F}_p[[x_1, \ldots, x_d]][x_1^{1/p^\infty}, \ldots, x_d^{1/p^\infty}]/(f_1, \ldots, f_m)\right)^\flat$
 $= \mathbb{F}_p[[x_1, \ldots, x_d]][x_1^{1/p^\infty}, \ldots, x_d^{1/p^\infty}]^{\wedge (f_1, \ldots, f_m)}$.

Once again, tilting gives a functor from the category of rings of characteristic $p$ to its full subcategory of perfect rings.

The study of singularities in characteristic $p$ through the lens of Frobenius is based upon the foundational result of Kunz asserting that a Noetherian ring $R$ of characteristic $p$ is regular if and only if the Frobenius endomorphism is flat [**Kun69**]. In order to leverage the intuition from this result and others like it in mixed characteristic, one must first revisit the situation in characteristic $p$ and remove the explicit dependence on Frobenius in some way. In the next section, we will introduce the notion of a perfectoid ring in mixed characteristic which serves as an analogue of a perfect ring. The following result can be seen as an extension of Kunz's result in terms of perfect rings, which both recovers Kunz's result by a novel proof and also guides our hand in formulating the analogous result in the mixed characteristic setting.

THEOREM 1.2 ([**BS17**, Corollary 11.35]). *If $R$ is a Noetherian ring with characteristic $p$, then $R$ is regular if and only if there exists a faithfully flat extension $R \longrightarrow A$ with $A$ perfect.*

The heart of the proof relies on a series of observations from [**AH97**] regarding the flat dimension of ideals generated by the arbitrary $p$-power roots of finitely many elements, which we recall here before proceeding further.

(a) Suppose $f$ is an element of a ring $A$ admitting a compatible system $\{f^{1/p^e}\}_{e\geq 0}$ of arbitrary $p$-power roots. Taking the horizontal colimits of the commutative diagram

$$\begin{array}{ccccccccc} A & \xrightarrow{f^{1-1/p}} & A & \xrightarrow{f^{1/p-1/p^2}} & A & \xrightarrow{f^{1/p^2-1/p^3}} & A & \longrightarrow & \cdots \\ {\scriptstyle f}\downarrow & & {\scriptstyle f^{1/p}}\downarrow & & {\scriptstyle f^{1/p^2}}\downarrow & & {\scriptstyle f^{1/p^3}}\downarrow & & \\ (f) & \subseteq & (f^{1/p}) & \subseteq & (f^{1/p^2}) & \subseteq & (f^{1/p^3}) & \longrightarrow & \cdots \end{array}$$

gives an induced mapping

$$\varinjlim \left( A \xrightarrow{f^{1-1/p}} A \xrightarrow{f^{1/p-1/p^2}} A \longrightarrow \cdots \right) \longrightarrow (f^{1/p^\infty})$$

that is clearly surjective. If additionally $A$ is reduced, it is straightforward to check this mapping is injective as well, from which it follows that $(f^{1/p^\infty})$ is a flat $A$-module.

(b) Suppose now $A \longrightarrow A'$ is a map of reduced rings, and $f$ is an element of a ring $A$ admitting a compatible system $\{f^{1/p^e}\}_{e\geq 0}$ of arbitrary $p$-power roots. We claim that $\mathrm{Tor}_i^A(A/(f^{1/p^\infty}), A') = 0$ for $i > 0$. Indeed, as the flat resolution $(f^{1/p^\infty}) \longrightarrow A$ of $A/(f^{1/p^\infty})$ can be used for this computation, one need only verify the vanishing of

$$\mathrm{Tor}_1^A(A/(f^{1/p^\infty}), A') = \ker\left( (f^{1/p^\infty}) \otimes_A A' \longrightarrow A' \right).$$

However, as tensor products are compatible with direct limits, the injectivity of

$$(f^{1/p^\infty}) \otimes_A A' = \varinjlim \left( A' \xrightarrow{f^{1-1/p}} A' \xrightarrow{f^{1/p-1/p^2}} A' \longrightarrow \cdots \right) \cong (f^{1/p^\infty})A' \subseteq A'$$

again follows using that $A'$ is reduced.

(c) Consider now an ideal of the form $(f_1^{1/p^\infty}, f_2^{1/p^\infty}, \ldots, f_m^{1/p^\infty})$, where each $f_i$ is an element of a ring $A$ admitting a compatible system $\{f_i^{1/p^e}\}_{e\geq 0}$ of arbitrary $p$-power roots, and furthermore each of the rings

$$A,\ A/(f_1^{1/p^\infty}),\ A/(f_1^{1/p^\infty}, f_2^{1/p^\infty}),\ \ldots,\ A/(f_1^{1/p^\infty}, f_2^{1/p^\infty}, \ldots, f_m^{1/p^\infty})$$

are reduced. Let $\mathcal{K}_\bullet(f_i^{1/p^\infty})$ denote the two term complex of flat $A$-modules

$$\mathcal{K}_1(f_i^{1/p^\infty}) := (f_i^{1/p^\infty}) \longrightarrow \mathcal{K}_0(f_i^{1/p^\infty}) := A$$

giving a flat resolution of $A/(f_i^{1/p^\infty})$, and set

$$\mathcal{K}_\bullet(f_1^{1/p^\infty}, \ldots, f_i^{1/p^\infty}) = \bigotimes_{j=1}^{i} \mathcal{K}_\bullet(f_j^{1/p^\infty}) = \bigotimes_{j=1}^{i} \left[ (f_j^{1/p^\infty}) \longrightarrow A \right].$$

Proceeding inductively, one may then conclude $\mathcal{K}_\bullet(f_1^{1/p^\infty}, \ldots, f_m^{1/p^\infty})$ is a flat resolution of $A/(f_1^{1/p^\infty}, f_2^{1/p^\infty}, \ldots, f_m^{1/p^\infty})$ of length $m$. Indeed, if we assume that the complex $\mathcal{K}_\bullet(f_1^{1/p^\infty},\dots,f_i^{1/p^\infty})$ is a flat resolution of $A/(f_1^{1/p^\infty},\dots,f_i^{1/p^\infty})$ for some $1 \le i < m$, it follows that

$$\begin{aligned} H_j\left(\mathcal{K}_\bullet(f_1^{1/p^\infty},\dots,f_{i+1}^{1/p^\infty})\right) &= H_j\left(\mathcal{K}_\bullet(f_1^{1/p^\infty},\dots,f_i^{1/p^\infty})\otimes_A \mathcal{K}_\bullet(f_{i+1}^{1/p^\infty})\right) \\ &= \operatorname{Tor}_j^A(A/(f_{i+1}^{1/p^\infty}), A/(f_1^{1/p^\infty},\dots,f_i^{1/p^\infty})) \end{aligned}$$

as flat resolutions can be used in Tor computations. Applying the observation from (b) above to $A \longrightarrow A/(f_1^{1/p^\infty},\dots,f_i^{1/p^\infty})$, we see that this Tor vanishes for $j>0$ and it follows together with (a) that $\mathcal{K}_\bullet(f_1^{1/p^\infty},\dots,f_i^{1/p^\infty})$ is a flat resolution as claimed. In particular, the flat dimension of $A/(f_1^{1/p^\infty},f_2^{1/p^\infty},\dots,f_m^{1/p^\infty})$ is bounded by $m$ (and in particular finite), *i.e.* $\operatorname{Tor}_j^A(M, A/(f_1^{1/p^\infty},f_2^{1/p^\infty},\dots,f_m^{1/p^\infty})) = 0$ for all $A$-modules $M$ and $j > m$.

Proof of Theorem 1.2. One direction here is straightforward: if $R$ is regular, Frobenius is flat so that $R \longrightarrow R_{\mathrm{perf}}$ is the colimit of flat extensions and hence also flat. As $R \longrightarrow R_{\mathrm{perf}}$ is also purely inseparable, the induced map on prime spectra is bijective, so the extension is faithfully flat.

For the reverse implication,[1] as a localization of a perfect ring is perfect, we may assume $(R,\mathfrak{m},k)$ is local and $A$ is a perfect faithfully flat $R$-algebra. If $x_1,\dots,x_d \in R$ are a system of parameters, we have $\sqrt{\mathfrak{m}A} = \sqrt{(x_1,\dots,x_d)A} = (x_1^{1/p^\infty},\dots,x_d^{1/p^\infty})$ and moreover each of the rings

$$A/(x_1^{1/p^\infty},\dots,x_i^{1/p^\infty}) = A/\sqrt{(x_1,\dots,x_i)A}$$

is reduced for $1 \le i \le d$ (and even still perfect). From the discussion above, it follows that $A/\sqrt{\mathfrak{m}A}$ has a resolution by flat $A$-modules of length $d$, and so in particular

$$0 = \operatorname{Tor}_{d+1}^A(A/\mathfrak{m}A, A/\sqrt{\mathfrak{m}A}) = \operatorname{Tor}_{d+1}^R(k, A/\sqrt{\mathfrak{m}A})$$

where the final equality uses that $R \longrightarrow A$ is flat (*e.g.* base change of a free resolution of $k$ over $R$ gives a free resolution of $k \otimes_R A = A/\mathfrak{m}A$ over $A$). On the other hand, when viewed as an $R$-module $A/\sqrt{\mathfrak{m}A}$ is annihilated by $\mathfrak{m}$ and hence is a vector space over $k$ – and necessarily non-zero as $A$ is faithfully flat over $R$. Writing $A/\sqrt{\mathfrak{m}A} \cong k^{\oplus \mathscr{I}}$ for an index set $\mathscr{I}$, we have

$$0 = \operatorname{Tor}_{d+1}^R(k, A/\sqrt{\mathfrak{m}A}) = \operatorname{Tor}_{d+1}^R(k, k^{\oplus\mathscr{I}}) = \operatorname{Tor}_{d+1}^R(k,k)^{\oplus\mathscr{I}}$$

and thus $\operatorname{Tor}_{d+1}^R(k,k) = 0$ so that $R$ is necessarily regular by the theorem of Auslander-Buchsbaum-Serre [**Eis95**, Corollary 19.5 and Theorem 19.12]. □

## 2. Perfectoid rings

Simply stated, our goal in this section is to introduce perfectoid rings, allowing us to extend the characterization of regular rings from Theorem 1.2 to mixed characteristic. While the general definition will take us some time to develop, the reader is encouraged to keep the following two basic cases of perfectoid rings in

[1] This implication can alternatively be seen to follow directly from Kunz's result. If $A$ is a perfect faithfully flat $R$-algebra, then $R \longrightarrow A \xrightarrow[\cong]{F} A$ is also faithfully flat. Since this composition also factors as $R \xrightarrow{F} R \longrightarrow A$, we have that $R \xrightarrow{F} R$ is flat as well [**Sta**, Lemma 039V]. By [**Kun69**], $R$ is then regular as desired.

mind throughout. In fact, in a sense that will subsequently be made precise in Proposition 2.3, every perfectoid ring can be understood in terms of the simpler cases below via a gluing procedure.

(1) If $A$ is a ring of characteristic $p$, then $A$ is perfectoid if and only if it is perfect.

(2) If $A$ is a ring that is $p$-torsion free, then $A$ is perfectoid if and only if:
   - it is $p$-complete, *i.e.* $A \longrightarrow A^{\wedge p} = \varprojlim A/(p^n)$ is an isomorphism, and consequently $p$-separated as well;
   - up to multiplication by a unit, $p$ admits a $p$-th root in $A$, *i.e.* there exist $\varpi \in A$ and $u \in A^\times$ with $p = u\varpi^p$.
   - the Frobenius endomorphism of $A/(p)$ is surjective with kernel generated by $\varpi$, *i.e.* $A/(\varpi) \xrightarrow{x \mapsto x^p} A/(p)$ is an isomorphism.

Examples of $p$-torsion free rings in mixed characteristic satisfying the latter criteria include

$$\mathbb{Z}_p[p^{1/p^\infty}]^{\wedge p},\ \mathbb{Z}_p[p^{1/p^\infty}, x_1^{1/p^\infty}, \ldots, x_d^{1/p^\infty}]^{\wedge p},$$

and

$$\mathbb{Z}_p[[x_1, \ldots, x_d]][p^{1/p^\infty}, x_1^{1/p^\infty}, \ldots, x_d^{1/p^\infty}]^{\wedge p}$$

and the reader is strongly encouraged to work through what happens after going modulo $p$ for each of these and to verify the above conditions hold. In particular, note that in each of these three examples one may simply take $u = 1$ and $\varpi = p^{1/p}$.

For rings of characteristic $p$, we have previously definited the tilt as the inverse limit over iterations of the Frobenius endomorphism. More generally, for a ring $A$ of arbitary characteristic we define the *tilt* $A^\flat := (A/(p))^\flat$ of $A$ as the tilt of its reduction modulo $p$. Perhaps the first indication of the utility of tilting in mixed characteristic is that there is a well-defined untilting or sharp map

$$\begin{array}{rcl} \sharp\colon A^\flat & \longrightarrow & A \\ \alpha = (a_0 \bmod p, a_1 \bmod p, a_2 \bmod p, \ldots) & \longmapsto & \alpha^\sharp := \varprojlim a_n^{p^n} \end{array}$$

provided $A$ is $p$-complete. The key observation is that whenever $a = \tilde{a} \bmod p$ for two elements $a, \tilde{a} \in A$, one has $a^{p^n} = \tilde{a}^{p^n} \bmod p^{n+1}$ for all $n \geq 0$, and it also follows that the sharp map modulo $p$ satisfies $\mathrm{pr}_0(\alpha) = a_0 \bmod p = \alpha^\sharp \bmod p$. In contrast to the situation in positive characteristic, the sharp map is multiplicative but not additive in mixed characteristic. Untilting is an extremely useful tool in practice with many applications. In particular, it allows one to show that the natural map of multiplicative monoids

$$\begin{array}{ccc} \varprojlim_{a \mapsto a^p} A & \longrightarrow & A^\flat \\ (a_0, a_1, a_2, \ldots) & \longmapsto & (a_0 \bmod p, a_1 \bmod p, \ldots) \\ a_{n+1}^p = a_n & & a_{n+1}^p = a_n \bmod p \end{array}$$

is an isomorphism with inverse

$$\begin{array}{ccc} \varprojlim_{a \mapsto a^p} A & \longleftarrow & A^\flat \\ (\alpha^\sharp, (\alpha^{1/p})^\sharp, (\alpha^{1/p^2})^\sharp, \ldots) & \longleftarrow & \alpha \end{array}$$

by leveraging both that $A^\flat$ is perfect and $\sharp$ is multiplicative. It follows that, for $p$-complete $A$, the image of $\sharp\colon A^\flat \longrightarrow A$ is precisely the subset of elements $f \in A$

admitting a compatible system $\{f^{1/p^e}\}_{e\geq 0}$ of arbitrary $p$-power roots. Note that, if $p \in (\varpi) \subseteq A$ and $A$ is also $\varpi$-complete, the argument above similarly gives $A^\flat \cong \varprojlim_{a\mapsto a^p} A \cong \varprojlim_{a\mapsto a^p} A/(\varpi)$.

As the definition of a perfectoid ring we shall present makes use of them, we now recall the (classical) construction and basic properties of $p$-typical Witt vectors. For an arbitrary ring $A$, the ring of Witt vectors $W(A) = \{(a_0, a_1, a_2, \ldots) \mid a_n \in A\}$ is the set of sequences of elements of $A$ together with an exotic addition and multiplication structure

$$\begin{aligned}(a_0,a_1,a_2,\ldots)+(b_0,b_1,b_2,\ldots) &= (a_0+b_0, a_1+b_1-\tfrac{(a_0+b_0)^p-a_0^p-b_0^p}{p},\ldots)\\ (a_0,a_1,a_2,\ldots)\cdot(b_0,b_1,b_2,\ldots) &= (a_0b_0, a_0^pb_1+a_1b_0^p+pa_1b_1,\ldots)\end{aligned}$$

determined by universal polynomials ensuring that for all $i \geq 0$ the mappings

$$\begin{aligned} w_i\colon W(A) &\longrightarrow A\\ (a_0,a_1,a_2,\ldots) &\longmapsto a_0^{p^i}+pa_1^{p^{i-1}}+p^2a_2^{p^{i-2}}+\cdots+p^ia_i\end{aligned}$$

are ring homomorphisms. As the formulas for addition and multiplication must work for any ring, it suffices to describe them over the (countably infinite) universal polynomial ring

$$\mathbb{Z}[X_0, Y_0, X_1, Y_1, \ldots]$$

over the integers. Writing

$$(X_0, X_1, X_2, \ldots) + (Y_0, Y_1, Y_2, \ldots) = (S_0, S_1, S_2, \ldots)$$

gives $X_0 + Y_0 = S_0$ as $w_0$ is additive. In order for $w_1$ to be additive as well, we then must have

$$(X_0^p + pX_1) + (Y_0^p + pY_1) = S_0^p + pS_1 = (X_0 + Y_0)^p + pS_1$$

and so

$$\begin{aligned} S_1 &= X_1 + Y_1 - \frac{(X_0+Y_0)^p - X_0^p - Y_0^p}{p}\\ &= X_1 + Y_1 - \sum_{i=1}^{p-1} \frac{(p-1)!}{(i!)((p-i)!)} X_0^i Y_0^{p-i} \in \mathbb{Z}[X_0, Y_0, X_1, Y_1].\end{aligned}$$

Continuing in this manner allows one to compute $S_n \in \mathbb{Z}[X_0, Y_0, X_1, Y_1, \ldots, X_n, Y_n]$ recursively (where one must argue the coefficients are necessarily integral along the way). Similarly, if

$$(X_0, X_1, X_2, \ldots) \cdot (Y_0, Y_1, Y_2, \ldots) = (T_0, T_1, T_2, \ldots)$$

we must have $T_0 = X_0Y_0$ as $w_0$ is multiplicative. To make $w_1$ multiplicative as well, we then must have

$$(X_0^p + pX_1) \cdot (Y_0^p + pY_1) = T_0^p + pT_1 = X_0^pY_0^p + pT_1$$

and so

$$T_1 = X_0^pY_1 + X_1Y_0^p + pX_1Y_1 \in \mathbb{Z}[X_0, Y_0, X_1, Y_1].$$

Note that, when deriving properties of the formulas for the ring structure of $W(A)$, it suffices to do so assuming $A$ is $p$-torsion free as we have here. In this case, the ghost map

$$\begin{aligned} w\colon W(A) &\longrightarrow \textstyle\prod_{\mathbb{Z}_{\geq 0}} A \\ \alpha &\longmapsto (w_0(\alpha), w_1(\alpha), w_2(\alpha), \cdots) \\ (a_0, a_1, a_2, \ldots) &\mapsto (a_0, a_0^p + pa_1, a_0^{p^2} + pa_1^p + p^2 a_2, \ldots) \end{aligned}$$

is an injection of rings; if moreover $p$ is a unit, the ghost map is an isomorphism. In contrast, we will largely be concerned with the Witt rings $W(A)$ where $A$ is characteristic $p$ (and moreoever perfect).

The construction of the Witt ring $W(A)$ (and similarly for the truncation rings $W_i(A)$ given by considering only the first $i$ components), together with the surjective ring homomorphism $\pi (= w_0)\colon W(A) \longrightarrow A$ such that

$$\pi((a_0, a_1, \ldots)) = a_0$$

is functorial in $A$. In particular, when $A$ has characteristic $p$, $W(A)$ has an induced Frobenius lift[2]

$$\begin{array}{ccc} W(A) & \xrightarrow[(a_0, a_1, a_2, \ldots) \mapsto (a_0^p, a_1^p, a_2^p, \ldots)]{F} & W(A) \\ \pi \downarrow & & \downarrow \pi \\ A & \xrightarrow{F} & A \end{array}$$

that is an isomorphism if $A$ is perfect. Note that the mappings above are well-defined and the above diagram[3] commutes even if $A$ is $p$-torsion free; however, in this situation, the horizontal arrows are merely maps of multiplicative monoids as the $p$-th power map is not additive. In contrast, the shifting map (denoted $V$ for the German Verschiebung)

$$\begin{aligned} V\colon W(A) &\longrightarrow W(A) \\ (a_0, a_1, a_2, \ldots) &\longmapsto (0, a_0, a_1, a_2, \ldots) \end{aligned}$$

is additive when $A$ has characteristic $p$ but not multiplicative. One checks (*e.g.* exploiting that $w$ is injective when $A$ is $p$-torsion free) that

$$V \circ F = F \circ V = \text{ (multiplication by) } p$$

and that $V$ is "$F$-linear" in the sense that $\alpha \cdot V(\beta) = V\left(F(\alpha) \cdot \beta\right)$ for all $\alpha, \beta \in W(A)$. In particular, we see that $p = VF(1) = (0, 1, 0, 0, \ldots) \in W(A)$, and similarly $p^n = (\underbrace{0, \ldots, 0}_{n}, 1, 0, \ldots) \in W(A)$ for $n \in \mathbb{Z}_{>0}$.

The ring of Witt vectors $W(A)$ over a perfect ring $A$ (necessarily of characteristic $p$) has particularly nice properties that characterize it completely. In this case, $W(A)$ is the unique $p$-complete and $p$-torsion free ring endowed with the surjection

[2]More generally, even when $A$ does not have characteristic $p$, the Witt ring $W(A)$ always has a lift of Frobenius modulo $p$. However, it is not given by raising the individual entries to $p$-th powers as shown here. While we will not make use of this fact, it is theoretically quite useful and can be used to construct the Witt ring functor as an adjoint. In other words, $W(A)$ is the largest $\delta$-ring admitting a ring map to $A$.

[3]When $A$ does not have characteristic $p$, the letter $F$ is generally reserved for the lift of Frobenius on $W(A)$ and not the $p$-th power map on components.

$\pi\colon W(A) \longrightarrow A$ where $\ker \pi = (p)$, *i.e.* with a given isomorphism $W(A)/(p) \cong A$. The key to showing uniqueness is to exploit properties of the Teichmüller map

$$[-]\colon A \cong A^\flat \cong W(A)^\flat \xrightarrow[\quad a \longmapsto [a] := (a,0,0,0,\ldots)\quad]{\sharp} W(A)\ ,$$

noting that any $p$-complete ring that yields $A$ upon killing $p$ will have a similar mapping (as $\sharp$ can be defined in this setting). By what we have seen above concerning $\sharp$, we have that $[rs] = [r][s]$ is multiplicative and the image of $[-]$ is precisely the set of elements of $W(A)$ admitting a compatible system of arbitrary $p$-power roots. In general, one has

$$[r]\cdot(s_0, s_1, s_2, \ldots) = (rs_0, r^p s_1, r^{p^2} s_2, \ldots).$$

and clearly $\pi([r]) = r$, so that $[-]$ is injective as well. Using the Teichmüller map and exploiting that $A$ is perfect, we can rewrite a Witt vector uniquely

$$(a_0, a_1, a_2, \ldots) = \sum_{n=0}^{\infty} V^n([a_n]) = \sum_{n=0}^{\infty} V^n F^n([a_n^{1/p^n}]) = \sum_{n=0}^{\infty} [a_n^{1/p^n}] p^n$$

as a power series in $p$, giving what are called the Teichmüller coordinates on the Witt ring. Note in particular that it follows easily from the above expansion that $\alpha = (a_0, a_1, \ldots) \in W(A)$ is then a unit if and only if $\pi(\alpha) = a_0 \in A$ is a unit.

DEFINITION 2.1. Suppose that $B$ is a perfect ring of characteristic $p$. We say that an element

$$d = (d_0, d_1, d_2, \ldots) \in W(B)$$

is *distinguished* if $B$ is $d_0$-complete and $d_1 \in B^\times$ is a unit. A ring $A$ is *perfectoid* if it has the form $A \cong W(B)/(d)$ for some perfect ring $B$ and a distinguished element $d \in W(B)$.

The first observation to be made is that, if $A \cong W(B)/(d)$ is a perfectoid ring, the perfect ring $B$ appearing in the presentation is necessarily isomorphic to $A^\flat$ and thus entirely determined by $A$ itself. To see this, note first that since $d = (d_0, d_1, \ldots) \in W(B)$ is distinguished, we may write

$$d = [d_0] + [d_1^{1/p}]p + [d_2^{1/p^2}]p^2 + \cdots = [d_0] + p\left([d_1^{1/p}] + [d_2^{1/p^2}]p + \cdots\right) = [d_0] + p\tilde{u}$$

where $\tilde{u} \equiv [d_1^{1/p}] \bmod p$ and so $\tilde{u} \in W(B)^\times$ is a unit. Thus, as $B$ is perfect,

$$A/(p) = W(B)/(d,p) = B/(d_0)$$

is a semi-perfect ring and we have

$$A^\flat = (A/(p))^\flat = (B/(d_0))^\flat = B^{\wedge d_0} = B$$

as $B$ is $d_0$-complete.

The second observation is that even the presentation of a perfectoid ring $A \cong W(A^\flat)/(d)$ as a quotient is entirely determined by $A$. Applying the sharp map $\sharp\colon A^\flat \longrightarrow A$ on Teichmüller coordinates induces a ring homomorphism

(2)

$$\theta\colon W(A^\flat) \xrightarrow[\quad (a_0, a_1, \ldots) = [a_0] + [a_1^{1/p}]p + \cdots \longmapsto a_0^\sharp + (a_1^{1/p})^\sharp p + \cdots\quad]{} A$$

called Fontaine's map, where we have leveraged again that $A$ is $p$-complete to make use of the sharp map. Let us first convince ourselves that this is indeed a map of rings. To check that it is, it suffices to verify it is additive and multiplicative modulo $p^{n+1}$ for each $n$. Note first that for $\alpha = (a_0, a_1, \ldots) \in W(A^\flat)$ we have

$$\begin{aligned} \theta(\alpha) &= \left((a_0^{1/p^n})^\sharp\right)^{p^n} + p\left((a_1^{1/p^n})^\sharp\right)^{p^{n-1}} + \cdots + p^n\left((a_1^{1/p^n})^\sharp\right) && \bmod p^{n+1} \\ &= w_n((a_0^{1/p^n})^\sharp, (a_1^{1/p^n})^\sharp, \ldots, (a_n^{1/p^n})^\sharp) && \bmod p^{n+1} \end{aligned}.$$

Now, appealing to the defining property of addition and multiplication of Witt vectors, we know that

$$\begin{array}{ccccc} W(A) & \xrightarrow{w_n} & A & \twoheadrightarrow & A/(p^{n+1}) \\ (x_0, x_1, x_2, \ldots) & \longrightarrow & & & x_0^{p^n} + px_1^{p^{n-1}} + \cdots p^n x_n \bmod p^{n+1} \end{array}$$

is a ring homomorphism. Moreover, since $x = x' \bmod p$ implies $x^{p^m} = (x')^{p^m} \bmod p^{m+1}$ for $x, x' \in A$ and $m \geq 0$, it is clear that this map only depends on the images of the $x_i$'s in $A/(p)$. In other words, there is an induced factorization

$$\begin{array}{ccc} (x_0, x_1, x_2, \ldots) & \qquad & W(A) \xrightarrow{w_n} A \\ \downarrow & & \downarrow \qquad\qquad \downarrow \\ (x_0 \bmod p, x_1 \bmod p, x_2 \bmod p, \ldots) & & W(A/(p)) \xrightarrow{\overline{w_n}} A/(p^{n+1}) \end{array}$$

through a ring map $\overline{w_n}\colon W(A/(p)) \longrightarrow A/(p^{n+1})$ with

$$\overline{w_n}(x_0 \bmod p, x_1 \bmod p, \ldots) = w_n(x_0, x_1, \ldots) \bmod p^{n+1}.$$

Thus, we may conclude

$$\begin{array}{ccccccc} W(A^\flat) & \xrightarrow{W(F^{-n})} & W(A^\flat) & \xrightarrow{W(\mathrm{pr}_0)} & W(A/(p)) & \xrightarrow{\overline{w_n}} & A/(p^{n+1}) \\ \alpha = (a_i) & \longmapsto & (a_i^{1/p^n}) & \longmapsto & (\mathrm{pr}_0(a_i^{1/p^n})) = ((a_i^{1/p^n})^\sharp \bmod p) & \longmapsto & \theta(\alpha) \bmod p^{n+1} \end{array}$$

is both additive and multiplicative as claimed, using that the Witt ring construction is functorial and thus the inverse $n$-iterated Frobenius $F^{-n}\colon A^\flat \longrightarrow A^\flat$ and projection $\mathrm{pr}_0\colon A^\flat \longrightarrow A/(p)$ give rise to corresponding ring maps $W(F^{-n})$ and $W(\mathrm{pr}_0)$ on Witt vectors, respectively. Fontaine's $\theta$ map yields a commutative diagram

$$\begin{array}{ccc} W(A^\flat) & \xrightarrow{\theta} & A \\ {\scriptstyle\pi}\downarrow & & \downarrow \\ A^\flat & \xrightarrow[\mathrm{pr}_0]{} & A/(p) \end{array}$$

and it is straightforward to see that $\theta$ is surjective if and only if $A/(p)$ is semi-perfect. In particular, this necessarily holds when $A$ is perfectoid. As $\theta$ is $W(A^\flat)$-linear, the kernel of $\theta$ is naturally identified with $(d)$, and thus the presentation of a perfectoid ring as $A \cong W(A^\flat)/(d)$ is essentially entirely determined by $A$ itself.

In the perfectoid ring $A \cong W(A^\flat)/(d)$ for $d = [d_0] + p\tilde{u}$ as above, setting $\varpi = \theta([d_0^{1/p}])$ and $u = -\theta(\tilde{u}^{-1}) \in A^\times$ gives $p = u\varpi^p$ in $A$. In other words, we have in $A$ that $\varpi$ is a $p$-th root of $p$ up to multiplication by a unit. This process can in turn be reversed, and gives a way to identify perfectoid rings in the wild.

ALTERNATE DEFINITION 2.2. A ring $A$ is perfectoid if and only if:

(i) $A$ is $p$-adically complete;
(ii) $A/(p)$ is semi-perfect, *i.e.* the Frobenius $A/(p) \xrightarrow{F} A/(p)$ is surjective;
(iii) the kernel of Fontaine's map $\theta\colon W(A^\flat) \longrightarrow A$ is principal;
(iv) there is a $p$-th root of $p$ in $A$ up to a unit, *i.e.* there is some $\varpi \in A$ and $u \in A^\times$ with $\varpi^p = up$.

Note that, condition (ii) implies that $\theta\colon W(A^\flat) \longrightarrow A$ is surjective. Since elements in the image of $\theta$ have the form

$$a_0^\sharp + (a_1^{1/p})^\sharp p + (a_2^{1/p^2})^\sharp p^2 + \cdots = a_0^\sharp + p\left((a_1^{1/p})^\sharp + (a_2^{1/p^2})^\sharp p + \cdots\right)$$

where $a_0, a_1, a_2, \ldots \in A^\flat$ by (2), it follows that every element of $A$ is necessarily in the image of the sharp map $\sharp\colon A^\flat \longrightarrow A$ up to a multiple of $p$. In particular, replacing $\varpi$ by a unit multiple in (iv), one may guarantee there is some $\varpi^\flat \in A^\flat$ with $(\varpi^\flat)^\sharp = \varpi$ or equivalently that $\varpi$ has a compatible system $\{\varpi^{1/p^e}\}_{e\geq 0}$ of arbitrary $p$-power roots in $A$. In the explicit formulation $A \cong W(A^\flat)/(d)$ for $d = [d_0] + p\tilde{u}$ as above with $\varpi = \theta([d_0^{1/p}]) = (d_0^{1/p})^\sharp$, we have $\varpi^\flat = d_0^{1/p}$ and $\varpi^{1/p^e} = ((d_0^{1/p^{e+1}})^\sharp)$. Conversely, to see the equivalence with our initial definition of a perfectoid ring, suppose that $A$ satisfies conditions (i)-(iv) above with $\varpi = \theta([d_0^{1/p}]) = (d_0^{1/p})^\sharp$ and $\theta(\tilde{u}) = -u$ for $d_0 \in A^\flat$ and $\tilde{u} \in W(A^\flat)$. By (iii), suppose that $e = (e_0, e_1, \ldots) \in W(A^\flat)$ generates the kernel of $\theta$. Since $\mathrm{pr}_0(e_0) = \mathrm{pr}_0(\pi(e)) = \theta(e) \bmod p$, we see $e_0$ is in the kernel of $\mathrm{pr}_0$ and so $A^\flat$ is $e_0$-complete and separated. Also, since $\varpi^p = up$ in $A$, we have that $\theta([d_0] + \tilde{u}p) = 0$ and so

$$[d_0] + \tilde{u}p = (d_0, \tilde{u}_0^p, \tilde{u}_1^p, \ldots) = (e_0 f_0, e_0^p f_1 + e_1 f_0^p, \ldots) = ef$$

for some $f \in W(A^\flat)$. Since $u \in A$ is a unit, from $\tilde{u}_0^p = e_0^p f_1 + e_1 f_0^p$ we see that $\mathrm{pr}_0(e_1 f_0^p) = \mathrm{pr}_0(\tilde{u}_0^p - e_0^p f_1) = (\mathrm{pr}_0(\pi(\tilde{u})))^p = (-u)^p \bmod p$ is a unit as well, and so it follows $e_1 \in A^\flat$ is a unit, $e$ is a distinguished element, and thus $A$ is perfectoid by the initial definition. In fact, this argument shows also that $f_0 \in A^\flat$ and $f \in W(A^\flat)$ are also units, so $d = [d_0] + p\tilde{u} = ef$ is an alternate choice of generator for the kernel of $\theta$.

With the perspective of the alternative definition in mind, let us return briefly to the two simpler cases of perfectoid rings from the very beginning of this section.

(1) A ring $A$ of characteristic $p$ is perfectoid if and only if $A$ is perfect. Indeed, one direction of the implication is immediate: if $A$ is perfect, then $p \in W(A^\flat)$ is a distinguished element and $W(A^\flat)/(p) = A^\flat = A$ is perfectoid. An easy way to see the reverse direction is to use that a perfectoid ring is necessarily reduced, which we will give an argument for below in Corollary 2.4.
(2) [**BMS15**, Section 3] A $p$-torsion free ring $A$ is perfectoid if and only if it is $p$-adically complete, there exist $\varpi \in A$ and $u \in A^\times$ with $p = u\varpi^p$, and $A/(\varpi) \xrightarrow{x \mapsto x^p} A/(p)$ is an isomorphism. Indeed, supposing these properties hold, we may ensure $\varpi$ has a compatible system of arbitary $p$-power roots by replacing it with a unit multiple. This gives $d_0 \in A^\flat$ with $\theta([d_0]) = d_0^\sharp = \varpi^p$; setting $\varpi^\flat = d_0^{1/p}$ and leveraging that $A/(\varpi) \xrightarrow{x \mapsto x^p} A/(p)$ is an isomorphism gives that $d_0 = (\varpi^\flat)^p$ generates the kernel of $\mathrm{pr}_0\colon A^\flat \longrightarrow A/(p)$. Since $\theta$ is surjective as $A$ is semi-perfect, we may also find a $\tilde{u} \in W(A^\flat)$ with $\theta(\tilde{u}) = -u^{-1}$. One then checks $\tilde{u}$ is a unit, and

so $d = [d_0] + \tilde{u}p \in W(A^\flat)$ is a distinguished element of the kernel of $\theta$. If $e \in W(A^\flat)$ is an arbitrary element of the kernel, $\pi(e) \in A^\flat$ maps to zero in $A/(p)$ and so is divisible by $d_0$. Thus, we can find $w, e' \in W(A^\flat)$ with $e - wd = pe'$. As $A$ is $p$-torsion free, we must have $e' \in \ker\theta$ and iterating this process gives that $d$ generates the kernel of $\theta$. Thus, $A = W(A^\flat)/(d)$ is perfectoid.

Interpolating between the two cases above involves an understanding of the $p$-torsion[4] $A[p] := \operatorname{Ann}_A p$ of an arbitrary perfectoid ring $A$, and is the key to the proof below.

PROPOSITION 2.3. *Suppose $A = W(A^\flat)/(d)$ is a perfectoid ring.*

- *$\overline{A} = A/A[p]$ is perfectoid and $p$-torsion free.*
- *$A/\sqrt{(p)}$ and also $\overline{A}/\sqrt{(p)}$ are perfect rings of characteristic $p$, hence also perfectoid.*
- *$A$ can be interpreted as the gluing of $\overline{A}$ and $A/\sqrt{(p)}$ along the common subscheme $\overline{A}/\sqrt{(p)}$. In other words, we have $A = \overline{A} \times_{\overline{A}/\sqrt{(p)}} A/\sqrt{(p)}$, and there is a short exact sequence*

$$0 \longrightarrow A \longrightarrow A/A[p] \oplus A/\sqrt{(p)} \longrightarrow \overline{A}/\sqrt{(p)} \longrightarrow 0.$$

PROOF. Set $\varpi = \theta([d_0^{1/p}]) = (d_0^{1/p})^\sharp$ as usual, so that $p = u\varpi^p$ with $u \in A^\times$ and the kernel of Frobenius on $A/(p)$ generated by $\varpi$. One verifies easily that $\sqrt{(p)} = (\varpi^{1/p^\infty})$, and so in particular $A/\sqrt{(p)}$ is perfect.

Next, note that $p$ is a non-zero divisor on $W(A^\flat)$, and we claim that $d = [d_0] + \tilde{u}p$ is as well. Suppose $e = (e_0, e_1, \ldots) \in W(A^\flat)$ with $de = 0$. Since $[d_0]e = -\tilde{u}pe$, we have that $[d_0]^i e \in (p^i)$ for $i \geq 1$, and so we see that the first $i - 1$ entries of

$$[d_0]^i e = (d_0^i e_0, d_0^{ip} e_1, d_0^{ip^2} e_2, \ldots)$$

must be zero. In particular, $e_i \in A^\flat[d_0^{(i+1)p^{i+1}}]$. Since $A^\flat$ is perfect however, it follows that $A^\flat[d_0] = A^\flat[d_0^m]$ for all $m \geq 1$, and so we have $e_i \in A^\flat[d_0] = A^\flat[d_0^{p^i}]$. It follows that $[d_0]e = (d_0 e_0, d_0^p e_1, \ldots) = 0$, hence $-\tilde{u}pe = 0$, and so $e = 0$ as $\tilde{u} \in W(A^\flat)^\times$.

Now, since we have that $p \in W(A^\flat)$ is a non-zero divisor, it follows that $\operatorname{Tor}_1^{W(A^\flat)}(A^\flat, A) = A[p]$ by using $W(A^\flat) \xrightarrow{p} W(A^\flat)$ to resolve the first entry. On the other hand, since $d \in W(A^\flat)$ is a non-zero divisor, we also see that $\operatorname{Tor}_1^{W(A^\flat)}(A^\flat, A) = A^\flat[d_0]$ by using $W(A^\flat) \xrightarrow{d} W(A^\flat)$ to resolve the second entry. Moreover, as $\operatorname{Tor}_1^{W(A^\flat)}(A^\flat, A)$ can also be computed by tensoring resolutions of both entries together, we therefore have isomorphisms

$$\begin{array}{ccc} & \dfrac{\{(\alpha,\beta) \in W(A^\flat) \oplus W(A^\flat) \mid d\alpha = p\beta\}}{\{(d\gamma, p\gamma) \mid \gamma \in W(A^\flat)\}} & \\ {}^{(\alpha,\beta)\mapsto\pi(\alpha)}\swarrow & & \searrow^{(\alpha,\beta)\mapsto\theta(\beta)} \\ A^\flat[d_0] & \xrightarrow[x \mapsto \theta(\tilde{u})^{-1} x^\sharp]{} & A[p]. \end{array}$$

[4]In what follows, we will repeatedly make use of the notation $R[d] := \operatorname{Ann}_d R$ to denote the kernel of multiplication by an element $d$ in a ring $R$.

In other words, every element of $a \in A[p]$ has the form $u'x^\sharp$ for some $x \in A^\flat[d_0]$. As a consequence, since $A^\flat[d_0] = A^\flat[d_0^{1/p}] = A^\flat[d_0^{1/p^m}] = A[d_0^{p^m}]$ for $m \geq 0$ as $A^\flat$ is perfect, we see that that $A[p] = A[\varpi] = A[\varpi^{1/p^m}] = A[\varpi^{p^m}]$ for all $m \geq 0$ where we have set $\varpi = \theta([d_0^{1/p}]) = (d_0^{1/p})^\sharp$ as usual. In particular, we also have $A[p] = A[p^\ell] = A[\varpi^{p\ell}]$ for all $\ell \geq 1$.

As $A[p] = A[p^2]$, it follows easily that $\overline{A} = A/A[p]$ is $p$-torsion free. To check that it is perfectoid, let us use the criteria from (2) above. Since $A$ is $p$-complete, the natural map $\overline{A} \longrightarrow \overline{A}^{\wedge p}$ is surjective, and so $A$ is $p$-complete provided it is $p$-separated. Suppose $a \in \bigcap_{m\geq 1}(A[p], p^m)$ in $A$. Then $pa \in \bigcap_{m \geq 1}(p^{m+1}) = 0$, so we see $a \in A[p]$ and $\overline{A}$ is $p$-complete as desired. Now, denoting by overline the passage to the quotient $A \longrightarrow \overline{A}$, the equation $p = u\varpi^p$ becomes $p = \overline{p} = \overline{u}\,\overline{\varpi}^p$ and we still have $\overline{u} \in \overline{A}^\times$. Thus, to see $\overline{A}$ is perfectoid, it suffices to check that the kernel of Frobenius on $\overline{A}/(p)$ is generated by $\overline{\varpi}$. To see this, suppose we have $a \in A$ with $a^p \in (A[p], p)$. Write $a^p = u'x^\sharp + vp$ for some $u' \in A^\times$ and $v \in A$. We have $u' = \pi(u)^\sharp$ mod $p$, we can modify our equation and write $a^p = (u_0x)^\sharp + v'p$ with $u_0 = \pi(u') \in (A^\flat)^\times$. Observe that $(u_0^{1/p}x^{1/p})^\sharp \in A[p]$ once more, and so we have $(a - (u_0^{1/p}x^{1/p})^\sharp)^p = v'p = 0 \bmod p$. It follows that $a - (u_0^{1/p}x^{1/p})^\sharp \in (\varpi)$, and hence the kernel of Frobenius on $\overline{A}/(p)$ is generated by $\overline{\varpi}$ and $\overline{A}$ is perfectoid as claimed.

Finally, if $a \in A[p] \cap (\varpi^{1/p^\infty})$, then we can write $a = b\varpi^{1/p^N}$ for some $N$ and $b \in A$. As $0 = pa = \varpi^p a = (\varpi)^{p+1/p^N}b$, we have $b \in A[\varpi^{p+1}] = A[p] = A[\varpi^{1/p^N}]$ whence $a = b\varpi^{1/p^N} = 0$. Thus, $A[p] \cap (\varpi^{1/p^\infty}) = 0$ and so there is a short exact sequence

$$0 \longrightarrow A \longrightarrow A/A[p] \oplus A/(\varpi^{1/p^\infty}) \longrightarrow A/(A[p], \varpi^{1/p^\infty}) \longrightarrow 0$$

observing that $A/(A[p], \varpi^{1/p^\infty}) = \overline{A}/(\overline{\varpi}^{1/p^\infty}) = \overline{A}/\sqrt{(p)}$ is also perfect. □

COROLLARY 2.4. *Perfectoid rings are reduced.*

PROOF. By the previous proposition, an arbitrary perfectoid ring is a subring of a product of a perfect ring of characteristic $p$ and a $p$-torsion free perfectoid ring. As perfect rings are reduced, it suffices to check simply that a $p$-torsion free perfectoid ring is reduced.

To that end, suppose that $A = W(A^\flat)/(d)$ is perfectoid and $p$-torsion free. Write $p = u\varpi^p$ for $u \in A^\times$ and $\varpi = \theta([d_0^{1/p}])$. If $a^p = 0$ in $A$, then we must have $a = \varpi a'$ for some $a' \in A$ since $\varpi$ generates the kernel of Frobenius on $A/(p)$. We have $0 = ua^p = p(a')^p$ and so $(a')^p = 0$ as $A$ is $p$-torsion free. Iterating this argument gives $a \in \bigcap_{m\geq 1}(\varpi^m) = 0$ and so $A$ is reduced as desired. □

We are now in a position to extend the characterization of regular rings from Theorem 1.2 to mixed characteristic.

THEOREM 2.5 ([**BIM19**]). *If $R$ is a Noetherian ring and $p$ belongs to the Jacobson radical of $R$, then $R$ is regular if and only if there exists a faithfully flat extension $R \longrightarrow A$ with $A$ perfectoid.*

PROOF OF THEOREM 2.5. Suppose first that $R$ is regular. Since $R$ is Noetherian, an arbitrary product of flat $R$-modules is flat [**Sta**, Proposition 05CZ]. Thus, via standard reductions, we may assume without loss of generality that $(R, \mathfrak{m}, k)$

is a complete local ring with $p \in \mathfrak{m}$ and $k = \overline{k}$ is an algebraically closed field of characteristic $p$. By the Cohen structure theorem, there are three cases to consider.

(a) If $R$ itself has characteristic $p$, then $R = k[[x_1, \ldots, x_d]]$.
(b) If $R$ has mixed characteristic and $p$ is part of a regular system of parameters (*i.e.* $R$ is unramified), then $R = W(k)[[x_2, \ldots, x_d]]$.
(c) Else if $R$ has mixed characteristic and $p \in \mathfrak{m}^2$ is not part of a regular system of parameters (*i.e.* $R$ is ramified), then $R = W(k)[[x_1, \ldots, x_d]]/(p - f)$ where $f \in (x_1, \ldots, x_d)^2$.

In all three cases, it is not hard to exhibit a faithfully flat map $\widehat{R} \longrightarrow A$ where $A$ is perfectoid in terms of the descriptions above. Explicitly, we may take the following construction in the respective situations.

(a) One may set $A = R_{\mathrm{perf}}$ as in Theorem 1.2.
(b) Taking $A = R[p^{1/p^\infty}, x_2^{1/p^\infty}, \ldots, x_d^{1/p^\infty}]^{\wedge p}$, we have seen above that $A$ is perfectoid. Let $A^{\mathrm{un}} = R[p^{1/p^\infty}, x_2^{1/p^\infty}, \ldots, x_d^{1/p^\infty}]$, which is a colimit of free $R$-modules and hence flat over $R$. Thus, $A^{\mathrm{un}}/pA^{\mathrm{un}}$ is flat over $R/(p)$. Moreover, as $A^{\mathrm{un}}$ is $p$-torsion free, we also have $\mathrm{Tor}_1^R(A^{\mathrm{un}}, R/(p)) = 0$. By [**Sta**, Lemma 0AGW], it follows that $A$ is faithfully flat over $R$ (alternatively, see [**Bha12**, Proposition 5.1]).
(c) Let $A$ be the $p$-adic completion of

$$W(k)[x_1^{1/p^\infty}, \ldots, x_d^{1/p^\infty}] \otimes_{W(k)[x_1, \ldots, x_d]} R.$$

Similarly as above, we have that $A$ is faithfully flat over $R$. In particular, $A$ remains $p$-torsion free. One can check directly that $A$ is perfectoid by verifying that there exist $\varpi \in A$ and $u \in A^\times$ with $p = u\varpi^p$, and that $A/(\varpi) \xrightarrow{x \mapsto x^p} A/(p)$ is an isomorphism. Indeed, considering $f = f(x_1, \ldots, x_d) \in W(k)[[x_1, \ldots, x_d]]$ as a power series, have that $p \nmid f$ and $f$ has no constant term. Given that the Frobenius lift $F \colon W(k) \longrightarrow W(k)$ is an isomorphism, let $F^{-1}f$ be the power series given by applying $F^{-1}$ to each of the coefficients of $f$, and set $\varpi = F^{-1}(f)(x_1^{1/p}, \ldots, x_d^{1/p}) \in W(k)[[x_1^{1/p}, \ldots, x_d^{1/p}]]$. It follows that $\varpi^p = f + ph$ for some $h \in W(k)[[x_1^{1/p}, \ldots, x_d^{1/p}]]$. As $f$ has no constant term, neither do $\varpi$ or $h$, whence in $A$ we have $\varpi^p = f + ph = p + ph = p(1+h)$ with $(1+h) \in A^\times$. Setting $u = (1+h)^{-1}$ gives $p = u\varpi^p$ in $A$ as claimed. Moreover, as $A/(p) = k[[x_1, \ldots, x_d]]_{\mathrm{perf}}/(f \bmod p)$ and $(\varpi \bmod p)^p = f \bmod p$ in the perfect ring $k[[x_1, \ldots, x_d]]_{\mathrm{perf}}$, it follows that $A/(\varpi) \xrightarrow{x \mapsto x^p} A/(p)$ is an isomorphism.

For the reverse implication, assume that $R \longrightarrow A$ is faithfully flat with $A$ perfectoid. Let $\mathfrak{m} = (p, f_2, \ldots, f_n)$ be an arbitary maximal ideal of $R$ with $k = R/\mathfrak{m}$. Leveraging the observations regarding the flat dimension of ideals generated by the arbitary $p$-power roots of finitely many elements once more, we have the following claim.

CLAIM 2.6. If $I_j = \sqrt{(p, f_2, \ldots, f_j)A}$ for each $j = 1, \ldots, n$, then the flat dimension of $A/I_j$ is at most $j$.

PROOF OF CLAIM 2.6. We have seen that $I_1 = \sqrt{(p)A} = (\varpi^{1/p^\infty})$ is a flat ideal as $A$ is reduced, so the claim is valid for $j = 1$. Set $\overline{A} = A/(\varpi^{1/p^\infty})$, which is a perfect ring of characteristic $p$. If $\overline{f_j}$ denotes the image of $f_j$ in $\overline{A}$, we have $A/I_j = \overline{A}/(\overline{f_2}^{1/p^\infty}, \ldots, \overline{f_j}^{1/p^\infty})$. We proceed to verify the claim by induction[5] on $j$. Assume that $j \geq 2$ and $M$ is an $A$-module. Consider the exact sequence

$$\mathrm{Tor}_i^A(M, A/I_{j-1}) \longrightarrow \mathrm{Tor}_i^A(M, A/I_j) \longrightarrow \mathrm{Tor}_{i-1}^A(M, I_j/I_{j-1}).$$

By the induction assumption, the left entry is zero for $i \geq j$. Since

$$\begin{aligned} I_j/I_{j-1} &= (\overline{f_j}^{1/p^\infty}) \cdot A/I_{j-1} \\ &= \varinjlim \left( A/I_{j-1} \xrightarrow{\overline{f_j}^{1-1/p}} A/I_{j-1} \xrightarrow{\overline{f_j}^{1/p-1/p^2}} A/I_{j-1} \longrightarrow \cdots \right), \end{aligned}$$

as $A/I_{j-1}$ is perfect and hence reduced, the right entry is zero for $i - 1 \geq j$ (as Tor is compatible with direct limits). Thus, it follows that the middle entry $\mathrm{Tor}_i^A(M, A/I_j) = 0$ for $i \geq j+1$ as desired. □

To conclude, we will modify the earlier argument from the proof of Theorem 1.2. It follows from Claim 2.6 that

$$0 = \mathrm{Tor}_{n+1}^A(A/\mathfrak{m}A, A/\sqrt{\mathfrak{m}A}) = \mathrm{Tor}_{n+1}^R(k, A/\sqrt{\mathfrak{m}A})$$

where the final equality uses that $R \longrightarrow A$ is flat (*e.g.* base change of a free resolution of $k$ over $R$ gives a free resolution of $k \otimes_R A = A/\mathfrak{m}A$ over $A$). On the other hand, when viewed as an $R$-module $A/\sqrt{\mathfrak{m}A}$ is annihilated by $\mathfrak{m}$ and hence is a vector space over $k$ – and necessarily non-zero as $A$ is faithfully flat over $R$. Writing $A/\sqrt{\mathfrak{m}A} \cong k^{\oplus \mathscr{I}}$ for an index set $\mathscr{I}$, we have

$$0 = \mathrm{Tor}_{n+1}^R(k, A/\sqrt{\mathfrak{m}A}) = \mathrm{Tor}_{n+1}^R(k, k^{\oplus \mathscr{I}}) = \mathrm{Tor}_{n+1}^R(k, k)^{\oplus \mathscr{I}}$$

and thus $\mathrm{Tor}_{n+1}^R(k,k) = \mathrm{Tor}_{n+1}^{R_\mathfrak{m}}(k,k) = 0$ so that $R_\mathfrak{m}$ is necessarily regular by the theorem of Auslander-Buchsbaum-Serre [**Eis95**, Corollary 19.5 and Theorem 19.12]. Varying over all of the maximal ideals $\mathfrak{m}$ of $R$, we have that $R$ is regular. □

We will see more examples of perfectoid rings in subsequent sections. However, to end this section, let us highlight an important result of Bhatt and Scholze on perfectoidization.

THEOREM 2.7 ([**BS22**]). *Suppose that $R \longrightarrow S$ is a finitely presented finite map such that $R$ is perfectoid. Then there exists a unique perfectoid $S$-algebra $S_{\mathrm{perfd}}$ through which all others must factor. In other words, if $S \longrightarrow A$ is an arbitrary ring map where $A$ is perfectoid, there is a canonical factorization $S \longrightarrow S_{\mathrm{perfd}} \longrightarrow A$.*

Put another way, the ring $S_{\mathrm{perfd}}$ is the smallest perfectoid $S$-algebra in the sense that it is initial in the category of perfectoid $S$-algebras. As the notation suggests, while the ring $R$ was necessary to guarantee the existence of $S_{\mathrm{perfd}}$, in fact $S_{\mathrm{perfd}}$ is independent of the initial ring $R$. Here are some additional facts to keep in mind about such perfectoidizations.

[5]Alternatively, as we know the flat dimension of $A/I_j$ over $\overline{A}$ is at most $j-1$ from the proof of Theorem 1.2, appealing to the (base-change for Tor) spectral sequence (see [**Sta**, Section 061Y, Example 15.62.2] or [**Wei94**, Theorem 5.6.6])

$$E^2_{p,q} = \mathrm{Tor}_p^{\overline{A}}(\mathrm{Tor}_q^A(M, \overline{A}), A/I_j) \Rightarrow \mathrm{Tor}_{p+q}^A(M, A/I_j)$$

for an $A$-module $M$ gives that $\mathrm{Tor}_i^A(M, A/I_j) = 0$ for $i \geq j+1$, *i.e.* the flat dimension of $A/I_j$ is at most $j$ over $A$.

- If $S$ has characteristic $p > 0$, then $S = S_{\mathrm{perf}}$. In other words, perfectoidization should be seen as a generalization of the notion of perfection to the mixed characteristic setting.
- If $R \twoheadrightarrow S$ is surjective, than so too is $R = R_{\mathrm{perfd}} \twoheadrightarrow S_{\mathrm{perfd}}$. As a particular example, if $S = R/(g_1, \ldots, g_n)$ and each $g_i$ admits a compatible system of $p$-power roots $\{g_i^{1/p^e}\}_{e \in \mathbb{Z}_{\geq 0}}$ in $R$, then $S_{\mathrm{perfd}} = (S/(g_1^{1/p^\infty}, \ldots, g_n^{1/p^\infty}))^{\wedge p}$.
- If $S$ is $p$-torsion free, then $S_{\mathrm{perfd}}$ is $p$-torsion free as well; see [**MST+22**].
- More generally, without such a presentation in terms of a perfectoid ring $R$, Bhatt and Scholze formulate a notion of the perfectoidization $S_{\mathrm{perfd}}$ in the derived setting.

## 3. Perfectoid BCM-algebras and the direct summand theorem

In recent years, there have been phenomenal advances in commutative algebra in mixed characteristic stemming from the introduction of perfectoid methods. In 2018, Peter Scholze was awarded the Field's Medal for "transforming arithmetic algebraic geometry over p-adic fields through his introduction of perfectoid spaces." Building on these techniques, Yves André was able to give an affirmative answer to the decades-old Direct Summand Conjecture of Hochster and show the existence of weakly functorial big Cohen-Macaulay algebras. Further seminal advances in this direction have been made by Bhargav Bhatt, particularly with regards to absolute integral closure. The goal of this section is to highlight and overview these advances.

Our starting point is the following result of André.

THEOREM 3.1 ([**And20**]). *Every Noetherian local ring $(R, \mathfrak{m}, k)$ with $p \in \mathfrak{m}$ admits a BCM $R$-algebra $B$.*

Here, the term BCM $R$-algebra is shorthand for a balanced big Cohen-Macaulay $R$-algebra, meaning that every system of parameters of $R$ is a regular sequence on $B$ and $B \neq \mathfrak{m}B$. The word big is used as a warning that $B$ is not assumed to be a finitely generated $R$-module, in which case extra care is needed when working with the Cohen-Macaulay condition. Given a (big) $R$-module $M$ with $M \neq \mathfrak{m}M$, one has the following implications (see [**BH93**, Section 8.5] and [**BMP+23**, Section 2.2]):

all systems of parameters are a regular sequence on $M$

$\Downarrow$

some system of parameters is a regular sequence on $M$

$\Downarrow$

$H^i_{\mathfrak{m}}(M) = 0$ for all $i < \dim(R)$.

(and, if $M = \widehat{M}$, the last condition implies the first.)

We will sketch a proof of Theorem 3.1 in mixed characteristic shortly, and will subsequently discuss a number of refinements. However, note that there are fundamentally important BCM $R$-algebras that are distinct from the constructions to be presented here. Recall that, given a domain $R$, its absolute integral closure $R^+$ is the integral closure of $R$ inside an algebraic closure of its fraction field. When $R$ is a complete local domain of equal characteristic $p > 0$, Hochster and Huneke [**HH92**] showed that one can take $B$ to be $R^+$ in Theorem 3.1 above; in mixed characteristic, Bhatt has analagously proven that we can take $B$ to be the $p$-adic completion of $R^+$ [**Bha20**]. The reader should also verify here that $(R^+)^{\wedge p}$ is in fact perfectoid, checking that the kernel of Frobenius modulo $p$ is generated by (a choice of) $p^{1/p}$.

In earlier sections, we have observed repeatedly the importance of exploiting elements with compatible systems of arbitary $p$-power roots. To that end, the crucial ingredient needed for the proof of Theorem 3.1 allows one to adjoin such roots to a perfectoid ring in a controlled manner.

THEOREM 3.2 (André's Flatness Lemma). [**And18, Bha18, Gab18, CS24**] *Suppose $A$ is perfectoid and $g \in A$. Then there exists a perfectoid $A$-algebra $A'$ where $A \longrightarrow A'$ is faithfully flat modulo $p^n$ for all $n \geq 1$ and so that $g$ has a compatible system $\{g^{1/p^e}\}_{e \in \mathbb{Z}_{\geq 0}}$ of arbitrary $p$-power roots in $A'$. Moreover, if $A$ is $p$-torsion free, then $A'$ is $p$-torsion free as well.*

We will assume this important result, which has also been fundamental for some recent advances in $p$-adic Hodge theory. The original proof given in [**And18**] (see also [**Bha18**]) relies on the geometry of perfectoid spaces, an alternative proof was given in [**BS22**] using prismatic cohomology, an elementary proof that only uses basic knowledge of perfectoid algebra was given in [**CS24**].

A SKETCH OF THE PROOF OF THEOREM 3.1 IN MIXED CHARACTERISTIC. By standard reductions (such as killing a minimal prime, completing, and extending the residue field), we may assume without loss of generality that $(R, \mathfrak{m}, k)$ is a Noetherian complete local domain with $\mathrm{char}(R) = 0$, $k$ perfect of characteristic $p$, and $\dim(R) = d$. Write $S = W(k)[[z_2, \ldots, z_n]]$ and $R = S/Q$ for a prime ideal $Q$ of $S$ with height $c$, noting that $p \notin Q$ and $n = c + d$. By prime avoidance, we can find elements $f_1, \ldots, f_c \in Q$ and $x_2, \ldots, x_d \in S$ so that $Q$ is minimal over $(f_1, \ldots, f_c)$, and $f_1, \ldots, f_c, p, x_2, \ldots, x_d$ is a system of parameters of $S$. Similarly, choose also $g \notin Q$ such that $gQ \subseteq \sqrt{(f_1, \ldots, f_c)}$.

Let $S_\infty = S[p^{1/p^\infty}, z_2^{1/p^\infty} \ldots, z_n^{1/p^\infty}]^{\wedge p}$, which is a perfectoid ring. Applying André's Flatness Lemma Theorem 3.2, we can find $S_\infty \longrightarrow S'_\infty$ faithfully flat modulo $p^n$ for all $n \geq 1$ where $S'_\infty$ is perfectoid and so that $f_1, \ldots, f_c, g$ all have compatible systems of arbitary $p$-power roots in $S'_\infty$. Moreover, as $S_\infty$ is $p$-torsion free, so too is $S'_\infty$ so that $\mathrm{Tor}_1^S(S/(p), S'_\infty) = \ker(S'_\infty \xrightarrow{\cdot p} S'_\infty) = 0$. Since $S'_\infty/(p)$ is also flat over $S/(p)$, it follows using [**Sta**, Lemma 0AGW] that $S'_\infty$ is flat over $S$ (as $S$ is Noetherian). Since $S$ is regular and hence Cohen-Macaulay, $f_1, \ldots, f_c, p, x_2, \ldots, x_d$ is a regular sequence in $S$, and hence also in $S'_\infty$. In particular, $f_1^{1/p^e}, \ldots, f_c^{1/p^e}, p, x_2, \ldots, x_d$ is also a regular sequence in $S'_\infty$ for any $e \geq 0$, implying $p, x_2, \ldots, x_d$ is a regular sequence on $S'_\infty/(f_1^{1/p^e}, \ldots, f_c^{1/p^e})$ for any $e \geq 0$. In general, a regular element on a module remains a regular element on the completion of the module with respect to that element, so that $(S'_\infty/(f_1^{1/p^e}, \ldots, f_c^{1/p^e}))^{\wedge p}$ remains $p$-torsion free. All together, this then gives that $p, x_2, \ldots, x_d$ is a regular sequence on $(S'_\infty/(f_1^{1/p^e}, \ldots, f_c^{1/p^e}))^{\wedge p}$ – which is perfectoid (and equals $(S'_\infty/(f_1, \ldots, f_c))_{\mathrm{perfd}}$).

Let $T = (S'_\infty/(f_1^{1/p^e}, \ldots, f_c^{1/p^e}))^{\wedge p}$, and set $T' = \mathrm{Hom}_T((g^{1/p^\infty}), T)$. Note that $T'$ is a $T$-algebra because $(g^{1/p^\infty}) = (g^{1/p^\infty})^2$ is an idempotent ideal, and we also have that $(g^{1/p^\infty})$ is flat as $T$ is perfectoid and hence reduced. This is precisely the setup of almost mathematics – the theory of which is due to Faltings, with a very general and detailed version presented in [**GR03**]. In what follows in these notes, we will give only a rough idea of how this material is used. The basic idea is to systematically treat modules killed by $(g^{1/p^\infty})$ as the zero-module; such modules are said to be $(g^{1/p^\infty})$-almost zero. Similarly, $T \longrightarrow T'$ is a $(g^{1/p^\infty})$-almost isomorphism – *i.e.* its kernel and cokernel are both annihilated by $(g^{1/p^\infty})$. This is verified by examining the long exact sequence

$$0 \longrightarrow \operatorname{Hom}_T(T/(g^{1/p^\infty}), T) \longrightarrow T \longrightarrow T' \longrightarrow \operatorname{Ext}^1_T((g^{1/p^\infty}), T) \longrightarrow \cdots.$$

Proceeding with our proof, let us first argue that $T'$ is also an $R$-algebra. Since $T$ is reduced (as it is perfectoid) and $(f_1, \ldots, f_c)T = 0$, we must have $\sqrt{(f_1, \ldots, f_c)T} = 0$. Because $gQ \in \sqrt{(f_1, \ldots, f_c)}$, this gives that $gQT = 0$. Using that $g \in S$ also has compatible $p$-power roots in $T$, this further gives that $(g^{1/p^\infty})QT = 0$ as well; if $y \in QT$ so that $gy = 0$, then also $gy^{p^e} = (g^{1/p^e}y)^{p^e} = 0$ hence $g^{1/p^e}y = 0$ as $T$ is reduced. Passing to $T'$ then gives that $QT' = 0$; for $t' \in T'$, viewing $t' \colon (g^{1/p^\infty}) \longrightarrow T$ as a $T$-linear map, we have $Qt'\big((g^{1/p^\infty})\big) = Qt'\big((g^{1/p^\infty})^2\big) = (g^{1/p^\infty})Qt'\big((g^{1/p^\infty})\big) \subseteq (g^{1/p^\infty})QT = 0$ and hence $Qt' = 0$. In particular, $T'$ is an algebra over $R = S/Q$. Better still, to see that we are near our goal, we next observe that $T'$ is essentially a $(g^{1/p^\infty})$-almost big Cohen-Macaulay $R$-algebra. Since $T \longrightarrow T'$ is a $(g^{1/p^\infty})$-almost isomorphism and $p, x_2, \ldots, x_d$ is a regular sequence on $T$, we have that:

- $p, x_2, \ldots, x_d$ are a $(g^{1/p^\infty})$-almost regular sequence on $T$, meaning $\frac{(p, x_2, \ldots, x_i) :_{T'} x_{i+1}}{(p, x_2, \ldots, x_i)}$ is annihilated by $(g^{1/p^\infty})$ (*i.e.* is $(g^{1/p^\infty})$-almost zero) for each $i$;
- $T'/(p, x_2, \ldots, x_d)T'$ is not annihilated by $(g^{1/p^\infty})$ (*i.e.* is not $(g^{1/p^\infty})$-almost zero).

The first condition is straightforward to verify using the almost isomorphism, the second condition requires a little work and can be found in [**And20**, 2.5].

The final step in the proof is to modify our almost big Cohen-Macaulay $R$-algebra so as to give an honest BCM $R$-algebra. Such a procedure has long been known in light of Hochster's work in the 1990's (see [**Hoc02**, Sections 3 and 4] or [**HM18**]). Here, we will rely instead on a clever trick due to Gabber [**Gab18**]. Set $B' = W^{-1} \prod_{\mathbb{N}} T'$ where $W$ is the multiplicative set generated by the element $(g, g^{1/p}, g^{1/p^2}, \ldots)$. The ingenious part here is the construction, after which it is essentially an elementary exercise (with a fair bit of notational bookkeeping) to verify that $p, x_2, x_3, \ldots, x_d$ is a regular sequence on $B'$ and $B'/(p, x_2, \ldots, x_d) \neq 0$. Lastly, setting $B = (B')^{\wedge \mathfrak{m}}$ gives the desired BCM $R$-algebra [**BH93**, Corollary 8.5.3]. □

REMARK 3.3. Suppose that $(R, \mathfrak{m}, k)$ is a complete local ring with $p \in \mathfrak{m}$. If $\{B_\lambda\}_{\lambda \in \Lambda}$ is a set of perfectoid BCM $R$-algebras, Ma and Schwede [**MS21**] have shown that one can find a perfectoid BCM $R$-algebra $B$ dominating all of the $B_\lambda$. Moreover, $B$ can be chosen compatibly with $\varinjlim B_\lambda$ provided the system is directed. In particular, this can be used to show existence of perfectoid BCM $R$-algebras dominating $R^+$. A BCM $R$-algebra dominating $R^+$ is called a BCM $R^+$-algebra for short, and similar domination and compatibility properties hold for perfectoid BCM $R^+$-algebras. Note, in particular, that perfectoid BCM $R^+$-algebras admit compatible systems of arbitrary $p$-power roots for all elements in $R$. We also record the following generalized version of Theorem 3.1.

THEOREM 3.4. *Any local homomorphism $R \longrightarrow R'$ of complete local domains of residue characteristic $p > 0$ fits into a commutative diagram*

$$\begin{array}{ccc} R & \longrightarrow & R' \\ \downarrow & & \downarrow \\ R^+ & \longrightarrow & R'^+ \\ \downarrow & & \downarrow \\ B & \longrightarrow & B' \end{array}$$

*with $B$ a perfectoid BCM $R^+$-algebra and $B'$ a perfectoid BCM $R'^+$-algebra* [**And20**]. *In fact, we can take $B$ and $B'$ to be the $p$-adic completions of $R^+$ and $R'^+$, respectively* [**Bha20, HH92**].

The next result, historically referred to as Hochster's Direct Summand Conjecture from its introduction in 1973 [**Hoc73**], involves the notion of a splinter ring. We say a Noetherian domain $R$ is a splinter provided every module-finite extension of rings $R \subseteq S$ splits as a map of $R$-modules, *i.e.* there exists $\phi \in \operatorname{Hom}_R(S, R)$ restricting to the identity on $R$ (or simply satisfying $\phi(1) = 1$). Equivalently still, $R$ is a splinter if and only if the natural map $R \longrightarrow R^+$ is pure (universally injective), *i.e.* it remains injective after tensoring with an arbitary $R$-module. The splinter condition is sensitive to the characteristic of the ring in question: for instance, while splinter rings are always normal, the reverse implication fails in general but holds if $\mathbb{Q} \subseteq R$. In contrast, when $(R, \mathfrak{m}, k)$ is local with $p \in \mathfrak{m}$, splinter rings are necessarily Cohen-Macaulay (using the previously mentioned result that $(R^+)^{\wedge p}$ is Cohen-Macaulay [**Bha20, HH92**]). Even basic properties of splinters can be quite subtle. For instance, for $(R, \mathfrak{m}, k)$ local Noetherian, the henselization of a splinter is always a splinter but the completion need not be unless $R \longrightarrow R^{\wedge \mathfrak{m}}$ is a regular morphism; see [**DT23**] for this result as well as additional background.

COROLLARY 3.5 (Direct summand theorem). [**And18**] *Regular Noetherian rings are splinters.*

PROOF. As a map $R \longrightarrow S$ where $S$ is a finite $R$-module splits if and only if it splits locally and after faithfully flat base change. Thus, it suffices to check that a complete regular local ring $(R, \mathfrak{m}, k)$ is a splinter. If $\mathbb{Q} \subseteq R$, then $R$ is a splinter as it is normal. Else, we have that $p \in \mathfrak{m}$. Suppose $(R, \mathfrak{m}, k) \longrightarrow (S, \mathfrak{n}, \ell)$ is module-finite local domain extension, so that $p \in \mathfrak{n}$ as well. By Theorem 3.1, take $B$ a BCM $S$-algebra. Since a system of parameters for $R$ is also a system of parameters for $S$, we have that $B$ is a BCM $R$-algebra as well. As $R$ is regular local, this implies $B$ is faithfully flat over $R$ (see [**HH92**, page 77, second paragraph] for a solution to this exercise). However, tensoring with the injective hull of the residue field and taking the Matlis dual, it is easy to see that a faithfully flat extension of a complete local ring is split (see [**HH95**, Lemma 2.1 (e)] for an argument, where [**HH95**, Section 2] also has an expanded version of this proof). □

## 4. BCM-rational and BCM-regular singularities

The goal of this section is to give an idea how to apply the machinery from previous sections to study singularities.

DEFINITION 4.1 ([**MS21**][6]). Let $(R, \mathfrak{m}, k)$ be a complete local Noetherian ring of dimension $d$ with $p \in \mathfrak{m}$, and let $B$ be a perfectoid BCM $R$-algebra.

- We say $R$ is *$BCM_B$-rational* if
  - $R$ is Cohen-Macaulay and
  - $H^d_{\mathfrak{m}}(R) \longrightarrow H^d_{\mathfrak{m}}(B)$ is injective, or equivalently $\operatorname{Hom}_R(B, \omega_R) \xrightarrow{\phi \mapsto \phi(1)} \omega_R$ is surjective.

  We say $R$ is *BCM-rational* if $R$ is BCM$_B$-rational for all perfectoid BCM $R$-algebras $B$.
- We say $R$ is *$BCM_B$-regular* if
  - $R$ is a normal $\mathbb{Q}$-Gorenstein domain and
  - $R \longrightarrow B$ splits as a map of $R$-modules, or equivalently $\operatorname{Hom}_R(B, R) \xrightarrow{\phi \mapsto \phi(1)} R$ is surjective.

  We say $R$ is *BCM-regular* if $R$ is BCM$_B$-regular for all perfectoid BCM $R$-algebras $B$.

It is easy to see that BCM-regular rings are BCM-rational. Indeed, considering the splitting of a map to a BCM $R^+$-algebra, it follows that BCM-regular rings are splinters – and as mentioned earlier, splinters are Cohen-Macaulay in mixed or positive characteristic since $(R^+)^{\wedge p}$ is a BCM $R$-algebra [**Bha20, HH92**]. For the second condition, one need only observe that applying local cohomology (or any functor) to a split injection remains a (split) injection. Note that it is an interesting conjecture (known in positive characteristic by [**Sin99**], open in mixed characteristic) as to whether BCM-regular is the same as BCM$_{(R^+)^{\wedge p}}$-regular in general, *i.e.* whether $(R^+)^{\wedge p}$ is a sufficiently large BCM $R$-algebra that it detects splittings to all BCM $R^+$-algebras; see also Question 4.5 below.

While BCM-rational rings are not BCM-regular in general, in the Gorenstein setting this implication holds so that the two notions are equivalent. Indeed, setting $\omega_R = R$, we see immediately that the second conditions in their respective definitions coincide; normality follows from the lemma below.

LEMMA 4.2. *If $R$ is BCM-rational, then $R$ is a normal domain.*

PROOF. Let $P$ be a minimal prime of $R$ such that $\dim R/P = \dim R$ and let $B$ be a perfectoid BCM $R/P$-algebra (and hence a BCM $R$-algebra). Since $R$ is BCM-rational, the composition $H^d_{\mathfrak{m}}(R) \longrightarrow H^d_{\mathfrak{m}}(R/P) \longrightarrow H^d_m(B)$ is injective, thus $H^d_{\mathfrak{m}}(R) \longrightarrow H^d_{\mathfrak{m}}(R/P)$ is injective. For every system of parameters $x_1, \ldots, x_d$ of $R$, consider the following diagram:

$$\begin{array}{ccc} H^d_{\mathfrak{m}}(R) & \hookrightarrow & H^d_{\mathfrak{m}}(R/P) \\ \uparrow & & \uparrow \\ R/(x_1, \ldots, x_d) & \longrightarrow & (R/P)/(x_1, \ldots, x_d)(R/P) \end{array}$$

[6] Our definition of BCM-rational singularities above slightly differs from [**MS21**] because we only consider rings of residue characteristic $p > 0$ and we only consider big Cohen-Macaulay algebras that are perfectoid. This will not affect our applications though since we can always work with perfectoid big Cohen-Macaulay algebras by Theorem 3.4. On the other hand, it seems natural to ask whether every big Cohen-Macaulay $R$-algebra $B$ map to a perfectoid big Cohen-Macaulay $R$-algebra $B'$. We do not know the answer to this question in mixed characteristic (if $R$ has characteristic $p > 0$, the question has an affirmative answer by taking $B' = B_{\text{perf}}$).

where the injectivity of the left vertical map is because $R$ is Cohen-Macaulay (by the definition of BCM-rational) and thus $x_1, \dots, x_d$ is a regular sequence on $R$. Chasing the diagram, we know that $R/(x_1, \dots, x_d) \longrightarrow (R/P)/(x_1, \dots, x_d)(R/P)$ is injective. Thus $P \subseteq (x_1, \dots, x_d)$ for every system of parameters $x_1, \dots, x_d$ of $R$. Hence $P \subseteq \cap_t(x_1^t, \dots, x_d^t) = 0$. This proves that $R$ is a domain.

Now let $R^{\mathrm{N}}$ denote the normalization of $R$, which is a complete normal domain. A perfectoid BCM $R^{\mathrm{N}}$-algebra $C$ is still a BCM $R$-algebra, and hence it follows that $H^d_{\mathfrak{m}}(R) \longrightarrow H^d_{\mathfrak{m}}(R^{\mathrm{N}})$ is injective. Suppose $\frac{y}{x_1} \in R^{\mathrm{N}}$ and $x_1, \dots, x_d$ is a system of parameters of $R$. The injectivity of $H^d_{\mathfrak{m}}(R) \longrightarrow H^d_{\mathfrak{m}}(R^{\mathrm{N}})$ yields $y \in \cap_t(x_1, x_2^t, \dots, x_d^t) = (x_1)$, thus $\frac{y}{x_1} \in R$. Thus $R$ is normal. □

In positive characteristic, Hochster and Huneke's celebrated theory of tight closure [**HH90**] introduced a number of important classes of singularities defined in terms of the Frobenius endomorphism. In particular, these include $F$-rational rings (those for which all ideals generated by partial systems of parameters are tightly closed) and $F$-regular rings (those for which all ideals in all localizations are tightly closed). Part of the motivation in defining (and naming) BCM-rational and BCM-regular singularities is to give analogous notions of singularity in mixed characteristic. Indeed, to that end, we record the following result.

THEOREM 4.3. *Let $(R, m, k)$ be a complete local Noetherian ring with* $\mathrm{char}(R) = p$.

(a) [**MS21**, Proposition 3.5] *$R$ is BCM-rational if and only if $R$ is $F$-rational.*
(b) [**MS21**, Corollary 6.23] *$R$ is BCM-regular if and only if $R$ is $\mathbb{Q}$-Gorenstein and $F$-regular.*[7]

Note that the importance of $F$-rational and $F$-regular rings stems in part from their connection to singularities used in the study of complex algebraic varieties. Via standard reduction to characteristic $p$ methods, $F$-rationality and $F$-regularity are linked with rational and Kawamata log terminal singularities, respectively. The following result, analagous to that of Smith for $F$-rationality [**Smi97**], furthers these connections.

THEOREM 4.4. *If $R$ is BCM-rational, then $R$ is pesudo-rational, that is, $R$ is normal, Cohen-Macaulay, and for every normal projective birational map $\pi\colon X \longrightarrow$* $\mathrm{Spec}(R)$ *we have $\pi_*\omega_X = \omega_R$.*

PROOF. We know $R$ is normal by Lemma 4.2 and Cohen-Macaulay by definition. Therefore to show $R$ is pseudo-rational, it suffices to prove $\pi_*\omega_X = \omega_R$. Let $X = \mathrm{Proj}\, R[Jt]$ for some ideal $J \subseteq R$ (since $X$ and $R$ are normal, we may replace $R[Jt]$ by its integral closure to assume $R[Jt]$ is normal). Let $f_1t, f_2t, \dots, f_nt$ be a generating set of $Jt$. The affine charts of $X$ can be identified with $\mathrm{Spec}\left(R[Jt][\frac{1}{f_it}]_0\right)$. Consider the Čech complex $C^\bullet = C^\bullet(f_1t, \dots, f_nt; R[Jt])$ and note that we have an exact sequence of complexes:

$$0 \longrightarrow [C^{\bullet \geq 1}]_0 \longrightarrow [C^\bullet]_0 \longrightarrow R[Jt]_0 \cong R \longrightarrow 0.$$

[7]As we are in the $\mathbb{Q}$-Gorenstein, there is no need to distinguish between the different flavors of $F$-regularity: weakly $F$-regular, $F$-regular, strongly $F$-regular,and even splinter are all the same [**Sin99**].

Note that $[C^\bullet]_0$ computes $\mathbf{R}\Gamma_{Jt}(R[Jt])_0$ and after shifting degree by one, $[C^{\bullet \geq 1}]_0$ computes $\mathbf{R}\Gamma(X, O_X) = \mathbf{R}\pi_* O_X$. Thus the above exact sequence of complexes yields an exact triangle after rotation:

$$\mathbf{R}\Gamma_{Jt}(R[Jt])_0 \longrightarrow R \longrightarrow \mathbf{R}\pi_* O_X \xrightarrow{+1} .$$

Applying $\mathbf{R}\Gamma_{\mathfrak{m}}(-)$ we get

$$\mathbf{R}\Gamma_{\mathfrak{m}+Jt}(R[Jt])_0 \longrightarrow \mathbf{R}\Gamma_{\mathfrak{m}}(R) \longrightarrow \mathbf{R}\Gamma_{\mathfrak{m}}\mathbf{R}\pi_* O_X \xrightarrow{+1} .$$

Taking cohomology we get a long exact sequence:[8]

$$\cdots \longrightarrow [H^d_{\mathfrak{m}+Jt}(R[Jt])]_0 \longrightarrow H^d_{\mathfrak{m}}(R) \longrightarrow H^d(\mathbf{R}\Gamma_{\mathfrak{m}}\mathbf{R}\pi_* O_X) \longrightarrow \cdots .$$

Now by Theorem 3.4, we have a commutative diagram:

$$\begin{array}{ccc} \widehat{R[Jt]_{\mathfrak{m}+Jt}} & \longrightarrow & R \\ \downarrow & & \downarrow \\ B & \longrightarrow & C \end{array}$$

where $B$ and $C$ are perfectoid BCM algebras for $\widehat{R[Jt]_{\mathfrak{m}+Jt}}$ and $R$, respectively. This induces a commutative diagram:

$$\begin{array}{ccc} H^d_{\mathfrak{m}+Jt}(R[Jt]) & \longrightarrow & H^d_{\mathfrak{m}}(R) \\ \downarrow & & \big\downarrow \\ 0 = H^d_{\mathfrak{n}}(B) & \longrightarrow & H^d_{\mathfrak{m}}(C) \end{array}$$

where the bottom left $0$ is because $B$ is a BCM $\widehat{R[Jt]_{\mathfrak{m}+Jt}}$-algebra and $\dim(R[Jt]_{m+Jt}) = d+1$, and right injectivity follows as $R$ is BCM-rational. Thus the diagram tells us that $H^d_{\mathfrak{m}+Jt}(R[Jt]) \longrightarrow H^d_{\mathfrak{m}}(R)$ is the zero map. The long exact sequence above then implies that $H^d_{\mathfrak{m}}(R) \longrightarrow H^d(\mathbf{R}\Gamma_{\mathfrak{m}}\mathbf{R}\pi_* O_X)$ is injective. By Grothendieck and local duality, this is saying that

$$\pi_*\omega_X \cong H^{-d}(\mathbf{R}\pi_*\omega_X^\bullet) \longrightarrow \omega_R$$

is surjective, where $\omega_X^\bullet$ is the normalized dualizing complex (here, we can simply define $\omega_X^\bullet := \pi^!\omega_R^\bullet$, where $\omega_R^\bullet = \omega_R[d]$ is the normalized dualizing complex for $R$; see [**Har66**, Chapter V]). As $\pi_*\omega_X \longrightarrow \omega_R$ is always injective since $X \longrightarrow \operatorname{Spec}(R)$ is birational, we know that $\pi_*\omega_X \cong \omega_R$ as wanted. □

BCM algebras have also been used to introduce a theory in mixed characteristic analagous to test ideals and parameter test submodules in positive characteristic. To give a rough idea, following [**MS21**] (*cf.* [**BMP⁺23, HLS24, MS18, BMP⁺24**]), for simplicity let $(R, \mathfrak{m}, k)$ be a complete normal $\mathbb{Q}$-Gorenstein Cohen-Macaulay local ring of dimension $d$ and residue characteristic $p > 0$. Let $B$ be a perfectoid BCM $R$-algebra. We have:

- the $\mathrm{BCM}_B$-test ideal $\tau_B(R) = \operatorname{im}\left(\operatorname{Hom}_R(B, R) \xrightarrow{\phi \mapsto \phi(1)} R\right) \subseteq R$, satisfying $\tau_B(R) = R$ if and only if $R$ is $\mathrm{BCM}_B$-regular, and;

[8]This is exactly the Sancho-de-Salas exact sequence [**SdS87**].

- the BCM$_B$-test submodule $\tau_B(\omega_R) = \operatorname{im}\left(\operatorname{Hom}_R(B, \omega_R) \xrightarrow{\phi \mapsto \phi(1)} \omega_R\right) \subseteq \omega_R$, satisfying $\tau_B(\omega_R) = \omega_R$ if and only if $R$ is BCM$_B$-rational.

QUESTION 4.5. Set $\tau_+(R) = \tau_{(R^+)^{\wedge p}}(R)$ and $\tau_+(\omega_R) = \tau_{(R^+)^{\wedge p}}(\omega_R)$. It is natural to ask:

(1.) Do we have that $\tau_+(R) = \tau_B(R)$ and $\tau_+(\omega_R) = \tau_B(\omega_R)$ for all perfectoid BCM $R^+$-algebras $B$? In other words, is $(R^+)^{\wedge p}$ a sufficiently large BCM $R$-algebra to compute all further BCM test ideals and submodules?

(2.) Do we have $\tau_+(R)_Q^{\wedge Q} \cong \tau_+(R_Q^{\wedge Q})$ and $\tau_+(\omega_R)_Q^{\wedge Q} \cong \tau_+(\omega_{R_Q^{\wedge Q}})$ for all $Q \in \operatorname{Spec}(R)$ that contain $p$? In other words, do the BCM test ideals and submodules for absolute integral closures commute with localization up to completion?

REMARK 4.6. In [**BMP**$^+$**24**], it is shown that $\tau_B(R)$ and $\tau_B(\omega_R)$ commute with localization up to completion (as in (2.)) for all sufficiently large $B$ under certain finiteness assumptions on $R$. Thus, under these assumptions, we have (1.) implies (2.). In this setting, it is also shown that $\tau_B(R)[\frac{1}{p}] = \mathcal{J}(R[\frac{1}{p}])$ and $\tau_B(\omega_R)[\frac{1}{p}] = \mathcal{J}(\omega_{R[\frac{1}{p}]})$; in other words, for sufficiently large $B$, the BCM$_B$-test ideal and submodule agree with the multiplier ideal and submodule respectively after inverting $p$.

EXAMPLE 4.7. We conclude this section with a list of some examples of BCM-regular rings.

- $R = \mathbb{Z}_p[[y_2, \ldots, y_n]]/(p^m + y_2^m + \cdots + y_n^m)$ is BCM-regular provided $m < n$ and $p \gg 0$ is sufficiently large (see [**MST**$^+$**22**]).
- $R = \mathbb{Z}_p[[x, z]]/(x^2 + p^2 z + z^3)$ is BCM-regular when $p > 5$. Indeed, this is an example of a rational double point in mixed characteristic of type $D_4$. In [**CRMP**$^+$**21**], it is shown that all rational double points with characteristic $(0, p > 5)$ are direct summands of regular rings – and hence BCM-regular in light of the Direct Summand Theorem. An alternative proof can also be found in [**BMP**$^+$**23**], where it is shown that all two-dimensional klt pairs with standard coefficients in mixed characteristic $(0, p > 5)$ are BCM-regular.
- Generalizing the example above, it follows from the Direct Summand Theorem that $\mathbb{Q}$-Gorenstein direct summands of regular rings are BCM-regular.
- $\mathbb{Q}$-Gorenstein log regular rings are BCM-regular, *i.e.* $\mathbb{Q}$-Gorenstein rings of the form $W(k)[[\mathscr{M}]]/(p - f)$ where $\mathscr{M}$ is a strongly convex saturated normal monoid and $f \in I_{\mathscr{M}} + (p^2) \subseteq W(k)[[\mathscr{M}]]$ for the ideal of monomials $I_{\mathscr{M}}$. See [**CLM**$^+$**23**, Proposition 5.3.5].

## 5. Perfectoid Hilbert-Kunz multiplicity and perfectoid signature

The goal of this section is to give an idea how to apply the machinery from previous sections to give analogs of Hilbert-Kunz multiplicity and $F$-signature in the mixed characteristic setting introduced in [**CLM**$^+$**23**]. For brevity, we shall assume a general familiarity with the definitions of these invariants (see, for example, [**Hun13**]). We will make use of the following notation throughout this section.

NOTATION 5.1.

- Let $(R, \mathfrak{m}, k)$ be a Noetherian complete local domain of dimension $d$ with $p \in \mathfrak{m}$ and $k$ a perfect field of characteristic $p$.
- By the Cohen-Gabber structure theorem, we can find a module finite local inclusion $(A, \mathfrak{m}_A, k) \hookrightarrow (R, \mathfrak{m}, k)$ that is generically separable where
  - (positive characteristic setting) $A = k[[x_1, \dots, x_d]]$ when $\operatorname{char}(R) = p$, or
  - (mixed characteristic setting) $A = W(k)[[x_2, \dots, x_d]]$ when $\operatorname{char}(R) = 0$ (in which case we set $x_1 = p$).
- Set $A_e = A[x_1^{1/p^e}, \dots, x_d^{1/p^e}]$ for $e \in \mathbb{Z}_{\geq 0}$, and

$$A_\infty^{\text{un}} = \varinjlim_e A_e = A[x_1^{1/p^\infty}, \dots, x_d^{1/p^\infty}].$$

  Put $A_\infty = (A_\infty^{\text{un}})^{\wedge p} = A[x_1^{1/p^\infty}, \dots, x_d^{1/p^\infty}]^{\wedge p}$, which we have seen is a perfectoid ring. Note that $A_\infty = A_{\text{perf}}$ in the positive characteristic setting.
- Let $R_{\text{perfd}}^{A_\infty} = (R \otimes_A A_\infty)_{\text{perfd}}$, which is well-defined by Theorem 2.7 as $R \otimes_A A_\infty$ is finite and finitely presented over the perfectoid ring $A_\infty$. Once more, note that $R_{\text{perfd}}^{A_\infty} = R_{\text{perf}}$ in the positive characteristic setting.

The tool needed to extract numerical invariants from the constructions above is the notion of normalized length first introduced by Faltings [**Fal02**] (see also [**GR03**]). For $M$ an $A_\infty^{\text{un}}$-module annihilated by a power of $\mathfrak{m}_A$, the normalized length $\lambda_\infty(M)$ is defined step-by-step as follows.

1. Suppose first that $M$ is finitely presented over $A_\infty^{\text{un}}$. In this case, we have that $M$ is defined over the (Noetherian local) ring $A_e$ for some $e \in \mathbb{Z}_{\geq 0}$. In other words, there is a finitely presented $A_e$-module $M_e$ annihilated by a power of $\mathfrak{m}_A$ so that $M = M_e \otimes_{A_e} A_\infty^{\text{un}}$; $M$ is then necessarily defined over $A_{e'}$ for all $e > e'$ as well. Set $\lambda_\infty(M) = \frac{1}{p^{ed}} \ell_{A_e}(M_e)$. This is well-posed (*i.e.* independent of $e$) as the extensions $A_e \longrightarrow A_{e+1}$ are all faithfully flat of constant rank $p^d$.
2. Next, suppose $M$ is merely finitely generated over $A_\infty^{\text{un}}$. Set

$$\lambda_\infty(M) = \inf\{\lambda_\infty(M') \mid M' \twoheadrightarrow M \text{ surjects onto } M, \text{ and } M' \text{ finitely presented}\}.$$

3. Lastly, for an arbitrary $A_\infty^{\text{un}}$-module annihilated by a power of $\mathfrak{m}_A$, set

$$\lambda_\infty(M) = \sup\{\lambda_\infty(M') \mid M' \subseteq M \text{ is a finitely generated submodule}\}.$$

It is straightforward to check that $\lambda_\infty(\_\_)$ is well-defined, and is additive on short exact sequences. Moreover, if $M$ is a finite length $A$-module, we have that $\lambda_\infty(M \otimes_A A_\infty) = \ell_A(M) \in \mathbb{R}_{\geq 0} \cup \{\infty\}$ essentially by construction.

Note that, even if $\lambda_\infty(M) = 0$, it may not be true that $M = 0$. However, in a sense we will now describe, this gives rise to yet another way of saying that a module is almost zero – but in a way that is distinct from the frameworks of almost mathematics we encountered earlier in the proof of Theorem 3.1. Consider the following examples which point to the underlying issue. It is a straightforward exercise from the definition to show that $\lambda_\infty\left(A_\infty/(\mathfrak{m}_A, x_i^{1/p^\infty})\right) = 0$ for any $i = 1, \dots, d$. In these examples, note that each of $A_\infty/(\mathfrak{m}_A, x_i^{1/p^\infty})$ are $(x_i^{1/p^\infty})$-almost zero; that is, while non-zero, they are almost zero with respect to different ideals. Similarly, suppose we are in the characteristic $p$ setting, so that $A_\infty = A_\infty^{un} = A_{\text{perf}}$.

We have that

$$\lambda_\infty\left(\frac{A_{\text{perf}}}{(\mathfrak{m}_A, f^{1/p^\infty})}\right) \leq \lambda_\infty\left(\frac{A_{\text{perf}}}{(\mathfrak{m}_A, f^{1/p^e})}\right) = \ell_{A_e}\left(\frac{A_e}{(\mathfrak{m}_A, f^{1/p^e})}\right).$$

Since the right hande side is $e_{\text{HK}}(A/(f))/p^e$ for $e \gg 0$ and so approaches 0 as $e \longrightarrow \infty$, we see $\lambda_\infty\left(A_{\text{perf}}/(\mathfrak{m}_A, f^{1/p^\infty})\right) = 0$.

In practice, given a setup for almost mathematics in $A_\infty^{un}$ as in the proof of Theorem 3.1, one can usually show that almost zero modules for that framework have zero normalized length. For instance, it is again straightforward from the definitions to check that $(p^{1/p^\infty})$-almost zero modules have zero normalized length; however, as the examples in the paragraph above illustrate, the reverse implication does not hold. Here is one way to address this issue from [**Shi07**]. Let $\nu\colon \operatorname{Frac}(A_\infty^{\text{un}}) \setminus \{0\} \longrightarrow \mathbb{Q}$ be a valuation with value group $\mathbb{Q}$ extending the $\mathfrak{m}_A$-adic valuation on $A$. An $A_\infty^{\text{un}}$-module is said to be $\nu$-almost zero if for any $m \in M$ and $\epsilon \in \mathbb{Q}_{>0}$ there exists an element $a \in A_\infty^{\text{un}}$ with $a \cdot m = 0$ and $\nu(a) < \epsilon$. It is not hard to show that $M$ is $\nu$-almost zero provided $\lambda_\infty(M) = 0$; see [**Shi07**, Proposition 2.15].

The following result demonstrates how to use normalized length in the positive characteristic setting to recover Hilbert-Kunz multiplicity $e_{\text{HK}}(J)$ of an ideal $J \subseteq R$ and the $F$-signature $s(R)$.

THEOREM 5.2. *Suppose we are in the positive characteristic setting of Notation* 5.1.

(a.) *If* $J \subseteq R$ *is an ideal with* $\ell_R(R/J) < \infty$, *then* $e_{\text{HK}}(J) = \lambda_\infty\left(R_{\text{perf}}/JR_{\text{perf}}\right)$.
(b.) *Let* $I_\infty = \left\{x \in R_{\text{perf}} \mid R \xrightarrow{1 \mapsto x} R_{\text{perf}} \text{ is not split}\right\}$. *Then* $s(R) = \ell_\infty(R_{\text{perf}}/I_\infty)$.

We shall give here a rough idea of the proof of this result. Before we begin, the reader is cautioned to observe first that $R_{\text{perf}}$ is *not* finitely generated as a $A_{\text{perf}}$-module – which makes using the normalized lengths in the statement more cumbersome to analyze.

PROOF SKETCH. Given that $R$ is generically separable over $A$, we have $R^{1/p^e} \otimes_{A_e} A_{\text{perf}} \subseteq R_{\text{perf}}$; moreoever, there exists an element $0 \neq g \in A$ such that $g^{1/p^e} R_{\text{perf}} \subseteq R^{1/p^e} \otimes_{A_e} A_{\text{perf}}$ for all $e$ by [**HH90**, Lemma 6.5]. Consider the surjections

$$\frac{R^{1/p^e} \otimes_{A_e} A_{\text{perf}}}{J\left(R^{1/p^e} \otimes_{A_e} A_{\text{perf}}\right)} \twoheadrightarrow \frac{R^{1/p^e} \otimes_{A_e} A_{\text{perf}}}{(JR_{\text{perf}}) \cap \left(R^{1/p^e} \otimes_{A_e} A_{\text{perf}}\right)}. \tag{3}$$

Limiting over the normalized lengths of the lefthand sides, we have

$$\begin{aligned}\lim_{e \longrightarrow \infty} \lambda_\infty\left(\frac{R^{1/p^e} \otimes_{A_e} A_{\text{perf}}}{J\left(R^{1/p^e} \otimes_{A_e} A_{\text{perf}}\right)}\right) &= \lim_{e \longrightarrow \infty} \frac{1}{p^{ed}} \ell_{A_e}\left(\frac{R^{1/p^e}}{JR^{1/p^e}}\right) \\ &= \lim_{e \longrightarrow \infty} \frac{1}{p^{ed}} \ell_R\left(\frac{R}{J^{[p^e]}}\right) = e_{\text{HK}}(J)\end{aligned}$$

which gives by (3) that

$$e_{\text{HK}}(J) \geq \lim_{e \longrightarrow \infty} \lambda_\infty\left(\frac{R^{1/p^e} \otimes_{A_e} A_{\text{perf}}}{(JR_{\text{perf}}) \cap \left(R^{1/p^e} \otimes_{A_e} A_{\text{perf}}\right)}\right). \tag{4}$$

On the other hand, using that $g^{1/p^e}R_{\text{perf}} \subseteq R^{1/p^e} \otimes_{A_e} A_{\text{perf}}$, we have

$$\begin{aligned}
&\lim_{e\longrightarrow\infty} \lambda_\infty\left(\frac{R^{1/p^e} \otimes_{A_e} A_{\text{perf}}}{(JR_{\text{perf}}) \cap (R^{1/p^e} \otimes_{A_e} A_{\text{perf}})}\right)\\
&\geq \lim_{e\longrightarrow\infty} \lambda_\infty\left(\frac{R^{1/p^e} \otimes_{A_e} A_{\text{perf}}}{J\left(R^{1/p^e} \otimes_{A_e} A_{\text{perf}}\right) :_{R^{1/p^e}\otimes_{A_e} A_{\text{perf}}} g^{1/p^e}}\right)\\
&= \lim_{e\longrightarrow\infty} \frac{1}{p^{ed}}\ell_R\left(\frac{R}{J^{[p^e]} :_R g}\right)\\
&= \lim_{e\longrightarrow\infty} \frac{1}{p^{ed}}\ell_R\left(\frac{R}{J^{[p^e]}}\right) - \lim_{e\longrightarrow\infty} \frac{1}{p^{ed}}\ell_R\left(\frac{R}{(J^{[p^e]}, g)}\right)\\
&= e_{\text{HK}}(J)
\end{aligned}$$

where the final equality uses that $\dim(R/(g)) = d-1$. Thus, we have equality in (4). As $R_{\text{perf}} = \bigcup_e(R^{1/p^e} \otimes_{A_e} A_{\text{perf}})$, we have that the righthand side of (4) agrees with $\lambda_\infty(R_{\text{perf}}/JR_{\text{perf}})$ giving (a.).

For (b.), let $I_e = \left\{r \in R \mid \xrightarrow{1\mapsto r^{1/p^e}} R^{1/p^e} \text{ is not split}\right\}$. A well-known characterization of the $F$-signature gives

$$s(R) = \lim_{e\longrightarrow\infty} \frac{1}{p^{ed}}\ell_R\left(\frac{R}{I_e}\right) = \lim_{e\longrightarrow\infty} \frac{1}{p^{ed}}\ell_{A_e}\left(\frac{R^{1/p^e}}{I_e^{1/p^e}}\right) = \lim_{e\longrightarrow\infty} \lambda_\infty\left(\frac{R^{1/p^e} \otimes_{A_e} A_{\text{perf}}}{I_e^{1/p^e} \otimes_{A_e} A_{\text{perf}}}\right).$$

Also, it is easy to see that $I_\infty \cap R^{1/p^e} = I_e^{1/p^e}$ when $R$ is $F$-split, giving that $I_\infty = \bigcup_e I_e^{1/p^e}$ and $I_e^{1/p^e} \otimes_{A_e} A_{\text{perf}} \subseteq I_\infty \cap \left(R^{1/p^e} \otimes_{A_e} A_{\text{perf}}\right)$ in general. Since $R_{\text{perf}} = \bigcup_e(R^{1/p^e} \otimes_{A_e} A_{\text{perf}})$, we have

$$\begin{aligned}
\lambda_\infty\left(\frac{R_{\text{perf}}}{I_\infty}\right) &= \lim_{e\longrightarrow\infty} \lambda_\infty\left(\frac{R^{1/p^e} \otimes_{A_e} A_{\text{perf}}}{I_\infty \cap \left(R^{1/p^e} \otimes_{A_e} A_{\text{perf}}\right)}\right)\\
&\leq \lim_{e\longrightarrow\infty} \lambda_\infty\left(\frac{R^{1/p^e} \otimes_{A_e} A_{\text{perf}}}{I_e^{1/p^e} \otimes_{A_e} A_{\text{perf}}}\right) = s(R).
\end{aligned}$$

For the reverse inequality, using that $g^{1/p^e}R_{\text{perf}} \subseteq R^{1/p^e} \otimes_{A_e} A_{\text{perf}}$, one first verifies that $I_\infty \cap \left(R^{1/p^e} \otimes_{A_e} A_{\text{perf}}\right) \subset (I_e :_R g)^{1/p^e} \otimes_{A_e} A_{\text{perf}}$ (see [**CLM+23**, Lemma 3.0.6] for details) after which the analagous argument from (a.) leveraging $\dim(R/(g)) = d-1$ applies. □

Inspired by the result above, we are now in a position to define the perfectoid Hilbert-Kunz multiplicity and perfectoid signature analagously in the mixed characteristic setting.

DEFINITION 5.3 ([**CLM+23**]). Suppose we are in the mixed characteristic setting of Notation 5.1.

- If $J \subseteq R$ is an ideal with $\ell_R(R/J) < \infty$, then the *perfectoid Hilbert-Kunz multiplicity* of $J$ is $e^{\underline{x}}_{\text{perfd}}(J) = \lambda_\infty\left(R^{A_\infty}_{\text{perfd}}/JR^{A_\infty}_{\text{perfd}}\right)$.
- Let $I_\infty = \left\{x \in R^{A_\infty}_{\text{perfd}} \mid R \xrightarrow{1\mapsto x} R^{A_\infty}_{\text{perfd}} \text{ is not split}\right\}$. The *perfectoid signature* of $R$ is $s^{\underline{x}}_{\text{perfd}}(R) = \lambda_\infty\left(R^{A_\infty}_{\text{perfd}}/I_\infty\right)$.

A priori, these definitions rely heavily not only on the choice of $A \subset R$, but also the choice of regular system of parameters $\underline{x} = x_1, \ldots, x_d$ of $A$ (which is used to give the filtration of $A_\infty^{\mathrm{un}} = \bigcup_e A_e$ in the definition of normalized length). The notation incorporates $\underline{x}$ in the definition above to emphasize this fact. It remains an important open question whether perfectoid Hilbert-Kunz multiplicity and perfectoid signature are independent of the choice of $\underline{x}$.

One of the most important properties of the $F$-signature in positive characteristic is that $s(R) > 0$ if and only if $R$ is (strongly) $F$-regular. Provided $R$ is normal and $\mathbb{Q}$-Gorenstein, we saw in the previous section that BCM-regular rings are an analog $F$-regular rings in mixed characteristic. The next result says that the perfectoid signature can be used in mixed characteristic to characterize BCM-regularity.

THEOREM 5.4. *Suppose we are in the mixed characteristic setting of Notation 5.1, and $R$ is normal and $\mathbb{Q}$-Gorenstein. Then $s^{\underline{x}}_{\mathrm{perfd}}(R) > 0 \iff R$ is BCM-regular.*

While the proof of this result is beyond the scope of these notes, based on our previous work we are well-positioned to give a heuristic argument for the more difficult implication $\Longrightarrow$. Throwing caution to the wind, let us content ourselves in being wildly imprecise in our use of almost mathematics. For simplicity, let us assume that $R$ is a Gorenstein ring and $\Delta \in R$ gives a generator for the socle modulo $(\underline{x})$. Suppose that $s^{\underline{x}}_{\mathrm{perfd}}(R) = 0$. Using standard arguments, one checks $I_\infty = (\underline{x})R^{A_\infty}_{\mathrm{perfd}} :_{R^{A_\infty}_{\mathrm{perfd}}} \Delta$. Since $\lambda_\infty \left(R^{A_\infty}_{\mathrm{perfd}}/I_\infty\right) = 0$, roughly speaking we have that $R^{A_\infty}_{\mathrm{perfd}}/I_\infty$ is almost zero. In other words, we have that $1 \in^a I_\infty = (\underline{x})R^{A_\infty}_{\mathrm{perfd}} :_{R^{A_\infty}_{\mathrm{perfd}}} \Delta$ or $\Delta \in^a (\underline{x})R^{A_\infty}_{\mathrm{perfd}}$ up to almost. Since $R^{A_\infty}_{\mathrm{perfd}}$ is an almost BCM $R$-algebra by [**BS22**], using Gabber's trick from the proof of Theorem 3.1 we can dominate $R^{A_\infty}_{\mathrm{perfd}}$ by an (honest) perfectoid BCM $R$-algebra $B$ so that $\Delta \in (\underline{x})B$ actually holds. In particular, $R \longrightarrow B$ cannot split (as $\Delta \notin (\underline{x})$) and so $R$ is not BCM-regular.

The perfectoid Hilbert-Kunz multiplicity and perfectoid signature in mixed characteristic share many other properties analagous properties with their positive characteristic counterparts. Below is a summary list of some key results; see [**CLM$^+$23**] for further detail.

- We have $e^{\underline{x}}_{\mathrm{perfd}}(\mathfrak{m}) \geq 1$ with equality if and only if $R$ is regular in analogy with [**WY00**].
- We have $s^{\underline{x}}_{\mathrm{perfd}}(R) \leq 1$ with equality if and only if $R$ is regular in analogy with [**HL02**, Theorem 16].
- If $I \subseteq J \subseteq R$ have finite colength, then $e^{\underline{x}}_{\mathrm{perfd}}(I) \geq e^{\underline{x}}_{\mathrm{perfd}}(J)$ with equality if and only if their full extended plus closures of $I^{\mathrm{epf}}$ and $J^{\mathrm{epf}}$ are the same. In other words, perfectoid Hilbert-Kunz multiplicity tests for full extended plus closure in the same manner that Hilbert-Kunz mulitiplicy tests for tight closure [**HH90**, Theorem 8.17]. Here, we have that $x \in I^{\mathrm{epf}}$ if and only if there exists a $0 \neq c \in R$ so that $c^{1/p^e}x \in (I, p^n)R^+$ for all $e, n > 0$; see [**HM21**].
- We have that $e^{\underline{x}}_{\mathrm{perfd}}(J)$ is at most the Hilbert-Samuel multiplicity $e(J)$ for all finite colength ideals $J \subseteq R$, with equality when $J$ is generated by a system of parameters for $R$. This is analagous to (but substantially harder to shown than) Lech's Lemma [**Mat89**, Theorem 14.12].

- We have $s^{\underline{x}}_{\mathrm{perfd}}(R) = \inf_{I \subsetneq J} \frac{e^{\underline{x}}_{\mathrm{perfd}}(I) - e^{\underline{x}}_{\mathrm{perfd}}(J)}{\ell_R(J/I)}$ in analogy with [**PT18**]. Note that a similar approach can also be used to give a perfectoid relative rational signature that detects BCM-rationality similar to [**HY22, ST24**].
- If $(R, \mathfrak{m}, k) \subseteq (S, \mathfrak{n}, l)$ is a module-finite split quasi-étale local domain extension, then

$$s^{\underline{x}}_{\mathrm{perfd}}(S) \cdot [l : k] = s^{\underline{x}}_{\mathrm{perfd}}(R) \cdot \mathrm{rank}_R(S)$$

in analogy with [**CRST18**], and with similar applications to those for $F$-signature to finiteness of local étale fundamental groups [**CRST18, BCRG$^+$19**] and bounding torsion in the class group [**Pol22, Mar22, CR22**].

QUESTION 5.5. We conclude by mentioning a few open questions.

1. As mentioned earlier, are perfectoid Hilbert-Kunz mulitiplicy and perfectoid signature independent of the choice of $\underline{x}$?
2. If $J \subseteq R$ is a finite colength ideal, is it true that $\frac{1}{d!}e(J) \leq e^{\underline{x}}_{\mathrm{perfd}}(J)$? Note that the analagous statement in positive characteristic is shown by exploiting the containment $J^{[p^e]} \subseteq J^{p^e}$.
3. What explicit computations of perfectoid Hilbert-Kunz multiplicity or perfectoid signature are possible? At present, a scant few very simple examples (perfectoid Hilbert-Kunz multiplicity of parameter ideals, perfectoid signature of finite quotients) can be extracted from the list of key results above.
4. Are perfectoid Hilbert-Kunz multiplicity of the maximal ideal or perfectoid signature compatible with localization?

## References


[AH97] Ian M. Aberbach and Melvin Hochster, *Finite Tor dimension and failure of coherence in absolute integral closures*, J. Pure Appl. Algebra **122** (1997), no. 3, 171–184, DOI 10.1016/S0022-4049(97)00049-2. MR1481086

[And18] Yves André, *La conjecture du facteur direct* (French, with French summary), Publ. Math. Inst. Hautes Études Sci. **127** (2018), 71–93, DOI 10.1007/s10240-017-0097-9. MR3814651

[And20] Yves André, *Weak functoriality of Cohen-Macaulay algebras*, J. Amer. Math. Soc. **33** (2020), no. 2, 363–380, DOI 10.1090/jams/937. MR4073864

[Bha12] Bhargav Bhatt, *Derived splinters in positive characteristic*, Compos. Math. **148** (2012), no. 6, 1757–1786, DOI 10.1112/S0010437X12000309. MR2999303

[Bha18] Bhargav Bhatt, *On the direct summand conjecture and its derived variant*, Invent. Math. **212** (2018), no. 2, 297–317, DOI 10.1007/s00222-017-0768-7. MR3787829

[Bha20] B. Bhatt, *Cohen-Macaulayness of absolute integral closures*, `arXiv:2008.08070`, 2020.

[BCRG$^+$19] Bhargav Bhatt, Javier Carvajal-Rojas, Patrick Graf, Karl Schwede, and Kevin Tucker, *Étale fundamental groups of strongly F-regular schemes*, Int. Math. Res. Not. IMRN **14** (2019), 4325–4339, DOI 10.1093/imrn/rnx253. MR3984070

[BIM19] Bhargav Bhatt, Srikanth B. Iyengar, and Linquan Ma, *Regular rings and perfect(oid) algebras*, Comm. Algebra **47** (2019), no. 6, 2367–2383, DOI 10.1080/00927872.2018.1524009. MR3957103

[BMP$^+$23] Bhargav Bhatt, Linquan Ma, Zsolt Patakfalvi, Karl Schwede, Kevin Tucker, Joe Waldron, and Jakub Witaszek, *Globally +-regular varieties and the minimal model program for threefolds in mixed characteristic*, Publ. Math. Inst. Hautes Études Sci. **138** (2023), 69–227, DOI 10.1007/s10240-023-00140-8. MR4666931

[BMP+24] B. Bhatt, L. Ma, Z. Patakfalvi, K. Schwede, K. Tucker, J. Waldron, and J. Witaszek, *Test ideals in mixed characteristic: a unified theory up to perturbation*, arXiv:2401.00615, 2024.

[BMS15] B. Bhatt, M. Morrow, and P. Scholze, *Integral p-adic Hodge theory—announcement*, Math. Res. Lett. **22** (2015), no. 6, 1601–1612, DOI 10.4310/MRL.2015.v22.n6.a3. MR3507252

[BS17] Bhargav Bhatt and Peter Scholze, *Projectivity of the Witt vector affine Grassmannian*, Invent. Math. **209** (2017), no. 2, 329–423, DOI 10.1007/s00222-016-0710-4. MR3674218

[BS22] Bhargav Bhatt and Peter Scholze, *Prisms and prismatic cohomology*, Ann. of Math. (2) **196** (2022), no. 3, 1135–1275, DOI 10.4007/annals.2022.196.3.5. MR4502597

[BH93] Winfried Bruns and Jürgen Herzog, *Cohen-Macaulay rings*, Cambridge Studies in Advanced Mathematics, vol. 39, Cambridge University Press, Cambridge, 1993. MR1251956

[CLM+23] H. Cai, S. Lee, L. Ma, K. Schwede, and K. Tucker, *Perfectoid signature, perfectoid hilbert-kunz multiplicity, and an application to local fundamental groups*, arXiv:2209.04046, 2023.

[CRMP+21] Javier Carvajal-Rojas, Linquan Ma, Thomas Polstra, Karl Schwede, and Kevin Tucker, *Covers of rational double points in mixed characteristic*, J. Singul. **23** (2021), 127–150, DOI 10.5427/jsing. MR4292622

[CRST18] Javier Carvajal-Rojas, Karl Schwede, and Kevin Tucker, *Fundamental groups of F-regular singularities via F-signature* (English, with English and French summaries), Ann. Sci. Éc. Norm. Supér. (4) **51** (2018), no. 4, 993–1016, DOI 10.24033/asens.2370. MR3861567

[CR22] Javier A. Carvajal-Rojas, *Finite torsors over strongly F-regular singularities*, Épijournal Géom. Algébrique **6** (2022), Art. 1, 30, DOI 10.46298/epiga.2022.7532. MR4391081

[CS24] Kęstutis Česnavičius and Peter Scholze, *Purity for flat cohomology*, Ann. of Math. (2) **199** (2024), no. 1, 51–180, DOI 10.4007/annals.2024.199.1.2. MR4681144

[DT23] Rankeya Datta and Kevin Tucker, *On some permanence properties of (derived) splinters*, Michigan Math. J. **73** (2023), no. 2, 371–400, DOI 10.1307/mmj/20205951. MR4584866

[Eis95] David Eisenbud, *Commutative algebra*, Graduate Texts in Mathematics, vol. 150, Springer-Verlag, New York, 1995. With a view toward algebraic geometry, DOI 10.1007/978-1-4612-5350-1. MR1322960

[Fal02] Gerd Faltings, *Almost étale extensions*, Astérisque **279** (2002), 185–270. Cohomologies $p$-adiques et applications arithmétiques, II. MR1922831

[Gab18] O. Gabber, *Observations made after the MSRI workshop on homological conjectures*, https://www.msri.org/workshops/842/schedules/23854/documents/3322/assets/31362.

[GR03] Ofer Gabber and Lorenzo Ramero, *Almost ring theory*, Lecture Notes in Mathematics, vol. 1800, Springer-Verlag, Berlin, 2003, DOI 10.1007/b10047. MR2004652

[HLS24] Christopher Hacon, Alicia Lamarche, and Karl Schwede, *Global generation of test ideals in mixed characteristic and applications*, Algebr. Geom. **11** (2024), no. 5, 676–711. MR4791070

[Har66] Robin Hartshorne, *Residues and duality*, Lecture Notes in Mathematics, No. 20, Springer-Verlag, Berlin-New York, 1966. Lecture notes of a seminar on the work of A. Grothendieck, given at Harvard 1963/64; With an appendix by P. Deligne. MR222093

[HM21] Raymond C. Heitmann and Linquan Ma, *Extended plus closure in complete local rings*, J. Algebra **571** (2021), 134–150, DOI 10.1016/j.jalgebra.2018.10.006. MR4200713

[HM18] Raymond Heitmann and Linquan Ma, *Big Cohen-Macaulay algebras and the vanishing conjecture for maps of Tor in mixed characteristic*, Algebra Number Theory **12** (2018), no. 7, 1659–1674, DOI 10.2140/ant.2018.12.1659. MR3871506

[Hoc73] M. Hochster, *Contracted ideals from integral extensions of regular rings*, Nagoya Math. J. **51** (1973), 25–43. MR349656

[Hoc02] Melvin Hochster, *Big Cohen-Macaulay algebras in dimension three via Heitmann's theorem*, J. Algebra **254** (2002), no. 2, 395–408, DOI 10.1016/S0021-8693(02)00086-8. MR1933876

[HH90] Melvin Hochster and Craig Huneke, *Tight closure, invariant theory, and the Briançon-Skoda theorem*, J. Amer. Math. Soc. **3** (1990), no. 1, 31–116, DOI 10.2307/1990984. MR1017784

[HH92] Melvin Hochster and Craig Huneke, *Infinite integral extensions and big Cohen-Macaulay algebras*, Ann. of Math. (2) **135** (1992), no. 1, 53–89, DOI 10.2307/2946563. MR1147957

[HH95] Melvin Hochster and Craig Huneke, *Applications of the existence of big Cohen-Macaulay algebras*, Adv. Math. **113** (1995), no. 1, 45–117, DOI 10.1006/aima.1995.1035. MR1332808

[HY22] Melvin Hochster and Yongwei Yao, *The F-rational signature and drops in the Hilbert-Kunz multiplicity*, Algebra Number Theory **16** (2022), no. 8, 1777–1809, DOI 10.2140/ant.2022.16.1777. MR4516193

[Hun13] Craig Huneke, *Hilbert-Kunz multiplicity and the F-signature*, Commutative algebra, Springer, New York, 2013, pp. 485–525, DOI 10.1007/978-1-4614-5292-8_15. MR3051383

[HL02] Craig Huneke and Graham J. Leuschke, *Two theorems about maximal Cohen-Macaulay modules*, Math. Ann. **324** (2002), no. 2, 391–404, DOI 10.1007/s00208-002-0343-3. MR1933863

[Kun69] Ernst Kunz, *Characterizations of regular local rings of characteristic $p$*, Amer. J. Math. **91** (1969), 772–784, DOI 10.2307/2373351. MR252389

[MS18] Linquan Ma and Karl Schwede, *Perfectoid multiplier/test ideals in regular rings and bounds on symbolic powers*, Invent. Math. **214** (2018), no. 2, 913–955, DOI 10.1007/s00222-018-0813-1. MR3867632

[MS21] Linquan Ma and Karl Schwede, *Singularities in mixed characteristic via perfectoid big Cohen-Macaulay algebras*, Duke Math. J. **170** (2021), no. 13, 2815–2890, DOI 10.1215/00127094-2020-0082. MR4312190

[MST+22] Linquan Ma, Karl Schwede, Kevin Tucker, Joe Waldron, and Jakub Witaszek, *An analogue of adjoint ideals and PLT singularities in mixed characteristic*, J. Algebraic Geom. **31** (2022), no. 3, 497–559, DOI 10.1090/jag/797. MR4484548

[Mar22] I. Martin, *The number of torsion divisors in a strongly $F$-regular ring is bounded by the reciprocal of $F$-signature*, Comm. Algebra **50** (2022), no. 4, 1595–1605, DOI 10.1080/00927872.2021.1986057. MR4391510

[Mat89] Hideyuki Matsumura, *Commutative ring theory*, 2nd ed., Cambridge Studies in Advanced Mathematics, vol. 8, Cambridge University Press, Cambridge, 1989. Translated from the Japanese by M. Reid. MR1011461

[Pol22] Thomas Polstra, *A theorem about maximal Cohen-Macaulay modules*, Int. Math. Res. Not. IMRN **3** (2022), 2086–2094, DOI 10.1093/imrn/rnaa154. MR4373232

[PT18] Thomas Polstra and Kevin Tucker, *F-signature and Hilbert-Kunz multiplicity: a combined approach and comparison*, Algebra Number Theory **12** (2018), no. 1, 61–97, DOI 10.2140/ant.2018.12.61. MR3781433

[SdS87] J. B. Sancho de Salas, *Blowing-up morphisms with Cohen-Macaulay associated graded rings*, Géométrie algébrique et applications, I (La Rábida, 1984), Travaux en Cours, vol. 22, Hermann, Paris, 1987, pp. 201–209. MR907914

[Shi07] Kazuma Shimomoto, *The Frobenius action on local cohomology modules in mixed characteristic*, Compos. Math. **143** (2007), no. 6, 1478–1492, DOI 10.1112/S0010437X07002825. MR2371377

[Sin99] Anurag K. Singh, **Q**-*Gorenstein splinter rings of characteristic $p$ are F-regular*, Math. Proc. Cambridge Philos. Soc. **127** (1999), no. 2, 201–205, DOI 10.1017/S0305004199003710. MR1735920

[ST24] Ilya Smirnov and Kevin Tucker, *The theory of $F$-rational signature*, J. Reine Angew. Math. **812** (2024), 1–58, DOI 10.1515/crelle-2024-0010. MR4767385

[Smi97] Karen E. Smith, *$F$-rational rings have rational singularities*, Amer. J. Math. **119** (1997), no. 1, 159–180. MR1428062

[Sta] T. Stacks Project Authors, *Stacks Project*.

[WY00] Kei-ichi Watanabe and Ken-ichi Yoshida, *Hilbert-Kunz multiplicity and an inequality between multiplicity and colength*, J. Algebra **230** (2000), no. 1, 295–317, DOI 10.1006/jabr.1999.7956. MR1774769

[Wei94] Charles A. Weibel, *An introduction to homological algebra*, Cambridge Studies in Advanced Mathematics, vol. 38, Cambridge University Press, Cambridge, 1994, DOI 10.1017/CBO9781139644136. MR1269324

Department of Mathematics, Purdue University, West Lafayette, Indiana 47907
*Email address*: ma326@purdue.edu

Department of Mathematics, University of Illinois at Chicago, Chicago, Illinois 60607
*Email address*: kftucker@uic.edu

FSC
www.fsc.org
MIX
Papier | Fördert gute Waldnutzung
FSC® C083411

Kolibri 360 GmbH
Ferdinand-Jühlke-Straße 7
99095 Erfurt, Deutschland
produktsicherheit@kolibri360.de